AF248621

Integrated Pest Management in Tropical Regions

Integrated Pest Management in Tropical Regions

Edited by

Carmelo Rapisarda

and

Giuseppe E. Massimino Cocuzza

CABI is a trading name of CAB International

CABI
Nosworthy Way
Wallingford
Oxfordshire OX10 8DE
UK

CABI
745 Atlantic Avenue
8th Floor
Boston, MA 02111
USA

Tel: +44 (0)1491 832111
Fax: +44 (0)1491 833508
E-mail: info@cabi.org
Website: www.cabi.org

T: +1 (617)682-9015
E-mail: cabi-nao@cabi.org

British Library Cataloguing-in-Publication Data

A catalogue record for this book is available from the British Library.

Library of Congress Cataloging-in-Publication Data

Names: Rapisarda, Carmelo, editor. | Cocuzza, Giuseppe E. Massimino, editor.
Title: Integrated pest management in tropical regions /
  edited by: Carmelo Rapisarda and Giuseppe E. Massimino Cocuzza.
Description: Boston, MA : CABI, [2018] | Includes bibliographical references and index.
Identifiers: LCCN 2017023566 (print) | LCCN 2017024622 (ebook) | ISBN 9781780648019 (ePDF) |
  ISBN 9781780648026 (ePub) | ISBN 9781780648002 (hbk : alk. paper)
Subjects: LCSH: Pests--Integrated control--Tropics.
Classification: LCC SB950.3. T73 (ebook) | LCC SB950.3. T73 I58 2018 (print) |
  DDC 632/.60913--dc23
LC record available at https://lccn.loc.gov/2017023566

ISBN: 978 1 78064 800 2 (hbk)
         978 1 78064 801 9 (e-book)
         978 1 78064 802 6 (e-pub)

Commissioning editor: Rachael Russell
Editorial assistant: Emma McCann
Production editor: Alan Worth

Typeset by AMA DataSet Ltd, Preston
Printed and bound by CPI Group (UK) Ltd, Croydon, CR0 4YY

Contents

Contributors

Tsedeke Abate, International Maize and Wheat Improvement Center (CIMMYT) – ICRAF House, United Nations Avenue, Gigiri, PO Box 1041, Village Market-00621, Nairobi, Kenya

Siti Ramlah A. Ali, Malaysia Palm Oil Board (MPOB), Biological Research Division, No. 6 Persiaran Institusi, Bandar Baru Bangi, 43000 Kajang, Selangor, Malaysia

Miguel A. Altieri, Department of Environmental Science, Policy and Management, University of California, Berkeley, 215 Mulford Hall, Berkeley, California 94720-3112, USA

Salvatore Bella, Council for Agricultural Research and Economic Analysis, Research Centre for Citrus and Mediterranean Crops (CREA-ACM), Corso Savoia 190, 95024 Acireale (CT), Italy

Daniel L. Coyne, International Institute of Tropical Agriculture (IITA), PO Box 30772-00100, Nairobi, Kenya

Mieke S. Daneel, ARC-Institute for Tropical and Subtropical Crops, PB X11208, Nelspruit, 1200, South Africa

Fábio Maximiano de Andrade Silva, Insecticide Resistance Action Committee (IRAC), Brazil and DuPont do Brasil S.A., Rodovia PLN 145, 943, Bairro Boa Esperança, Paulínia-SP, Brazil

José Gilberto de Moraes, Departamento de Entomologia e Acarologia, ESALQ, USP, 13418-900, Piracicaba, São Paulo, Brazil

Thomas Dubois, The World Vegetable Center, Eastern and Southern Africa, PO Box 10, Duluti, Arusha, Tanzania

Odair Aparecido Fernandes, Departamento de Fitossanidade, FCAV, Universidade Estadual Paulista – UNESP, 14884-900, Jaboticabal, São Paulo, Brazil

François-Régis Goebel, CIRAD, Unité de Recherche AIDA, 34398 Montpellier cedex 5, France

Shoil M. Greenberg, (retired), Kika de la Garza Subtropical Agricultural Research Center, Agricultural Research Service, United States Department of Agriculture, Weslaco, Texas, 12472 Brookline Street, Carmel, Indiana 46032, USA

Devid Guastella, Agrisudafrica (Pty) Ltd, Springvallei Road D601, PO Box 31, Franklin 4706, South Africa

Abdelhaq Hanafi, International Fund for Agricultural Development (IFAD), Via Paolo di Dono 44, 00142 Rome, Italy

Luko Hilje, Professor Emeritus, Tropical Agricultural Research and Higher Education Center (CATIE), Turrialba, Costa Rica

Norman Kamarudin, Malaysia Palm Oil Board (MPOB), Biological Research Division, No. 6 Persiaran Institusi, Bandar Baru Bangi, 43000 Kajang, Selangor, Malaysia

Fred Kanampiu, International Institute of Tropical Agriculture (IITA), PO Box 30772-00100, icipe-campus, Kasarani, Nairobi, Kenya

Nitin Kulkarni, Forest Entomology Division, Tropical Forest Research Institute, PO-RFRC, Jabalpur – 482021, India

James Legg, International Institute of Tropical Agriculture (IITA), PO Box 34441, Dar es Salaam, Tanzania

George Mahuku, International Institute of Tropical Agriculture (IITA), Regional Hub for Eastern Africa, Plot 25, Light Industrial Area, Coca Cola Road, Mikocheni, PO Box 34441, Dar es Salaam, Tanzania

Zulkefli Masijan, Malaysia Palm Oil Board (MPOB), Biological Research Division, No. 6 Persiaran Institusi, Bandar Baru Bangi, 43000 Kajang, Selangor, Malaysia

Giuseppe E. Massimino Cocuzza, Dipartimento di Agricoltura, Alimentazione e Ambiente, Università degli Studi, via Santa Sofia 100, 95123 Catania, Italy

Ramle Moslim, Malaysia Palm Oil Board (MPOB), Biological Research Division, No. 6 Persiaran Institusi, Bandar Baru Bangi, 43000 Kajang, Selangor, Malaysia

Urbano Nava-Camberos, Facultad de Agricultura y Zootecnia, Universidad Juárez del Estado de Durango, Gómez Palacio, Durango, México

Clara I. Nicholls, Department of Environmental Science, Policy and Management, University of California, Berkeley, 215 Mulford Hall, Berkeley, California 94720-3112, USA

Amin Nikpay, Department of Plant Protection, Sugarcane and By-products Development Company, Salman Farsi Unit, Ahwaz, Iran

Joshua Okonya, International Potato Center (CIP), PO Box 22274, Kampala, Uganda

Megha N. Parajulee, Texas A&M University System, Texas A&M AgriLife Research, 1102 East FM 1294, Lubbock, Texas 79403, USA

Silvana V. Paula-Moraes, Entomology and Nematology Department, University of Florida, West Florida Research and Education Center, Jay, Florida 32565, USA

Carmelo Rapisarda, Dipartimento di Agricoltura, Alimentazione e Ambiente, Università degli Studi, Via Santa Sofia 100, 95123 Catania, Italy

Alexandre Specht, Embrapa Cerrados, BR-040, Km 18, Planaltina-DF, Brazil

Edison R. Sujii, Empresa Brasileira de Pesquisa Agropecuária (EMBRAPA), Centro Nacional de Pesquisa em Recursos Genéticos e Biotecnologia (CENARGEN), Parque Estação Biológica, Brasilia, DF, Brazil

Mohd Basri Wahid, Malaysia Palm Oil Board (MPOB), Biological Research Division, No. 6, Persiaran Institusi, Bandar Baru Bangi, 43000 Kajang, Selangor, Malaysia

Vitalis Wafula Wekesa, Department of Biological Science and Technology, Technical University of Kenya, PO Box 52428-00200, Nairobi, Kenya

Everlyne Wosula, International Institute of Tropical Agriculture (IITA), Regional Hub for Eastern Africa, Plot 25, Light Industrial Area, Coca Cola Road, Mikocheni, PO Box 34441, Dar es Salaam, Tanzania

1   Introduction

Carmelo Rapisarda* and Giuseppe E. Massimino Cocuzza

*Dipartimento di Agricoltura, Alimentazione e Ambiente,
Università degli Studi, Catania, Italy*

1.1 Tropics and Subtropics

The Tropics, geographically limited in latitude by the Tropic of Cancer (to the north) and the Tropic of Capricorn (to the south), are characterized by limited seasonal differences, with a mean warm to high temperature and a high humidity level almost all year round, at most with difference between a dry and a rainy season (McGregor and Nieuwolt, 1998). Plant diversity and biology are influenced by these peculiar climatic conditions and herbivores may develop almost continuously throughout the year in these regions, showing homodynamic cycles and high biodiversity, whatever their trophic habits.

Slightly similar features are shown by the Subtropics, which extend from the Tropics to the temperate regions (to about 40° latitude) and are characterized by warm to hot summers and cool to mild winters, thus with a well-defined seasonality but with almost rare frost (Rohli and Vega, 2015). This relative similarity between Tropics and Subtropics allows frequently tropical crops to be cultivated also in subtropical areas, with only gradual smooth changes of cultural contexts with increasing latitude and some overlap of agroecological environments, including the species composition of pests, their population dynamics and phytosanitary importance.

As a consequence of climatic change, dispersal of organisms is becoming increasingly frequent, especially to more northern latitudes, causing deep changes to global biodiversity (Engel *et al.*, 2011). Thus, we are witnessing an increasing number of invasions of subtropical environments by typical tropical species, which can sometimes move even up to temperate regions (Levine and D'Antonio, 2003; Parker *et al.*, 2006; Levine, 2008; Roques *et al.*, 2010). For their high displacement capacity, also of anthropogenic origin, crop pests significantly respond to this trend and important examples may be found in the context of various crop types, such as vegetables, on which the Tomato leaf miner has rapidly invaded nearly all subtropical and most temperate areas of Europe, Africa and Asia (Desneux *et al.*, 2010); or citrus, always characterized for being susceptible to being colonized by exotic species but recently threatened, in the Mediterranean region, by two tropical vectors of dangerous pathogens, such as the Black citrus aphid *Toxoptera citricidus* (Kirkaldy), which is the main vector of the Citrus Tristeza Virus, and the African citrus psyllid *Trioza erytreae* (Del Guercio), vector of the Citrus huanglongbing or greening disease (Hermoso de Mendoza *et al.*, 2008; Massimino Cocuzza *et al.*, 2017). These continuous movements of pests, and their constant invasion of new

* Corresponding author e-mail: rapicar@unict.it

areas where they were previously absent, require continuous updates both in basic knowledge about their biology and epidemiology as well as in more purely phytosanitary issues related to control strategies aimed at managing their populations in different ecological contexts.

1.2 Pest Control and IPM

All over the world, the health status and production of agricultural and forest plants are affected by several limiting factors, of both biotic and non-biotic origin. The word pests usually indicates every kind of insect, mite or pathogen that directly or indirectly injures a crop or a forest. With a broad interpretation of the term, every wild plant that reduces availability of water or nutrients to cultivated plants also meets the definition of a pest.

Pests have competed with humans since the birth of agricultural activities. Pest control, aimed at reducing their damage to food crops, started already from the very ancient civilizations of the world to explore ways to kill or simply repel noxious organisms on cultivated plants (Flint and van den Bosch, 1981). Use of chemicals has been the base of pest control for a long time, initially through the use of substances of inorganic origin (e.g. sulfur) or extracted from living organisms, mainly plants (e.g. *Pyrethrum*). A radical change of chemical control strategies occurred soon after the Second World War, with the spread of synthetic organic insecticides, which led to an intensification of the chemical pressure in agroecosystems and the consequent technical, environmental and health secondary effects (Arias-Estévez *et al.*, 2008; Aktar *et al.*, 2009; Damalas and Eleftherohorinos, 2011; Gill *et al.*, 2012; Kohler and Triebskorn, 2013; Lin *et al.*, 2013).

Following a growing awareness of the negative effects of chemical pesticides on the environment as well as on animal (including humans) health, the concept of Integrated Pest Management (IPM) was developed around the middle of the 20th century, as an alternative to chemical control involving the use of different and combined tactics, aimed at keeping pest populations below the levels at which they cause economic injuries (Stern *et al.*, 1959; Kogan, 1998). In contrast with previous strategies, that viewed the crop as the centre of interest, IPM may be considered as an ecosystem-based philosophy for pest control, aiming at achieving a long-term prevention of pests or their damage through a combination of techniques. Cultural practices, especially the use of resistant varieties (but also a rational management of water and fertilizer supply), and biological control (in its broad sense of habitat manipulation) constitute two important elements of IPM. Mechanical and physical control have also an important role in IPM, by blocking pests or making the environment unsuitable for them. Chemicals can also be used in IPM programmes, of course, but only after monitoring has assessed relevant pest numbers and damage, as well as caring to select pesticides and apply them so as to avoid risks to human health or to beneficial and non-target organisms, and to minimize secondary effects on the environment.

IPM programmes have become a relevant issue in the agricultural and environmental policies in many countries worldwide. Taking a cue from noting how conventional pest control approaches give rise to unsustainable production systems, these programmes have been increasingly implemented in parallel with the global growth of awareness on the need for a greater sustainability of economic and social development. A broad knowledge of plant protection disciplines, completed by skills in different sectors of crop production, such as climatology, plant genetics or soil science, is needed for planning a system-oriented IPM strategy, where the target is not the pest to be reduced but the ecological balance to be maintained. Therefore, research is a key point for implementing any kind of IPM programme and training is essential to transfer the results of research to farmers and technicians.

1.3 Integrated Pest Management in the Tropics

During the last decades, Integrated Pest Management approaches have been extensively and successfully implemented over nearly all temperate areas, especially North America, Europe and Australia; in some of these areas, they are even promoted and sometimes even made compulsory by law. For instance, in Europe, current legislation is sharply reducing the use of synthetic chemical pesticides and promoting research and application of IPM strategies (Lefebvre *et al.*, 2015). Comparatively, the role of IPM within crop protection grew slowly in tropical regions and possibilities for its application have been only partially explored. In these areas, the peculiar effects of environmental conditions on pest biology are constantly faced, but also constraints to IPM implementation are represented by a limited understanding of the basic agroecological factors which are the primary support to the development of modern pest control programmes (Hilje *et al.*, 2003).

In most tropical countries, agrochemicals still remain an essential component of agricultural practices (Carvalho *et al.*, 1998; Aktar *et al.*, 2009; Schiesari *et al.*, 2013; Pretty and Pervez Bharucha, 2015) and it is likely that they will continue to play a key role also in the foreseeable future, in spite of regional policies and specific supporting programmes, which have greatly increased the use of IPM strategies during recent decades. The wide diffusion of both the basic knowledge on biotic and non-biotic factors influencing agroecosystems and the technical skills to manage biological balances, aimed at minimizing the risk of infestation by pests, are indispensable starting points for a rationalization of crop protection even in tropical regions, helping to achieve high standards of both environmental and economic sustainability. Therefore, a constant update on results of ongoing worldwide research on pests of major tropical crops and on techniques for controlling their damage is essential to enable improvement processes that could lead to a higher quality of life for both farmers and consumers. To disseminate widely IPM practices among farmers, the role of political institutions is of crucial importance, through a regulatory action to limit the use of pesticides to those with lower environmental impact (Pretty and Pervez Bharucha, 2015). Moreover, economic investments should be increased to spread IPM practices, as already happens in many tropical and subtropical countries (Pretty and Pervez Bharucha, 2015).

Apart from a huge number of existing papers, focusing on precise crops, or on limited geographical areas, or on results of the research on specific pests or pest groups, a first comprehensive contribution aiming to update on the possibilities of Integrated Pest Management in tropical regions was provided about 20 years ago by Mengech *et al.* (1995). More recently, new and updated insights have been given for IPM in specific cropping systems, such as in the case of tropical vegetable crops (Muniappan and Heinrichs, 2016). The present volume responds to the need for assessing and updating the available techniques aimed at sustainably reducing pest damage on crops, within the context of present climatic dynamism and global movement of organisms, which are improving pest problems almost everywhere. It also aims at updating a large-scale overview of IPM applications in the main tropical food crops (such as cereals, legumes, root and tuber crops, sugarcane, vegetables, banana and plantain, citrus, oil palm, tea, cocoa, coffee) but also in fibre crops (such as cotton) and tropical forests. In the first part, a review is also reported of the basic agroecological framework representing the IPM foundations and of the emerging technologies in chemical and biological methods, which are the pillar of pest control in tropical (and subtropical) crops.

In addition to universities and research institutions (public and private), this volume may be of interest also to individual and/or associated farmers, agricultural companies, technicians, NGOs working at various levels in the field of sustainable agricultural development and, more generally, to all who are technically or politically involved in the implementation of sustainable crop protection programmes in developing countries of the tropical (and subtropical) world.

References

Aktar, M.W., Sengupta, D. and Chowdhury, A. (2009) Impact of pesticides use in agriculture: their benefits and hazards. *Interdisciplinary Toxicology* 2 (1), 1–12.

Arias-Estévez, M., López-Periago, E., Martínez-Carballo, E., Simal-Gándara, J., Mejuto, J.C. and García-Río, L. (2008) The mobility and degradation of pesticides in soils and the pollution of groundwater resources. *Agriculture, Ecosystems & Environment* 123, 247–260.

Carvalho, F.P., Nhan, D.D., Zhong, C., Tavares, T. and Klaine, S. (1998) Tracking pesticides in the tropics. *IAEA Bulletin* 40 (3), 24–30.

Damalas, C.A. and Eleftherohorinos, I.G. (2011) Pesticide exposure, safety issues, and risk assessment indicators. *International Journal of Environmental Research and Public Health* 8 (12), 1402–1419.

Desneux, N., Wajnberg, E., Wyckhuys, K.A.G., Burgio, G., Arpaia, S. *et al.* (2010) Biological invasion of European tomato crops by *Tuta absoluta*: ecology, geographic expansion and prospects for biological control. *Journal of Pest Science* 83, 197–215.

Engel, K., Tollrian, R. and Jeschke, J.M. (2011) Integrating biological invasions, climate change and phenotypic plasticity. *Communicative & Integrative Biology* 4(3), 247–250.

Flint, M.L. and van den Bosch, R. (1981) A history of pest control. In: Flint, M.L. and van den Bosch, R. (eds) *Introduction to Integrated Pest Management*. Plenum Press, New York, pp. 51–81.

Gill, R.J., Ramos-Rodriguez, O. and Raine, N.E. (2012) Combined pesticide exposure severely affects individual- and colony-level traits in bees. *Nature* 491 (7422), 105–108.

Hermoso de Mendoza, A., Alvarez, A., Michelena, J.M., Gonzales, P. and Cambra, M. (2008) *Toxoptera citricida* (Kirkaldy) (Hemiptera, Aphididae) and its natural enemies in Spain. *IOBC/WPRS Bulletin* 38, 225–232.

Hilje, L., Araya, C.M. and Valverde, B.E. (2003) Pest management in mesoamerican agroecosystems. In: Vandermeer, J.H. (ed.) *Tropical Agroecosystems*. CRC Press, Boca Raton, Florida, pp. 59–93.

Kogan, M. (1998) Integrated pest management: historical perspectives and contemporary developments. *Annual Review of Entomology* 43, 243–270.

Kohler, H.R. and Triebskorn, R. (2013) Wildlife ecotoxicology of pesticides: can we track effects to the population level and beyond? *Science* 341 (6147), 759–765.

Lefebvre, M., Langrell, S.H. and Gomez-y-Paloma, S. (2015) Incentives and policies for integrated pest management in Europe: a review. *Agronomy for Sustainable Development* 35, 27–45.

Levine, J.M. (2008) Biological invasions. *Current Biology* 18 (2), R57–R60.

Levine, J.M. and D'Antonio, C.M. (2003) Forecasting biological invasions with increasing international trade. *Conservation Biology* 17, 322–326.

Lin, P.C., Lin, H.J., Liao, Y.Y., Guo, H.R. and Chen, K.T. (2013) Acute poisoning with neonicotinoid insecticides: a case report and literature review. *Basic & Clinical Pharmacology & Toxicology* 112 (4), 282–286.

Massimino Cocuzza, G.E., Urbaneja, A., Hernández-Suárez, E., Siverio, F., Di Silvestro, S.A., Tena A. and Rapisarda, C. (2017) A review on *Trioza erytreae* (African citrus psyllid), now in mainland Europe, and its potential risk as vector of huanglongbing (HLB) in citrus. *Journal of Pest Science* 90 (1), 1–17.

McGregor, G.R. and Nieuwolt, S. (1998) *Tropical Climatology: An Introduction to the Climates of the Low Latitudes*, 2nd edn. John Wiley & Sons, Chichester.

Mengech, A.N., Saxena, K.N. and Gopalan, H.N.B. (eds) (1995) *Integrated Pest Management in the Tropics: Current Status and Future Prospects*. John Wiley & Sons, Chichester.

Muniappan, R. and Heinrichs, E.A. (eds) (2016) *Integrated Pest Management of Tropical Vegetable Crops*. Springer, Dordrecht, The Netherlands.

Parker, J.D., Burkepile, D.E. and Hay, M.E. (2006) Opposing effects of native and exotic herbivores on plant invasions. *Science* 311, 1459–1461.

Pretty, J. and Pervez Bharucha, Z. (2015) Integrated Pest Management for sustainable intensification of agriculture in Asia and Africa. *Insects* 6, 152–182.

Rohli, R.V. and Vega, A.J. (2015) *Climatology*, 3rd edn. Jones & Bartlett Learning, Burlington, Massachusetts.

Roques, A., Kenis, M., Lees, D., Lopez-Vaamonde, C., Rabitsch, W., Rasplus, J.-C. and Roy, D. (2010) Alien terrestrial arthropods of Europe. *Biorisk* 4 (1) (special issue), Pensoft, Sofia-Moscow.

Schiesari, L., Waichman, A., Brock, T., Adams, C. and Grillitsch, B. (2013) Pesticide use and biodiversity conservation in the Amazonian agricultural frontier. *Philosophical Transactions of the Royal Society of London, Series B, Biological Sciences* 368 (1619), 20120378. DOI:10.1098/rstb.2012.0378

Stern, V.M., Smith, R.F., van den Bosch, R. and Hagen, K.S. (1959) The integrated control concept. *Hilgardia* 29 (2), 81–101.

2 Agroecological Foundations for Pest Management in the Tropics: Learning from Traditional Farmers

Miguel A. Altieri* and Clara I. Nicholls

*Department of Environmental Science, Policy and Management,
University of California, Berkeley, USA*

2.1 Introduction

The integrity of tropical ecosystems is at risk as demand for food and other resources from industrialized countries increases. More than ever, tropical agroecosystems face unrelenting intensification and expansion. For decades, the production of agricultural export commodities has represented a major source of foreign income for many tropical countries (Gomiero, 2016; Altieri *et al.*, 2017). Two strategies have been used in the tropics to increase agricultural production: clearing natural habitats to plant new crops or intensifying output from existing crop lands. It is not new that crops such as coffee, cacao, oil palm, rice and soybean, which encompass a range of product types (oils, grain and fruits) grown mainly for export in areas of rich biodiversity, have been produced and expanded involving deforestation and biodiversity loss (Donald, 2004; Altieri *et al.*, 2015). On the other hand, Green Revolution approaches to enhance production via high-yielding varieties accompanied by agrochemical inputs has left an immense ecological and social footprint (Altieri, 2004; Gomiero, 2016). Moreover, excess inputs have compounded pest problems, as in the case of rice in Asia, where heavy use of nitrogen fertilizer has increased pest reproductive potential in rice (Altieri and Nicholls, 2004, 2008). The nitrogen-rich hybrid rice plants seem to create favourable conditions for brown plant hopper outbreaks in areas with large plantings of the hybrids (Altieri, 2005; Watanabe *et al.*, 2013).

In the last two decades, the expansion of monocultures has accelerated radically due to the increased acreage devoted to transgenic crops and biofuel plantations (Gomiero, 2016). For example, in Malaysia and Indonesia, oil palm plantations that are rapidly expanding for biodiesel production are a poor substitute for native tropical forests. They support few species of conservation importance, and affect biodiversity in adjacent habitats through fragmentation, edge effects and pollution (Fletcher *et al.*, 2011). In Brazil alone, Roundup Ready® soybean occupies an area no less than 45 million ha, and in Argentina, more than 95% of all soybean acreage is transgenic (James, 2013). Such simplification can have serious ecological implications for pest management, such as in the case in four US Midwest states, where recent biofuel-driven growth in maize and soybean planting resulted in lower landscape diversity,

---

* Corresponding author e-mail: agroeco3@berkeley.edu

decreasing the supply of pest natural enemies to maize and soybean fields and reducing biocontrol services by 24% (Altieri and Nicholls, 2008; Altieri *et al.*, 2015). This loss of biocontrol services cost soybean and maize producers in these states an estimated US$58 million per annum in reduced yield and increased pesticide use (Landis *et al.*, 2008; Altieri *et al.*, 2015).

Massive increases in production areas and intensifying yields per hectare have reached a dead-end, as managing monocultures with high external inputs has resulted in massive loss of natural habitats and biodiversity, pesticide and nitrate pollution, and overall environmental degradation (Woodhouse, 2010). The ecological futility of promoting mechanized monocultures in tropical areas of overwhelming biotic intricacy where pests flourish year-round and nutrient leaching is a major constraint has been amply demonstrated (Ewel, 1986; Altieri and Nicholls, 2004). The modern agricultural strategy is totally opposite to the way small farmers have practised traditional agriculture for centuries, and whom were bypassed by agricultural modernization and have not relied on agrochemicals to sustain production (Dewalt, 1994; Denevan, 1995).

In most tropical areas, small farmers have used intercropping and agroforestry systems which involve mixtures of annual crops and/or perennial trees grown on the same piece of land at the same time, with enough proximity for ecological interactions to occur between component crops and associated biota. These systems constitute an effective agroecological strategy for introducing more biodiversity into agroecosystems, and the resulting increased crop diversity enhances a number of ecosystem services provided to farmers (Chang, 1977; Altieri and Nicholls, 2004; Altieri *et al.*, 2017).

A more reasonable approach to design more pest-resilient tropical farming systems is to imitate natural ecosystems, as most traditional farmers do, rather than struggle to impose horticultural simplicity in ecosystems that are inherently complex (Altieri, 2002). Ewel (1986) argues that successional ecosystems can be particularly appropriate templates for the design of sustainable tropical agroecosystems. Building on the experience and time-tested systems of small farmers and the contributions of modern agroecology, principles can be derived for agroecosystem design emphasizing the development of cropping systems that confer associational resistance to pests, thus reducing agroecosystem vulnerability while providing biological stability and productivity (Altieri, 2004).

2.2 Traditional Farming: Lessons for Pest Management

For many agroecologists, a starting point in the development of new agricultural systems are the very systems that traditional farmers have developed and/or inherited throughout centuries (Altieri and Toledo, 2005). Such complex farming systems, adapted to local conditions, have helped small farmers to sustainably manage harsh environments and to meet their subsistence needs, without depending on mechanization, chemical fertilizers, pesticides or other technologies of modern agricultural science (Altieri *et al.*, 2017). Guided by an intricate knowledge of nature, traditional farmers have nurtured biologically and genetically diverse smallholder farms with a robustness and a built-in resilience necessary to adjust to rapidly changing climates, pests and diseases, and more recently to globalization, technological penetration and other modern trends (Clawson, 1985; Khan and Pickett, 2008). Indigenous farmers tend to combine various production systems as part of a typical household resource management scheme. Much research on the features of these systems suggests that a series of factors and characteristics listed in Table 2.1 underlie the sustainability of multiple use systems (Altieri, 2004; Altieri and Nicholls, 2008).

A salient feature of traditional farming systems is its high level of biodiversity deployed in the form of polycultures, agroforestry and other complex cropping

Table 2.1. Socio-ecological features that underlie the sustainability and resilience of traditional farming systems (Koohafkan and Altieri, 2016).

Deep knowledge of plants, animals, soils and how to manage them.

Multiple use strategies of the landscapes, including management of mosaics of fields, fallows, forest remnants, etc.

Farms are small in size with a continuous production serving subsistence and local market demands.

Diversified farm systems based on several cropping systems, featuring mixtures of crops, trees and/or animals with rich varietal and other genetic diversity.

Maximum and effective use of local resources and low dependence on off-farm inputs.

High net energy yield because energy inputs are relatively low.

Labour is skilled and complementary, drawn largely from the household or community relations.

Dependency on animal traction and manual labour shows favourable energy input/output ratios.

Heavy emphasis on recycling of nutrients and materials.

Building on natural ecological processes (e.g. succession) rather than struggling against them.

systems (Altieri *et al.*, 2012). Guided by an acute observation of nature, many traditional farmers have intuitively mimicked the structure of natural systems with their cropping arrangements. Examples of such biomimicry abound and below we describe three examples of biodiverse farming systems and their implications for insect pest management (Nicholls and Altieri, 2004; Altieri *et al.*, 2017).

2.2.1 The rice–fish–duck systems in China

Over 90% of the world's rice is produced and consumed in the Asia-Pacific Region. Asian farmers account for 87% of the world's total rice production. In many river basins, paddy cultivation is a main provider of livelihoods, as rice is the major staple food of the people in such regions. Almost all social and cultural activities of millions of rural people are directly or indirectly related to rice seasons and landscapes. Each and every part of the rice plant has economic and social significance. Rice has many uses – medicinal, food, animal feed, cosmetics, rituals, etc. Grain is the staple food for humans, husk, and bran are used as livestock feeds, straw is used in thatching the rural huts, bedding materials for livestock, and mainly as fodder for cattle during the dry season. Straw is also used as fuel in cooking and as binding materials for mud

plastering of houses, etc. (Hanks, 1992; Das *et al.*, 2015).

Rice culture has been followed by the people of the Asian continent, and has been guided by their wisdom for centuries. The components of traditional rice-based farming systems are location-specific and based on the farmers' choices and resources available to them. Although variable, the common components of rice systems are rice, fish, livestock (cow, buffalo, goat), poultry, fruits, vegetables and fruit trees. Soil- and water-conservation measures and composting are an integral part of such systems. There is wide diversity of rice varieties, including sticky, aromatic, glutinous, scented, coloured, short, long, etc. and each one has a particular use (Hanks, 1992; Das *et al.*, 2015). Many farmers mix local rice cultivars and improved varieties, thus enhancing genetic diversity at the field level. Studies by plant pathologists provide evidence suggesting that indeed genetic heterogeneity reduces the vulnerability of monocultured crops to disease (Altieri, 2004). Mixing of crop species and or varieties can delay the onset of diseases by reducing the spread of disease carrying spores, and by modifying environmental conditions so that they are less favourable to the spread of certain pathogens (Altieri, 2008). Recent research in China, where four different mixtures of rice varieties grown by farmers from 15 different townships over

3000 ha, suffered 44% less blast incidence and exhibited 89% greater yield than homogeneous fields without the need to use fungicides (Zhu *et al.*, 2000).

In China alone, there are about 75 million farmers who still practise rice farming methods that are over 1000 years old. Many Chinese traditional rice paddies include fish, ducks, weeds, plankton, photosynthetic bacteria, aquatic insects, benthos, rice pests, water mice, water snakes, birds, and other soil and water microbes (Altieri *et al.*, 2017). In addition, farmers plant up to ten different species of indigenous vegetables in the field borders of the terrace fields, where also at least 62 forest species thrive; 21 of these are used as food and 53 for medicinal and herbal purposes (Koohafkan and Altieri, 2016). These rice-based farming systems support a variety of beneficial interactions: the various species of fish (*Tilapia nilotica* and *Cyprinus carpio*) consume insect pests (mainly leaf hoppers and leaf rollers) that attack the rice plant, as well as weeds that choke rice plants and rice leaves infected by sheath blight disease, thus reducing the need for pesticides. These systems exhibit a lower incidence of insect pests and plant diseases when compared to monoculture rice farming. Further, the fish oxygenate the water and move the nutrients, thereby benefiting the rice. *Azolla* species proliferate fixing nitrogen (243–402 kg/ha) some of which (17–29%) is used by the rice (Zheng and Deng, 1998). The ducks consume the *Azolla* before it covers the whole surface and triggers eutrophication, in addition to consuming snails and weeds. By consuming biomass, the fish and ducks reduce the methane emissions otherwise produced by decomposing vegetation by up to 30%, as compared to conventional farming. Clearly, the complex and diverse food webs of microbes, insects, predators and associated crop plants promote a number of ecological as well as social and economic services, beneficial to the local rural communities (Fig. 2.1) (Altieri and Nicholls, 2008; Altieri *et al.*, 2017).

In traditional systems of rice production, organic fertilizers are used extensively, and this is how rice has been cultivated

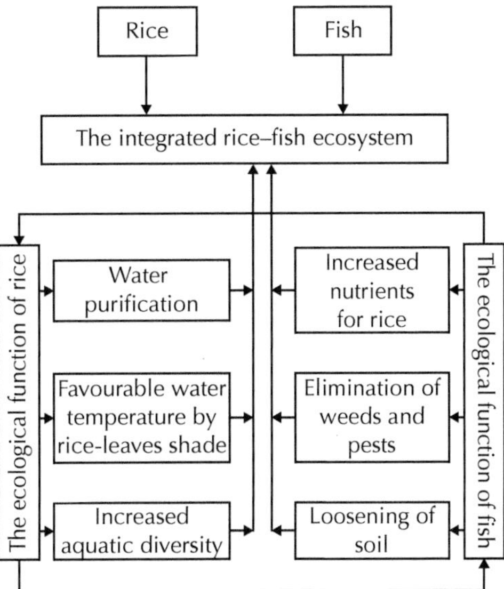

Fig. 2.1. Interactions among different components in Chinese rice–fish–duck agricultural systems (Zheng and Deng, 1998).

continuously for centuries without any impairment of fertility. In fact, the fertility of paddy fields tended to increase over time. Under modern high-input production systems, since the middle of the 1980s farmers and scientists have noticed a decline in fertilizer efficiency, particularly in areas where rice has been intensively cultivated for some time (Hanks, 1992). Several possible reasons for this degradation of the resource base have been suggested: changes in the N-supplying capacity of the soil, possibly because of repeated applications of rice straw which has a high carbon content; a build-up of soil pests such as nematodes specific to rice from continuous monocropping; deterioration from water-logging of soil given intensive irrigation, etc. (Altieri and Nicholls, 2008). Whatever the reason(s), this decrease in fertilizer efficiency means that current yields cannot be maintained without increasing inputs (Lobell *et al.*, 2009). The best solution seems to be to return to more traditional diversified cropping systems, in which at best rice is grown in rotation with other crops, preferably legumes (Altieri and Nicholls, 2004).

2.2.2 Insect pest suppression in polycultures

Intercropping is widely practised in Latin America, Asia and Africa, by smallholders as a means of increasing crop production per unit land area, with limited capital investment and minimal risk of total crop failure (Lithourgirdis *et al.*, 2011; Altieri *et al.*, 2017). Polycultures are estimated to still provide as much as 15–20% of the world's food supply. In Latin America, farmers grow 70–90% of their beans with maize, potatoes and other crops, whereas maize is intercropped on 60% of its growing areas in the region (Francis, 1986). Eighty-nine per cent of cowpeas in Africa are intercropped and the total percentage of cropped land actually devoted to intercropping varies from a low of 17% for India to a high of 94% in Malawi (Lithourgidis *et al.*, 2011; Massawa *et al.*, 2016; Mwamlima *et al.*, 2016). In these traditional multiple cropping systems, productivity in terms of harvestable products per unit area can range from 20 to 60% higher than under sole cropping with the same level of management (Kass, 1978; Vandermeer, 1989; Altieri *et al.*, 2017).

Polycultures involve spatial diversification of cropping systems (intercropping, agroforestry systems, etc.) allowing the cultivation of two or more crops simultaneously on the same field, with or without row arrangements (Vandermeer, 1989; Altieri *et al.*, 2012). Intercropping systems may involve mixtures of annual crops with other annuals, annuals with perennials, or perennials with perennials. In intercropping systems, plant species are grown in close proximity so that beneficial interactions occur between them. Intercropping provides insurance against crop failure and allows lower inputs through reduced fertilizer and pesticide requirements, thus reducing production costs and minimizing environmental impacts (Altieri and Liebman, 1986; Lithourgidis *et al.*, 2011).

It is accepted by many entomologists that inter(species) and intra(genetic) specific diversity reduces crop vulnerability to insect pests (Altieri *et al.*, 2015). There is a large body of literature documenting that diversification of cropping systems (variety mixtures, polycultures, agroforestry systems, etc.) often leads to reduced herbivore populations (Altieri, 2002; Altieri and Nicholls, 2004; Risch *et al.*, 1983). Two hypotheses have been offered to explain such reductions. The *natural enemy hypothesis* predicts that there will be a greater abundance and diversity of natural enemies of pest insects in polycultures than in monocultures (Altieri and Nicholls, 2004; Letourneau *et al.*, 2011). Predators tend to be polyphagous and have broad habitat requirements, so they would be expected to encounter a greater array of alternative prey and microhabitats in a heterogeneous environment. Several studies support the natural enemy hypothesis. In tropical maize/bean/squash systems, Letourneau (1987) studied the importance of parasitic wasps in mediating the differences in pest abundance between simple and complex crop arrangements. A squashfeeding caterpillar, *Diaphania hyalinata* (Lepidoptera: Pyralidae), occurred at low densities on intercropped squash in tropical Mexico. Part of the effect of the associated maize and bean plants may have been to render the squash plants less apparent to ovipositing moths. Polyculture fields also harboured greater numbers of parasitic wasps than did squash monocultures. Malaise trap captures of parasitic wasps in monoculture consisted of half the number of individuals caught in mixed culture. The parasitoids of the target caterpillars were also represented by higher numbers in polycultures throughout the season (Altieri and Liebman, 1986). Not only were parasitoids more common in the vegetationally diverse, traditional system, but the parasitization rates of *D. hyalinata* eggs and larvae on squash were higher in polycultures. Approximately 33% of the eggs in polyculture samples over the season were parasitized, and only 11% of eggs in monocultures. Larval samples from polycultures showed an incidence of 59% parasitization for *D. hyalinata* larvae, whereas samples from monoculture larval specimens were 29% parasitized (Letourneau, 1987).

The *resource concentration hypothesis* is based on the fact that insect populations can be influenced directly by the concentration and spatial dispersion of their food plants. Many herbivores, particularly those with narrow host ranges, are more likely to find and remain on hosts that are growing in dense or nearly pure stands and which are thus providing concentrated resources and homogeneous physical conditions (Andow, 1991; Altieri, 1999). One study that supports this hypothesis (Risch, 1981) looked at the population dynamics of six chrysomelid beetles in monocultures and polycultures of maize/bean/squash (*Cucurbita pepo*). In polycultures containing at least one non-host plant (maize), the number of beetles per unit was significantly lower, relative to the numbers of beetles on host plants in monocultures. Measurement of beetle movements in the field showed that beetles tended to emigrate more from polycultures than from host monocultures (Altieri and Liebman, 1986). Apparently, this was due to several factors: (i) beetles avoided host plants shaded by maize, (ii) maize stalks interfered with flight movements of beetles, and (iii) as beetles moved through polycultures they remained on non-host plants for a significantly shorter time than on host plants. There were no differences in rates of parasitism or predation of beetles between systems (Risch, 1981). A second study examined the effects of plant diversity on the cucumber beetle, *Acalymma vittata* (Bach, 1980). Population densities were significantly greater in cucumber (*Cucumis sativus*) monocultures than in polycultures containing cucumber and two non-host species. Bach (1980) also found greater tenure time of beetles in monocultures than in polycultures. She also determined that these differences were caused by plant diversity per se, and not by differences in host plant density or size. Thus they do not reveal if differences in numbers of herbivores between monocultures and polycultures are due to diversity or rather to the interrelated and confounding effects of plant diversity, plant density, and host plant patch size (Altieri, 2004).

Over the last 40 years, much research has been devoted to evaluate the effects of crop diversity on densities of herbivore pests and has tried to prove one or both hypothesis. An early review by Risch *et al.* (1983) summarized 150 published studies of the effect of diversifying an agroecosystem on insect pest abundance. Some 198 herbivore species were examined in these studies: 53% of these species were found to be less abundant in the more diversified system, 18% were more abundant in the diversified system, 9% showed no difference, and 20% showed a variable response (Altieri and Nicholls, 2004; Lithourgidis *et al.*, 2011). Andow (1991) analysed results from 209 studies involving 287 pest species, and found that, compared with monocultures, the population of pest insects was lower in 52% of the studies (i.e. 149 species) and higher in 15% of the studies (i.e. 44 species) (Nicholls *et al.*, 2016). Of the 149 pest species with lower populations in intercrops, 60% were monophagous and 28% polyphagous. The population of natural enemies of pests was higher in the intercrop in 53% of the studies and lower in 9%. The reduction in pest numbers was almost twice for monophagous insects (53.5% of the case studies showed lowered numbers in polycultures) than for polyphagous insects (33.3%) (Andow, 1991).

In a meta-analysis of 21 studies comparing pest suppression in polycultures versus monocultures, Tonhasca and Byrne (1994) found that polycultures significantly reduced pest densities by 64%. In a later meta-analysis, Letourneau *et al.* (2011) found a 44% increase in abundance of natural enemies (148 comparisons), a 54% increase in herbivore mortality, and a 23% reduction in crop damage on farms with species-rich vegetational diversification systems than on farms with species-poor systems. Unequivocally, earlier reviews and recent meta-analyses suggest that diversification schemes generally achieve significant positive outcomes including natural enemy enhancement, reduction of herbivore abundance, and reduction of crop damage, from a combination of bottom-up and top-down effects (Kremen and Miles, 2012).

Work in Kenya by scientists at the International Center of Insect Physiology and Ecology (ICIPE) added a new dimension by considering the chemical ecology of these systems (Kahn and Pickett, 2008). A habitat management system was developed to control the stem borer, using two kinds of crops that are planted together with maize: a plant that repels borers (the push) and another that attracts (pulls) them (Khan *et al.*, 1998). The plant chemistry responsible for stem borer control involves the release of attractive volatiles from the trap plants and repellent volatiles from the intercrops. Two of the most useful trap crops that pull in the borers' natural enemies such as the parasitic wasp (*Cotesia sesamiae*), are Napier grass and Sudan grass, both important fodder plants; these are planted in a border around the maize (Cook *et al.*, 2007). Two excellent borer-repelling crops, which are planted between the rows of maize, are molasses grass, which also repels ticks, and the leguminous silverleaf (*Desmodium*), which in addition can suppress the parasitic weed *Striga* by a factor of 40 compared to maize monocrop. *Desmodium*'s N-fixing ability increases soil fertility leading to a 15–20% increase in maize yield (Khan *et al.*, 1998).

The push–pull strategy, was adopted by more than 10,000 households in 19 districts in Kenya, 5 districts in Uganda, and 2 districts in Tanzania, helping participating farmers to increase their maize yields by an average of 20% in areas where only stem borers are present and by more than 50% in areas where both stem borers and *Striga* are problems (Khan and Pickett, 2008). Participating farmers in the breadbasket of Trans Nzoia reported a 15–20% increase in maize yield. In the semi-arid Suba district, plagued by both stem borers and *Striga* a substantial increase in milk yield has occurred in the last four years, with farmers now being able to support grade cows on the fodder produced by *Desmodium* and other plants. When farmers plant maize, Napier grass and *Desmodium* together, a return of US$2.30 for every dollar invested is made, as compared to only $1.40 obtained by planting maize as a monocrop (Cook *et al.*, 2007).

2.2.3 Insect prevalence in agroforestry systems

Agroforestry is an intensive land-management system that combines trees and/or shrubs with crops and/or livestock on a landscape level to achieve optimum benefits from biological interactions. The few reviews on pest management in agroforestry (Schroth *et al.*, 2000; Rao *et al.*, 2000) stipulate that the high plant diversity associated with agroforestry systems provide some level of protection from pest and disease outbreaks (Altieri and Nicholls, 2007).

Shade from trees may markedly reduce pest density in understorey intercrops. The effect of shade on pests and diseases in agroforestry has been studied quite intensively in cocoa and coffee systems undergoing transformation from traditionally shaded crop species to management in unshaded conditions (Altieri and Nicholls, 2007). In cocoa plantations, insufficient overhead shade favours the development of numerous herbivorous insect species, including thrips (*Selenothrips rubrocinctus*) and mirids (*Sahlbergella, Distantiella*, etc.). Even in shaded plantations, these insects concentrate at spots where the shade trees have been destroyed, e.g. by wind (Beer *et al.*, 1997). Bigger (1981) found an increase in the numbers of Lepidoptera, Homoptera, Orthoptera and the mirid *Sahlbergella singularis* and a decrease in the number of Diptera and parasitic Hymenoptera from the shaded towards the unshaded part of a cocoa plantation in Ghana (Altieri and Nicholls, 2008).

In coffee, the effect of shade on insect pests is less clear than in cocoa, as the leaf miner (*Leucoptera meyricki*) is reduced by shade, whereas the coffee berry borer (*Hypothenemus hampei*) may increase under shade (Altieri and Nicholls, 2007). Similarly, unshaded tea suffers more from attack by thrips and mites, such as the red spider mite (*Oligonychus coffeae*) and the pink mite (*Acaphylla theae*), whereas heavily shaded and moist plantations are more severely damaged by mirids (*Helopeltis* spp.) (Guharay *et al.*, 2000). For a low elevation dry coffee zone in Central America, shade should be managed between 35 to 65%, as shade promotes leaf retention in the dry season and reduces *Cercospora coffeicola*, weeds and *Planococcus citri* (Staver *et al.*, 2001; Altieri *et al.*, 2015). The complete elimination of shade trees from coffee systems can have an enormous impact on the diversity and density of arthropods, especially ants. Studying the ant community in a gradient of coffee plantations going from systems with high density of shade to shadeless plantations, Perfecto and Vandermeer (1996) reported a significant decrease in ant diversity, with important implications for pest control as a diverse ant community can offer more safeguards against pest outbreaks. This canopy layer provides plantations with a forest-like vegetation structure that can help maintain biodiversity (Gonthier *et al.*, 2013). Ant biodiversity is high in many coffee plantations with a forest-like vegetation structure and ants attack and prey on many coffee pests, including the coffee berry borer (CBB) (*Hypothenemus hampei*) (Philpott and Armbrecht, 2006). *Azteca instabilis* F. Smith is a competitively dominant ant that aggressively patrols arboreal territories in high densities and it has been found that it impacts the berry borer (Gonthier *et al.*, 2013).

The manipulation of ground cover vegetation in tropical plantations can significantly affect tree growth by altering nutrient availability, soil physics, and moisture, and the prevalence of weeds, plant pathogens, and insect pests and associated natural enemies (Altieri *et al.*, 2017). A number of entomological studies conducted in these systems indicate that plantations with rich floral undergrowth exhibit a significantly lower incidence of insect pests than clean cultivated orchards, mainly because of an increased abundance and efficiency of predators and parasitoids, or other effects related to habitat changes (Altieri and Nicholls, 2004, 2007). In the Solomon Islands,

O'Connor (1950) recommended the use of a cover crop in coconut groves to improve the biological control of coreid pests by the ant *Oecophylla smaragdina subnitida*. In Ghana, coconut gave light shade to cocoa and supported, without apparent crop loss, high populations of *Oecophylla longinoda*, keeping the cocoa crop free from cocoa capsids (Leston, 1973). Wood (1971) reported that in Malaysian oil palm (*Elaeis guineensis*) plantations, heavy ground cover, irrespective of type, reduced damage to young trees caused by rhinoceros beetle (*Oryctes rhinoceros*).

2.3 Conclusions

The persistence of millions of hectares under traditional agriculture in the form of rice terraces, polycultures, agroforestry systems, etc., document a successful indigenous agricultural adaptation strategy to difficult environments and comprises a tribute to the 'creativity' of peasants throughout the developing world (Koohafkan and Altieri, 2016; Altieri *et al.*, 2015). These microcosms of traditional agriculture offer promising models for other areas as they promote biodiversity, thrive without agrochemicals, and sustain year-round yields (Denevan, 1995; Altieri, 2004). Only recently have applied ecologists, agronomists and pest management specialists recognized the virtues of diversified traditional agroecosystems whose sustainability lies in the complex ecological models they follow. The study of traditional agroecosystems and the ways in which peasants maintain and use biodiversity can speed the emergence of agroecological principles, which are urgently needed to develop more sustainable agroecosystems and agrobiodiversity conservation strategies both in the industrial and developing countries (Altieri and Toledo, 2005). In fact, such studies have already helped some agroecologists to create novel farm designs that considerably reduce pest problems and thus obviate the use of pesticides (Malezieux, 2012). A key challenge has involved the translation of the ecological principles underlying traditional farms into practical strategies for pest management. Nevertheless, there are some noteworthy examples such as the deployment of variety mixtures of local rice with hybrids in Yumman, China to reduce blast incidence (Zhu *et al.*, 2000; Altieri and Nicholls, 2003a), the push–pull system in Africa for stem borer control (Khan *et al.*, 1998) and the design of pest-suppressive multistrata shade-grown coffee systems in Central America to simultaneously reduce *Cercospora coffeicola*, weeds and *Planococcus citri* (Staver *et al.*, 2001).

Following are several practical suggestions for pest management emerging from lessons learned by studying traditional farming systems:

- Agroecosystems should mimic the diversity and functioning of local ecosystems thus exhibiting tight nutrient cycling, complex structure and enhanced biodiversity (Altieri and Nicholls, 2017). The expectation is that such agricultural mimics, like their natural models, can be productive, pest resistant and conservative of nutrients and biodiversity. Ewel (1986) argues that natural plant communities have several traits (pest suppression among them) that would be desirable to incorporate into agroecosystems. Thus, the prevalent coevolved natural secondary plant associations of an area should provide the model for the design of multi-species crop mixtures.
- Increase species diversity at the landscape and field level, as this promotes fuller use of resources (nutrients, radiation, water, etc.), protection from pests and compensatory growth (Altieri and Nicholls, 2004). Many researchers have highlighted the importance of various spatial and temporal plant combinations to facilitate complementary resource use or to provide intercrop advantage, such as in the case of legumes facilitating the growth of cereals by supplying extra nitrogen (Vandermeer, 1989; Altieri and Nicholls, 2004). Compensatory growth

is another desirable trait as, if one species succumbs to pests, weather or harvest, another species fills the void, maintaining full use of available resources. Crop mixtures also minimize risks, especially by creating the sort of vegetative texture that suppresses specialist pests (Altieri and Nicholls, 2004).

- Enhance longevity through the addition of perennials that contain a thick canopy, thus providing continual cover that protects the soil. Constant leaf fall builds organic matter and allows uninterrupted nutrient circulation. Dense, deep root systems of long-lived woody plants is an effective mechanism for nutrient capture offsetting the negative losses through leaching (Beer *et al.*, 1997; Altieri and Nicholls, 2007).
- Impose a fallow to restore soil fertility through biomass accumulation and biological activation, and to reduce agricultural pest populations as life cycles are interrupted with a rotation of fallow vegetation and crops (Altieri and Nicholls, 2004).
- Enhance additions of organic matter by including legumes, biomass-producing plants and incorporating animals. Accumulation of organic matter is key for activating soil biology, improving soil structure and macroporosity and elevating the nutrient status of soils (Altieri and Nicholls, 2007).
- Lower applications of synthethic N fertilizers, as many studies documenting lower abundance of several insect herbivores in low-input systems have partly attributed such reductions to the lower nitrogen content in organically farmed crops. In Japan, density of immigrants of the plant hopper species *Sogatella furcifera* was significantly lower and the settling rate of female adults and survival rate of immature stages of ensuing generations were generally lower in organic compared to conventional rice fields. Consequently, the density of plant hopper nymphs and adults in the ensuing generations

was found to decrease in organically farmed fields (Altieri and Nicholls, 2003b).

- Increase landscape diversity by having in place a mosaic of agroecosystems representative of various stages of succession. Improved pest control is also linked to spatial heterogeneity at the landscape level (Altieri and Nicholls, 2008). The species pool in the surrounding landscape and the distance of crop from natural habitat are important for the conservation of enemy diversity and, in particular, the conservation of poorly-dispersing and specialized enemies (Tscharntke *et al.*, 2007). Structurally complex landscapes with high habitat connectivity may enhance the probability of pest regulation. In addition, risk of complete failure is spread among, as well as within, the various farming systems.

References

Altieri, M.A. (1999) The ecological role of biodiversity in agroecosystems. *Agriculture, Ecosystems and Environment* 74, 19–31.

Altieri, M.A. (2002) Agroecology: the science of natural resource management for poor farmers in marginal environments. *Agriculture, Ecosystems and Environment* 93, 1–24.

Altieri, M.A. (2004) Linking ecologists and traditional farmers in the search for sustainable agriculture. *Frontiers in Ecology and the Environment* 2, 35–42.

Altieri, M.A. (2005) The myth of coexistence: why transgenic crops are not compatible with agroecologically based systems of production. *Bulletin of Science, Technology and Society* 25, 287–288.

Altieri, M.A. (2008) Small farms as a planetary ecological asset: five key reasons why we should support the revitalization of small farms in the global south. Available at: www.plantingjustice.org/resources/food-justice-research/support-small-farms (accessed 25 July 2017).

Altieri, M.A. and Liebman, M. (1986) Insect, weed, and plant disease management in multiple cropping systems. In: Francis, C.A. (ed.) *Multiple Cropping Systems*. Macmillan, New York.

Altieri, M.A. and Nicholls, C.I. (2003a) Agroecology: rescuing organic agriculture from a specialized industrial model of production and distribution. *Ecology and Farming* 34, 24–26.

Altieri, M.A. and Nicholls, C.I. (2003b) Soil fertility management and insect pests: harmonizing soil and plant health in agroecosystems. *Soil and Tillage Research* 72, 203–211.

Altieri, M.A. and Nicholls, C.I. (2004) *Biodiversity and Pest Management in Agroecosystems*. Haworth Press, New York.

Altieri, M.A. and Nicholls, C.I. (2007) Ecologically based pest management in agroforestry systems. In: Daizy, R.B., Ravinder, K.K., Shibu, J. and Harminder, P.S. (eds) *Ecological Basis of Agroforestry*. CRC Press, Boca Raton, London and New York, pp. 95–107.

Altieri, M.A. and Nicholls, C.I. (2017) Agroecology: a brief account of its origins and currents of thought in Latin America. *Agroecology and Sustainable Food Systems* 41, 231–237.

Altieri, M.A. and Toledo, V.M. (2005) Natural resource management among small-scale farmers in semi-arid lands: building on traditional knowledge and agroecology. *Annals of Arid Zone* 44, 365–385.

Altieri, M.A., Funes-Monzote, F.R. and Petersen, P. (2012) Agroecologically efficient agricultural systems for smallholder farmers: contribution to food sovereignty. *Agronomic Sustainable Development* 32, 1–13.

Altieri, M.A., Nicholls, C.I., Henao, A. and Lana, M.A. (2015) Agroecology and the design of climate change-resilient farming systems. *Agronomy for Sustainable Development* 35, 869–890.

Altieri, M.A., Nicholls, C.I. and Montalba, R. (2017) Technological approaches to sustainable agriculture at a crossroad: an agroecological perspective. *Sustainability* 9, 349. DOI:10.3390/su9030349

Andow, D.A. (1991) Vegetational diversity and arthropod population response. *Annual Review of Entomology* 36, 561–586.

Bach, C.E. (1980) Effects of plant diversity on the population dynamics of a specialist herbivore, the striped cucumber beetle, *Acalymma vittata* (Fab.). *Ecology* 61, 1515–1530.

Beer, J., Muschler, R., Kass, D. and Somarriba, E. (1997) Shade management in coffee and cacao plantations. In: Nair, P.K.R. and Latt C.R. (eds) *Forestry Sciences: Directions in Tropical Agroforestry Research*. Kluwer Academic Publishers, Dordrecht, The Netherlands, pp. 139–164.

Bigger, M. (1981) Observations on the insect fauna of shaded and unshaded Amelonado cocoa. *Bulletin of Entomological Research* 71, 107–119.

Chang, J.H. (1977) Tropical agriculture: crop diversity and crop yields. *Economic Geography* 53, 241–254.

Clawson, D.L. (1985) Harvest security and intraspecific diversity in traditional tropical agriculture. *Economic Botany* 39, 56–67.

Cook, S.M., Khan, Z.R. and Pickett, J.A. (2007) The use of push-pull strategies in Integrated Pest Management. *Annual Review of Entomology* 52, 375–400.

Das, A., Ramkrushn, G.I., Yadav, G.S, Layek, J., Debnath, C., Choudhury, B.U., Mohaptara, K.P., Ngachan, S.V. and Das, S. (2015) Capturing traditional practices of rice based farming systems and identifying interventions for resource conservation and food security in Tripura, India. *Applied Ecology and Environmental Sciences* 3, 100–107.

Denevan, W.M. (1995) Prehistoric agricultural methods as models for sustainability. *Advanced Plant Pathology* 11, 21–43.

Dewalt, B.R. (1994) Using indigenous knowledge to improve agriculture and natural resource management. *Human Organization* 5, 23–31.

Donald, P.F. (2004) Biodiversity impacts of some agricultural commodity production systems. *Conservation Biology* 18, 17–38.

Ewel, J.J. (1986) Designing agricultural ecosystems for the tropics. *Annual Review of Ecology and Systematics* 17, 245–271.

Fletcher, R.J., Robertson, B.A., Evans, J., Doran, P.J., Alavalapati, J.R. and Schemske, D.W. (2011) Biodiversity conservation in the era of biofuels: risks and opportunities. *Frontiers in Ecology and the Environment* 9, 161–168. DOI:10.1890/090091

Francis, C.A. (1986) *Multiple Cropping Systems*. Macmillan, New York.

Gomiero, T. (2016) Soil degradation, land scarcity and food security: reviewing a complex challenge. *Sustainability* 8, 281. DOI:10.3390/su8030281

Gonthier, D.J., Ennis, K.K., Philpott, S.M., Vandermeer, J. and Perfecto, I. (2013) Ants defend coffee from berry borer colonization. *BioControl* 58, 815. DOI:10.1007/s10526-013-9541-z

Guharay, F., Monterrey, J., Monterroso, D. and Staver, C. (2000) *Manejo integrado de plagas en el cultivo de café*. CATIE, Managua, Nicaragua.

Hanks, L. (1992) *Rice and Man: Agricultural Ecology in Southeast Asia*. University of Hawaii Press, Honolulu, Hawaii.

James, C. (2013) *Global Status of Commercialized Biotech/GM Crops*. ISAAA Briefs 46, Ithaca, New York.

Kass, D.C.L. (1978) Polyculture cropping systems: review and analysis. *Cornell International Agricultural Bulletin* 32.

Khan, Z. and Pickett, J. (2008) Push-pull strategy for insect pest management. In: Capinera, J.L. (ed.) *Encyclopedia of Entomology*. Springer, Dordrecht, The Netherlands, pp. 3074–3082.

Khan, Z.R., Ampong-Nyarko, K., Hassanali, A. and Kimani, S. (1998) Intercropping increases parasitism of pests. *Nature* 388, 631–632.

Koohafkan, P. and Altieri, M.A. (2016) *Forgotten Agricultural Heritage: Reconnecting Food Systems and Sustainable Development*. Earthscan Food and Agriculture Series, Routledge, London.

Kremen, C. and Miles, A. (2012) Ecosystem services in biologically diversified versus conventional farming systems: benefits, externalities, and trade-offs. *Ecology and Society* 17, 40. DOI:10.5751/ES-05035-170440

Landis, D.A., Gardiner, M.M., van der Werf, W. and Swinton, S.M. (2008) Increasing corn for biofuel production reduces biocontrol services in agricultural landscapes. *Proceedings of the National Academy of Sciences* 105, 20552–20557.

Leston, D. (1973) The ant mosaic: tropical tree crops and the limiting of pests and diseases. *Proceedings of the National Academy of Sciences* 19, 311.

Letourneau, D.K. (1987) The enemies hypothesis: tritrophic interaction and vegetational diversity in tropical agroecosystems. *Ecology* 68, 1616–1622.

Letourneau, D.K., Armbrecht, I., Salguero Rivera, B., Montoya Lerma, J., Jimenez Carmona, E. *et al.* (2011) Does plant diversity benefit agroecosystems? A synthetic review. *Ecological Applications* 21(1), 9–21.

Lithourgidis, A.S., Dordas, C.A., Damalas, C.A. and Vlachostergios, D.N. (2011) Annual intercrops: an alternative pathway for sustainable agriculture. *Australian Journal of Crop Science* 5, 396–410.

Lobell, D.B., Cassman, K.G. and Field, C.B. (2009) Crop yield gaps: their importance, magnitudes, and causes. *Annual Review of Environment and Resources* 34, 179–204.

Malezieux, E. (2012) Designing cropping systems from nature. *Agronomy for Sustainable Development* 32, 15–29.

Massawa, P.I., Mtei, K.M., Munishi, L.K. and Ndakidemi, P.A. (2016) Existing practices for soil fertility management through cereals-legume intercropping systems. *World Research Journal of Agricultural Sciences* 3, 80–91.

Mwamlima, L.H., Kabambe, V.H., Nyirenda, G.K.C. and Mhango, W.G. (2016) Effects of intercropping systems and foliar pesticides applied to control cotton (*Gossypium hirsutum* L.) pests on incidences of cowpea (*Vigna unguiculata* L. Walp) pests. *Agricultural Science Research Journal* 6, 313–321.

Nicholls, C.I. and Altieri, M.A. (2004) Designing species-rich, pest-suppressive agroecosystems through habitat management. In: Rickerl, D. and Francis, C. (eds) *Agroecosystems Analysis.* American Society of Agronomy, Madison, Wisconsin, pp. 49–62.

Nicholls, C.I., Altieri, M.A. and Vazquez, L. (2016) Agroecology: principles for the conversion and redesign of farming systems. *Journal of Ecosystem & Ecography* 55. DOI:10.4172/2157-7625

Perfecto, I. and Vandermeer, J.H. (1996) Microclimatic changes and the indirect loss of ant diversity in a tropical agroecosystem. *Oecologia* 108, 577–582.

Philpott, S.M. and Armbrecht, I. (2006) Biodiversity in tropical agroforests and the ecological role of ants and ant diversity in predatory function. *Ecological Entomology* 31, 369–377.

Rao, M.R., Singh, M.P. and Day, R. (2000) Insect pest problems in tropical agroforestry systems: contributory factors and strategies for management. *Agroforestry Systems* 50(3), 243–277.

Risch, S.J. (1981) Insect herbivore abundance in tropical monocultures and polycultures: an experimental test of two hypotheses. *Ecology* 62, 1325–1340.

Risch, S.J., Andow, D. and Altieri, M.A. (1983) Agroecosystem diversity and pest control: data, tentative conclusions, and new research directions. *Environmental Entomology* 12, 625–629.

Schroth, G., Krauss, U., Gasparotto, L., Aguilar, J., Duarte, A. and Vohland, K. (2000) Pests and diseases in agroforestry systems of the humid tropics. *Agroforestry Systems* 50, 199–241.

Staver, C., Guharay, F., Monterroso, D. and Muschler, R.G. (2001) Designing pest-suppressive multistrata perennial crop systems: shade grown coffee in Central America. *Agroforestry Systems* 53, 151–170. DOI:10.1023/A:1013372403359

Tonhasca, A. and Byrne, D.N. (1994) The effects of crop diversification on herbivorous insects: a meta-analysis approach. *Ecological Entomology* 19 (3), 239–244.

Tscharntke, T., Bommarco, R., Clough, Y., Crist, T.O., Kleijn, D., Rand, T.A., Tylianakis, J.M., van Nouhuys, S. and Vidal, S. (2007) Conservation biological control and enemy diversity on a landscape scale. *Biological Control* 43, 294–309.

Vandermeer, J. (1989) *The Ecology of Intercropping.* Cambridge University Press, Cambridge, UK.

Watanabe, K., Koji, S., Hidaka, K. and Nakamura, K. (2013) Abundance, diversity, and seasonal population dynamics of aquatic Coleoptera and Heteroptera in rice fields: effects of direct seeding management. *Environmental Entomology* 42(5), 841–850. DOI:10.1603/EN13109

Wood, B.J. (1971) Development of integrated control programs for pests of tropical perennial crops in Malaysia. In: Huffaker, C.B. (ed.) *Biological Control.* Plenum Press, New York, pp. 422–457.

Woodhouse, P. (2010) Beyond industrial agriculture? Some questions about farm size, productivity and sustainability. *Journal of Agrarian Change* 10, 437–453.

Zheng, Y. and Deng, G. (1998) Benefits analysis and comprehensive evaluation of rice – fish-duck symbiotic model. *Chinese Journal of Eco-Agriculture* 6(1), 48–51.

Zhu, Y., Fen, H., Wang, Y., Li, Y., Chen, J., Hu, L. and Mundt, C.C. (2000) Genetic diversity and disease control in rice. *Nature* 406, 718–772.

3 Options and Challenges for Pest Control in Intensive Cropping Systems in Tropical Regions

Silvana V. Paula-Moraes[1,*], Fábio Maximiano de Andrade Silva[2] and Alexandre Specht[3]

[1]Entomology and Nematology Department, West Florida Research and Education Center, University of Florida, Jay, USA; [2]Insecticide Resistance Action Committee Paulínia, Brazil; [3]Embrapa Cerrados, Planaltina, Brazil

3.1 Introduction

The Food and Agriculture Organization (FAO) estimates that, by mid-century, global demand for food production will be 60% higher than it is currently, and there will be a need to provide food for an additional 2 billion people (FAO, 2012). Thus the demand to develop more hectares of arable land has been highlighted, particularly with respect to the potential for farmland expansion in sub-Saharan Africa and Latin America (Morris *et al.*, 2012).

In Brazil, the successful agricultural transformation of a broad savanna, commonly called Cerrado, is an example of the adoption of a green revolution programme. The region, which covers 2 million km² of central Brazil and represents about 23% of the country's land surface (Ratter and Ribeiro, 1996), went through a transformation from non-agricultural regions to intensive cropping systems, especially for the production of grains and fibre. This transformation happened from 1975 until the beginning of the 1980s, and many governmental programmes were launched with the intent of stimulating development of the Brazilian savanna through subsidies for agriculture (Amabile and Barcelos, 2008).

Another important government policy was a robust investment in research. Contributions from the work of governmental institutions, such as Embrapa, state companies and universities in Brazil played, and continue to play, an important role in the increase of agricultural and cattle production in the savanna.

Challenges associated with poor soils and the availability of local cultivars were the first barriers to be overcome. Moreover, knowledge of the environmental peculiarities of the region and the economic and social characteristics was not available. The challenge of low fertility drove the development of correction techniques, such as appropriate soil fertilization and the selection of grain and pasture varieties tolerant to aluminium. The development of soil management techniques, such as tillage and more recently, no-tillage systems, in addition to the use of appropriate farm implements, contributed to maintaining the physical properties of soils, increasing water infiltration and reducing the risk of

* Corresponding author e-mail: paula.moraes@ufl.edu

erosion. The integration of crop–livestock systems at different levels of intensification, and the possibility of including forestry inputs, represent alternative strategies to improve food production. Furthermore, diversified systems that integrate livestock and forestry have been advocated as an approach to increase ecosystem services. Climatic studies and irrigation techniques were also developed and allowed better understanding and management of the distribution of rainfall in the region. Socioeconomic factors were analysed, and currently the farm expansion promoted a total farm revenue increase from 19.2% in 1985 to 28.7% in 1995/96 and 33.2% in 2006 (IBGE, 2006).

However, a challenge that remains for intensive cropping systems in the tropics, and particularly in the Brazilian savanna, is the occurrence of pests, which can compromise agricultural activity. Stern *et al.* (1959) listed the manipulation of the environment, primarily the establishment of monocultures, as one of the causes of pest outbreaks in agroecosystems. The establishment of lower economic thresholds in intensive cropping systems, which implies an extremely low pest tolerance, should be also considered (Raupp *et al.*, 1988). A final challenge is the globalization of markets and associated traffic of numerous commodities. The modern global economy has several pathways for the introduction of invasive species, which threaten agroecosystems. In savanna areas, with conditions that are favourable for year-round pest survival and establishment, the risk of invasive species is even more problematic.

3.2 Current Status of Pest Problems and Management in the Intensive Cropping Systems of the Brazilian Savanna

There are several examples of pests that reach high population densities and demand management in the large maize, soybean and cotton production areas of the Brazilian savanna. Pest groups, especially from Lepidoptera, Coleoptera and Hemiptera, present

significant risk to the agricultural economy. The level of importance of the specific pest depends on several factors, such as the ecological attributes of the pest and the specific area of the savanna region, phenology stage of the crop, etc.

In Lepidoptera, several species from the Superfamily Noctuoidea are a critical pest group, considering the damage capacity of the larval stage. They favour conditions in savanna areas, developing several generations per year and have the ability to disperse and colonize new host plants in the succession of crops in the agricultural landscape. In this group, species in the Erebidae family from the genera *Alabama* Grote, 1895 and *Anticarsia* Hübner, 1818, have significant economic importance. Another pest group belongs to the Noctuidae family, of which members of the genera *Agrotis* Ochsenheimer, 1816, *Chloridea* Duncan & Westwood, 1841, *Chrysodeixis* Hübner, 1821, *Helicoverpa* Hardwick, 1965, *Peridroma* Hübner, 1821, *Rachiplusia* Hampson, 1913 and *Spodoptera* Guenée, 1852 have economic importance. All these genera have species with key pest status in maize (e.g *Spodoptera frugiperda* (J.E. Smith, 1797)), soybean (e.g. *Chrysodeixis includens* (Walker, 1858)), and cotton (e.g. *Chloridea virescens* (Fabricius, 1777)), among other species which have secondary or subeconomic pest status (Table 3.1). Moreover, the detection of *Helicoverpa armigera* (Hübner, 1808) in Brazil (Czepak *et al.*, 2013; EMBRAPA, 2013; Specht *et al.*, 2013; Sosa-Gómez *et al.*, 2016) causing economic impact, particularly in the Savanna region, represents an important new pest problem since its introduction.

Considering the phenology of crops and the time of pest occurrence of members from the Noctuoidea superfamily, Zahiri *et al.* (2011) divided the group into four 'pest clades': (i) cutworms (*Agrotis* latu sensu and *Peridroma*); (ii) semiloopers (Plusiinae); (iii) armyworms (*Spodoptera* spp.); and (iv) budworms and earworms (Heliothines).

Cutworms typically cause economic impact in the beginning of the plant development, and can compromise the crop

Table 3.1. Species from Noctuidae: Noctuinae (Prodeniini), Plusiinae and Heliothinae present in the Brazilian savanna associated maize, soybean and cotton (Silva *et al.*, 1968; Lafontaine and Poole, 1991; Coto *et al.*, 1995; Pogue, 2002; Pastrana, 2004; Specht *et al.*, 2013).

Subfamily/species	Cotton	Maize	Soybean
Noctuinae (Prodeniinae)			
Agrotis ipsilon (Hufnagel, 1766)	X	X	X
Agrotis malefida Guenée, 1852	X	X	X
Feltia repleta (Walker, 1857)	X	X	X
Feltia subterranea (Fabricius, 1794)	X	X	X
Peridroma saucia (Hübner, 1808)	X	X	X
Spodoptera albula (Walker, 1857)	X	X	X
Spodoptera androgea (Stoll, 1782)		X	
Spodoptera cosmioides (Walker, 1858)	X	X	X
Spodoptera dolichos (Fabricius, 1794)	X	X	X
Spodoptera eridania (Stoll, 1782)	X	X	X
Spodoptera frugiperda (J.E. Smith, 1797)	X	X	X
Plusiinae			
Autoplusia egena (Guenée, 1852)	Malvaceae	X	X
Ctenoplusia oxygramma (Geyer, 1832)	X		X
Chrysodeixis includens (Walker, 1858)	X	X	X
Rachiplusia nu (Guenée, 1852)	X	X	X
Trichoplusia ni (Hübner, 1803)	X	X	X
Heliothinae			
Helicoverpa armigera (Hübner, 1809)	X	X	X
Helicoverpa zea (Boddie, 1850)	X	X	X
Chloridea virescens (Fabricius, 1777)	X	X	X

stand. Cutworm larvae have underground habits and roll up when disturbed. During the night, larvae come to the surface to feed, and damage is caused by cutting the stems of the plants at or near the ground surface. Generally, the damage occurs in maize, soybean or cotton up to 45 days after crop emergence (Baudino, 2004; Lafontaine, 2004; Kitching and Rawlins, 1999; Pogue, 2004; Angulo *et al.*, 2008; Specht *et al.*, 2013). In Brazil, the most common genera are *Agrotis*, *Feltia* Walker, 1856 and *Peridroma*.

Semilooper larvae occur during all crop phenological stages of several dicotyledonous crops such as cotton and soybean. However, it can also occur in maize when regional populations are high. *Trichoplusia ni* (Hübner, 1803) is a cosmopolitan and polyphagous species of many vegetables, but can also cause economic damage to cotton and soybean. *Chrysodeixis includens* is considered one of the most important pests in the Brazilian savanna, feeding on soybean and cotton. *Rachiplusia nu* (Guenée, 1852) is another pest in the semi-looper clade, but is of localized economic importance in the south and southeast regions of Brazil.

The armyworm clade is polyphagous and feeds during all crop phenological stages. *Spodoptera frugiperda* is the most important species in the Brazilian savanna and has a broad host range and huge ecological plasticity, with reported presence in different host phenotypes (Machado *et al.*, 2008). *Spodoptera dolichos* (Fabricius, 1794), likewise a cutworm, can also feed during the initial crop development stages. *Spodoptera albula* (Walker, 1857), *S. eridania* (Stoll, 1782) and *S. cosmioides* (Walker, 1858) are well known by the common name 'black caterpillar', and have been increasing in economic importance in Brazilian savanna regions. These species feed on maize leaves, reproductive tissues of cotton and soybean pods.

Budworm and earworm larvae include species with a preference to feed on plant shoots and reproductive tissue. Even though species of this clade are polyphagous, they do have preferred hosts, such as the preference of *Helicoverpa zea* (Boddie, 1850) for maize and the preference of *Chloridea virescens* (Fabricius, 1777) for cotton and soybean. In the case of *Helicoverpa armigera* (Hübner, 1808), there were reports of economic damage caused by this species in several crops during the crop season of 2012/2013. One hypothesis to explain the recent occurrence of significant economic damage to multiple hosts is the behaviour of an invasive species in a new environment, where the population built through time (Sosa-Gómez *et al.*, 2016) and resulted in an outbreak, which caused significant impact to several crops cultivated in Brazil. However, field reports and observations in the Brazilian savanna indicate that *H. armigera* has a preference for cotton, soybean and sorghum (Cunningham and Zalucki, 2014).

Insecticides are a common tactic in most intensive systems of maize, soybean and cotton because of their fast action, ease of application, and relative economic viability, at least in the short term. Several classes of insecticides have been adopted, and synthetic pyrethroids (SPs) and organophosphates (OPs) are most frequently used. The newest class of synthetic insecticides that have been used on a large scale to control insect pests in the Brazilian savanna are neonicotinoids, especially imidacloprid and thiamethoxan in combination with synthetic pyrethroids. However, neonicotinoids are options for the control of piercing-sucking insects, such as aphids and whiteflies, and are not effective against Lepidoptera. During the last few years, with the growth of importance of Lepidoptera in soybean and cotton, including the new scenario of *H. armigera* in Brazil, new chemical groups such as Diamides (chlorantraniliprole, flubendiamide and cyantraniliprole), Pyrroles (chlorfenapyr), Oxadiazines (indoxacarb), Diacylhydrazines (methoxyfenozide) have increased in use. These new chemical classes have increased in use because of several reasons such as high efficacy, control spectrum (across different species of Lepidoptera), flexibility of use (different size of larva and broad window of use during the crop cycle) and increased residual (more than 7–10 days of control). Virus (NPV) and *Bacillus thuringiensis* formulations (the insecticidal proteins they produce) have been also adopted on a large scale to control lepidopterans.

In soybean production in the Brazilian savanna, the use of chemical control has been the basis of pest control against the pest groups of Lepidoptera, stink bug and, most recently, whitefly, mites and thrips. Overall, the recommendations highlight the following aspects.

1. Application of appropriate products to appropriate pest stages (e.g. Diamides, IGR, Oxadiazines and Pyrroles typically work best on early stages of *A. gemmatalis* Hübner, 1818, *C. includens*, *H. armigera* and *Spodoptera* spp).
2. Preference for selective products which preserve pest natural enemies, because they can delay or reduce the infestation of a new pest population.
3. Products are often most appropriate for specific phenological stages of a crop, so an understanding of the residual, translaminar effect, systemic activity, ingestion or contact activity and residual can assist the grower in proper product use and application timing.
4. Spray intervals adjustment to account for pest biology (e.g. pests with short life cycles may undergo a rapid population increase, so more than one application at a shorter intervals might be appropriate to effectively manage the pest).
5. Product rotation to avoid high selection pressure which increases the risk of resistance.
6. Adequate insecticide coverage to maximize product performance.

7. Use of recommended product label rates and avoidance of sub-dosage or super-dosage.
8. Correct calibration and maintenance of application equipment.
9. Avoidance of tank mixtures of pesticides with the same mode of action.
10. Avoidance of insecticides which have existing resistance problems.

Since the crop season of 2013/2014, plants of soybean with insecticidal proteins from *Bacillus thuringiensis* (Bt) expressing the *Cry*1Ac toxin have been available in Brazil (Table 3.2). Expectation of massive adoption of the technology was confirmed in the crop season of 2015/2016. The reasons for rapid farmer acceptance can be explained by the high level of control provided by Bt technology for important lepidopteran pest species in the Brazilian savanna, such as *H. armigera*, *C. includens*, *A. gemmatalis* and *C. virescens* (MacRae *et al.*, 2005; Bernardi *et al.*, 2012; Yano *et al.*, 2015).

Maize production in the Brazilian savanna, and indeed the entire country, has been challenged in the last years. With the introduction of Bt traits, pest scouting became less practised because all transgenic technologies at that time exhibited a high level of control of the main pest, *S. frugiperda*, in maize. However, with the reported failure of *Cry*1F (Farias *et al.*, 2014) and cross-resistance to *Cry*1Ab (Bernardi *et al.*, 2015), it is necessary to consider a new management platform for Lepidoptera in maize, particularly in the case of *S. frugiperda*. One of the first challenges is to change farmer behaviour. Farmers became used to the benefits of Bt transgenic technology, but should continue to practise IPM and adopt best practices, such as:

1. scout for insects and apply insecticide only after reaching recommended thresholds;
2. use thresholds for Bt traits;
3. incorporate refuge to increase technology long term durability;
4. use crop rotation as the primary strategy to reduce pest populations in the next cropping cycle;
5. rotate technologies (e.g. seed treatment or fallow) to reduce pest populations;
6. use insecticides with high efficacy for the target pest;
7. use insecticides appropriate for the crop phenological stage;
8. use insecticide rotation;
9. give preference to products that preserve beneficial organisms;
10. use recommended label rates for each product and avoid use sub-dosage or super-dosage; and
11. avoid insecticides which have existing resistance problems.

In the case of cotton, during the last 5–10 years the cropping system has been through several changes, primarily associated with planting time and new technologies. However, any recommendation in cotton is a challenge, because the pest complex associated with cotton, in some situations, occurs at the same time. Co-occurring examples include aphids, thrips, boll weevil (*Anthonomus grandis* Boheman, 1843), *H. armigera*, *S. frugiperda* and whitefly (*Bemisia tabaci* (Gennadius, 1889)). In addition, broader planting times and overlapping presence of cotton in the field in different regions of the Brazilian savanna expose the crop to increasing risk of insect injury. The Bt traits in cotton (Table 3.2) have been adopted on a large scale, with variable levels of efficacy. Especially with

Table 3.2. *Bacillus thuringiensis* toxins available in Brazil, in maize, soybean and cotton Bt traits, targeting Lepidoptera species.

Crop	*Cry*1Ac	*Cry*1A.105	*Cry*1F	*Cry*1Ab	*Cry*2Ae	*Cry*2Ae	Vip 3Aa20
Maize		x	x	x	x		
Soybean	x						
Cotton	x		x	x	x	x	x

respect to the presence of *Cry*1F resistance in *S. frugiperda* (Bernardi *et al.*, 2014). In addition, boll weevil is one of the main pests in Brazilian cotton and can jeopardize all cotton IPM programmes.

Based on the cotton pest complex and the aspects previously mentioned, it is much more difficult to have a single recommendation for cotton. For this reason, insect pest management recommendations in cotton should be based on: (i) crop phenology; (ii) pest infestation at different crop phenological stages; (iii) pest location; (iv) pest stage; and (v) product efficacy. The basis for the most effective insecticide application on cotton includes pest monitoring and the use of action thresholds, and by using this information, the chances for effective pest management and plant protection are maximized.

3.3 Challenges and Considerations for Pest Management in the Intensive Cropping Systems of the Brazilian Savanna

Integrated Pest Management (IPM) (Kogan, 1998; Pedigo and Rice, 2009) is a globally recognized approach to manage pests in agroecosystems, which represents a replacement of calendar-based scheduling of insecticide application. The components of the IPM approach embrace the concepts of maintaining a pest population below the economic injury level (EIL) via the economic threshold (ET) and adoption of multiple management tactics, as much as possible, in an environmentally compatible way.

Criticism exists which states that the EIL and ET concepts are only based on a chemical approach (Higley and Pedigo, 1996). Zalucki *et al.* (2009) describe two different schools of thought in IPM: the 'sample, spray and pray' approach, which is based on rational pesticide use by developing intervention thresholds to be used in conjunction with sampling schedules that determine pest densities; and the second approach, which supports the idea of an 'understanding of the agroecosystem', with great emphasis on natural enemies. This second school of thought has not been well defined and has not been frequently applied. Similarly, Morse (2009) states there are two approaches in the adoption of IPM in developing countries. One approach puts IPM in terms of elimination of the need for the pesticide (strategic approach). The other approach emphasizes the reduction of pesticide pressure on the environment (tactical approach).

In the Brazilian savanna, chemical control is a common tactic in most intensive cropping systems, with increasing insecticide use in crops such as soybean. However, recommendations for IPM has played a role in crop protection in Brazil (e.g. Cruz *et al.*, 1997, 1999; Bueno *et al.*, 2011a, 2012; Viana and Mendes, 2012; Figueiredo *et al.*, 2015; Miranda *et al.*, 2015). Concepts such as the EIL and ET, when available and validated, make it possible to identify and differentiate species that have pest status from organisms that do not cause economic loss, and reduce overuse of insecticides (Bueno *et al.*, 2013). Thus, the establishment and adoption of the specific EIL and ET for the specific pest/crop scenario is required, followed by the possibility of curative tactics and the understanding that there are no universal solutions for pest management.

Moreover, it should be accepted that IPM it is not a simple concept. The complexity of knowledge required for developing and establishing an IPM programme for specific pest and crop requires several years of research, followed by a validation process. These steps allow the extrapolation of research results to technological answers in the field, but sometimes do not eliminate the complexity that effective management involves.

An example of this complexity is the level of development of EILs, ETs and sampling plans. This is challenging not only in the Brazilian savanna, but globally. Despite their critical importance, Peterson (1996) reports in a review that in 43 commodities only 100 arthropod pest species had research-based EILs. Castle and Naranjo (2009) reported in a review of 1970 to 2009 that 105 relevant papers were published in the *Journal of Economic Entomology*, and from this total, only 60 addressed the development of EILs, ETs and sampling plans. In half of them, the thresholds developed did not come with a respective sampling plan for their effective use and adoption. In the same way, Wearing (1988) interviewed more than 150 researchers and extension service professionals in Europe, USA, and Australia/New Zealand and identified the 'lack of simple monitoring methods' as a major technical barrier to the implementation of IPM. This problem cuts across regions and crops. Even in the cases where the statistical aspects of sampling plan methodology are addressed, the problem of sampling plan complexity often persists and results in modifications or deleterious adjustments of the sampling protocol when it is applied in the field (Naranjo *et al.*, 1997; Grieshop *et al.*, 1988).

3.3.1 Development and validation of EIL, ET and sampling plans

The EIL is defined as the lowest density of the pest population which causes economic damage (Stern *et al.*, 1959). Several contributions have been made to this concept (Stone and Pedigo, 1972; Pedigo *et al.*, 1986; Higley and Pedigo, 1996), and the result is a current cost–benefit equation (Pedigo and Rice, 2009). The basic EIL equation proposes a balance between the losses associated with pest damage and loss prevention from pest management (Higley and Peterson, 1994).

The concept of the EIL incorporates aspects of the injury/damage relationship of the pest and crop, and the dynamics of the cost of management and market value of the product. A complex aspect of the EIL is the distinction between injury and damage. Injury is the deleterious effect of the pest on the plant tissue (Pedigo and Rice, 2009), and damage is the economic loss in quantity or quality of the crop (Pedigo *et al.*, 1986). In order to establish this relationship between injury per insect

and the consequent damage caused by a pest in a specific crop, substantial data are required.

New IPM approaches have been proposed which are relevant to the discussion of how feasible IPM can be in the intensive maize, soybean and cotton cropping systems of the Brazilian savanna. The incorporation of environmental costs (Higley and Wintersteen, 1992) in the EIL is one approach, which proposes the selection of the least environmentally damaging pesticide. However, more information about the level of environmental risk of insecticides currently in the market is necessary to apply this approach.

Another potential contribution to the EIL concept is the multiple-species EIL (Higley and Peterson, 1994; Hunt *et al.*, 2003). The idea is to group pest species feeding on the same tissue, in the same way, and at the same time in an injury guild (Higley and Peterson, 1994). Bueno *et al.* (2011b) evaluated an insect-injury equivalency for soybean in Brazil, considering *A. gemmatalis* as the standard equivalent species, and comparing the consumption of *C. includens*, *S. eridania*, *S. cosmioides* and *S. frugiperda*. The results indicated a significant difference in insect-injury equivalence of *S. cosmioides*, which was nearly double that of *A. gemmatalis*. The conclusion was that the insect-injury equivalent should be two for *S. cosmioides* and one for all other listed species. Nevertheless, the authors highlight the demand for more studies considering the changes in the insect-injury equivalency as a function of insect density.

The economic threshold (ET) is another important concept, and one which has practical application in the field. The ET represents the moment when the pest population must be controlled in order that it does not reach the EIL (Stern *et al.*, 1959), and is based on time (Pedigo and Rice, 2009). The value of ET is typically set below the EIL. Several approaches to ET have been proposed, which incorporate mortality factors or pest survival information (Ostile and Pedigo, 1987; Barrigossi *et al.*, 2003; Paula-Moraes *et al.*, 2013), and establish the ET in a pest stage other than that for which the EIL is defined.

Besides the EIL and ET, a sampling plan is another pillar of IPM (Kogan, 1998). The development of sampling plans based on a probabilistic foundation allows analysis and inference about the pest population density under evaluation (Young and Young, 1998).

Sampling plan design requires several types of information associated with the appropriate sampling unit, the number of observations that will be necessary, and the precision associated with the pest population density estimate. The probability distribution of the pest, together with information of its biology, ecology and behaviour are useful to determine the number of samples in pest sampling protocols (Binns, 1994). There are formulas that allow a precise estimate of pest density population, based on the probability distribution of the pest. In some cases, when estimating a pest population at the field level, it is necessary to ignore the randomization of the samples. The collection of samples follows predetermined patterns and the variance in the estimate will not depend on the chance of the draw, but will depend on the sampling plan and spatial distribution of the pest in the crop (Morris, 1960; Legg and Moon, 1994). There are also differences in an insect's spatial pattern at different insect life stages, and is common in species with holometabolous development (Pedigo and Zeiss, 1996).

3.3.2 Natural biological control in intensive cropping systems in the Brazilian savanna

Biological control has been promoted as an additional tactic to chemical control, and is utilized in soybean, maize and cotton production in Brazil. Several methods of augmentative biological control have been adopted. As previously cited, the presence of *H. armigera* in Brazilian intensive cropping systems drove the adoption of the NPV virus and formulations of *Bacillus thuringiensis*, as well as the release of the parasitoid *Trichogramma pretiosum*.

However, there is a demand to improve biological control, especially focusing on conservation of the existing natural enemies in crops of the Brazilian savanna. Based on the reduction of pesticide pressure approach (Morse *et al.*, 2009; Zalucki *et al.*, 2009), determining the presence of native beneficial arthropods, and ways to increase their effectiveness in this specific Brazilian agricultural landscape, is required.

A main assumption of natural biological control in the agroecosystem states that herbivore density decreases in complex landscapes and with a diversity of non-crop habitats. However, during the main cropping season in the summer, the mosaic of crops in agricultural areas of the Brazilian savanna (particularly Mato Grosso and Bahia states) declines and predominantly becomes an 'ocean' of soybean, with surrounding crops of maize and cotton and a few other field crops in smaller areas. Hence, the general recommendation to provide undisturbed sites near or in strips within crops is not feasible for some farms. First, because the increase of the natural enemies is variable and dependent on the natural enemy taxa and the landscape structure (Gardiner *et al.*, 2009). Second, because of the large scale at which most farmers of the region operate, the field boundaries will not always act as an efficient natural enemy reservoir.

More studies should be directed at the characteristics of the landscape mosaic that influence the design of options to improve natural biological control in the intensive cropping systems in the Brazilian savanna. Special attention should focus on the influence of the proportion of the habitat edge to interior, patch area, patch quality, patch diversity, and microclimate in relation to pest and beneficial insect populations (Hunter, 2002). Conservation practices to improve favourable conditions for natural biological control should also be employed. Practices such as decreasing pesticide application through the development and adoption of EILs and ETs, use of more selective products, timing of pesticide application, and providing untreated areas in the agricultural landscape (i.e. refugia) should be considered in order to provide conditions that shelter and enhance survival of natural enemies (Roubos *et al.*, 2014). Considering the large operational scale of many farmers in the Brazilian savanna, a firm commitment from farmers in these high input systems to favour natural biological control can have a large landscape positive impact.

3.3.3 Farmer risk aversion

The concept of risk aversion is based on the resistance to a set of outcomes with a known probability distribution. Empirical evidences demonstrate that most individuals are risk averse.

In general, farmers in the Brazilian savanna accept the concept of sampling plans. In intensive maize, soybean and especially cotton cropping systems the idea of employing correct timing of scouting activities to determine pest population density sounds attractive. On the other hand, farmers are more resistant to the concepts of the EIL and ET. The easy explanation for this aversion is the convenience of chemical control, which represents a cheap alternative with faster control action than other tactics of management that are demanding of time, labour and money. However, the farmer concerns associated with the risk of pest control adoption (e.g. insecticide application) once the pest damage has already occurred, should be taken into consideration.

It also continues to be difficult for farmers to understand the difference between perceived pest risk and actual pest risk. Part of this difficulty is because concepts such as the ET involve more uncertainty (including risk) than the traditional technology of chemical control. The financial risk associated with economic dynamics, scale of large crop areas and operations, and other factors cause farmers to lack confidence in the ET as a well-timed application parameter to achieve effective pest control.

A rigorous diagnostic study of regional demands and validation of recommendations at field level should be conducted in order to address farmer concerns and help them accept and implement IPM concepts. Specifically, validation of the well-timed sampling and ET use on large-scale farms is required to achieve farmer acceptance. This type of validation should reduce the belief that reducing the use of chemical control necessarily results in loss of production and/or quality. In addition, it is important to document the economic yield loss compensation from using IPM and its relationship to gross margins, as compared to traditional chemical control (Picanço *et al.*, 2004; Féménia and Letort, 2016). Moreover, precise identification of local obstacles for the adoption of IPM should be employed followed by educational programmes targeting high-tech farmers in the Brazilian savanna.

3.3.4 Bt transgenic technology in the Brazilian savanna

Currently, Bt transgenic technology has been adopted on a large scale in commodity crops in several countries around the world (James, 2015). Pest exposure to Bt toxins expressed in plant tissues throughout most of the crop season provides a high level of pest control. The benefits of Bt crops are associated with the reduction of insecticide sprays against Lepidoptera species and suppression of the target pest. An example of these benefits is the suppression of *Ostrinia nubilalis* (Hübner, 1796) in the American Corn Belt (Hutchison *et al.*, 2010). In this case, the area-wide suppression of *O. nubilalis* resulting from widespread use of Bt field maize also reduced *O. nubilalis* pressure in non-Bt maize, including organic maize and sweetcorn.

In Brazil, Bt transgenic traits targeting lepidopteran pests are available in maize, cotton and, since the crop season of 2014/2015, in soybean (Table 3.2). Their adoption has significant economic relevance considering the high pressure of pests under the tropical conditions of the Brazilian savanna, the possibility to interrupt the population growth of polyphagous pests, and the possibility of widespread suppression of target pests.

Doubts have been raised about the survival of the IPM framework with the rapid and extensive adoption of Bt technology, as Bt technology decreases the pest population density with high level of efficiency and is adopted as a therapeutic measure (Hunt and Paula-Moraes, 2010). These characteristics can be considered a negation of all concepts and tenets of IPM. Moreover, it could reinforce farmer risk aversion and a low tolerance of pest presence in intensive maize, soybean and cotton cropping systems. However, this technology is not a solution for all pests associated with maize, soybean and cotton in the Brazilian savanna. For example, sucking insects, such as aphids, are not targets of this technology. Even in the group of Lepidoptera, there are many species for which current Bt proteins are not effective, and a decrease of conventional insecticide applications could change the status of formally secondary pests. For example, the Bt trait (*Cry*1Ac) currently available in the market provides poor control against the *Spodoptera* pest complex (Bernardi *et al.*, 2014). The result could be an increase of species such as *S. eridania* and *S. cosmioides* population densities in soybean.

Even in the cases where Bt technology provides effective control to a specific target species, IPM remains relevant. An example is *Cry*1Ac in cotton, which targets *Helicoverpa* spp. The toxin expression declines at the end of the cotton crop cycle (Siebert *et al.*, 2009), so control of this species could be compromised. Hence, a refined sampling plan that considers the surviving larvae, rather than egg evaluation, is required in *Cry*1Ac cotton fields. Moreover, it is necessary to redesign the EIL and ET for this pest in cotton, since some late season pest management (e.g. insecticide application) may be necessary (Naranjo *et al.*, 2008).

Another concern about the widespread adoption of Bt technology, which reinforces the continued need for IPM, is the risk of the evolution of Bt-resistant pest populations of target pests of Bt toxins. Worldwide, field-evolved resistance has been observed in 5 out of 13 pest species examined (Tabashnik *et al.*, 2013). Several different strategies to delay the evolution of resistance of target pests to Bt crops are discussed in the literature, and these strategies are generally referred to as insect resistance management (IRM). Current IRM for transgenic crops embraces the consideration of insect genetics, adoption of two-toxin pyramids, since no presenting a risk of cross resistance (Bernardi *et al.*, 2015), and the cultivation of refuges of non-Bt plants, all of which play fundamental roles in decreasing the risk of insect resistance evolution to Bt technology.

The IRM principles are based on: (i) resistance in insects is typically recessive or incompletely dominant (e.g. Tabashnik *et al.*, 1992, 1994; Gould, 1998; McGaughey and Beeman, 1998); (ii) the high dose expressed in Bt plants is a 25-fold toxin concentration necessary to kill homozygous susceptible larvae and most heterozygotes (defined in EPA, 1998) (e.g. Roush, 1989; Gould, 1994) and; (iii) the refuge is composed of non-Bt plants or other susceptible hosts of the target pest that provide unselected susceptible individuals that will mate with homozygote resistant individuals that escape from high dose exposure (e.g. Tabashnik, 1994; Tabashnik *et al.*, 2013). However, the extensive knowledge about the genetic, ecological and operational factors that influence insect resistance evolution is critical and more information is required, particularly in the Brazilian savanna.

In Brazil, there is no legal framework for IRM. However, in 2014 the Brazilian Agriculture Ministry established a technical group to discuss and propose recommendations for IRM for Bt maize, soybean and cotton (MAPA, 2014). Scientists, industry and representatives from maize, soybean and cotton commodity groups form this group. The idea is to exchange and discuss information about the recommendation and correct adoption of Bt crops in Brazil. In the same way, the technical group of the Insecticide Resistance Action Committee (IRAC) recently coordinated an industry effort to discuss IRM in Brazil. It was possible to align recommendations considering the specific demands of IPM and IRM in Bt and non-Bt crops (IRAC, 2016). Moreover, the structured refuge percentage recommendations for non-Bt maize, soybean and cotton were aligned in Brazil. However, the adoption of refuge in intensive cropping systems specifically in the Brazilian savanna has been the topic of much discussion. Aspects related to refuge size, spatial and temporal refuge patterns and mandatory rules for its adoption are still hot discussion topics.

More research is required to validate various aspects of IRM considering the particularities of agriculture in the Brazilian savanna. The effect of the spatial and temporal scale of the landscape with Bt crops (Table 3.2) in context with the target pest population dynamics should be investigated. Factors related to tropical conditions, such as pest survival, number of generations, larval and adult movement, mating behaviour, host plant range, number of crop seasons, and others factors require study. In this way, it will be possible to provide more comprehensive recommendations for IRM in the Brazilian savanna.

In summary, even with the problems that may result from the use of Bt technology, including the recent report of *S. frugiperda* resistance to *Cry*1F in Brazil (Farias *et al.*, 2014), Bt technology is a valuable pest management tool targeting economic species of Lepidoptera in intensive maize, soybean and cotton cropping systems in the Brazilian savanna. Indeed, there is the expectation that this technology may result in the area-wide reduction of pest population densities, such as *Helicoverpa* spp. and *C. includens*, in a spatial and temporal scale, and so bring the possibility of infestation reduction in non-Bt hosts in the landscape. However, the concerns associated with the widespread use of transgenic technology reinforce the need for it to be part of a new IPM framework which is complementary to ongoing IRM in the intensive cropping systems of the Brazilian savanna.

3.4 Final Considerations

There will be many challenges in the Brazilian savanna in the coming years, such as the following.

1. Real adoption of state-of-the-art IPM, with acceptance of the concepts of maintaining a pest population below the economic injury level (EIL) via the economic threshold (ET) and adoption of multiple management tactics, as much as possible, in an environmentally compatible way. Funding for research to provide more information and validation of recommendations for the management of pests in intensive cropping systems should be prioritized.
2. Studies, considering the spatial and temporal dynamics of the agriculture landscape to establish the best management practices concerning the use of native beneficial arthropods as a natural control, as well as augmentative control.
3. Education and communication to the growers about benefits of all technology available, based on how to use, how to preserve and how to sustain the Brazilian savanna as a food producer in the coming years.
4. Integration of all knowledge, in a IRM framework, to use Bt traits for a lengthy period to improve/maintain yield, to keep lower population of target pest complex, and to help crop management.

Brazilian savanna will continue be a very important grain and fibre producer for a long time, but will need a huge effort among researchers, growers, industry and governmental programmes to provide enough information to keep the region as an important food supplier for many years.

Acknowledgements

The authors acknowledge Dr Thomas E. Hunt of the University of Nebraska, Lincoln for preliminary review of this chapter.

References

Amabile, R.F. and Barcelos, A.O. (2008) Produção agropecuária e florestal: demandas para a pesquisa. In: *IX Simpósio Nacional do Cerrado e II Simpósio Internacional Savanas Tropicais*. Brasília-DF, Brasília, Brazil, pp. 53–66.

Angulo, A.O., Olivares, T.S. and Weigert, G.T. (2008) *Estados inmaduros de lepidópteros nóctuidos de importancia agrícola y forestal en Chile y claves para su identificación (Lepidoptera: Noctuidae)*, 3rd edn. Universidad de Concepción, Concepción, Chile.

Barrigossi, J.A.F., Hein, G.L. and Higley, L.G. (2003) Economic injury levels and sequential sampling plans for Mexican bean beetle (Coleoptera: Coccinellidae) on dry beans. *Journal of Economic Entomology* 96, 1160–1167.

Baudino, E. (2004) Presencia y distribución temporal del complejo de orugas cortadoras (Lepidoptera: Noctuidae) en pasturas de alfafa (*Medicago sativa* L.) del área fisiográfica Oriental de la provincia de La Pampa, Argentina. *Revista de la Facultad Agronomía* 15, 31–42.

Bernardi, D., Salmeron, E., Horikoshi, R.J., Bernardi, O., Dourado, P.M. *et al.* (2015) Cross-resistance between *Cry*1 proteins in fall armyworm (*Spodoptera frugiperda*) may affect the durability of current pyramided Bt maize hybrids in Brazil. *PLOS ONE* 10, e014013.

Bernardi, O., Malvestiti, G.S., Dourado, P.M., Oliveira, W.S., Martinelli, S. *et al.* (2012) Assessment of the high-dose concept and level of control provided by MON 87701×MON 89788 soybean against *Anticarsia gemmatalis* and *Pseudoplusia includens* (Lepidoptera: Noctuidae) in Brazil. *Pest Management Science* 68, 1083–1091.

Bernardi, O., Sorgatto, R.J., Barbosa, A.D., Domingues, F.A., Dourado, P.M. *et al.* (2014) Low susceptibility of *Spodoptera cosmioides*, *Spodoptera eridania* and *Spodoptera frugiperda* (Lepidoptera: Noctuidae) to genetically-modified soybean expressing *Cry*1Ac protein. *Crop Protection* 58, 33–40.

Binns, M.R. (1994) Sequential sampling for classifying pest status. In: Pedigo, L. and Buntin, G. (eds) *Handbook of Sampling Methods for Arthropods in Agriculture*. CRC, Boca Raton, Florida, pp. 137–174.

Bueno, A.F., Batistela, M.J., Bueno, R.C.O.F., França-Neto, J.B., Nishikawa, M.A.N. and Filho, A.L. (2011a) Effects of integrated pest management, biological control and prophylactic use of insecticides on the management and sustainability of soybean. *Crop Protection* 30, 937–945.

Bueno, A.F., Panizzi, A.R., Corrêa-Ferreira, B.S., Hoffmann-Campo, C.B., Sosa-Gómez, D.R. *et al.* (2012) Histórico e evolução do manejo integrado de pragas da soja no Brasil. In: Hoffmann-Campo, C.B., Moscadi, F. and Corrêa-Ferreira, B.S. (eds) *Soja: manejo integrado de insetos e outros artrópodes-praga*. Embrapa, Brasília, Brazil.

Bueno, A.F., Paula-Moraes, S.V., Gazzoni, D.L. and Pomari, A.F. (2013) Economic thresholds in soybean-integrated pest management: old concepts, current adoption, and adequacy. *Neotropical Entomology* 42, 439–447.

Bueno, R.C.O.F., Bueno, A.F., Moscardi. F., Parra, J.R.P. and Hoffmann-Campo, C.B. (2011b) Lepidopteran larvae consumption of soybean foliage: basis for developing multiple-species economic thresholds for pest management decisions. *Pest Management Science* 67, 170–174.

Castle, S.J. and Naranjo, S.E. (2009) Sampling plans, selective insecticides and sustainability: the case for IPM as informed pest management. *Pest Management Science* 65, 1321–1328.

Coto, D., Saunders, J.L., Vargas, S.C.L. and King, B.S. (1995) *Plagas invertebradas de cultivos tropicales con enfasis en America Central – un inventário*. Série Técnica: Manual Técnico 12, CATIE, Turrialba, Costa Rica.

Cruz, I., Valicente, F.H., Santos do, J.P., Waquil, J.M. and Viana, P. (1997) *Manual de identificação de pragas da cultura do milho*. EMBRAPA-CNPMS, Sete Lagoas, Brazil.

Cruz, I., Viana, P.A. and Waquil, J.M. (1999) *Manejo das pragas iniciais de milho mediante o tratamento de sementes com inseticidas sistêmicos*. EMBRAPA-CNPMS. Circular Técnica n. 31, EMBRAPA-CNPMS, Sete Lagoas, Brazil.

Cunningham, J.P. and Zalucki, M.P. (2014) Understanding heliothine (Lepidoptera: Heliothinae) pests: What is a host plant? *Journal of Economic Entomology* 107, 881–896.

Czepak, C., Albernaz, K.C., Vivan, L.M., Guimarães, H.O. and Carvalhais, T. (2013) First reported occurrence of *Helicoverpa armigera* (Hübner) (Lepidoptera: Noctuidae) in Brazil. *Pesquisa Agropecuária Tropical* 43, 110–113.

Embrapa (2013) *Nota técnica sobre resultados do trabalho inicial de levantamento da lagarta do gênero* Helicoverpa – *detecção da espécie* Helicoverpa armigera *no Brasil.* Embrapa Cerrados, Planaltina, Brazil.

EPA (1998) *Insect Resistance Management Fact Sheet for* Bacillus thuringiensis (*Bt*). Environmental Protection Agency. Available at: https://archive.epa.gov/pesticides/biopesticides/web/html/bt_corn_refuge_2006.html (accessed 6 November 2017).

FAO (2012) *The State of Food and Agriculture: Investing in Agriculture for a Better Future.* FAO, Rome, Italy. Available at: www.fao.org/docrep/017/i3028e/i3028e.pdf (accessed 28 July 2017).

Farias, J.R., Andow, D.A., Horikoshi, R.J., Sorgatto, R.J., Fresia, P. *et al.* (2014) Field-evolved resistance to *Cry*1F maize by *Spodoptera frugiperda* (Lepidoptera: Noctuidae) in Brazil. *Crop Protection* 64, 150–158.

Féménia, F. and Letort, É. (2016) *How to Achieve Significant Reduction in Pesticide Use? An Empirical Evaluation of the Impacts of Pesticide Taxation Associated to a Change in Cropping Practice.* Working Paper Smart-Lereco, 16-02.

Figueiredo, M.L.C., Cruz, I., Silva, R.B. and Foster, J.E. (2015) Biological control with *Trichogramma pretiosum* increases organic maize productivity by 19.4%. *Agronomy for Sustainable Development* 35, 1175–1183.

Gardiner, M.M., Fiedler, A.K., Costamagna, A.D. and Landis, D.A. (2009) Integrated conservation biological control into IPM systems. In: Radcliffe, E.B., Hutchison, W.D. and Cancelado, R.E. (eds) *Integrated Pest Management: Concepts, Tactics, Strategies and Case Studies.* Cambridge University Press, New York, pp. 151–178.

Gould, F. (1994) Potential and problems with high-dose strategies for pesticidal engineered crops. *Biocontrol Science and Technology* 4, 451–461.

Gould, F. (1998) Sustainability of transgenic insecticidal cultivars: integrating pest genetics and ecology. *Annual Review of Entomology* 43, 701–726.

Grieshop, J.I., Zalom, F.G. and Miyao, G. (1988) Adoption and diffusion of integrated pest management innovations in agriculture. *Bulletin of Entomological Society of America* 34, 72–78.

Higley, L.G. and Pedigo, L.P. (1996) The EIL concept. In: Higley, L.G. and Pedigo, L. (eds) *Economic Threshold for Integrated Pest Management.* University of Nebraska Press, Lincoln, Nebraska, pp. 9–21.

Higley, L.G. and Peterson, R.K.D. (1994) Initiating sampling programs. In: Pedigo, L. and Buntin, G. (eds) *Handbook of Sampling Methods for Arthropods in Agriculture.* CRC Press, Boca Raton, Florida, pp. 119–136.

Higley, L.G. and Wintersteen, W. (1992) A novel approach to environmental risk assessment of pesticides as a basis for incorporating environmental costs into economic injury levels. *American Entomology* 38, 34–38.

Hunt, T.E. and Paula-Moraes, S.V. (2010) IPM in the age of transgenic crops: Are IPM and transgenic technology compatible or in conflict? In: *XXIII Congresso Brasileiro de Entomologia*, Setembro 26–30, Natal, RN, Brazil.

Hunt, T.E., Higley, L.G. and Haile, F.J. (2003) Imported longhorned weevil (Coleoptera: Curculionidae) injury to soybean: physiological response and injury guild-level economic injury levels. *Journal of Economic Entomology* 96, 1168–1173.

Hunter, M.D. (2002) Landscape structure, habitat fragmentation, and the ecology of insects. *Agricultural and Forest Entomology* 4, 159–166.

Hutchison, W., Burkness, E., Mitchell, P., Moon, R., Leslie, T. *et al.* (2010) Areawide suppression of European corn borer via transgenic maize reaps savings to non-Bt maize growers. *Science* 330, 220–225.

IBGE (2006) Instituto Brasileiro de Geografia e Estatística. Available at: htpp://ibge.gov.br (accessed 18 April 2016).

IRAC (2016) *Insect Resistance Management Guidelines for Bt Soybean, Cotton and Corn in Brazil.* Available at: www.irac-br.org/outros/Recomendações (accessed 18 April 2016).

James, C. (2015) *Global Status of Commercialized Biotech/GM Crops: 2015.* ISAAA Brief No. 51. ISAAA, Ithaca, New York.

Kitching, I.J. and Rawlins, J.E. (1999) The Noctuoidea. In: Kristensen, N.P. (ed.) *Lepidoptera: Moths and Butterflies. Volume 1: Evolution, Systematics and Biogeography* (Handbook of Zoology/ Handbuch der Zoologie). Walter de Gruyter, Berlin and New York, pp. 355–401.

Kogan, M. (1998) Integrated pest management: historical perspectives and contemporary developments. *Annual Review of Entomology* 43, 243–270.

Lafontaine, J.D. (2004) Noctuoidea, Noctuinae, Agrotini. In: Hodges, R.W., Davis, D.R., Ferguson, D.C., Munroe, E.G. and Powell, J.A. (eds) *The Moths of America North of Mexico*. Allen Press, Lawrence, KS, Fasc. 25-1.

Lafontaine, J.D. and Poole, R.W. (1991) Fascicle 25.1 – Noctuoidea, Noctuidae (Part) Plusiinae. In: Hodges, R.W. (ed.) *The Moths of America North of México*. The Wedge Entomological Research Foundation, Lawrence, Kansas.

Legg, D.E. and Moon, R.D. (1994) Bias and variability in statistical estimates. In: Pedigo, L.P. and Buntin, G.D. (eds) *Handbook of Sampling Methods for Arthropods in Agriculture*. CRC Press, Boca Raton, Florida, pp. 99–118.

Machado, C.A., Wunder, M., Baldissera, V.D., Oliveira, J.V., Fiuza, L.M. and Nagoshi, R.N. (2008) Molecular characterization of host strains of *Spodoptera frugiperda* (Lepidoptera: Noctuidae) in Southern Brazil. *Annals of the Entomological Society of America* 101, 619–626.

MacRae, T.C., Baur, M.E., Boethel, D.J., Fitzpatrick, B.J., Gao, A.G. *et al.* (2005) Laboratory and field evaluations of transgenic soybean exhibiting highdose expression of a synthetic *Bacillus thuringiensis cry1Ac* gene for control of Lepidoptera. *Journal of Economic Entomology* 95, 577–587.

MAPA (2014) Normative No. 950, September 24, 2014. Available at: www.agricultura.gov.br/legislacao (accessed 18 April 2016).

McGaughey, W. and Beeman, R. (1998) Resistance to *Bacillus thuringiensis* in colonies of India meal moth and almond moth (Lepidoptera: Pyralidae). *Journal of Economic Entomology* 81, 28–33.

Miranda, J.E., Rodrigues, S.M.M., Albuquerque, F.A. de, Silva, C.A.D. da, Almeida, R.P. de and Ramalho, F. de S. (2015) *Guia de identificação de pragas do algodoeiro* (Série Documentos, 225). Embrapa, Campina Grande, Brazil.

Morris, M., Binswanger, H., Byerlee, D. and Staatz, J.A. (2012) Breadbasket for Africa: farming in the Guinea Savanna Zone. *Solutions* 3, 44–49.

Morris, R.F. (1960) Sampling insect populations. *Annual Review of Entomology* 5, 243–264.

Morse, S. (2009) IPM, ideals and realities in developing countries. In: Radcliffe, E.B., Hutchison, W.D. and Cancelado, R.E. (eds) *Integrated Pest Management: Concepts, Tactics, Strategies and Case Studies*. Cambridge University Press, New York, pp. 458–470.

Naranjo, S.E., Diehl, J.W. and Ellsworth, P.C. (1997) Sampling whiteflies in cotton: validation and analysis of enumerative and binomial plans. *Environmental. Entomology* 26, 777–788.

Naranjo, S.E., Ruberson, J.R., Sharma, H.C., Wilson, L. and Wu, K. (2008) The present and future role of insect-resistant genetically modified cotton in IPM. In: Romeis, J., Shelton, A.M. and Kennedy, G.G. (eds) *Integration of Insect-Resistant Genetically Modified Crops within IPM Programs*. Springer, Dordrecht, The Netherlands, pp. 159–194.

Ostlie, K.R. and Pedigo, L.P. (1987) Incorporating pest survivorship into economic thresholds. *Bulletin of Entomological Society of America* 33, 98–102.

Pastrana, J.A. (2004) *Los Lepidópteros Argentinos: sus plantas hospedadoras y otros substratos alimenticios*. Sociedad Entomológica Argentina, Buenos Aires, Argentina.

Paula-Moraes, S.V., Hunt, T.E., Wright, R.J., Hein, G.L. and Blankenship, E.E. (2013) Western bean cutworm survival and the development of economic injury levels and economic thresholds in field corn. *Journal of Economic Entomology* 106, 1274–1285.

Pedigo, L.P. and Rice, M.E. (2009) *Entomology and Pest Management*. Prentice-Hall, Upper Saddle River, New Jersey.

Pedigo, L.P. and Zeiss, M.R. (1996) *Analyses in Insect Ecology and Management*. Iowa State University Press, Ames, Iowa.

Pedigo, L.P., Hutchins, S.H. and Higley, L.G. (1986) Economic injury levels in theory and practice. *Annual Review of Entomology* 31, 341–368.

Peterson, R.K.D. (1996) The status of economic-decision level development. In: Higley, L.G. and Pedigo, L.P. (eds) *Economic Thresholds for Integrated Pest Management*. University of Nebraska Press, Lincoln, Nebraska, pp. 151–178.

Picanço, M.C., Paula, S.V., Moraes Junior, A.R., Oliveira, I.R., Semeão, A.A. and Rosado, J.F. (2004) Impactos financeiros da adoção de manejo integrado de pragas na cultura do tomateiro. *Acta Scientiarum Agronomy* 26, 245–252.

Pogue, M.G. (2002) A world revision of the genus *Spodoptera* Guenée (Lepidoptera: Noctuidae). *Memoirs of the American Entomological Society* 43, 1–202.

Pogue, M.G. (2004) A new synonym of *Helicoverpa zea* (Boddie) and differentiation of adult males of *H. zea* and *H. armigera* (Hübner) (Lepidoptera: Noctuidae: Heliothinae). *Annals of the Entomological Society of America* 97, 1222–1226.

Ratter, J.A. and Ribeiro, J.F. (1996) Biodiversity of the flora of the cerrado. In: Pereira, R.C. and Nasser, L.C.B. (eds) *Proceedings of the VIII Simposio sobre o cerrado, 1st International Symposium on Tropical Savannas.* Embrapa, CPAC, Planaltina-DF, Brazil, pp. 3–6.

Raupp, M., Davidson, A., Koehler, C.S., Sadof, C.S. and Reichelderfer, C.S. (1988) Decision-making considerations for aesthetic damage caused by pests. *Bulletin of the Entomological Society of America* 34, 27–32.

Roubos, C.R., Rodriguez-Saona, C. and Isaacs, R. (2014) Mitigating the effects of insecticides on arthropod biological control at field and landscape scales. *Biological Control* 75, 28–38.

Roush, R.T. (1989) Designing resistance management programs: How can you choose? *Pesticide Science* 26, 423–441.

Siebert, M.W., Patterson, T.G., Gilles, G.J., Nolting, S.P., Braxton, L.B. *et al.* (2009) Quantification of Cry1Ac and Cry1F *Bacillus thuringiensis* insecticidal proteins in selected transgenic cotton plant tissue types. *Journal of Economic Entomology* 102, 1301–1308.

Silva, A.G.A., Gonçalves, C.R., Galvão, D.M., Gonçalves, A.J.L., Gomes, J. *et al.* (1968) *Quarto catálogo dos insetos que vivem nas plantas do Brasil, seus parasitos e predadores. Parte II, 1º tomo. Insetos, hospedeiros e inimigos naturais.* Ministério da Agricultura, Rio de Janeiro, Brazil.

Sosa-Gomez, D.R., Specht, A., Paula-Moraes, S.V., Lima, A.L., Yano, S.A.C. *et al.* (2016) Timeline distribution of the *Helicoverpa armigera* in Brazil. *Revista Brasileira de Entomologia* 60, 101–104.

Specht, A., Sosa-Gómez, D.R., Paula-Moraes, S.V. and Yano, S.A.C. (2013) Identificação morfológica e molecular de *Helicoverpa armigera* (Lepidoptera: Noctuidae) e ampliação de seu registro de ocorrência no Brasil. *Pesquisa Agropecuária Brasileira* 48, 689–692.

Stern, M., Smith, R.F., van den Bosch, R. and Hagen, K.S. (1959) The integrated control concept. *Hilgardia* 29, 330–350.

Stone, J.D. and Pedigo, L.P. (1972) Development and economic-injury level of the green cloverworm on soybean in Iowa. *Journal of Economic Entomology* 65, 197–201.

Tabashnik, B.E. (1994) Evolution of resistance to *Bacillus thuringiensis. Annual Review of Entomology* 39, 47–94.

Tabashnik, B.E., Schwartz, J., Finson, N. and Johnson, M. (1992) Inheritance of resistance to *Bacillus thuringiensis* in diamondback moth (Lepidoptera: Plutellidae). *Journal of Economic Entomology* 85, 1046–1055.

Tabashnik, B.E., Brevault, T. and Carriere, Y. (2013) Insect resistance to Bt crops: lessons from the first billion acres. *Nature Biotechnology* 31, 510–521.

Viana, P.A. and Mendes, S.M. (2012) *Guia de inseticidas para a cultura do milho* (Série Documentos). Embrapa Milho e Sorgo, Sete Lagoas, Brazil.

Wearing, C.H. (1988) Evaluating the IPM implementation process. *Annual Review of Entomology* 33, 17–38.

Yano, S.A., Specht, A., Moscardi, F., Carvalho, R.A., Dourado, P.M. *et al.* (2015) High susceptibility and low resistance allele frequency of *Chrysodeixis includens* (Lepidoptera: Noctuidae) field populations to Cry1Ac in Brazil. *Pest Management Science* 72 (8), 1578–1584.

Young, L.J. and Young, J.H. (1998) *Statistical Ecology: A Population Perspective.* Kluwer Academic Publishers, Boston, Massachusetts.

Zahiri, R., Kitching, I.J., Lafontaine, J.D., Mutanen, M., Kaila, L. *et al.* (2011) A new molecular phylogeny offers hope for a stable family-level classification of the Noctuoidea (Lepidoptera). *Zoological Scripta* 40, 158–173. DOI:10.1111/j.1365-3113.2011.00607.x

Zalucki, M.P., Adamson, D. and Furlong, M.J. (2009) The future of IPM: Whither or wither? *Australian Journal of Entomology* 48, 85–96.

4   Biological Pest Control in the Tropics

Odair Aparecido Fernandes[1,*], José Gilberto de Moraes[2] and Vitalis Wafula Wekesa[3]

[1]Departamento de Fitossanidade, Universidade Estadual Paulista, Jaboticabal, Brazil; [2]Departamento de Entomologia e Acarologia, Piracicaba, Brazil; [3]Department of Biological Science and Technology, Technical University of Kenya, Nairobi, Kenya

4.1 Introduction

The use of natural enemies to control arthropod pests has long been important in the tropics and successful cases have already been reported elsewhere (Prado, 1991; Parra *et al.*, 2002; Neuenschwander *et al.*, 2003; Pinto *et al.*, 2006; Alves and Lopes, 2008; Sampaio *et al.*, 2008; Bueno, 2009). Both arthropods and microorganisms have been used as biological control agents by growers to target pests in annual, semi-annual and perennial crops. Undoubtedly, microorganisms such as *Bacillus thuringiensis* Berliner and *Metarhizium anisopliae* (Metsch.) Sorokin are by far the most used biological control agents. However, despite the increasing awareness about food safety, growers still rely heavily on chemical pesticides to reduce pest problems in tropical regions.

Compared to the pesticide industry, the market share of biological control products is still small, corresponding to only about 2% of the worldwide total sales (Blum, 2002). Consequently, the pesticide market worldwide is still very important and has reached US$270–300 billion a year (Zhang *et al.*, 2011). Currently, available commercial formulations and application techniques of microbial biopesticides are similar to chemical pesticides and this might explain their acceptance and worldwide use by growers. Therefore, there are still extensive demands and opportunities for biological control in the tropics.

Tropical regions have great biodiversity, and for this reason they are potential sources of natural enemies. However, biodiversity in these regions is still underexplored for its potential use in biological control (Alves and Lopes, 2008). Moreover, new regulations related to biodiversity conservation and avoidance of biopiracy have imposed strong restrictions on the exploration of biodiversity, including biological agents.

In the recent past, the incorporation of genetically modified crops as a new tool for pest control in integrated pest management (IPM) programmes resulted, in some cases, in a reduction in pesticide use. Therefore, side effects due to such use were also reduced, allowing natural enemy populations to develop and to contribute to the improvement of biological control in agroecosystems.

In this review, we aim to explore the current status of biological control programmes and discuss future trends and challenges in the tropics. We understand that there are still many opportunities for implementing biological control

---

* Corresponding author e-mail: oafernandes@fcav.unesp.br

programmes, either small or large scale, which include research on natural enemies, revision of current regulations, improvement of formulation, packaging and methods of application, and enhancement of awareness campaigns to growers, stakeholders and the general public.

4.2 Concepts

Under an applied point of view, biological control refers to the suppression of pest populations by the use of natural enemies, which include predators, parasitoids, entomopathogens, competitors and antagonists (De Bach, 1974). There have been some debates in the past two decades about the convenience of broadening this definition to include genes and genetic products, such as genetically modified organisms, in the concept of biological control (Garcia *et al.*, 1988; Eilenberg *et al.*, 2001). However, we maintain here the concept of biological control as originally defined by De Bach.

Strategies used in biological control programmes are usually termed conservation, augmentation and classical biological control, and natural enemies used in these strategies are usually referred to as biological control agents.

Improvement of the efficiency of natural enemies can be sought by conserving and/or improving the ecosystem to allow them to find shelter and or developmental resources. These activities compose what is referred to as conservation biological control.

Moreover, improvement can also be attempted by massive production and release of natural enemies in the ecosystem. In this strategy, termed augmentation, relatively few individuals (inoculative release) or many individuals (inundative release) are released, depending upon the interest on promoting pest control either for long-term or short-term basis. The choice of the type of release to be adopted depends upon the different aspects, as, for example, the length of time allowed for the natural enemies to promote the control (if short, as for annual crops, inundative release would be preferred), the value of the crop to be protected (if expensive, the preference would be for inundative release), and where the stability of the crop environment is desired (inoculative release is preferred).

When potentially effective natural enemies are not found in the ecosystem where control of the pest is considered, these can be introduced from a different region. Known as classical biological control, this strategy has been mostly used for the control of invasive pests, especially hemipteran pests, including scales, mealybugs, and aphids that usually live in colonies and that are slow-moving or totally sessile (Sampaio *et al.*, 2008). However, successful classical biological control has also been adopted for pests of many other crops in annual and perennial systems.

The three biological control strategies have been used in tropical countries. However, biological control programmes based on inundative release and introduction of exotic natural enemies are often the most used.

4.3 Current Status of Biological Control in the Tropics

Biological control activities are frequently conducted in tropical countries in relation to research or practical use. However, practical use has been greatly influenced by the development of other control tools, demand of international markets, reduced efficiency of agrochemicals (especially due to the development of resistance), and new research results. Also, in the recent years, there has been increased awareness and education in the area of biological control. This has been made possible mainly among growers who practise agriculture in the form of extensive mixed cropping or polyculture, as exemplified in several African countries. Most African farmers practise 'small-plot' farming with several crops planted at the same time in the same plot. The main pest control practices with polyculture are cultural control measures like crop rotation and removal of plant residues, often by burning.

On the other hand, in one particular case, the implementation of new tools seems to have disrupted the use of biological control. This refers to the adoption of new (genetically modified) soybean varieties tolerant to herbicides associated with other agrochemicals (Petter *et al.*, 2007) and reduced use of biological control agents. Target markets in this case (internal and foreign markets) are not restrictive in relation to the adopted practices of pest control. Conversely, in other cases, importing countries have imposed stringent pest control practices, which have promoted the use of biological control. This has been the case with different goods produced in the tropics mainly for exportation, especially to Europe (Duarte Cueva, 2012).

4.3.1 Classical biological control

Comprehensive revisions about projects concerning this type of strategy were published by Hong *et al.* (1999) and Waterhouse and Sands (2001) for southeast Asia and Oceania, Parra *et al.* (2002) for Brazil, Greathead (2003) and Neuenschwander *et al.* (2003) for Africa, and Aguirre Gil *et al.* (2013) for Latin America, with special reference to Peru.

As often is the case, activities related to classical biological control have been conducted almost exclusively by governmental departments and organizations, which quite often suffer from limited availability of funds. Thus, efforts related to this type of strategy are usually more limited than efforts dedicated to other strategies.

The actual implementation of projects concerning classical biological control has become progressively more difficult, especially because of more restrictive legislations concerning the several phases of this type of strategy. In an attempt to ensure potential benefits from the prevailing biodiversity, in many tropical countries, permits are now required for the collection and for the shipment of prospective natural enemies to other countries. Also, shipment of natural enemies for biological control

purposes has become difficult, because some transporters simply refuse to carry biological control agents, in part, possibly for fear of transgressing the laws of the involved countries. In addition, official permits issued by receiving countries are also required for the introduction of the organisms. The main concern in this case is to prevent possible disruption of the local ecological balance, by the introduced natural enemies. Obtaining those permits is quite often complicated and time-consuming. All of these difficulties have discouraged the execution of potentially successful projects. Wilson *et al.* (2000) and Cock *et al.* (2010) discussed how stringent legislation adopted by countries may hamper the development of biological control.

Despite these difficulties, classical biological control has been successful in tropical countries. In the recent past, most outstanding projects involved the biological control of two cassava pests in Africa, namely, the cassava green mite, *Mononychellus tanajoa* (Bondar) (Acari: Tetranychidae), and the cassava mealybug, *Phenacoccus manihoti* (Matile-Ferrero) (Hemiptera: Pseudococcidae), in the 1970s–1990s. The former was controlled with the introduction of predaceous Phytoseiidae mites from South America (Yaninek and Hanna, 2003) and the latter, with the introduction of the parasitoid *Anagyrus lopezi* (De Santis) (Hymenoptera: Encyrtidae), also from South America (Neuenschwander, 2003).

A related and very promising project has been conducted in Thailand and neighbouring countries, where the fearful cassava mealybug was recently introduced (Winotai *et al.*, 2010). *Anagyrus lopezi* was then introduced from the Republic of Benin to Thailand (Winotai *et al.*, 2010; Parsa *et al.*, 2012). The obtained results so far are very promising, and could represent a new successful biological control programme. The parasitoid has seemingly become established and is now spreading from the initial sites of release.

Several recent cases of successful projects have been mentioned for Asia. One of these refers to the beetle *Brontispa longissima* Gestro (Coleoptera: Hispidae), a serious pest of coconut palms. This species was introduced from countries of southeast Asia, where it seems to have originated, to other Asian countries. Under the auspices of FAO, a major project was launched. As part of that project, the parasitoid *Asecodes hispinarum* Boucek (Hymenoptera: Eulophidae) was introduced from Samoa to several Asian countries, with promising results having so far been reported for Vietnam, where the parasitoid became established and has seemingly reduced the damage caused by the pest (Liebregts and Chapman, 2004). In the same continent, effective control of the papaya mealybug *Paracoccus marginatus* Williams and Granara de Willink (Hemiptera: Pseudococcidae) by *Acerophagus papayae* Noyes and Schauff, *Anagyrus loecki* Noyes, and *Pseudleptomastix mexicana* Noyes and Schauff (Hymenoptera: Encyrtidae) imported from Mexico and of the solenopsis mealybug *Phenacoccus solenopsis* Tinsley (Hemiptera: Pseudococcidae) by the fortuitously introduced *Aenasius bambawaley* Hayat (Hymenoptera: Encyrtidae) was achieved (Muniappan, 2011).

Ongoing projects involving mite pests refer to the introduction of predatory mites from Brazil to the Republic of Benin for the control of the coconut mite, *Aceria guerreronis* Keifer (Acari: Eriophyidae); from La Réunion Island to Brazil (Moraes *et al.*, 2012) for the control of the palm mite, *Raoiella indica* Hirst (Acari: Tenuipalpidae) and *Euseius stipulatus* Athias-Henriot (Acari: Phytoseiidae) from southern Europe to Peru (Aguirre-Gil *et al.*, 2013) for the control of the citrus rust mite, *Phyllocoptruta oleivora* (Ashmed) (Acari: Eriophyidae). On the other hand, ongoing projects involving insect pests refer to the introduction of parasitoids from the USA to Brazil for the control of the citrus leafminer, *Phyllocnistis citrella* Stainton (Lepidoptera: Gracilariidae) on citrus (Gravena, 2011) and of Psyllaephagus bliteus (Hymenoptera: Encyrtidae), apparently introduced along with its host, for the control of the red gum lerp psyllid *Glycaspis brimblecombei* Moore (Hemiptera: Aphalaridae) on eucalyptus (Wilcken *et al.*, 2010).

Another recent success story of pest control using introduced natural enemies involves the potato tuber moth *Phthorimaea operculella* Zeller (Lepidoptera: Gelechiidae) using the parasitoids *Copidosoma koehleri* Blanchard (Hymenoptera: Encyrtidae) and *Apanteles subandinus* Blanchard (Hymenoptera: Braconidae) from South America to Africa (Greathead, 1971; Weber, 2013).

4.3.2 Augmentation

Activities related to this strategy have been mostly conducted by private organizations, given the interest in the commercialization of the natural enemies. Of course, the actual practical use of this strategy depends first of all on the cost of the adoption. This certainly encourages producers of natural enemies to seek ways of producing them at costs compatible with that of other available control measures.

Releases in extensive crops

Use of this strategy in tropical countries was until recently restricted to a few crops grown in large areas. Extensive use has been made of the fungus *Metarhizium anisopliae* (Metschnikoff) for the control of the spittlebugs *Notozulia entreriana* (Berg) and *Deois flavopicta* (Stål) (Hemiptera: Cercopidae) in Brazil for many years in pastures (Alves and Lopes, 2008). Similarly, this fungus has also been largely used to control *Mahanarva posticata* (Stål) (Hemiptera: Cercopidae) especially in northeastern Brazil, where many sugarcane growers mass produce this biological control agent for periodic releases.

The most popular programme has been the use of parasitoids for the biological control of the sugarcane borer, *Diatraea saccharalis* (Fabricius) (Lepidoptera: Crambidae) in Brazil (Botelho and Macedo, 2002). In fact, initially this referred to a case of classical biological control with the introduction of the larval parasitoid *Cotesia flavipes* (Cameron) (Hymenoptera: Braconidae) from Asia to the Americas in the 1960s, followed by the launching of an extensive programme of mass production by growers and commercial producers of natural enemies (Fig. 4.1). This very effective measure has been widely used for the control of this pest in 3 million ha of sugarcane in Brazil (Parra, 2011), which represents around 30% of the total area covered with this crop. In this country, the level of damage was reduced from 7% of infestation (percentage of infested internodes) to 2% after the introduction of the parasitoid in 1974 (Parra *et al.*, 2002). Biological control programmes using this parasitoid were also developed in other South and Central American countries such as Colombia, Guatemala and Venezuela.

Similarly, *Chilo partellus* (Swinhoe) (Lepidoptera: Crambidae), a serious pest of maize and sorghum in Africa, is under control by *C. flavipes* introduced from Pakistan and India (Overholt, 1998). Together with habitat management through push–pull strategy, stem borer populations have been tremendously reduced (Khan *et al.*, 1997). Varietal mixtures are a viable strategy for subsistence agriculture in Africa and is also very much revered. The use of this resource is currently being investigated in the manipulation of the habitats to the point of being widespread among smallholder farming in pest management.

More recently, commercial laboratories have mass-produced species of *Trichogramma* Westwood (Hymenoptera: Trichogrammatidae) for periodic releases to control *D. saccharalis* eggs (Parra, 2010). Species of *Trichogramma* have also been used to control the tomato leafminer, *Tuta absoluta* (Meyrick) (Lepidoptera: Gelechiidae) in

Fig. 4.1. The larval parasitoid, *C. flavipes*, represents a successful biological control programme of the sugarcane borer, *D. saccharalis*, in the tropics. The host is mass reared and manually offered to parasitoid females for oviposition (A and B). After completing the larval period, *C. flavipes* larvae leave the host (C) to pupate. Cocoons are kept in plastic cups and released in sugarcane fields (D). Photos by O.A. Fernandes.

northeastern Brazil. In combination with other control practices, such as applications of *B. thuringiensis*, this has reduced the use of broad-spectrum pesticides and enforcement of a mandatory three-month crop-free period (Haji *et al.*, 2002). Mass production of parasitoids has been an important business in Brazil (Parra, 2010). A long list of organisms of the same and of many other groups for use in organic farming in Chile was published by Certificadora Chile Orgánico/Fundación para la Inovación Agraria (2005).

Since the 1990s, a gradual reduction in the use of burning to harvest sugarcane has been imposed by Brazilian law enforcers to the sugarcane industry. The harvest of green sugarcane results in a large amount of litter (leaves and stalk tips) on the soil surface, which can reach 18 t/ha. Litter accumulation has favoured the development of the sugarcane spittlebug *Mahanarva* (Stål) spp. (Hemiptera: Cercopidae) and outbreaks of other soil-dwelling insect pests. As the sugarcane industry had already relied on biological control programmes (for sugarcane borer), the adoption of another ecologically sound tool was facilitated. The fungi

M. anisopliae and *Beauveria bassiana* (Bals.) Vuill were incorporated in the pest management system. Both fungi are produced using mainly rice as substrate (Alves and Lopes, 2008). As the insect feeds on the superficial roots, application directed towards the base of the plants using terrestrial sprayers is recommended. However, applications by aeroplane are possible during cloudy or rainy days because rain can wash out the spores deposited on the leaves to the base of the plants and soil where the spittlebugs are (Fig. 4.2).

One of the major examples of the use of a virus for the control of a pest insect was developed in the tropics. It refers to the use of *Baculovirus anticarsia* (AgMNVP), commercially produced for the control of the velvetbean caterpillar, *Anticarsia gemmatalis* (Hübner) (Lepidoptera: Erebidae (a new family for Anticarsia gemmatalis) on soybean (Sosa-Gomez *et al.*, 2008; Moscardi *et al.*, 2011). This is also a major crop in South America, where it covers an estimated 46.5 million ha (FAO, 2012). Just in Brazil, it has been calculated that *B. anticarsia* was used in an area of approximately

Fig. 4.2. Aerial application of the entomopathogen *Metarhizium anisopliae* for biological control of the spittle bug. To assure that applied spores are washed off the leaves and reach the nymphs at the base of sugarcane stalks, application should be followed by rain. Infected *Mahanarva* sp. adults are shown. Photos by O.A. Fernandes.

2 million ha in the 1980s–1990s, reducing by about 50% the use of insecticide for the control of that pest (L. Morales, personal communication). Given the extensive use of biological control in this crop, another programme was successfully implemented (Corrêa-Ferreira, 2002) on this agroecosystem in Brazil involving the use of the egg parasitoid *Trissolcus basalis* (Woll.) (Hymenoptera: Platygastridae) for the control of stink bugs (Hemiptera: Pentatomidae).

However, recently, traditional varieties of soybean have been widely replaced by genetically modified varieties. These have been introduced worldwide in production systems to improve plants to withstand pest incidences. Maize, cotton, canola and soybean are the main cultivated genetically modified crops. They can express tolerance to herbicides and/or insecticide proteins. Comprehensive risk assessment studies have demonstrated that genetically modified (GM) plants expressing proteins have no effect on the population of natural enemies (Naranjo, 2009). In fact, in some studies, natural enemy populations can even be favoured, due to reduction in pesticide use (Lu *et al.*, 2012). In GM maize expressing *B. thuringiensis* proteins, Fernandes *et al.* (2007) observed no reduction on generalist predators.

On the other hand, the use of genetically modified crops expressing tolerance to herbicides can facilitate the use of other pesticides that are harmful to natural enemy populations. After the introduction of the genetically modified soybean varieties to allow the use of the herbicide glyphosate after germination, growers now usually consider that it is more economical to combine the application of that herbicide with the use of agrochemicals to control pests and diseases (Petter *et al.*, 2007), instead of using the herbicide in combination with biological control. This has resulted in a major reduction of the use of *B. anticarsia* in Brazil, with a major impact on companies involved in the commercial production of that virus. In addition, the more widespread use of fungicides (to control the Asian soybean rust, introduced in Brazil in the early 2000s) has also reduced the epizootics of naturally occurring fungi such as *Nomuraea rileyi* (Farlow) Samson and favoured the occurrence of outbreaks of lepidopteran pests (Sosa-Gómez *et al.*, 2003). Thus, biotechnological tools may interfere directly or indirectly with biological control programmes.

Brazil has been the world leader in pesticide use, whose market reached US$8 billion in 2011 (Freitas Jr, 2011). The value of the main commodities has been high in the past years (late 2000s and early 2010s), paying off the use of extra inputs to increase yield and/or reduce losses caused by pests. This usually leads to low tolerance to pest population by growers who adopt and rely on pesticides as the primary control tool.

Most of the important agricultural pests in Africa are thought to be native and are closely associated with their range of natural enemies. This makes biological control in Africa unique as compared to other geographic areas such as Oceania, where the most important pests are exotic. Where pests are predominantly exotic, classical biological control is widely adopted, since natural enemies are introduced from the places of origin of the pests. Thus, in Africa, most pests can be controlled by augmentative biological control or by the conservation of natural enemies.

Use of entomopathogens has also evolved in biological pest management in Africa. One of the pioneering works refers to the control of locusts and grasshoppers using a commercial product called Green Muscle under an international project called Lutte Biologique contre les Locustes et les Sauteriaux (LUBILOSA). Green Muscle was based on the fungus *M. anisopliae* var. *acridium* (Lomer *et al.*, 2001). The success of LUBILOSA project led to other projects involving entomopathogens that have targeted termites (Maniania *et al.*, 2002), banana weevils (Nankinga and Moore, 2000) and fruit flies (Ekesi *et al.*, 2002). Several other insects are being targeted by entomopathogens, including diamond back moth, as well as thrips and whitefly. These entomopathogens are now a widespread tool in pest management and are being applied in the management of pests in both traditional and commercial crops.

Releases in small crops

Augmentation of natural enemies has only recently been extensively used in the tropics for the control of pests on crops grown in restricted areas. In part, this has occurred because of the increase in the acreage used for protected crops in tropical countries, in particular, for the cultivation of high-value ornamentals and horticultural crops. In open fields, biological control has been adopted on ornamentals and strawberry and is based primarily on periodic releases of natural enemies.

Many of the crops grown in small areas have high economic return, often implying the adoption of low economic damage level for each pest, leading growers to invest more in control practices. This trend has often led to the development of pest resistance due to the excessive use of agrochemicals and the resultant difficulties in controlling the pests. Growers' consciousness about the positive results of biological control initiatives in crops grown in other parts of the world has led them to contact research institutions in an attempt to make these techniques available. The recent improvement in communication, especially the use of the internet, has apparently played a role in this process.

Of considerable interest is also the control of the two spotted spider mite, *Tetranychus urticae* Koch (Acari: Tetranychidae), on ornamentals and strawberry, with the use of predatory mites, mainly in Argentina, Brazil, Colombia and Kenya (Fig. 4.3).

Ongoing research referring to periodic releases involves the mass production of entomopathogenic fungi, *Sporothrix insectorum* (Hoog and Evans) and *Isaria fumosorosea* (sin. *Paecilomyces fumosoroseus*), and green lacewings, *Chrysoperla cincta* (Schneider) (Neuroptera: Chrysopidae) for the control of the lacebug *Leptopharsa heveae* Drake and Poor (Hemiptera: Tingidae) in central Brazil (C.H. Scomparin, personal communication).

Mass production and periodic releases of natural enemies have also been conducted on citrus in Peru on small as well as in relatively extensive crops. In that country, citrus production for exportation is a major agricultural business. In order to comply with requirements of European consumers, the citrus rust mite is controlled by an association of control practices, involving the use of the predatory mite of the genus *Euseius*, as well as the application of other control methods.

These aspects have attracted the interest of companies of different sizes, from large multinational corporations to microenterprises, to establish in different countries. These are now found in many tropical countries, especially in the American continent and in Asia. As a result of the growing business possibilities, those companies have organized themselves in associations that look after their commercial interest. Associations of producers of biological control agents are found in tropical countries such as Brazil (ABCBIO, n.d.) and Peru (Duarte Cueva, 2012), though there is also the International Biocontrol Manufacturers' Association (IBMA) which is a worldwide association of biocontrol industries involved in the production of microorganisms, macroorganisms, semiochemicals and natural pesticides for crop protection and public health (www.ibma-global.org).

Another new impetus concerning the use of biological control refers specifically to the upsurge of frequent outbreaks of

Fig. 4.3. *Phytoseiulus macropilis* (Banks) (Acari: Phytoseiidae), a predator of the two spotted mite, *Tetranychus urticae* Koch (Acari: Tetranychidae). Photo by PROMIP, Brazil.

whiteflies of the genus *Bemisia* (Hemiptera: Aleyrodidae) in the last 20–30 years throughout the world and on many different crops. This has led to the search and finding of effective predators and to the development of techniques to mass produce them at costs that allow their practical use (van Lenteren, 2007). These techniques have often been developed by private companies, which do not have interest in publishing details about them. Also, the more intensive problems with the incidence of thrips (Thysanoptera: Thripidae) and fungus gnats (Diptera: Sciaridae), especially in ornamentals, in Brazil and Colombia has led to the use of effective predatory mites for their control (Gerson *et al.*, 2003).

Despite the new impetus to biological control, a major constraint in some countries refers to the very restrictive legislation, in which the commercialization and use of biological control agents are treated in a similar manner as agrochemicals. In this sense, the commercialization of all types of mass-produced natural enemies requires previous registration, which may be sufficiently expensive or complicated as to discourage the small companies from dedicating to the production of those organisms. On the one hand, the rigour favours the maintenance of high-quality products; while on the other hand, it sometimes hampers the establishment of new initiative of small, local entrepreneurs in favour of well-established, large companies. These requirements do not exist in some countries with longstanding experience on the use of biological control, as in the United States of America, in what refers to the use of predators or parasitoids. However, there are some exceptions. In Peru, for example, the national initiatives to engage in the mass production of natural enemies are protected by the state. Small businesses receive several benefits from the national government, making them more competitive in relation to foreign companies wanting to operate in the country (Duarte Cueva, 2012). Thus, natural enemies in Peru are mass-produced by a series of private laboratories supported by SENASA (Peruvian National Service for Agrarian Health).

4.3.3 Conservation

Activities related to this topic are mainly attributed to the use of more selective pesticides, so as to prevent negative impact from the use of chemicals on natural enemies present in agroecosystems. This is discussed more fully in the next section.

In relation to other forms of conservation, of particular interest has been a recent biological control programme of rats in sugarcane. The rat *Sigmodon hirsutus* (Burmeister) has been found in sugarcane fields and wet pastures of Central America, Mexico and Venezuela (IUCN, 2013). This species is known to be preyed upon (Delgado and Cataño, 2004) by barn owls, *Tyto alba* (Scopoli). This led growers from Guatemala to distribute artificial nests within sugarcane plantations in order to retain owls, so as to reduce damage by rats. This has been reported as very effective (Falla *et al.*, 2012) and is part of the sugarcane IPM programme in Central America.

4.4 Relationship of Biological Control and IPM

The management of insect pests does not rely on a single control practice. Usually a variety of tactics are integrated to maintain pests at levels that are acceptable to growers. Applying multiple control tactics minimizes the chance that insects will adapt to any one tactic of pest control, making it very desirable to commercial growers. However, this could be partly the obstacle in adopting IPM strategies, because it requires a lot of knowledge in making combinations of control tactics. The goal of IPM has never been to eliminate all pests; some pests are tolerable and even essential, as they serve as food for natural enemies. Rather, the aim is to reduce pest populations to non-damaging numbers. The IPM control tactics include pest-resistant plants, and cultural, physical, mechanical, biological and chemical control. Therefore, biological control is just one of the components of IPM. But insecticides are still the main tool used to control insect pests even in IPM systems. Not only growers in the tropics but also those elsewhere rely on such compounds for pest management. However, integration of biological control and insecticides are only possible under certain circumstances.

1.   Insecticides are selective to natural enemies: selectiveness can be achieved by inherent properties of the natural enemy to stand the toxicity of the compound or specificity of the compound towards the target pest.
2.   Insecticides are selectively applied to reduce the risk to natural enemies; product formulation, application method, and time of application can be changed to decrease the likelihood of natural enemies coming into contact with the insecticides and thus reduce mortality.
3.   Other agrochemicals (e.g. fungicides and herbicides) may also affect natural enemies. Consequently, their effect should also be taken into consideration in pest management.

Generally, the information about selectivity is not required to be present in the product's label in most countries, making the selection of those products that are compatible with natural enemies a very difficult task. The International Organization for Biological Control and Integrated Control of Noxious Animals and Plants (IOBC) has been working along with researchers to enhance the knowledge on pesticide toxicity to natural enemies through the working group on pesticides and beneficial organisms.

Several strategies have been developed and implemented (for example, in Africa) such as educating farmers, habitat management through push–pull technology, plant health management, and selection of planting material as a result of small-plot farming. And because of this type of setting, farmers have adopted practical IPM options, with little or no reliance on pesticides. The result of this approach is that Africa's environment remains unsaturated with pesticides and there is a wide range of natural enemies, as well as resistant cultivars.

Biological control should be the basis of any IPM programme. Therefore, strategies to enhance the effectiveness of naturally occurring biological control agents should always be encouraged to increase adoption by growers. However, as pointed out by Kogan (1998), most IPM programmes still rely on chemical control as the main corrective strategy to reduce pest problems.

4.5 Future Trends and Concluding Remarks

Pest problems continue to increase all over the world. One of the main reasons seems to be the process of globalization and hence a broadening of the international market of agricultural goods (Muniappan, 2011). This process may facilitate the spread of pests around the world, including into, out of and between tropical countries. The rate of introduction of new pest species has been noted to be much higher in the tropics than in temperate regions (Fish *et al.*, 2010). For instance, at least four major mite pests have been introduced in tropical America, including the Caribbean islands, since the last few years of the 20th century, namely, *Steneotarsonemus spinki* Smiley (Almaguel *et al.*, 2000), *R. indica* Hirst (Etienne and Flechtmann, 2006), *Aceria litchi* Keifer (Raga *et al.*, 2010) and *Aceria tosichella* Keifer (Navia *et al.*, 2006). Within that same period, exogenous insects such as the Old World bollworm, *Helicoverpa armigera* (Hübner) (Lepidoptera: Noctuidae), has been introduced to the same continent (Czepak *et al.*, 2013). New insect pests recently introduced to Asia include mainly species of mealybugs and whiteflies (Muniappan, 2011; Parsa *et al.*, 2012). This calls for efficient quarantine measures to be implemented, in order to reduce the chances for increased pest problems in the tropics.

The results of research work conducted by different government institutions and private organizations in the tropics and elsewhere have made possible the increase in the use of augmentation for traditional as well as for new pest problems. Of major importance has been the development of new techniques allowing the reduction of the production cost of natural enemies, making their use competitive with costs of chemical control.

In the tropics, favourable climatic conditions and high biodiversity would be expected to favour the development of biological control. For example, natural epizootics caused by a diversity of fungi are quite frequent in most countries (Alves and Lopes, 2008). However, one of the main causes for the relatively low use of biological control in the tropics seems to be mainly related to the low investment in research and/or regulation on restrictions to avoid biopiracy and protect biodiversity.

However, for several countries, investments in research are very difficult because of the limitation of available funds, and in this case, international support would be very appropriate. However, the recent international economic crises have negatively affected the forecast of biological control in the near future. International research centres have faced significant downsizing and budget constraints. In this regard, the important role played in the recent past by institutions like the International Institute of Tropical Agriculture (IITA) and the International Center for Tropical Agriculture (CIAT) concerning the biological control of major pests in Africa and in the Americas, such as the cassava green mite, has been severely affected. Thus, it is necessary to revisit the current situation of these centres and to provide support to allow them to resume their role as regional leaders on biological control.

In an attempt to promote the use of biological control, some work has been conducted for the discovery of new effective biological control agents. Examples of those efforts include the work conducted in Brazil to determine native predatory mites to be used for the control of mites and small insects that spend part or all of their life in the soil (Castilho *et al.*, 2009a,b). The participation of taxonomists is of fundamental importance for this type of work, and often that is one of the main limitations for its success. In any case, considering that many of the tropical countries are located in regions of high biodiversity, it seems convenient to foster this type of work in these regions, where the chances to find promising natural enemies are expected to be higher than in temperate countries.

New candidates for use as biological control agents could be of use in tropical countries as well as in countries of other regions. It is striking to note that there is a very low number of commercialized natural enemies of pest organisms around the world in comparison with the expected number of species of promising groups. More intensive exploration and bioprospecting in tropical countries may allow the discovery of a range of natural enemies perhaps more fitting than the ones already in use, concerning their efficiency, suitability for mass production and specificity.

Acknowledgements

Thanks are due to Sofia Jiménez Jorge and Roberto Trincado, for the information about the status of biological control in Peru and Chile, respectively.

References

ABCBIO (n.d.) Brazilian Association of Biological Control Companies. Available at: www.abcbio.org. br (accessed 13 January 2014).

Aguirre-Gil, O.J., Jorge, S.J. and Busoli, A.C. (2013) Controle biológico clássico na América Latina: o caso do Peru. In: Busoli, A.C., Alencar, J.R.D.C.C., Fraga, D.F., Souza, L.A., Souza, B.H.S. and Grigolli, J.F.J. (eds) *Tópicos em Entomologia Agrícola VI*. Gráfica Multipress Ltd, Jaboticabal, Brazil, pp. 67–75.

Almaguel, L., Hernandez, J., de la Torre, P.E., Santos, A., Cabrera, R.I. *et al.* (2000) Evaluación del comportamiento del acaro *Steneotarsonemus spinki* (Acari: Tarsonemidae) en los estúdios de regionalización desarrollados en Cuba. *Fitosanidad* 4, 15–19.

Alves, S.B. and Lopes, R.B. (eds) (2008) *Controle microbiano de pragas na América Latina: avanços e desafios*. Fealq, Piracicaba, Brazil.

Blum, B. (2002) Blocked opportunities for bio-control. *Pesticides News* 57, 18.

Botelho, P.S. and Macedo, N. (2002) *Cotesia flavipes* para o controle de *Diatraea saccharalis*. In: Parra, J.R.P., Botelho, P.S.M., Corrêa-Ferreira, B.S. and Bento, J.M.S. (eds) *Controle biológico no Brasil: parasitóides e predadores*. Manole, Barueri, Brazil, pp. 409–425.

Bueno, V.H.P. (2009) *Controle biológico de pragas*. Editora UFLA, Lavras, Brazil.

Castilho, R.C., Moraes, G.J., Silva, E.S. and Silva, L.O. (2009a) Predation potential and biology of *Protogamasellopsis posnaniensis* Wisniewski & Hirschmann (Acari: Rhodacaridae). *Biological Control* 48, 164–167.

Castilho, R.C., Moraes, G.J., Silva, E.S., Freire, R.A.P. and Eira, F.C. (2009b) The predatory mite *Stratiolaelaps scimitus* as a control agent of the fungus gnat *Bradysiamato grossensis* in commercial production of the mushroom *Agaricus bisporus*. *International Journal of Pest Management* 55, 181–185.

Certificadora Chile Orgánico/Fundación para la Innovación Agraria (2005) *Catalogo de insumos para el control de plagas y enfermedades en agricultura orgánica en Chile*. Certificadora Chile Orgánico/Fundación para la Innovación Agraria, Santiago, Chile.

Cock, M.J., van Lenteren, J.C., Brodeur, J., Barratt, B.I., Bigler, F. *et al.* (2010) Do new access and benefit sharing procedures under the Convention on Biological Diversity threaten the future of biological control? *BioControl* 55 (2), 199–218.

Corrêa-Ferreira, B.S. (2002) *Trissolcus basalis* para o controle de percevejos da soja. In: Parra, J.R.P., Botelho, P.S.M., Corrêa-Ferreira, B.S. and Bento, J.M.S. (eds) *Controle biológico no Brasil: parasitóides e predadores*. Manole, Barueri, Brazil, pp. 449–476.

Czepak, C., Albernaz, K.C., Vivan, L.M., Guimarães, H.O. and Carvalhais, T. (2013) Primeiro registro de ocorrência de *Helicoverpa armigera* (Hübner) (Lepidoptera: Noctuidae) no Brasil. *Pesquisa Agropecuária Tropical* 43 (1), 110–113.

De Bach, P. (1974) *Biological Control by Natural Enemies*. Cambridge University Press, London.

Delgado, V.C.A. and Cataño, B.E.J.F. (2004) Diet of the barn owl (*Tyto alba*) in the lowlands of Antioquia, Colombia. *Ornitologia Neotropical* 15, 413–415.

Duarte Cueva, F. (2012) El control biológico como estrategia para apoyar las exportaciones agrícolas no tradicionales en Perú: un análisis empírico. *Contabilidad y Negocios* 7 (14), 81–100. Available at: http://revistas.pucp.edu.pe/index.php/contabilidadyNegocios/article/view/3881 (accessed 29 July 2017).

Eilenberg, J., Hajek, A. and Lomer, C. (2001) Suggestions for unifying the terminology in biological control. *BioControl* 460 (4), 387–400.

Ekesi, S., Maniania, N.K. and Lux, S.A. (2002) Mortality in three African tephritid fruit fly puparia and adults caused by the entomopathogenic fungi, *Metarhizium anisopliae* and *Beauveria bassiana*. *Biocontrol Science and Technology* 12, 7–17.

Etienne, J. and Flechtmann, C.H. (2006) First record of *Raoiella indica* (Hirst, 1924) (Acari: Tenuipalpidae) in Guadeloupe and Saint Martin, West Indies. *International Journal of Acarology* 32 (3), 331–332.

Falla, C., Márquez, M. and Lemus, J.M. (2012) *Características bioecológicas de la lechuza* (Tyto alba) *como depredador dentro del manejo integrado de la rata de campo*. In: Cengicaña Memoria: presentation de resultados de investigación – zafra 2011/2012. Cengicaña, pp. 187–194. Available at: www.cengicana.org/files/20150902101636279.pdf (accessed 29 July 2017).

FAO (2012) FAOSTAT. Food and Agriculture Organization of the United Nations. Available at: www.faostat.fao.org (accessed 12 June 2013).

Fernandes, O.A., Faria, M.R., Martinelli, S., Schmidt, F.G.V., Carvalho, V.F. and Moro, G.L. (2007) Short-term assessment of Bt maize on non-target arthropods in Brazil. *Scientia Agricola* 64, 249–255.

Fish, J., Chiche, Y., Day, R., Efa, N., Witt, A. *et al.* (2010) Mainstreaming gender into prevention and management of invasive species. Global Invasive Species Programme (GISP), Washington DC, and Nairobi. Available at: http://issg.org/cii/Electronic%20references/pii/references/mainstreaming_gender_into_prevention_and_management_of_invasive_species.PDF (accessed 29 July 2017).

Freitas Jr., G. (2011) Brasil deve passar os EUA em venda de defensivos agrícolas. *Valor Econômico*, 22 July. Available at: www.valor.com.br/arquivo/899865/brasil-deve-passar-os-eua-em-venda-de-defensivos-agricolas#ixzz2YPGFacKV (accessed 13 January 2014).

Garcia, R., Caltagirone, L.E. and Gutierrez, A.P. (1988) Comments on a redefinition of biological control. *BioScience* 38 (10), 692–694.

Gerson, U., Smiley, R.L. and Ochoa, R. (2003) *Mites (Acari) for Pest Control.* Blackwell Science, Oxford, UK.

Gravena, S. (2011) History of pest control in the Brazilian citrus. *Citrus Research and Technology* 32 (2), 85–92.

Greathead, D.J. (1971) *A Review of Biological Control in the Ethiopian Region.* Commonwealth Institute of Biological Control, Technical Communication No. 5. Commonwealth Agricultural Bureau, Farnham, UK.

Greathead, D.J. (2003) Historical overview of biological control in Africa. In: Neuenschwander, P., Borgemeister, C. and Langewald, J. (eds) *Biological Control in IPM Systems in Africa.* CAB International, Wallingford, UK.

Haji, F.N.P., Prezotti, L., Carneiro, J.S. and Alencar, J.A. (2002) *Trichogramma pretiosum* para o controle de pragas no tomateiro industrial. In: Parra, J.R.P., Botelho, P.S.M., Corrêa-Ferreira, B.S. and Bento, J.M.S. (eds) *Controle biológico no Brasil: parasitóides e predadores.* Manole, Barueri, Brazil, pp. 377–394.

Hong, L.W., Sastroutomo, S.S., Caunter, I.G., Ali, J., Yeang, L.K. *et al.* (eds) (1999) *Symposium on Biological Control in the Tropics.* CAB International, Selangor, Malaysia.

IUCN (2013) *The IUCN Red List of Threatened Species.* Available at: www.iucnredlist.org/details/136426/0 (accessed 12 June 2013).

Khan, Z.R., Ampong-Nyarko, K., Chiliswa, P., Hassanali, A., Kimani, S. *et al.* (1997) Intercropping increases parasitism of pests. *Nature* 388, 631–632.

Kogan, M. (1998) Integrated pest management: historical perspectives and contemporary developments. *Annual Review of Entomology* 43 (1), 243–270.

Liebregts, W. and Chapman, K. (2004) Impact and control of the coconut hispine beetle, *Brontispa longissima* Gestro (Coleoptera: Chrysomelidae). In: *Report of the Expert Consultation on Coconut Beetle Outbreak in APPPC Member Countries.* RAP Publication 2004/29, 26–27 October. FAO, Bangkok, pp. 19–25.

Lomer, C.J., Bateman, R.P., Johnson, D.L., Langewald, J. and Thomas, M.B. (2001) Biological control of locusts and grasshoppers. *Annual Review of Entomology* 46, 667–702.

Lu, Y., Wu, K., Jiang, Y., Guo, Y. and Desneux, N. (2012) Widespread adoption of Bt cotton and insecticide decrease promotes biocontrol services. *Nature* 487, 362–365.

Maniania, N.K., Ekesi, S. and Songa, J.M (2002) Managing termites in maize cropping systems with entomopathogenic fungus, *Metarhizium anisopliae. Insect Science and Its Application* 21, 41–46.

Moraes, G.J., Castro, M.M.G., Kreiter, S., Quilici, S., Gondim Jr, M.G.C. and Sá, L.A.N. (2012) Search for natural enemies of *Raoiella indica* Hirst in Réunion Island (Indian Ocean). *Acarologia* 52, 129–134.

Moscardi, F., Souza, M.L., Castro, M.E.B., Moscardi, M.L. and Szewczyk, B. (2011) Baculovirus pesticides: present state and future perspectives. In: Ahmad, I., Ahmad, F. and Pichtel, J. (eds) *Microbes and Microbial Technology: Agricultural and Environmental Applications.* Springer, New York, pp. 415–445.

Muniappan, R. (2011) Recent invasive hemipterans and their biological control in Asia. In: *5th Meeting of the Asian Cotton Research and Development Network,* Lahore, Pakistan. Available at: www.icac.org/tis/regional_networks/asian_network/meeting_5/documents/papers/PapMuniappanR.pdf (accessed 13 January 2014).

Nankinga, C.M and Moore, D. (2000) Reduction of banana weevil populations using different formulations of the entomopathogenic fungus *Beauveria bassiana. Biocontrol Science and Technology* 10, 645–657.

Naranjo, S.E. (2009) Impacts of Bt crops on non-target invertebrates and insecticide use patterns. *Perspectives in Agriculture, Veterinary Science, Nutrition and Natural Resources* 4, 1–23.

Navia, D., Truol, G., Mendonça, R.S. and Sagadin, M. (2006) *Aceria tosichella* Keifer (Acari: Eriophyidae) from wheat streak mosaic virus-infected wheat plants in Argentina. *International Journal of Acarology* 32 (2), 189–193.

Neuenschwander, P. (2003) Biological control of cassava and mango mealybugs in Africa. In: Neuenschwander, P., Borgemeister, C. and Langewald, J. (eds) *Biological Control in IPM Systems in Africa.* CAB International, Wallingford, UK, pp. 45–59.

Neuenschwander, P., Borgemeister, C. and Langewald, J. (2003) *Biological Control in IPM Systems in Africa.* CAB International, Wallingford, UK.

Overholt, W.A. (1998) Biological control. In: Polaszek, A. (ed.) *African Cereal Stemborers: Economic Importance, Taxonomy, Natural Enemies and Control.* CAB International, Wallingford, UK, pp. 349–362.

Parra, J.R.P. (2010) Egg parasitoids commercialization in the new world. In: Consoli, F.L., Parra, J.R.P. and Zucchi, R.A. (eds) *Egg Parasitoids in Agroecosystems with Emphasis on* Trichogramma. Springer, London, pp. 373–388.

Parra, J.R.P. (2011) Controle biológico de pragas no Brasil: histórico, situação atual e perspectivas. *Ciência & Ambiente* 43, 7–18.

Parra, J.R.P., Botelho, P.S.M., Corrêa-Ferreira, B.S. and Bento, J.M.S. (2002) *Controle biológico no Brasil: parasitóides e predadores.* Manole, Barueri, Brazil.

Parsa, S., Kondo, T. and Winotai, A. (2012) The cassava mealybug (*Phenacoccus manihoti*) in Asia: first records, potential distribution, and an identification key. *PLOS ONE* 7 (10), e47675. DOI:10.1371/journal.pone.0047675

Petter, F.A., Procópio, S.O., Cargnelutti Filho, A., Barroso, A.L.L., Pacheco, L.P. and Bueno, A.F. (2007) Associações entre o herbicida glyphosate e inseticidas na cultura da soja Roundup Ready®. *Planta daninha* 25 (2), 389–398.

Pinto, A.S., Nava, D.E., Rossi, M.M. and Malerbo-Souza, D.T. (2006) *Controle biológico de pragas na prática.* CP 2, Piracicaba, Brazil.

Prado, E. (1991) *Artrópodos y sus enemigos naturales asociados a plantas cultivadas en Chile.* Serie Boletín Técnico no. 169. Instituto de Investigaciones Agropecuaria, Santiago, Chile.

Raga, A., Mineiro, J.L.D.C., Sato, M.E., Moraes, G.J. and Flechtmann, C.H.W. (2010) First report of *Aceria litchii* (Keifer) (Prostigmata: Eriophyidae) on litchi trees in Brazil. *Revista Brasileira de Fruticultura* 32 (2), 628–629.

Sampaio, M.V., Bueno, V.H.P., Silveira, L.C.P. and Auad, A.M. (2008) Biological control of insect pests in the tropics. In: Del Claro, K. *et al.* (eds) *Encyclopedia of Life Support Systems (EOLSS).* Developed under the auspices of UNESCO, Oxford, UK, pp. 1–36. Available at: http://www.eolss.net/sample-chapters/C20/E6-142-TA-04.pdf (accessed 7 June 2017).

Sosa-Gómez, D.R., Delpin, K.E., Moscardi, F. and Nozaki, M.H. (2003) The impact of fungicides on *Nomuraea rileyi* (Farlow) Samson epizootics and on populations of *Anticarsia gemmatalis* Hübner (Lepidoptera: Noctuidae) on soybean. *Neotropical Entomology* 32 (2), 287–291.

Sosa-Gómez, D.R., Moscardi, F., Santos, B., Alves, L.F.A. and Alves, S.B. (2008) Produção e uso de vírus para o controle de pragas na América Latina. In: Alves, S.B. and Lopes, R.B. (eds) *Controle microbiano de pragas na América Latina: avanços e desafios.* Fealq, Piracicaba, Brazil, pp. 49–68.

van Lenteren, J.C. (2007) Biological control for insect pests in greenhouses: an unexpected success. In: Vincent, C., Goettel, M.S. and Lazarovits, G. (eds) *Biological Control: A Global Perspective.* CAB International, Wallingford, UK, pp. 105–117.

Waterhouse, D.F. and Sands, D.P.A. (2001) *Classical Biological Control of Arthropods in Australia.* ACIAR Monograph No. 77. Australian Centre for International Agricultural Research, Bruce, Australia.

Weber, D.C. (2013) Biological control of potato insect pests. In: Giordanengo, P., Vincent, C. and Alyokhin, A. (eds) *Insect Pests of Potato: Global Perspectives on Biology and Management.* Elsevier, Amsterdam, pp. 399–437.

Wilcken, C.F., Sá, L.A.N., Dal Pogetto, M.H.F.A., Couto, E.B., Ferreira Filho, P.J. and Firmino-Winckler, D.C. (2010) Rearing system of red gum lerp psyllid (*Glycaspis brimblecombei*) (Hemiptera: Psyllidae) and its parasitoid (*Psyllaephagus bliteus*) (Hymenoptera: Encyrtidae) for biological control in eucalyptus plantations. *Documentos técnicos IPEF* 2 (2), 1–23.

Wilson, C.G. and McFadyen, R.E.C. (2000) Biological control in the developing world: safety and legal issues. *Proceedings of the X International Symposium on Biological Control of Weeds, 4–14 July 1999.* Montana State University, Bozeman, Montana, pp. 505–511.

Winotai, A., Goergen, G., Tamo, M. and Neuenschwander, P. (2010) Cassava mealybug has reached Asia. *Biocontrol News and Information* 31 (2), 10N–11N.

Yaninek, S. and Hanna, R. (2003) Cassava green mite in Africa – a unique example of successful classical biological control of a mite pest on a continental scale. In: Neuenschwander, P., Borgemeister, C. and Langewald, J. (eds) *Biological Control in IPM Systems in Africa.* CAB International, Wallingford, UK, pp. 61–75.

Zhang, W.J., Jiang, F.B. and Ou, J.F. (2011) Global pesticide consumption and pollution: with China as a focus. *Proceedings of the International Academy of Ecology and Environmental Sciences* 1 (2), 125–144.

5   Integrated Pest Management in Tropical Cereal Crops

George Mahuku[1,*], Everlyne Wosula[1] and Fred Kanampiu[2]
*[1]International Institute of Tropical Agriculture, Dar es Salaam, Tanzania;
[2]International Institute of Tropical Agriculture, Kasarani, Nairobi, Kenya*

5.1 Introduction

The supply of food – especially grains – in developing countries should increase by about 70% by 2050 if the approximately 9.7 billion people who are expected to be living then are going to be food-secure (Godfray *et al.*, 2010; FAO, 2017a). Annual cereal production will need to rise to about 3 billion tons from the current 2.1 billion (Alexandratos and Bruinsma, 2012; FAO, 2017a). This ambitious goal can be achieved through an increase in yield of the major grains and by lowering the crop losses caused by pests. Cereals are major staple food crops that are cultivated on approximately 75% of the arable land (484 million ha) in the tropics and contributing about 1.71 billion tons (65.7%) of the total world cereal production (FAO, 2014). The four most important cereals in the tropics are rice, maize, sorghum and pearl millet (Bragg *et al.*, 2016). The winter cereals, wheat and barley, are also grown in the tropics, but to a limited extent, and largely at high altitudes. As opportunities for expanding irrigation and productive arable land are limited, improved pest management is an important strategic component for increasing available supplies of food, especially in developing countries. This chapter will highlight strategies for managing cereal pests that are most suitable for smallholder farmers, who in tropical countries produce more than 70% of the cereals. Examples highlighting major biotic constraints for each major cereal will be given, followed by an in-depth analysis of opportunities for harnessing IPM for increased cereal productivity.

5.1.1 Integrated management of pests in tropical cereals

Integrated pest management (IPM) is a systematic plan that brings together different pest-control tactics into one programme. The primary objective is to keep pest intensity below an economic injury threshold and prevent reductions in crop yield and quality (Hill, 2008). In an IPM programme, use of pesticides is reduced and emphasis is placed on using cultural, biological, genetic, physical, regulatory and mechanical control methods. The goal is to prevent pests from reaching economically damaging levels (Ehler, 2006). Success of an IPM programme depends on careful observation, a thorough knowledge of the pest and the damage caused, and an understanding of all available control options.

---

* Corresponding author e-mail: G.Mahuku@cgiar.org

5.1.2 Biotic constraints to cereal production

Crop losses to biotic stresses (weeds, insect pests and pathogens) vary among crops and regions, and this results primarily from differences in host reaction to the interaction (Oerke and Dehne, 2004). On average, 15.8% of maize is lost to pests, and this ranges from 9% to 31%, depending on pest type (Gibbon *et al.*, 2007). In rice, average yield losses are higher than maize, at 22%, but vary for different constraints, with diseases (25% loss) contributing more than insects (20% loss) (Diagne *et al.*, 2013). Sorghum and millets are generally grown in dry and marginal lands and, as such, losses from pests are low, averaging 6.9% and ranging from 0.3 to 17% (Oerke and Dehne, 2004; Dhaliwal *et al.*, 2010).

5.2 Maize

Maize (*Zea mays* L.) is one of the most important cereal crops in the world and, together with rice and wheat, provides at least 30% of the food calories to more than 4.5 billion people in 94 developing countries (Shiferaw *et al.*, 2011). It is estimated that maize is being cultivated on approximately 8.12 million ha with annual production of 19.77 million tons (Dass *et al.*, 2008; FAO, 2014). Currently, maize covers 25 million ha in sub-Saharan Africa, largely in smallholder systems that produce 38 million metric tonnes, primarily for food (Smale *et al.*, 2011). The highest amounts of maize consumed are in southern Africa at 85 kg/capita/year; and contributing more than 40% of total calories (Shiferaw *et al.*, 2011).

5.2.1 Losses by insect pests and diseases

Many pests constrain maize production in the tropics (Ehler, 2006), and the intensity of damage varies according to type of disease or frequency of pest occurrence, prevailing environmental conditions, host genotype, and time (growth stage of the crop) of

infection (Oerke and Dehne, 2004; Dhaliwal *et al.*, 2010). As shown in Table 5.1, an estimated 54% of attainable yield is lost annually to diseases (16%), animals and insects (20%) and weeds (18%) in Africa alone (Oerke *et al.*, 1994; Oerke, 2006). Similar losses have been observed for Central and South America (48%) and Asia (42%) (Oerke *et al.*, 1994). Therefore, efforts to reduce losses from diseases and insect pests offer tremendous opportunities for increasing and stabilizing maize productivity.

5.2.2 Important diseases of maize in the tropics

More than 100 pathogens infect maize, but only a fraction cause economic damage in a specific location (White, 1999). Some major diseases of maize in the tropics and the estimated yield losses are listed in Table 5.2. Some diseases are of regional importance such as tar spot complex (TSC) and corn stunt complex (CSC) in Central America (Mahuku and Kumar, 2017); maize streak virus (MSV), and the parasitic weed Striga (*Striga asiatica* (L.) Kuntze and *S. hermonthica* (Delile) Benth.) in Africa (Gethi *et al.*, 2005; Karavina, 2014; Khan *et al.*, 2016). Some emerging diseases, such as maize lethal necrosis (MLN), have devastated maize production in eastern Africa (Mahuku *et al.*, 2015).

Table 5.1. Estimated yield losses for maize from different biotic constraints.

Region	Attainable Yield (t×10⁶)[a]	Per cent losses associated with		
		Diseases	Animals/ Insects	Weeds
Africa	74	16	20	18
C/S America	98	12	17	19
Asia	203	12	18	12
North America	258	9	11	11
Europe	68	7	9	9

[a]Based on 1988–1990 data summarized by Oerke *et al.*, 1994.

Table 5.2. Some important diseases and insect pests of maize in tropical countries and estimated yield losses.

Common name	Scientific name	Estimated yield loss (%)	References
Diseases			
Northern corn leaf blight	*Setosphaeria turcia* ((Luttrell) Leonard & Suggs)	0–66	Pataky *et al.*, 1998
Grey leaf spot	*Cercospora zeae* (Tehon & E.Y. Daniels)	5–30	Ward *et al.*, 1999
Southern corn rust	*Puccinia polysora* (Underwood)	0–50	Castellanos *et al.*, 1998
Common rust	*Puccinia sorghi* (Schweinitz)	12–61	Dey *et al.*, 2012
Post flowering stalk rots	*Fusarium verticillioides* ((Saccardo) Nirenberg), *Macrophomina phaseolina* ((Tassi) Goidànich), *Colletotrichum graminicola* ((Cesati) G.W. Wilson), *Stenocarpella maydis* ((Berkeley) B. Sutton), *Gibberella zeae* ((Schweinitz) Petch)	10–42	Khokhar *et al.*, 2014
Tar spot complex	*Phyllachora maydis* (Maublanc) and *Monographella maydis* (E. Müller & Samuels)	0–75	Bajet *et al.*, 1994
Corn stunt complex	*Spiroplasma kunkelii* (Whitcomb), Maize bushy stunt phytoplasma, Maize rayado fino virus	0–50	Bradfute *et al.*, 1981
Fusarium ear rots	*Fusarium verticillioides* (Saccardo)	5–15	Chen *et al.*, 2016
Maize lethal necrosis	Maize chlorotic mottle virus and Sugarcane mosaic virus	0–100	Mahuku *et al.*, 2015
Maize streak	Maize streak virus	0–90	Karavina, 2014
Field and storage pests			
Stem borers	*Chilo partellus* (Swinhoe), *Busseola fusca* (Fuller), *Sesamia calamistis* (Hampson), *Eldana saccharina* (Walker), *Mussidia nigrivenella* (Ragonot)	9–80	De Groote, 2001; Kfir *et al.*, 2002
Armyworms	*Spodoptera exempta* (Walker) and *Spodoptera frugiperda* (J. E. Smith)	7–100	Dal Pogetto *et al.*, 2012
Maize weevil	*Sitophilus zeamais* (Motschulsky) and *Sitophilus oryzae* (L.)	4–30	De Groote *et al.*, 2013; Suleiman and Rosentrater, 2015
Larger grain borer	*Prostephanus truncatus* (Horn)	9–45	Suleiman and Rosentrater, 2015
Termites		10–30	Wood and Cowie, 1988; Sekamatte *et al.*, 2003

5.2.3 Integrated management of maize diseases

Effective management of maize diseases involves the selection and use of appropriate techniques that prevent or suppress disease development to a tolerable level (Maloy, 2005). Techniques for reducing initial pathogen inoculum include tillage practices, crop rotations and other agronomic practices, and reducing rate of disease development. The latter can be done through host resistance, choice of shorter-season hybrids and early planting, ensuring optimum plant density, irrigation and soil fertility, as well as the judicious use of fungicides (Ward and Nowell, 1998; Nutter and Guan, 2001). Proper disease management requires correct identification of the pathogen and symptoms, as well as knowledge of the impact of the pathogen/disease (Ward and Nowell, 1998; Nutter and Guan, 2001). The cornerstone of an integrated approach will be the development of high-yielding resistant hybrids grown in rotation with non-host crops.

Reduction of initial inoculums

Quarantine is very effective in minimizing introduction of a potential pest or pathogen to new areas or country where the pest is currently absent (Waage and Mumford, 2008). This can be done at country/regional and/or continental level and should be supported by pest risk analysis (Beed, 2014). For maize, seeds are inspected before leaving and entering a country or before being moved from a disease endemic region to a disease free region within a country to prevent spread of pest/pathogen to new areas. For example, strict quarantine measures have been effective in limiting the introduction and spread of maize chlorotic mottle virus (MCMV) and MLN into southern and western Africa (Mahuku *et al.*, 2015). However, for quarantine to be effective, it should be supported by continual surveillance, appropriate diagnostic tools, and regional, continental and international cooperation to monitor known quarantine pests and pathogens simultaneously.

Eradication is a technique that reduces pathogen inoculum between seasons, making sure that the amount of pathogen present is not sufficient to cause significant disease or affect the plant's development and hence yield reduction (Maloy, 2005). Sanitation methods such as cleaning tools used in infected fields, removal of infected maize plant debris that will act as a source of inoculum in the next season, roguing diseased maize plants, eliminating weeds and other alternative hosts which serve as reservoir for viruses, crop rotation and control of vectors are methods employed in eradication (Webster *et al.*, 2004; Maloy, 2005). Rotating maize with MCMV non-host crops, such as Irish potatoes, sweet potatoes, cassava, beans, onions, vegetables and garlic has been used to minimize the impact of MLN (Wangai *et al.*, 2012).

Reducing the rate of infection

Avoidance aims at preventing contact between host plant and the pathogen by planting in fields with no history of the disease, providing adequate plant spacing to avoid crowding and plant injury, as well as inhibiting the use of recycled maize seeds by using certified seeds. Planting on the onset of the main rainy season and not during the short rainy season to create a break in maize planting seasons will also reduce the population of vectors and result in a low rate of infection and disease severance.

Plant protection involves protection of the host (maize) from invading pathogens and can be achieved by modifying plant nutrient base (the use of manure and fertilizers) and environment. For example, MLN viruses cannot be controlled using chemicals, but chemicals can be used to kill vectors that transmit/spread these viruses. Several insecticides, formulated either as granules or spray applications can be used to manage vectors (e.g. aphids, rootworms, thrips) that transmit MLN causing viruses. However, the use of chemicals is not adequate for managing plant virus diseases (Perring *et al.*, 1999), and also insects might develop resistance (Satapathy, 1998).

Host plant resistance has proven to be the most reliable, effective, environmentally friendly and economical way of controlling maize diseases (Pratt *et al.*, 2003). It is especially attractive to smallholder farmers because once the technology is developed it is packed and disseminated as seed; therefore, it is practical, cost-effective, and environmentally friendly. For cereals, the use of host plant resistance is regarded as the only realistic pest management strategy, especially for smallholder farmers who cannot afford chemical control options. Resistance is available to almost all the major diseases of maize; the big challenge to breeders is to incorporate these genes into elite but susceptible cultivars (Pratt *et al.*, 2003; Mahuku and Kumar, 2017). CIMMYT and the International Institute for Tropical Agriculture (IITA) have been working closely with scientists from national maize programmes in developing countries to develop maize varieties and hybrids with resistance to major maize diseases (Pratt *et al.*, 2003; CGIAR, 2012). Several varieties with resistance to MSV, GLS, TLB, TSC, common rust and ear rots are available and have been deployed (Mahuku and Kumar, 2017).

Integrated management of selected maize diseases

Maize Streak Virus (MSV) was reported first from east Africa, and has now extended to several other African countries (Alegbejo *et al.*, 2002; Karavina, 2014). The virus is transmitted by *Cicadulina* spp. leafhoppers. *Cicadulina mbila* (Naudé) is the most prevalent vector, and it transmits the virus for most of its life after feeding on an infected plant. Losses from MSV can be 100%, if infection occurs early in the disease cycle on susceptible hybrids. Severe infection causes stunting; plants will not develop cobs and can die prematurely (Karavina, 2014). Several cereal crops and wild grasses host virus and vectors. MSV is managed using resistant maize hybrids (Barrow, 1993; Alegbejo *et al.*, 2002; Karavina, 2014). To date, several MSV-tolerant cultivars have been released throughout the sub-Saharan region (Karavina, 2014). In Zimbabwe, for example, SeedCo. (a private company) has released the cultivars SC403, SC411, SC621, SC713 and SC719 with acceptable levels of tolerance to MSV and these are being marketed in several countries in the region (SeedCo., 2010–2011).

Grey Leaf Spot (GLS) is caused by *Cercospora zeae-maydis* (Tehon & E.Y. Daniels) and *Cercospora zeina* (Crous & U. Braun). Disease development is favoured by extended periods of leaf wetness and cloudy conditions, and can result in severe leaf senescence after flowering and poor grain fill (Ward *et al.*, 1999). Managing GLS relies on host resistance and several sources of resistance are available and these have been incorporated and deployed in resistant varieties and hybrids (Ward *et al.*, 1999). As resistance to GLS is controlled by minor genes, complete resistance cannot be achieved (Maroof *et al.*, 1996; Benson *et al.*, 2015). For this reason, cultural practices are deployed with tolerant varieties for sufficient GLS management. Extended period of leaf surface moisture is critical for GLS development, as such, avoiding scheduling irrigation during late afternoon or early evening, especially after outbreaks have already occurred significantly contributes to adequate GLS management. GLS overwinters in crop residue in the field, thus destroying plant debris after harvest, removal of crop residues or deep ploughing to reduce the amount of initial inoculum, and crop rotation with non-host plants have successfully been used to manage GLS (Ward *et al.*, 1999). The disease can also be managed by fungicides, which are routinely used in seed production (Ward *et al.*, 1997). Though very important and effective in managing GLS, fungicides are rarely used by smallholder farmers.

Turcicum Leaf Blight (TLB), caused by *Exserohilum turcicum* ((Passerini) K.J. Leonard & Suggs), can lead to complete burning of the foliage, resulting in more than 70% yield reduction when infection occurs prior to silking and conditions are optimum for disease development (Perkins and Pedersen, 1987; Reddy *et al.*, 2013). Host resistance is the most efficient and cost-effective means for managing TLB and several sources of resistance have been identified and are being used to develop improved varieties (Welz and Geiger, 2000). A high level of resistance characterizes some hybrids, such as SC627, Longe 2H, Longe 6H, Longe 7h and Longe 8H. Cultural practices, such as crop rotations with non-cereal crops (like sunflower, soybean), burying infected debris soon after harvest are effective in reducing initial inoculum and subsequent disease pressure (Reddy *et al.*, 2013). Application of fungicides is effective in managing TLB, but this is rarely used by smallholder farmers.

Maize Lethal Necrosis (MLN) is a new viral disease of maize in Africa that is caused by synergistic interaction of MCMV and any of the viruses belonging to potyviridae family. In Africa, MLN results from co-infection of maize plants by MCMV and Sugarcane mosaic virus (SCMV). Loss of maize productivity to MLN in Kenya has been estimated at 0.5 million tons per year, or 23% of the average annual production equivalent to US$188 million (De Groote *et al.*, 2016). No commercial maize varieties are resistant (Mahuku *et al.*, 2015). When infection occurs early, plants are killed, while later infection results in poorly filled grain that is prone to ear rots. Presently, management strategy is based on prevention of the introduction of the disease through quarantine, careful control of plant material and early destruction of diseased plants. Rotation with legumes has been proposed to break the disease cycle (Kiruwa *et al.*, 2016).

5.2.4 Insect pests of maize

Maize is attacked by about 139 insect pests with varying degree of damage under field and storage conditions (Dhaliwal *et al.*, 2010; Dhillon *et al.*, 2014). Insect pests reduce maize production by directly attacking roots (rootworms, wireworms, white grubs, and seed-corn maggots), leaves (aphids, armyworms, stem borers, thrips, spider mites and grasshoppers), stalks (stem borers and termites), ears and tassels (stem borers, earworms, adult rootworms and armyworms), and grain during storage (grain weevils and grain borers) (Kfir *et al.*, 2002; Hill, 2008).

Stem borers

Stem borers are the most important field pests in maize cultivation in the tropics (De Groote, 2001; Kfir *et al.*, 2002). Five species of stem borers (*Chilo partellus* (Swinhoe), *Busseola fusca* (Fuller), *Sesamia calamistis* (Hampson), *Eldana saccharina* (Walker) and *Mussidia nigrivenella* (Ragonot) are the dominant pests (De Groote, 2001; Kfir *et al.*, 2002; Culliney, 2014). Damage is from feeding by the larvae and yield losses of up to 88% – depending on the cultivar, plant developmental stage at infestation and prevailing environmental conditions – have been reported (Kfir *et al.*, 2002).

INTEGRATED MANAGEMENT OF STEM BORERS The most success in stem borer management has been obtained from using cultural control strategies, early planting and implementing the 'push–pull' technique (Khan *et al.*, 2000; Dhillon *et al.*, 2014). Cultural control is the first line of defence against pests and includes techniques such as destruction of crop residues, intercropping, crop rotation, manipulation of planting dates, and tillage methods. Farmer cooperation is essential for these control measures to be effective, because insects emerging from untreated fields can infest adjacent crops. Destroying larvae in old stalks to reduce the first generation of adult population is very effective in limiting damage of new maize crops. This is achieved through tillage to bury infested stalks deeply into the soil; discing to break stems and expose larvae to adverse weather conditions, birds, rodents, ants, spiders, and other natural enemies; and burning infested old stalk and crop residues are effective in destroying the pest (Kfir *et al.*, 2002).

Planting early ensures that the most vulnerable crop stage does not coincide with periods of peak insect activity (Dhillon *et al.*, 2014). The 'push–pull' strategy involves combined use of intercropping and trap crop systems (Khan *et al.*, 2014; Pickett *et al.*, 2014). Stem borers are attracted to highly susceptible trap plants (pull) and are driven away from the maize crop by repellent intercrops (push) (Hassanali *et al.*, 2008; Pickett *et al.*, 2014). Napier grass and Sudan grass are used as trap plants, whereas molasses grass (*Melinis minutiflora* P. Beauv.) and silverleaf desmodium (*Desmodium uncinatum* (Jacq.) DC.) repel ovipositing stem borers (Khan *et al.*, 2014). In addition, molasses grass produces volatile compounds that attract the stem borer natural enemy, *Cortesia sesamiae* (Cameron), thus leading to increased parasitism of stem borer larvae (Kfir *et al.*, 2002).

HOST PLANT RESISTANCE Although several sources of stem borer resistance have been identified, most of the maize varieties and hybrids on the market do not have adequate resistance (Kumar, 1997; Dhillon *et al.*,

2014). This is because the mechanisms, inheritance and nature of gene action for resistance to stem borers are poorly understood making it difficult to breed for resistance. Transgenic maize expressing the *Bacillus thuringiensis* (Bt) gene have been effective in controlling Lepidopteran pests, but this can only be used in countries that have embraced cultivation of genetically modified crops (Christou *et al.*, 2006). Bt maize has not been widely adopted by smallholder farmers due to regulatory problems (Symth, 2017).

Armyworms

The African fall armyworm, *Spodoptera exempta* (Walker), is a major widespread migratory insect pest that is a perennial threat to cereal production over much of eastern and southern Africa (Grzywacz *et al.*, 2014). In 2007/08, severe armyworm outbreaks in Ethiopia affected >279,000 hectares of cropland (USAID, 2008). Outbreaks of a similar scale occurred in southern Africa in 2012/13, when in Zambia alone armyworm were reported in seven of the country's ten provinces and more than 96,000 hectares of maize and pasture were infested, affecting close to 73,000 farmers (USAID, 2013). Fall armyworm *Spodoptera frugiperda* (J.E. Smith), an invasive species, was reported on the African continent for the first time in 2016 (Goergen *et al.*, 2016). Recently, this pest was reported for the first time in southern Africa (Malawi, Mozambique, Namibia, South Africa, Zambia and Zimbabwe) and it is causing considerable damage to maize (FAO, 2017b). The fall armyworm is a voracious pest and, given its polyphagous nature, it is expected that its accidental introduction in the African continent will constitute a lasting threat to several cereal crops.

INTEGRATED MANAGEMENT OF ARMYWORM
Armyworms are mainly controlled using contact insecticides such as dimethoate or similar organophosphorous insecticide sold under many different brand names (Adamczyk *et al.*, 1999). Transgenic maize cultivars expressing the Cry1F toxin are effective against armyworms, but are currently not cultivated in Africa. Moreover, reports about increasing cases of fall armyworm resistance to Cry1F (Storer *et al.*, 2010) show there is a need to develop alternative control options including the use of nucleopolyhedroviruses (NPV), endophytic entomopathogenic fungi and insect biological control agents (Grzywacz *et al.*, 2014).

Termites

Termites are becoming important maize pests in many tropical countries (Rouland-Lefèvre, 2010; Dhillon *et al.*, 2014; Bragg *et al.*, 2016). Maize is attacked by several species of termites and the damage can be seen especially during drought seasons or in areas where rainfall is scarce. Termites can destroy the roots causing lodging of the stem. Destruction continues even on fallen plants. Attacks at the early stage may cause 100% yield loss. Damage after physiological maturity will lead to grains of poor quality because after lodging, cobs are exposed to contamination (Sileshi *et al.*, 2005). Termites are mainly controlled using chemical insecticides (Riekert and Van den Berg, 2003).

Aflatoxins in cereals and their management

Aflatoxins are highly toxic and carcinogenic mycotoxins that frequently contaminate several cereal crops grown in warm agricultural areas across the globe (Shephard, 2008; Liu and Wu, 2010). Aflatoxin contamination is widespread in maize and sorghum and can end up in milk from animals fed with contaminated feed (Shephard, 2008; Udomkun *et al.*, 2017). Several *Aspergillus* species possess the ability to produce aflatoxins although the major causal agent of contamination globally is *Aspergillus flavus* Link (Klich, 2007). Consumption of foods containing high aflatoxin concentrations can cause acute health effects, such as liver cirrhosis and death (CDC, 2004), while sub-lethal chronic exposure may cause cancer and is associated with immune system suppression, and impaired food conversion, interference with micronutrient metabolism, increased incidence and severity of infectious diseases, as well as retarded child growth and decrease in human and animal productivity (Williams *et al.*, 2004; Liu and Wu, 2010; Chan-Hon-Tong *et al.*, 2013). Women may expose their unborn child to aflatoxins during pregnancy and through breastfeeding, if they consume aflatoxin contaminated foods (Chan-Hon-Tong *et al.*, 2013). An estimated 4.5 billion people in developing countries are exposed to aflatoxins (CAST, 2003; Williams *et al.*, 2004).

INTEGRATED MANAGEMENT OF AFLATOXINS Aflatoxin contamination is a complex process that starts in the field and persists in storage (Lillehoj *et al.*, 1980; Williams, 2006). Deployment of good agricultural practices (GAP) is the most effective and economical strategy for achieving 'aflatoxin safe' crops and foods. Several pre-harvest and post-harvest management strategies have been recommended for the reduction of aflatoxin accumulation (Hell *et al.*, 2008). These include cultural practices, biological control of aflatoxin-producing fungi and proper post-harvest handling (Lillehoj *et al.*, 1980; Jones, 1987; Hell *et al.*, 2008; Bandyopadhyay *et al.*, 2016).

PRE-HARVEST MANAGEMENT OF AFLATOXINS Pre-harvest aflatoxin contamination can be minimized by ploughing to bury crop debris that provides a food source for *A. flavus*, selection of appropriate planting date (to take advantage of periods of rainfall and avoid end-season drought effects), seed dressing with systemic fungicides or bio-control agents, maintaining good plant density in the fields, removal of premature dead plants, managing weeds, pest and diseases and proper fertilizer application (Lillehoj *et al.*, 1980; Jones, 1987; Hell *et al.*, 2008). These practices minimize proliferation of aflatoxin producing *A. flavus.*

RESISTANT CROP VARIETIES Aflatoxin contamination flares when plants are grown under stressful conditions (Bandyopadhyay *et al.*, 2016). Use of crop varieties with tolerance to drought and insect pests, and resistance to major biotic stress, will minimize stressing plants and contribute towards minimizing aflatoxin contamination. Resistance to aflatoxin contamination exists in maize populations, but this is complex and is controlled by multiple genes (Warburton and Williams, 2014). Progress has been made in selecting maize inbred lines with resistance to aflatoxin accumulation (Windham and Williams, 2002; Warburton and Williams, 2014). However, despite all efforts, the level of resistance in available maize hybrids is not yet adequate to prevent unacceptable aflatoxin contamination.

BIO-CONTROL OF AFLATOXIN Biological control of aflatoxins is considered as the most promising strategy for pre-harvest control of aflatoxin as it does not demand much of the farmer's time (Bandyopadhyay *et al.*, 2016). It employs the ability of non-toxigenic *A. flavus* strains to effectively out-compete toxigenic strains for the same ecological niche (Cotty, 2006). Strains of atoxigenic *A. flavus* immobilized on heat killed carrier (i.e. wheat, sorghum or barley grain) that also serves as a nutrient source is broadcast in the field (Bandyopadhyay *et al.*, 2016). The biocontrol formulation provides atoxigenic *A. flavus* with both reproductive and dispersal advantages over resident aflatoxin-producers (Cotty *et al.*, 2008). Timing of bicontrol application is very crucial for success; normally it is carried out before resident *Aspergillus* populations begin to increase, 2–3 weeks before crop flowering, and this allows for effective displacement of aflatoxin producers (Atehnkeng *et al.*, 2014; Bandyopadhyay *et al.*, 2016). Application of the biological control has consistently been shown to reduce aflatoxin contamination by more than 80% and the effect carries into storage (Atehnkeng *et al.*, 2014).

POST-HARVEST MANAGEMENT OF AFLATOXINS Although pre-harvest control strategies are emphasized for control of aflatoxin, these should be augmented by post-harvest strategies (Table 5.3). The crop should be properly dried to safe moisture levels (10–13%) before storage, to reduce and prevent fungal growth in storage (Hell *et al.*, 2008). Naked

Table 5.3. Strategies for integrated management of aflatoxin in cereals for improved food safety and health.

Stage	Actions
Pre-harvest	Timing of planting; Crop variety used; Genotype of seed planted; Irrigation, insecticides; Biological control through competitive exclusion; Timing of harvesting
Post-harvest: drying and storage	Hand sorting; Drying on mats; Sun drying; Storing bags on wooden pallets or elevated platforms; Use of insecticides and hermetic storage structures; Rodent control

cobs or grain should be dried off the ground and on tarpaulins or raised platforms. Solar dryers have been introduced for faster and efficient maize drying under a controlled environment that offers improved sanitation (Sharma *et al.*, 2009; Ogunkoya *et al.*, 2011). To increase adoption of solar-drying technology, affordable, low maintenance solar dryers are required for smallholder farmers (Sharma *et al.*, 2009; Ogunkoya *et al.*, 2011).

IMPROVED STORAGE STRUCTURES Use of controlled atmosphere storage (hermetic) with high CO_2 and low O_2 has been shown to inhibit *A. flavus* growth and reduce aflatoxin production in staple grains (Anankware *et al.*, 2012; De Groote *et al.*, 2013). When used together with grain sorting to remove damaged grain, hermetic storage is very effective in minimizing aflatoxin contamination in storage (Chulze, 2010). Several storage technologies are available, including hermetic storage bags such as super grain bags and metal silos that are suitable for smallholder farmers (Anankware *et al.*, 2012; De Groote *et al.*, 2013).

5.3 Sorghum and Millets

Sorghum and millets (a diverse group of small-grain annual cereal grasses including pearl millet, foxtail millet, finger millet and several others) are particularly important for smallholder farmers on drought-prone marginal lands. In sub-Saharan Africa, sorghum and millets are typically grown as the primary food crop in dry rainfed systems on poor soils with minimal synthetic inputs (Belton and Taylor, 2004; Reynolds *et al.*, 2015). In contrast, in south Asia sorghum and millet crops are increasingly irrigated and given higher input as they are grown for market sale in sequence and rotation with other crops, mainly pulses and oilseeds (Reynolds *et al.*, 2015). Of the small grains, sorghum is the more commonly grown. The area under sorghum in sub-Saharan Africa increased by 82% from 1984 to 2014 (FAO-STAT, 2014). By 1994, sorghum production in the tropics was 55.2 million tons, with Africa accounting for 54% of production; while millet production was 27.2 million tons, with Asia contributing 54% of the quantity (FAO, 2014).

5.3.1 Diseases and insect pests of sorghum and millets

Yields of sorghum and millets are generally low (<500 kg/ha), and this is attributed to several factors, including genetics, environment, weeds, diseases and pests (Chandrashekar and Satyanarayana, 2006; Rurinda *et al.*, 2014). On average, about 17% of sorghum and millet yield is lost annually to weeds, 8% to insect pests and 5% to diseases (Reddy and Usha, 2004). Important diseases of sorghum include several leaf diseases (sorghum leaf blight, anthracnose, sooty stripe, leaf rust, grey leaf spot, downy mildew, and several bacterial diseases), grain diseases (such as head smut, and false smut, root and stalk diseases, insect pests), and weeds (such as *Striga*) (Reddy and Usha, 2004; Chandrashekar and Satyanarayana, 2006). Diseases of economic importance to millet cultivation include rust (*Puccinia substriata* Ellis & Bartholomew), ergot (*Claviceps fusiformis* Loveless), leaf blast (*Pyricularia grisea* Cavara), foot rot (*Sclerotium rolfsii* Saccardo), smut (*Tolyposporidium penicillariae* Brefeld), leaf spot (*Helmithosporium* sp.), and downy mildew (*Sclerospora graminicola* (Saccardo) J. Schröter) (Lubadde, 2014). *Striga hermonthica* (Del.) Benth. is an important weed pest in east Africa. Insects of economic importance to millet cultivation include armyworms, stem-fly (*Atherigona miliaceae* Malloch), stripe borer, pink borer flea beetle and millet head miner, *Heliocheilus albipunctella* (de Joannis) (Youm and Owusu, 1998).

Integrated management of selected sorghum and millet diseases

In smallholder farming systems in tropical countries, sorghum and millets are grown in dry marginal areas that are prone to drought. Therefore major diseases are the ones that affect the panicle and do well under dry conditions. Insect pests, including stem borers and gall midges are very important as they cause significant economic losses. Use of host resistance is the preferred pest management option for the smallholder farmers (Ejeta, 2007a). Resistance to most biotic problems is available and is widely deployed and used (Chandrashekar and Satyanarayana, 2006). However, as these crops are considered orphan crops, most farmers use their own seed and thus do not have access to improved technology. Awareness creation and extension are needed to bring the technology to farmers.

Sorghum leaf blight, caused by *Exserohilum turcicum* ((Luttr.) K.J. Leonard & Suggs), is a foliar disease common under conditions with heavy dew. Leaf blights are effectively controlled using resistant cultivars (Hennessy *et al.*, 1990; Ejeta, 2007a) and through rotation to non-susceptible crops. Although fungicides that are effective against foliar diseases are available, these are rarely used in managing sorghum and millet diseases in the tropics.

Head smut, caused by the soilborne fungus *Sporisorium reilianum* ((J.G. Kühn) Langdon & Fullerton), is a serious disease of sorghum and millets. When infected, some hybrids are dwarfed and will tiller profusely. As head smut spores can remain viable for years in the soil, crop rotation and

fungicides are not effective in managing the diseases. Hence host resistance is the only effective strategy for managing head smut. Sources of resistance are available and have been incorporated into preferred cultivars (Zou, 2010).

Pearl millet downy mildew, caused by *Sclerospora graminicola* ((Sacc.) J. Schröt.), is a serious disease in India and Africa with losses of at least 30% reported on susceptible varieties (Singh, 1995). Two types of spores, sporangia on the leaves and oospores on all plant parts, can survive in the soil and can be spread long distances in soil blown by the wind. Management is dependent on hybrids bred for resistance (Breese *et al.*, 2000), and treatment of the seeds with fungicides, most commonly metalaxyl (Williams and Singh, 1981). However, use of resistant varieties is the most suitable management strategy for smallholder farmers.

Rust of pearl millet (*Puccinia substriata* Ellis & Bartholomew) causes substantial losses in grain yield, especially if infection is early. Spores carried by the wind spread the rust, and the pathogen can survive in soil, on plant debris, volunteer pearl millet and alternative hosts (Khairwal *et al.*, 2007; Lubadde *et al.*, 2014). Management strategy includes crop rotation, removal of weeds, cultivation of tolerant varieties and destruction of the crop remains after harvest. Fungicides are not economically viable unless crops are grown for commercial purposes (Khairwal *et al.*, 2007).

Sorghum downy mildew (*Peronosclerospora sorghi* (W. Weston & Uppal) C.G. Shaw) is predominantly a soilborne disease, whose thick-walled oospores can survive for several years in the soil before infecting young plants. Oospores can also be carried over in seed. Infected plants fail to produce grain. Host-plant resistance has been a very effective method to control sorghum downy mildew and numerous resistant lines have been identified (Rashid *et al.*, 2013). The fungicide metalaxyl is effective in reducing the incidence of sorghum downy mildew when applied as a seed dressing (Williams and Singh, 1981). The main control methods are the use of a resistant variety combined with a fungicide seed treatment (Hash and Witcombe, 2002; Rashid *et al.*, 2013).

Integrated management of selected sorghum and millet insect pests

Millet stem borer (*Coniesta ignefusalis* Hampson) is a major pest of millet in the Sahelian and sub-Saharan regions (Ajayi, 1990). The larvae tunnel into stems leading to lodging, dead hearts, poor grain development and yield reduction. The use of chemicals is rarely justified due to difficulty in timing of application and cost. A combination of cultural practices, such as early planting, intercropping, the 'push–pull' system, and managing crop residues are the most effective approaches for controlling the pest (Ajayi, 1990; Khan *et al.*, 2014).

Sorghum midge (*Stenodiplosis sorghicola* (Coquillett)) is one of the most important pests of sorghum. The larvae feed on developing seeds resulting in malformation of the grain and empty or chaffy heads. The pest can effectively be managed by a combination of resistant varieties (Sharma *et al.*, 1993; Tao *et al.*, 2003), and cultural control measures, such as planting early and planting varieties that flower (Sharma, 1985). Chemical sprays are not very effective as the pest spends most of its life cycle protected inside the spikelets.

Sorghum stem borers (*Busseola* spp., *Chilo* spp., *Eldana* spp. and *Sesamia* spp.) are major pests of sorghum (Nwanze and Mueller, 1989). Their larvae feeds by digging the internal tissue of the plant stem, causing the weakening of the plant. The pest can be controlled through a combination of cultural practices, most notably intercropping and the 'push–pull' system as outlined under maize pests (Kfir *et al.*, 2002; Pickett *et al.*, 2014; Khan *et al.*, 2016).

Ear head caterpillar (*Helicoverpa armigera* (Hübner)) may attack sorghum from head emergence to early grain fill. The insect feeds on the developing grain, resulting in poorly filled and shrivelled seeds (Reddy

and Usha, 2004; Gandhi and Balikai, 2013). Ear head caterpillar is best managed using a naturally occurring nuclear polyhedrosis virus (NPV). There are several cost-effective, commercially formulated NPV products on the market. In addition, natural enemies of *Helicoverpa* are very effective in managing the pest (Reddy and Usha, 2004).

5.4 Rice

Rice is the principal food grain consumed by almost half of the world's population (Khush, 2005), making it the most important food crop currently produced. Rice is grown on over 163 million ha in more than 110 countries, and occupies almost one-fifth of the total world cropland under cereals (FAO, 2014). Most of the world's rice production is from irrigated and rain-fed lowland rice fields in warm and humid environments that are conducive for insect pest proliferation (Pathak and Khan, 1994). Rice yield losses in tropical countries have been estimated at 10.9% from diseases, 12.9% from insect pests and 9.2% from weeds (Gianessi, 2014). The estimates caused by diseases (16.6%) were highest in Asia, while those caused by weeds (11.9%) were highest in Africa (Gianessi, 2014).

5.4.1 Important diseases of rice

Many pathogens affect rice productivity in tropical countries worldwide (Webster and Gunnell, 1992). These are classified into: (i) **major pathogens** that include blast fungus (*Magnaporthe oryzae* B.C. Couch), Rice yellow mottle virus (RYMV) and the bacterium responsible for leaf blight (*Xanthomonas oryzae* pv. *oryzae* (Ishiyama) Swings *et al.*); (ii) **secondary pathogens** responsible for brown spot (*Bipolaris oryzae* (Breda de Haan) Shoemaker), leaf scald (*Gerlachia oryzae* (Hashioka & Yokogi) W. Gams) and sheath blight (*Rhizoctonia solani* J.G. Kühn); (iii) **minor pathogens** responsible for false smut (*Ustilaginoides virens* (Cooke) Takahashi), narrow brown spot

(*Cercospora jansenea* (Raciborski) Constantinescu), sheath rot (*Sarocladium oryzae* (Sawada) W. Gams & D. Hawksworth), bakanae disease (*Fusarium moniliforme* J. Sheldon), bacterial leaf streak (*Xanthomonas oryzae* pv. *oryzicola* (Fang *et al.*) Swings *et al.*) and grain discoloration (caused by a complex of fungi) (Séré *et al.*, 2013).

Integrated management of rice diseases

Host plant resistance is the most important tool for managing rice diseases, especially for smallholder farmers in developing countries (Buddenhagen, 1983). The use of resistant varieties is very much welcomed by resource-poor farmers because it does not require additional cost and it is environmentally friendly. Rice varieties resistant to rice blast (Bonman and Mackill, 1988), bacterial blight (Mew *et al.*, 1992), rice tungro (Azzam and Chancellor, 2002) and brown spot (Ou, 1985) are widely used. Examples of integrated rice disease management for the major diseases are given below, taking host resistance as the main component of the strategy.

Rice blast (*Magnaporthe grisea* (T.T. Hebert) M.E. Barr) affects all aerial parts of the plant, causing their death up to tillering, or reducing grain yield and quality on mature plants. Yield loses ranging from 9% to 80% have been reported, depending on cultivar, prevailing environmental conditions and stage of infection (Bonman *et al.*, 1992). An effective control of rice blast can be achieved by using tolerant or resistant varieties, fractioning the distribution of nitrogen fertilizers, satisfying the water needs of the plants, destroying the crop residues and using certified and fungicide treated seeds. Sources of resistance to rice blast are available (Bonman *et al.*, 1992; Bonman and Mackill, 1988), and these have been used to develop resistant/tolerant varieties for different localities (Fukuoka and Okuno, 2001; Suh *et al.*, 2009). Rice varieties with stacked resistance genes (gene pyramiding) have been used to extend the usefulness of host resistance to manage rice blast (Séré *et al.*, 2013). Multi-lines composed of a mixture of

varieties with different blast resistance genes have routinely been used to manage the disease (Wolfe, 1985; Zhu *et al.*, 2000). Although effective, chemical control is rarely used for managing blast under field conditions. Rather, fungicides, such as benomyl and edifenphos, tricyclazol, kitazin and thiophanate-methyl are commonly used in nurseries (Séré *et al.*, 2013).

Rice yellow mottle disease (Rice yellow mottle virus (RYMV)) is responsible for major epidemics and yield loss in lowland irrigated rice (Kouassi *et al.*, 2005). Leaves turn yellow or orange with green streaks, plants become stunted, tiller number is reduced and panicles produce unfilled or sterile grain (Kouassi *et al.*, 2005). Yield losses ranging from 10% to 90% have been reported in susceptible cultivars and when infection occurs early (Taylor, 1989). RYMV is spread by beetles and grasshoppers and perhaps also other insects and mites, through leaf-to-leaf and root-to-root contact, and on harvest implements (Koudamiloro *et al.*, 2015). Use of host resistance is the most efficient method for managing the diseases. Tolerant varieties are available and are routinely deployed in areas where the disease is a problem (Pinto, 1999; Salaudeen, 2014). The use of tolerant varieties in conjunction with cultural techniques, e.g. removal of rice ratoons, grasses and sedges that are alternative hosts of both virus and insects before planting and destruction of crop residues after harvest is very effective in managing the disease (Sorho *et al.*, 2005). Managing insect vectors in the nursery and weeds in fields surrounding the nurseries is important for controlling vectors of important viruses.

Bacterial leaf blight of rice (*Xanthomonas oryzae* pv. *oryzae* (Ishiyama) Swings *et al.*) kills seedlings and destroys the leaves of older plants. The disease is extremely serious worldwide and has emerged as a major problem in irrigated crops (Mizukami and Wakimoto 1969; Séré *et al.*, 2013). Yield losses ranging from 2.7% to 41% have been reported in Africa (Awoderu *et al.*, 1991). Wild hosts maintain the disease between crops and spread occurs through irrigation and floodwaters as well as by wind, rain and seed. Management requires use of pathogen-free seed or seed coming from pathogen-free crops. Resistant or tolerant varieties should be planted. Resistance to bacterial blight is available and more than 30 resistance genes have been characterized (Sun *et al.*, 2004). Gene pyramiding has been used as a strategy to develop rice varieties with stable and durable resistance to bacterial blight (Huang *et al.*, 1997; Datta *et al.*, 2002). Cultural practices, such as burning crop residues after harvesting heavily infected fields, destroying the surrounding weeds that serve as a reservoir of the pathogen, good drainage and removal of volunteer seedlings and proper use of nitrogen fertilizer have been effective in managing bacterial blight (Verdier *et al.*, 2012; Séré *et al.*, 2013).

5.4.2 Insect pests of rice

Over 100 species of insect pests attack rice and about 20 of these can cause economic damage (Pathak and Khan, 1994; Muralidharan and Pasalu, 2006), through either direct feeding and/or transmission of viruses. Important insect pests include stem borers, gall midge (*Orseolia oryzae* Wood-Mason), brown planthopper (*Nilaparvata lugens* (Stål)), leaf folder (*Cnaphalocrocis medinalis* (Guenée)), and green leafhopper (*Nephotettix virescens* (Distant)). Yield losses due to pests have been estimated at about 20% (Pathak and Khan, 1994). The stem borers are the most serious pests of rice and they infest plants from seedling stage to maturity. In Asia, the most destructive and widely distributed are yellow stem borer, *Scirpophaga incertulas* (Walker) and striped stem borer *Chilo suppressalis* (Walker) which are responsible for an annual damage of 5–10% (Pathak and Khan, 1994). In Africa, sorghum stem borer *Chilo partellus* (Swinhoe), *C. diffusilineus* (de Joannis), white stem borer *Maliarpha separatella* Ragonot, and African pink borer *Sesamia calamistis* Hampson are serious rice pests (Nwilene *et al.*, 2013). In South America, *Diatraea saccharalis* (Fabricius) is the most widespread species, followed by *Rupela albinella* Cramer and *Elasmopalpus lignosellus* (Zeller) (Pathak and Khan, 1994).

Integrated management of rice insect pests

YELLOW STEM BORER Feeding larvae of this insect cause deadhearts at the vegetative stages and whiteheads at the reproductive stages. Deadhearts cause a curl of the central leaf that remain unfolded, turns brownish and dries out, and do not support the panicle. Whereas whiteheads cause non-emergence to the panicle, that remains empty and white. Yields can be reduced by as much as 23%. Yellow stem borer can effectively be controlled using cultural practices but most are effective only if carried out through community-wide cooperation, while others are effective on a single field (Singh *et al.*, 2014). Practices that can be carried out on a single field include using optimal rates of nitrogen fertilizer in split applications. Applying slag increases the silica content of the crop, making it more resistant. Since the eggs of *S. incertulas* are laid near the tip of the leaf blade, the widespread practice of clipping the seedlings before transplanting greatly reduces the carry-over of eggs from the seedbed to the transplanted fields. Majority of the larvae, including those remaining in the stubble, can be removed by harvesting at ground level, burning or removing the stubble, decomposing the stubble with low rates of calcium cyanide, ploughing and flooding (Pathak and Khan, 1994). In several countries, delayed seeding and transplanting have been effective in evading first-generation moths. Crop rotation with non-graminaceous crops significantly reduces the borer population.

Most biological control of stem borers in tropical Asia and Africa comes from indigenous predators, parasites, and entomopathogens (Nwilene *et al.*, 2013). The success and the development of stable IPM systems depend primarily on the conservation of these valuable organisms. Over 100 species of stem borer parasitoids have been identified. The species belonging to the genera *Telenomus*, *Tetrastichus* and *Trichogramma* are the most important egg parasitoids. Egg masses are also the food of several predators, e.g. the longhorned grasshopper *Conocephalus longipennis* (Haan), Coccinellid beetles *Micraspis crocea* (Mulsant), *Harmonia octomaculata* (Fabricius) and carabid beetles such as *Ophionea* spp. prey on young newly emerged larvae before they penetrate the stem (Pathak and Khan, 1994).

Although a lot of rice germplasm (17,000 varieties) has been screened for resistance to stem borers, high levels of resistance have not been found; only partial resistance is available (Nwilene *et al.*, 2013). In addition, varieties are rarely resistant to all stem borer species, thus complicating the use of host resistance to manage the pest. Some wild rice germplasm, such as *Oryza officinalis* (Well ex Watt) and *Oryza ridleyi* (Hook. f.) have very high levels of resistance to stem borers; however, this resistance has so far not been transferred to cultivated rice (Pathak and Khan, 1994). Efforts are underway to transfer this resistance to cultivated rice (Pathak and Khan, 1994; Datta *et al.*, 2002). Stem borers are difficult to control with insecticides because after hatching, the larvae are exposed only for a few hours before they penetrate the stem (Pathak and Khan, 1994; Nwilene *et al.*, 2013).

AFRICAN RICE GALL MIDGE (*ORSEOLIA ORYZIVORA* HARRIS AND GAGNÉ) Larvae of the African rice gall midge feed on young shoots (tillers) of rice, causing long cylindrical galls and prevent growth of tillers, leading to severe yield reductions (Nacro *et al.*, 1996; Nwilene *et al.*, 2013). The most noticeable symptom of infestation is silvery white galls, also known as 'silver shoots' or 'onion leaf galls', unique to the gall midge. Breeding for resistance or tolerance to midge attack shows greatest promise (Omoloye *et al.*, 2002; Bragg *et al.*, 2016). An improved variety, Cisadane, is highly tolerant to natural midge infestations (Omoloye *et al.*, 2002). More sources of resistance have been found and these are being incorporated into high-yielding and adapted varieties. Other options include managing alternative hosts outside the cropping season, synchronizing wet season planting of rice to limit midge colonization of early planted rice crops. A combination of natural control, through encouragement of parasitic wasps, and planting of resistant or tolerant varieties is the most effective method for managing this damaging pest (Nwilene *et al.*, 2013).

5.5 Storage Pests

Cereal crops play a major role in smallholder farmers' livelihoods in sub-Saharan Africa (SSA), with maize being the most important food and cash crop for millions of rural farm families in the region (Cairns *et al.*, 2013). Therefore, proper storage significantly contributes to food security (Tefera, 2012). On average, an estimated 20–30% of cereal produce, amounting to more than US$4 billion annually is lost to storage pests (FAO, 2010). In maize, losses as high as 45% have been reported (Tefera, 2012). Millet and sorghum grains are relatively resistant to post-harvest pest attacks, and under good storage conditions, they can be kept for two or three years with relatively little damage, even in the absence of pesticides.

The main storage insect pests of maize are the maize weevil *Sitophilus zeamais* (Motschulsky), the larger grain borer (LGB), *Prostephanus truncatus* (Horn), angoumois grain moth *Sitotroga cereallela* (Olivier) and the lesser grain weevil *Sitophilus oryzae* (L.); these collectively cause an estimated 20–30% loss (Abass *et al.*, 2014; Tefera, 2012). LGB has emerged as the most important storage pest of maize. The pest was accidentally introduced into Africa (Tanzania, east Africa, west Africa) during the late 1970s, where it has spread rapidly (Farrell and Haines, 2002; Tefera, 2012; Abass *et al.*, 2014). The pest makes long-term storage of maize impossible, thus affecting food security and livelihoods of many smallholder farmers (Tefera, 2012; Abass *et al.*, 2014).

5.5.1 Integrated management of cereal storage pests

Several strategies are currently being used to manage storage pests of maize (Midega *et al.*, 2016), but the most effective is the use of storage pesticides. Correct identification of the pest is essential as efficacy of insecticides vary depending on pest species (Midega *et al.*, 2016). For example pyrethroid pesticides can effectively control the lesser grain borers (*Rhyzopertha dominica* (Fabricius)), but not maize weevil (*Sitophilus zeamais*) and rice weevil (*S. oryzae*), while *R. dominica* can be controlled by organophosphate insecticides (Lorini and Filho, 2006).

Storage hygiene

Poor storage hygiene can lead to the perpetuation of storage problems from one season to the other. Proper cleaning and maintaining good store hygiene are essential to good storage-pest management. Drying the harvested grains to safe storage moisture content, aeration and cooling, followed by suitable packaging in sanitized insect-proof containers can prevent grain loss.

Physical control

Altering the physical environment variables such as temperature, relative humidity/grain moisture content and composition of atmospheric gases can effectively be used to manage storage pests. Inert dusts including sand and other soil components, Diatomaceous earth, silica aerogel, non-silica dusts (e.g. rock phosphate) and clays (e.g. kaolin) kill storage insects through dehydration (Fields and Muir, 1996). Diatomaceous earths are the fossil remains of aquatic plankton that are as effective as the synthetic conventional insecticide, but are non-toxic to humans and animals (Fields and Muir, 1996; Stathers *et al.*, 2008).

Hermetic storage

A new technology using thick plastic bags 'Super Bags' that does not allow gaseous exchange (hermetic), thus suffocating the pests by depleting oxygen is gaining prominence for managing storage pests in smallholder settings (Abass *et al.*, 2014). One such method currently under extensive promotion is the use of 'triple bagging' technique that was developed as an effective hermetic storage method in Cameroon (Anankware *et al.*, 2012). It provides the use of two polyethylene inner bags of 80μ and one external bag, more durable and resistant. The control of insects given by the bags is very effective and with a duration of 3–4 years.

Chemical treatment

The judicious use of synthetic insecticides offers farmers storing grain a potent means of protection against storage pests. The best example of this has been the campaigns against LGB, in which shelling of maize cobs, the admixture of an insecticidal cocktail (mixture of organophosphorus and synthetic pyrethroid), and storage in sacks or other containers has limited grain losses. Fumigation uses gaseous pesticides to suffocate or poison the pests and this is applied mainly in commercial storage facilities such as warehouses or silos. Currently, phosphine is the most common fumigant used for stored crop protection worldwide despite reported failures in some countries due to insect resistance (Nguyen *et al.*, 2015). However, synthetic insecticides are falling out of favour for environmental and health reasons, and the future is likely to rest more on other approaches such as good hygiene, hermetic stores, and the application of alternatives to synthetic insecticides, such as diatomaceous earths.

Crop variety

The choice of grain variety is a critical initial step in preventing losses during storage. There have been efforts to breed maize varieties with increased resistance to storage pests over many years (Kumar, 2002), but to date these have not resulted in crops with both desired agronomic characteristics and the required resistance.

5.6. Striga

Striga is a parasitic weed that seriously constrains the productivity of staples such as maize, sorghum, millet and upland rice in sub-Saharan Africa (Ejeta, 2007b). The weed survives by siphoning-off water and nutrients from the crop for its own growth and impairs normal host growth via three processes: competition for nutrients, impairment of photosynthesis (Joel, 2000) and a phytotoxic effect within days of attachment to the hosts (Frost *et al.*, 1997; Gurney *et al.*, 1999). There are about 23 species of *Striga* in Africa, out of which *Striga hermonthica* (Del.) Benth. and *Striga asiatica* (L.) Kuntze, are the most important (Gressel *et al.*, 2004; Gethi *et al.*, 2005). Roughly 300 million people in sub-Saharan Africa (SSA) are adversely affected by *Striga* (Ejeta, 2007b). *Striga* infests nearly 100 million hectares, that is more than 40% of arable land in SSA (Lagoke *et al.*, 1991) and causes yield losses ranging from 20% to 80% and even total crop failure in severe infestation (Kanampiu *et al.*, 2002a; Khan *et al.*, 2016). Unfortunately, the problem of *Striga* is continuing to extend to new areas in SSA as farmers abandon heavily infested fields for new ones (Gressel *et al.*, 2004; Khan *et al.*, 2016).

5.6.1 Integrated management of *Striga*

Striga seeds can remain dormant and viable in the soil for up to 20 years (Khan *et al.*, 2016). Effective control of *Striga* should target reducing the seed bank in the soil, preventing new seed production and spread from infested to non-infested soils, and improving soil fertility (Ejeta, 2007b). *Striga* infestation can be controlled through heavy application of nitrogen fertilizer (Igbinosa *et al.*, 1996), crop rotation (Oswald and Ransom, 2001), use of trap crops that are not susceptible to parasitism by *Striga* such as legumes (Gbehounou and Adango, 2003), chemical stimulants (Worsham *et al.*, 1959) to abort seed germination, hoeing and hand-pulling (Ransom, 1996), herbicide application (Oswald, 2005), imazapyr herbicide-resistant maize (IR-maize) (Kanampiu *et al.*, 2001, 2002a, 2003), and use of resistant/tolerant crop varieties (Showemimo *et al.*, 2002).

Host-crop resistance

Relatively good progress in identifying resistance/tolerance in maize and sorghum has been achieved (Ejeta and Butler, 1993; Kim, 1994). Resistance is mainly quantitative, i.e. as the level of *Striga* infestation and virulence increases the resistance will eventually break down. *Striga* tolerance/resistance (STR) maize varieties have been developed and are being grown widely in west Africa (Menkir *et al.*, 2010), and east Africa (De Groote *et al.*, 2008). Recently maize inbred lines and hybrids with polygenic field resistance and the IR-genes have been developed (Menkir *et al.*, 2010). These hybrids sustained less damage and yield loss under *S. hermonthica* infestation and supported fewer emerged parasites than the susceptible hybrid check (Menkir *et al.*, 2007).

Hand weeding

Although it seems to be a straightforward approach to interrupt the growth cycle of *Striga*, easy to practice and understand, it is not very effective and farmers are reluctant to employ it. One reason is that *Striga* emerges 5–6 weeks after planting and it takes another 3 weeks until the plants are big enough to be uprooted. At that time the farmer has already done the 'normal' weeding of the crop, which means coming back to weed *Striga* not only once but several times, as *Striga* continues to emerge until a few weeks before harvest (Oswald, 2005). Second, *Striga* not only absorbs water and nutrients from its host-crop but also exerts a potent phytotoxic effect on the host-crop, which means that although *Striga* is weeded, it has already done considerable damage to the crop (Ransom, 1996). Third, *Striga* densities are often so high that hand-weeding is extremely time-consuming. Nevertheless, it remains an integral part of an integrated *Striga* control approach to minimize mature plants and replenishing the seed bank.

Herbicides

There are a number of herbicides available for controlling pre-flowering *Striga* (Langston and English, 1990), but these are largely unavailable to smallholder farmers mainly because of cost. Seed-dressing of imazapyr-resistant maize gets direct action on *Striga* seed. This causes *Striga* plants, which attach to the maize roots or around coated seeds, to immediately die. The maize remains *Striga*-free for the first weeks after planting and this considerably increases yield (Kanampiu *et al.*, 2002b). Development of imazapyr and pyrithiobac seed coatings for the control of *Striga* offers an effective means of controlling *Striga* with smaller amounts of herbicide than is used in spray applications (Berner *et al.*, 1997; Abayo *et al.*, 1998; Kanampiu *et al.*, 2002b, 2003). As little as 30 g/ha imazapyr seed coating applied to imidazolinone resistant (IR) maize seed before or at the time of *Striga* attachment to the maize root will prevent the phytotoxic effect of *Striga*. Imazapyr that is not absorbed by the maize seedling diffuses into the surrounding soil, thus killing ungerminated *Striga* seeds. Maize varieties that have been converted to herbicide resistance are available for use by smallholder farmers (Menkir *et al.*, 2010). This technology reduces yield loss to less than 20%, depletes the *Striga* seed bank in the soil so subsequent *Striga* numbers are less the following year, is cost-effective, and is compatible with existing cropping systems.

Crop rotations

Striga seed banks can be reduced by inducing suicidal germination with trap crops or through natural demise (Berner *et al.*, 1995; Khan *et al.*, 2016). Rotating susceptible cereal crops with crops that are not parasitized by *Striga* has long been advocated as a simple way of avoiding *Striga*-related losses. Rotating with trap crops that induce the germination of *Striga* but are not themselves parasitized is an effective way to reduce levels of *Striga* seeds in the soil. Cotton, sunflower and soybean are effective in reducing *Striga* seed bank and improve cereal yields (Teka, 2014). Crop rotation is probably the most effective way to reduce *Striga* infestation and increase cereal yields considering the limited resource base of smallholder farmers in SSA (Oswald and Ransom, 2001).

Intercropping

An important prerequisite to success in developing *Striga* control in Africa is an understanding of the cropping systems used. Multiple cropping predominates in smallholder farmers in the tropics (Akobundu, 1991). Intercropping cereals with legumes, such as cowpea, groundnut, green gram, dolichos bean and soybean has been shown to reduce the number of *Striga* plants that mature in an infested field (Carson, 1989; Carsky *et al.*, 1994). Effect of intercrops on *Striga* under intercropping might be acting as trap crops, stimulating suicidal *Striga* germination or altering the microclimate of the crop's canopy and soil surface to interfere with *Striga* germination and development (Parker and Riches, 1993).

The push–pull technology has been used to effectively manage *Striga* in sorghum and maize cropping systems. The 'push–pull' technology is based on a stimulo-deterrent concept (Miller and Cowles, 1990). In this strategy, maize is intercropped with a stem borer moth-repellent plant, *Desmodium uncinatum* (Jacq.) DC., while an attractant host plant, Napier grass (*Pennisetum purpureum* Schumach), is planted as a trap plant around this intercrop. Volatiles produced by the *Desmodium* repel the host-seeking moths, while those produced by the Napier grass are attractive to them (Chamberlain *et al.*, 2006). In Kenya, the above-mentioned forage legume, *D. uncinatum*, intercropped with maize has been found to reduce infestation by its allelopathic root exudates that stimulate germination of *S. hermonthica* seeds and concurrently inhibit growth of its radicle (Tsanuo *et al.*, 2003). The system has been modelled into the 'push–pull' technology for control of *Striga* and stem borers (Khan *et al.*, 2014). Though novel, adoption of the system is constrained by the use of *Desmodium* as a fodder crop and not directly as a food crop (Odhiambo *et al.*, 1994).

5.7 Challenges to IPM

Although IPM is viewed as the best strategy for managing pests and diseases of cereal crops, adoption by resource-limited farmers has been slow. A huge gap still exists that has limited the adoption and implementation of IPM strategies, including the following.

- Inadequate knowledge about actual losses from pests and the real as well as potential gains from pest management. Improving the estimation of cereal crop losses and the costs and benefits of reducing crop losses will help in adoption of IPM strategies by smallholder farmers.
- IPM requires collective action within a farming community (Parsa *et al.*, 2014). The recognition that pest management is most effective when implemented collectively at the country and regional level precedes IPM itself, and gave rise to the development of area-wide pest management (Knipling, 1960) and metapopulation theory (Levins, 1969). Success of such a programme requires buy-in from all people in the community. For example, smallholder farmers in Kenya were advised to plant maize only in one season to minimize the effect of MLN. However, some of the farmers planted maize in both the first and second season and this served as a virus reservoir and impeded proper management of MLN.
- For farmers, especially in developing countries, IPM is time-consuming and complicated; given the multiple demands of farm production, farmers cannot be expected to carry out the integration of multiple, suppressive tactics for all classes of pests (Ehler, 2006).

5.8 Conclusion

In most developing tropical countries, cereals are a major staple food crop and as such, there is no incentive for farmers to implement IPM strategies. The revenue from implementing IPM strategies is not huge enough to justify the time and resources that go into IPM. In such a scenario, successful pest management should rely on the use of host resistance. This strategy is very attractive to farmers as they do not have to invest more for the benefit accrued. However, improved insect pests, diseases and weeds management through IPM, relying primarily on interventions supporting crop health and discouraging pest outbreaks (such as through intercropping and use of 'push–pull' systems to attract and trap pests (Khan *et al.*, 2016)), have seen growing effectiveness and acceptance among farmers. More investments from governments to support extension services will be required for IPM to take root.

References

Abass, A.B., Ndunguru, G., Mamiro, P., Alenkhe, B., Mlingi, N. *et al.* (2014) Post-harvest food losses in a maize-based farming system of semi-arid savannah area of Tanzania. *Journal of Stored Products Research* 57, 49–57.

Abayo, G.O., English, T., Kanampiu, F.K., Ransom, J.K. and Gressel, J. (1998) Control of parasitic witchweeds (*Striga spp.*) on corn (*Zea mays*) resistant to acetolactate synthase inhibitors. *Weed Science* 46, 459–466.

Adamczyk, J.J., Leonard, B.R. and Graves, J.B. (1999) Toxicity of selected insecticides to fall armyworms (Lepidoptera: Noctuidae) in laboratory bioassay studies. *Florida Entomologist* 82, 230–236.

Ajayi, O. (1990) Possibilities for integrated control of the millet stem borer, *Acigona ignefusalis* Hampson (Lepidoptera: Pyralidae) in Nigeria. *International Journal of Tropical Insect Science* 11, 109–117.

Akobundu, I.O. (1991) Integrated weed management for *Striga* control in cropping systems in Africa. In: Kim, S.K. (ed.) *Combating Striga in Africa. Proceedings International Workshop organised by IITA, ICRISAT and IDRC, 22–24 August 1988.* IITA, Ibadan, Nigeria, pp. 122–125.

Alegbejo, M.D., Olejede, S.O., Kashina, B.D. and Abo, M.E. (2002) Maize streak mastrevirus in Africa: distribution, transmission, epidemiology, economic significance and management strategies. *Journal of Sustainable Agriculture* 19, 35–45.

Alexandratos, N. and Bruinsma, J. (2012) *World Agriculture Towards 2030/2050: The 2012 Revision.* ESA working paper No. 12-03, FAO, Rome. Available at: www.fao.org/docrep/016/ap106e/ap106e.pdf (accessed 11 March 2017).

Anankware, P.J., Fatunbi, A.O., Afreh-Nuamah, K., Obeng-Ofori, D. and Ansah, A.F. (2012) Efficacy of the multiple-layer hermetic storage bag for biorational management of primary beetle pests of stored maize. *Academic Journal of Entomology* 5, 47–53.

Atehnkeng, J., Ojiambo, P.S., Cotty, P.J. and Bandyopadhyay, R. (2014) Field efficacy of a mixture of atoxigenic *Aspergillus flavus* Link: Fr vegetative compatibility groups in preventing aflatoxin contamination in maize (*Zea mays* L.). *Biological Control* 72, 62–70.

Awoderu, V.A., Bangura, N. and John, V.T. (1991) Incidence, distribution and severity of bacterial disease on rice in West Africa. *Tropical Pest Management* 37, 113–117.

Azzam, O. and Chancellor, T.C. (2002) The biology, epidemiology, and management of rice tungro disease in Asia. *Plant Disease* 86, 88–100.

Bajet, N.B., Renfro, B.L. and Carrasco, J.M.V. (1994) Control of tar spot of maize and its effect on yield. *International Journal of Pest Management* 40, 121–125.

Bandyopadhyay, R., Ortega-Beltran, A., Akande, A., Mutegi, C., Atehnkeng, J. *et al.* (2016) Biological control of aflatoxins in Africa: current status and potential challenges in the face of climate change. *World Mycotoxin Journal* 9, 771–789.

Barrow, M.R. (1993) Increasing maize yields in Africa through the use of maize streak virus resistant hybrids. *African Crop Science Journal* 1, 139–144.

Beed, F.D. (2014) Managing the biological environment to promote and sustain crop productivity and quality. *Food Security* 6, 169–186.

Belton, P.S. and Taylor, J.R. (2004) Sorghum and millets: protein sources for Africa. *Trends in Food Science & Technology* 15, 94–98.

Benson, J.M., Poland, J.A., Benson, B.M., Stromberg, E.L. and Nelson, R.J. (2015) Resistance to gray leaf spot of maize: genetic architecture and mechanisms elucidated through nested association mapping and near-isogenic line analysis. *PLOS Genetics* 11, e1005045.

Berner, D.K., Kling J.G. and Singh, B.B. (1995) *Striga* research and control: a perspective from Africa. *Plant Disease* 79, 652–660.

Berner, D.K., Ikie, F.O. and Green, J.M. (1997) ALS-inhibiting herbicide seed treatments control *Striga hermonthica* in ALS-modified corn (*Zea mays*). *Weed Technology* 11, 704–707.

Bonman, J.M. and Mackil, D.J. (1988) Durable resistance to rice blast disease. *Oryza* 25, 103–110.

Bonman, J.M., Khush, G.S. and Nelson, R.J. (1992) Breeding rice for resistance to pests. *Annual Review of Phytopathology* 30, 507–528.

Bradfute, O.E., Tsai, J.A. and Gordon, D.T. (1981) Corn stunt spiroplasma and viruses associated with a maize disease epidemic in southern Florida. *Plant Disease* 65, 837–841.

Bragg, D.E., Rondon, S.I., Gavloski, J., Shankar, U. and Abrol, D.P. (2016) Integrated pest management in tropical cereal crops. In: Abrol, O. (ed.) *Integrated Pest Management in the Tropics.* New India Publishing Agency, New Delhi, pp. 249–273.

Breese, W.A., Hash, C.T., Devos, K.M. and Howarth, C.J. (2000) Pearl millet genomics and breeding for resistance to downy mildew. In: Leslie, J.F. (ed.) *Sorghum and Millets Diseases.* Wiley-Blackwell, Hoboken, New Jersey, pp. 243–246.

Buddenhagen, I.W. (1983) Disease resistance in rice. In: Lamberti, F. (ed.) *Durable Resistance in Crops.* Plenum Press, New York, pp. 401–428.

Cairns, J.E., Hellin, J., Sonder, K., Araus, J.L., MacRobert, J.F. *et al.* (2013) Adapting maize production to climate change in sub-Saharan Africa. *Food Security* 5, 345–360.

Carsky, R.J., Singh, L. and Ndikawa, R. (1994) Suppression of *Striga hermonthica* on sorghum using a cowpea intercrop. *Experimental Agriculture* 30, 349–358.

Carson, A.G. (1989) Effect of intercropping sorghum and groundnuts on density of *Striga hermonthica* in the Gambia. *Tropical Pest Management* 35, 130–132.

CAST (2003) Mycotoxins: risks in plant, animal, and human systems. In: Richard, J.L. and Payne, G.A. (eds) *Council for Agricultural Science and Technology Task Force Report No. 139*, Ames, Iowa. Available at: https://www.cast-science.org/publications/?mycotoxins_risks_in_plant_animal_and_human_systems&show=product&productID=2905

Castellanos, S., Hallauer, A.R. and Cordova, H.S. (1998) Relative performance of testers to identify elite lines of corn (*Zea mays* L.). *Maydica* 43, 217–226.

CDC (2004) Outbreak of aflatoxin poisoning – eastern and central provinces, Kenya, January–July 2004. Centers for Disease Control and Prevention. *Morbidity and Mortality Weekly Report* 53, pp. 790–793.

CGIAR (2012) *MAIZE Annual Report 2012*. CGIAR/CIMMYT, Mexico. Available at: http://libcatalog.cimmyt.org/download/cim/98018.pdf (accessed 13 March 2017).

Chamberlain, K., Khan, Z.R., Pickett, J.A. Toshova, T. and Wadhams, L.J. (2006) Diel periodicity in the production of green leaf volatiles by wild and cultivated host plants of stemborer moths, *Chilo partellus* and *Busseola fusca*. *Journal of Chemical Ecology* 32, 565–577.

Chandrashekar, A. and Satyanarayana, K.V. (2006) Disease and pest resistance in grains of sorghum and millets. *Journal of Cereal Science* 44, 287–304.

Chan-Hon-Tong, A., Charles, M.A., Forhan, A., Heude, B. and Sirot, V. (2013) Exposure to food contaminants during pregnancy. *Science of the Total Environment* 458, 27–35.

Chen, J., Shrestha, R., Ding, J., Zheng, H., Mu, C. *et al.* (2016) Genome-wide association study and QTL mapping reveal genomic loci associated with Fusarium ear rot resistance in tropical maize germplasm. *G3* 6, 3803–3815.

Christou, P., Capell, T., Kohli, A., Gatehouse, J.A. and Gatehouse, A.M.R. (2006) Recent developments and future prospects in insect pest control in transgenic crops. *Trends in Plant Science* 11, 302–308.

Chulze, S.N. (2010) Strategies to reduce mycotoxin levels in maize during storage: a review. *Food Additives & Contaminants* 27, 651–657.

Cotty, P.J. (2006) Biocompetitive exclusion of toxigenic fungi. In: Barug, D., Bhatnagar, D., van Egmond, H.P., van der Kamp, J.W., van Osenbruggen, W.A. *et al.* (eds) *The Mycotoxin Factbook*. Wageningen Academic Publishers, Wageningen, The Netherlands, pp. 179–197.

Cotty, P.J., Probst, C. and Jaime-Garcia, R. (2008) Etiology and management of aflatoxin contamination. In: Leslie, J.F., Bandyopadhyay, R. and Visconti, A. (eds) *Mycotoxins: Detection Methods, Management, Public Health, and Agricultural Trade*. CAB International, Wallingford, UK, pp. 287–299.

Culliney, T.W. (2014) Crop losses to arthropods. In: Pimentel, D. and Peshin, R. (eds) *Integrated Pest Management*. Springer, Dordrecht, The Netherlands, pp. 201–225.

Dal Pogetto, M.H.F.A., Prado, E.P., Gimenes, M.J., Christovam, R.S., Rezende, D.T. *et al.* (2012) Corn yield reduction of insecticidal sprayings against fall army worm *Spodoptera frugiperda* (Lepidoptera: Noctuidae). *Journal of Agronomy* 11, 17–21.

Dass, S., Jat, M.L., Singh, K.P. and Rai, H.K. (2008) Agro-economic analysis of maize-based cropping systems in India. *Indian Journal of Fertilisers* 4, 53–62.

Datta, K., Baisakh, N., Thet, K.M., Tu, J. and Datta, S. (2002) Pyramiding transgenes for multiple resistance in rice against bacterial blight, yellow stem borer and sheath blight. *Theoretical and Applied Genetics* 106, 1–8.

De Groote, H. (2001) Maize yield losses from stem borers in Kenya. *Insect Science and Its Application* 22, 89–96.

De Groote, H., Wangare, L., Kanampiu, F., Odendo, M., Diallo, A. *et al.* (2008) The potential of a herbicide resistant maize technology for Striga control in Africa. *Agricultural Systems* 97, 83–94.

De Groote, H., Kimenju, S.C., Likhayo, P., Kanampiu, F., Tefera, T. *et al.* (2013) Effectiveness of hermetic systems in controlling maize storage pests in Kenya. *Journal of Stored Products Research* 53, 27–36.

De Groote, H., Oloo, F., Tongruksawattana, S. and Das, B. (2016) Community-survey based assessment of the geographic distribution and impact of maize lethal necrosis (MLN) disease in Kenya. *Crop Protection* 82, 30–35.

Dey, U., Harlapur, S.I., Dhutraj, D.N., Suryawanshi, A.P., Badgujar, S.L. *et al.* (2012) Spatiotemporal yield loss assessment in corn due to common rust caused by *Puccinia sorghi* Schw. *African Journal of Agricultural Research* 7, 5265–5269.

Dhaliwal, G.S., Jindal, V. and Dhawan, A.K. (2010) Insect pest problems and crop losses: changing trends. *Indian Journal of Ecology* 37, 1–7.

Dhillon, M.K., Kalia, V.K. and Gujar, G.T. (2014) Insect-pests and their management: current status and future need of research in quality maize. In: Chaudhary, D.P., Kumar, S. and Langyan, S. (eds) *Maize: Nutrition Dynamic and Novel Uses.* Springer, New Delhi, pp. 95–104.

Diagne, A., Alia, D.Y., Amovin-Assagba, A., Wopereis, M.C.S., Saito, K. *et al.* (2013) Farmer perceptions of the biophysical constraints to rice production in sub-Saharan Africa, and potential impact of research. In: Wopereis, M.C.S., Johnson, D.E., Ahmadi, N., Tollens, E. and Jalloh, A. (eds) *Realizing Africa's Rice Promise,* CAB International, Wallingford, UK, pp. 46–68.

Ehler, L.E. (2006) Integrated pest management (IPM): definition, historical development and implementation, and the other IPM. *Pest Management Science* 62, 787–789.

Ejeta, G. (2007a) Breeding for resistance in sorghum: exploitation of an intricate host–parasite biology. *Crop Science* 47, S-216.

Ejeta, G. (2007b) The Striga scourge in Africa: a growing pandemic. In: Ejeta, G. and Gressel, J. (eds) *Integrating New Technologies for Striga Control: Towards Ending the Witch-Hunt,* World Scientific, Toh Tuck Link, Singapore, pp. 145–158.

Ejeta, G. and Butler, L.G. (1993) Host plant resistance to *Striga.* In: *International Crop Science I, Crop Science Society of America,* Madison, Wisconsin, pp. 561–569.

FAO (2010) *Reducing Post-Harvest Losses in Grain Supply Chains in Africa: Lessons Learned and Practical Guidelines.* Available at: www.fao.org/3/a-au092e.pdf (accessed 12 March 2017).

FAO (2017a) *The Future of Food and Agriculture – Trends and Challenges.* Available at: www.fao.org/3/a-i6583e.pdf (accessed 12 March 2017).

FAO (2017b) Plant pests and diseases. Available at: www.fao.org/emergencies/emergency-types/plant-pests-and-diseases/en (accessed 12 March 2017).

FAOSTAT (2014) Food and Agriculture Organization of the United Nations (FAO), Rome. Available at: http://faostat.fao.org (accessed 23 November 2016).

Farrell, G. and Haines, C.P. (2002) The taxonomy, systematics and identification of *Prostephanus truncates* (Horn). *Integrated Pest Management Reviews* 7, 85–90.

Fields, P.G. and Muir, W.E. (1996) Physical control. In: Subramanyam, B. and Hagstrum, D.W. (eds) *Integrated Management of Insects in Stored Products.* Marcel Dekker Inc., New York, pp. 195–221.

Frost, D.L., Gurney, A.L., Press, M.C. and Scholes, J.D. (1997) *Striga hermonthica* reduces photosynthesis in sorghum: the importance of stomatal limitations and a potential role for ABA? *Plant, Cell & Environment* 20, 483–492.

Fukuoka, S. and Okuno, K. (2001) QTL analysis and mapping of *pi21,* a recessive gene for field resistance to rice blast in Japanese upland rice. *Theoretical and Applied Genetics* 103, 185–190.

Gandhi, B.K. and Balikai, R.A. (2013) Estimation of crop loss due to earhead caterpillar *Helicoverpa armigera* (Hubner) under artificial condition in sorghum hybrid CSH-16. *Vegetos – An International Journal of Plant Research* 26, 45–49.

Gbehounou, G. and Adango, E. (2003) Trap crops of *Striga hermonthica*: in vitro identification and effectiveness in situ. *Crop Protection* 22, 395–404.

Gethi, J.G., Smith, M.E., Mitchell, S.E. and Kresovich, S. (2005) Genetic diversity of *Striga hermonthica* and *Striga asiatica* populations in Kenya. *Weed Research* 45, 64–73.

Gianessi, L.P. (2014) *Importance of Pesticides for Growing Rice in Latin America.* Available at: https://croplife.org/wp-content/uploads/pdf_files/Case-Study-112-Rice-in-Latin-America.pdf (accessed 12 March 2017).

Gibbon, D., Dixon, J. and Flores, D. (2007) *Beyond Drought Tolerant Maize: Study of Additional Priorities in Maize.* Report to Generation Challenge Program. CIMMYT Impacts, Targeting and Assessment Unit. CIMMYT, Mexico City.

Godfray, C., Beddington, J.R., Crute, I.R., Haddad, L., Lawrence, D. *et al.* (2010) Food security: the challenge of feeding 9 billion people. *Science* 327, 812–818.

Goergen, G., Kumar, P.L., Sankung, S.B., Togola, A. and Tamò, M. (2016) First report of outbreaks of the fall armyworm *Spodoptera frugiperda* (J.E. Smith) (Lepidoptera, Noctuidae), a new alien invasive pest in West and Central Africa. *PLOS ONE* 11, e0165632.

Gressel, J., Hanafi, A., Head, G., Marasas, W., Obilana, A.B. *et al.* (2004) Major heretofore intractable biotic constraints to African food security that may be amendable to novel biotechnological solutions. *Crop Protection* 23, 661–689.

Grzywacz, D., Stevenson, P.C., Mushobozi, W.L., Belmain, S. and Wilson, K. (2014) The use of indigenous ecological resources for pest control in Africa. *Food Security* 6, 71–86.

Gurney, A.L., Press, M.C. and Scholes, J.D. (1999) Infection time and density influence the response of sorghum to the parasitic angiosperm *Striga hermonthica*. *New Phytologist* 143, 573–580.

Hash, C.T. and Witcombe, J.R. (2002) Gene management and breeding for downy mildew resistance. In: Leslie, J.F. (ed.) *Sorghum and Millets Diseases*. Wiley-Blackwell, Hoboken, New Jersey, pp. 27–36.

Hassanali, A., Herren, H., Khan, Z.R., Pickett, J.A. and Woodcock, C.M. (2008) Integrated pest management: the push–pull approach for controlling insect pests and weeds of cereals, and its potential for other agricultural systems including animal husbandry. *Philosophical Transactions of the Royal Society of London. Series B: Biological Sciences* 363, 611–621.

Hell, K., Fandohan, P., Bandyopadhyay, R., Kiewnick, S., Sikora, R. *et al.* (2008) Pre-and post-harvest management of aflatoxin in maize: an African perspective. In: Leslie, J.F., Bandyopadhyay, R. and Visconti, A. (eds) *Mycotoxins: Detection Methods, Management, Public Health and Agricultural Trade*. CAB International, Wallingford, UK, pp. 219–229.

Hennessy, G.G., De Milliano, W.A.J. and McLaren, C.G. (1990) Influence of primary weather variables on sorghum leaf blight severity in southern Africa. *Phytopathology* 80, 943–945.

Hill, D.S. (2008) *Pests of Crops in Warmer Climates and Their Control*. Springer, Dordrecht, The Netherlands.

Huang, N., Angeles, E.R., Domingo, J., Magpantay, G., Singh, S. *et al.* (1997) Pyramiding of bacterial resistance genes in rice: marker-aided selection using RFLP and PCR. *Theoretical and Applied Genetics* 95, 313–320.

Igbinosa, I., Cardwell, K.F. and Okonkwo, S.N.C. (1996) The effect of nitrogen on the growth and development of giant witchweed, *Striga hermonthica* Benth.: effect on cultured germinated seedlings in host absence. *European Journal of Plant Pathology* 102, 77–86.

Joel, D.M. (2000) The long-term approach to parasitic weed control: manipulation of specific developmental mechanisms of the parasite. *Crop Protection* 19, 753–758.

Jones, R.K. (1987) The influence of cultural practices on minimizing the development of aflatoxin in field maize. In: Zuber, M.S., Lillehoj, E.B. and Renfro, B.L. (eds) *Aflatoxin in Maize: A Proceedings of the Workshop*. CIMMYT, Mexico City, pp. 136–144.

Kanampiu, F.K., Ransom, J.K. and Gressel, J. (2001) Imazapyr seed dressings for Striga control on acetolactate synthase target-site resistant maize. *Crop Protection* 20, 885–895.

Kanampiu, F.K., Friesen, D. and Gressel, J. (2002a) CIMMYT unveils herbicide-coated maize seed technology for striga control. *Haustorium* 42, 1–3.

Kanampiu, F.K., Ransom, J.K., Friesen, D. and Gressel, J. (2002b) Imazapyr and pyrithiobac movement in soil and from maize seed coats to control Striga in legume intercropping. *Crop Protection* 21, 611–619.

Kanampiu, F.K., Kabambe, V., Massawe, C., Jasi, L., Friesen, D. *et al.* (2003) Multi-site, multi-season field tests demonstrate that herbicide seed-coating herbicide-resistance maize controls Striga spp. and increases yields in several African countries. *Crop Protection* 22, 697–706.

Karavina, C. (2014) Maize streak virus: a review of pathogen occurrence, biology and management options for smallholder farmers. *African Journal of Agricultural Research* 9, 2736–2742.

Kfir, R., Overholt, W.A., Khan, Z.R. and Polaszek, A. (2002) Biology and management of economically important lepidopteran cereal stem borers in Africa. *Annual Review of Entomology* 47, 701–731.

Khairwal, I.S., Rai, K.N., Diwakar, D., Sharma, Y.K., Rajpurohit, B.S. *et al.* (2007) *Pearl Millet: Crop Management and Seed Production Manual*. ICRISAT, Patancheru, India.

Khan, Z.R., Pickett, J.A., Van den Berg, J., Wadhams, L.J. and Woodcock, C.M. (2000) Exploiting chemical ecology and species diversity: stemborer and Striga control for maize and sorghum in Africa. *Pest Management Science* 56, 957–962.

Khan, Z.R., Midega, C.A., Pittchar, J.O., Murage, A.W., Birkett, M.A. *et al.* (2014) Achieving food security for one million sub-Saharan African poor through push–pull innovation by 2020. *Philosophical Transactions of the Royal Society of London. Series B: Biological Sciences* 369, 20120284.

Khan, Z.R., Midega, C.A., Hooper, A. and Pickett, J. (2016) Push–pull: chemical ecology-based integrated pest management technology. *Journal of Chemical Ecology* 42, 689–697.

Khokhar, M.K., Hooda, K.S., Sharma, S.S. and Singh, V. (2014) Post flowering stalk rot complex of maize: present status and future prospects. *Maydica* 59, 226–242.

Khush, G.S. (2005) What it will take to feed 5.0 billion rice consumers in 2030. *Plant Molecular Biology* 59, 1–6.

Kim, S.K. (1994) Genetics of maize tolerance of *Striga hermonthica*. *Crop Science* 34, 900–907.

Kiruwa, F.H., Feyissa, T. and Ndakidemi, P.A. (2016) Insights of maize lethal necrotic disease: a major constraint to maize production in East Africa. *African Journal of Microbiology Research* 10, 271–279.

Klich, M.A. (2007) *Aspergillus flavus*: the major producer of aflatoxin. *Molecular Plant Pathology* 8, 713–722.

Knipling, E. (1960) Use of insects for their own destruction. *Journal of Economic Entomology* 53, 415–420.

Kouassi, N.K., N'guessan, P., Albar, L., Fauquet, C.M. and Brugidou, C. (2005) Distribution and characterization of Rice yellow mottle virus: a threat to African farmers. *Plant Disease* 89, 124–133.

Koudamiloro, A., Nwilene, F.E., Togola, A. and Akogbeto, M. (2015) Insect vectors of Rice yellow mottle virus. *Journal of Insects*. DOI:10.1155/2015/721751

Kumar, H. (1997) Resistance in maize to *Chilo partellus* (Swinhoe) (Lepidoptera: Pyralidae): an overview. *Crop Protection* 16, 243–250.

Kumar, H. (2002) Resistance in maize to the larger grain borer, *Prostephanus truncatus* (Horn) (Coleoptera: Bostrichidae). *Journal of Stored Products Research* 38, 267–280.

Lagoke, S.T.O., Parkinson, V. and Agunbiade R.M. (1991) Parasitic weeds and control methods in Africa. In: Kim, S.K. (ed.) *Combating Striga in Africa. Proceedings of an International Workshop on Striga*. IITA, Ibadan, Nigeria, pp. 3–14.

Langston, M.A. and English, T.J. (1990) Vegetative control of witchweed and herbicide evaluation of techniques. In: Sand, P.F., Eplee, R.E. and Westbrooks, R.G. (eds) *Witchweed Research and Control in the United States*. Weed Science Society of America, Champaign, Illinois, pp. 108–113.

Levins, R. (1969) Some demographic and genetic consequences of environmental heterogeneity for biological control. *Bulletin of the Entomological Society of America* 15, 237–240.

Lillehoj, E.B., Kwolek, F.W., Horner, E.S., Widstrom, N.W., Josephson, L.M. *et al.* (1980) Aflatoxin contamination of preharvest corn: role of *Aspergillus flavus* inoculum and insect damage. *Cereal Chemistry* 57, 255–257.

Liu, Y. and Wu, F. (2010) Global burden of aflatoxin-induced hepatocellular carcinoma: a risk assessment. *Environmental Health Perspectives* 118, 818–824.

Lorini, I. and Filho, A.F. (2006) Integrated pest management strategies used in stored grain in Brazil to manage pesticide resistance. In: Lorini, I., Bacaltchuk, B., Beckel, H., Deckers, D., Sundfeld, E. *et al.* (eds) *Proceedings of the 9th International Working Conference on Stored Product Protection*. São Paulo, Brazil.

Lubadde, G. (2014) Genetic analysis and improvement of pearl millet for rust resistance and grain yield in Uganda. Doctoral dissertation. University of KwaZulu-Natal Pietermaritzburg, Scottsville, South Africa.

Mahuku, G. and Kumar, L. (2017) Rapid response to disease outbreaks in maize cultivation: the case of maize lethal necrosis. In: Watson, D. (ed.) *Achieving Sustainable Cultivation of Maize*, vol. 2. Burleigh Dodds Science Publishing, Cambridge, UK.

Mahuku, G., Lockhart, B.E., Wanjala, B., Jones, M.W., Kimunye, J.N. *et al.* (2015) Maize lethal necrosis (MLN): an emerging threat to maize-based food security in sub-Saharan Africa. *Phytopathology* 105, 956–965.

Maloy, O.C. (2005) *Plant Disease Management. The Plant Health Instructor*. Available at: www.apsnet.org/edcenter/intropp/topics/Pages/PlantDiseaseManagement.aspx (accessed 12 March 2017).

Maroof, M.A., Yue, Y.G., Xiang, Z.X., Stromberg, E.L. and Rufener, G.K. (1996) Identification of quantitative trait loci controlling resistance to gray leaf spot disease in maize. *Theoretical and Applied Genetics* 93, 539–546.

Menkir, A., Badu-Apraku, B., Yallou, C.G., Kamara, A.Y. and Ejeta, G. (2007) Breeding maize for broad-based resistance to *Striga hermonthica*. In: Ejeta, G. and Gressel, J. (eds) *Integrating New Technologies for Striga Control: Towards Ending the Witch-Hunt*, World Scientific, Toh Tuck Link, Singapore, pp. 99–114.

Menkir, A., Adetimirin, V.O., Yallou, C.G. and Gedil, M. (2010) Relationship of genetic diversity of inbred lines with different reactions to *Striga hermonthica* (Del.) Benth and the performance of their crosses. *Crop Science* 50, 602–611.

Mew, T.W., Vera Cruz, C.M. and Medalla, E.S. (1992) Changes in race frequency of *Xanthomonas oryzae* pv. *oryzae* in response to rice cultivars planted in the Philippines. *Plant Disease* 76, 1029–1032.

Midega, C.A., Murage, A.W., Pittchar, J.O. and Khan, Z.R. (2016) Managing storage pests of maize: farmers' knowledge, perceptions and practices in western Kenya. *Crop Protection* 90, 142–149.

Miller, J.R. and Cowles, R.S. (1990) Stimulo-deterrent diversion: concept and its possible application to onion maggot control. *Journal of Chemical Ecology* 16, 3197–3212.

Mizukami, T. and Wakimoto, S. (1969) Epidemiology and control of bacterial leaf blight of rice. *Annual Review of Phytopathology* 7, 51–72.

Muralidharan, K. and Pasalu, I.C. (2006) Assessments of crop losses in rice ecosystems due to stem borer damage (Lepidoptera: Pyralidae). *Crop Protection* 25, 409–417.

Nacro, S., Heinrichs, E.A. and Dakouo, D. (1996) Estimation of rice yield losses due to the African rice gall midge, *Orseolia oryzivora* Harris and Gagne. *International Journal of Pest Management* 42, 331–334.

Nguyen, T.T., Collins, P.J. and Ebert, P.R. (2015) Inheritance and characterization of strong resistance to phosphine in *Sitophilus oryzae* (L.). *PLOS ONE* 10, e0124335.

Nutter, F.W. and Guan, J. (2001) Disease losses. In: Maloy, O.C and Murray, T.D. (eds) *Encyclopedia of Plant Pathology*. John Wiley and Sons, Inc., New York, pp. 340–351.

Nwanze, K.F. and Mueller, R.A.E. (1989) Management options for sorghum stem borers for farmers in the semi-arid tropics. In Nwanze, K.F. (ed.) *International Workshop on Sorghum Stem Borers, 17–20 November 1987*. ICRISAT, Patancheru, India. pp. 105–113.

Nwilene, F.E., Nacro, S., Tamò, M., Menozzi, P., Heinrichs, E.A. *et al.* (2013) Managing insect pests of rice in Africa. In: Wopereis, M.C.S., Johnson, D.E., Ahmadi, N., Tollens, E. and Jalloh, A. (eds) *Realizing Africa's Rice Promise*, CAB International, Wallingford, UK, pp. 229–240.

Odhiambo, G.D. and Ransom, J.K. (1994) Preliminary evaluation of long-term effects of trap cropping on Striga. In: Pieterse, A.H., Verkleij, J.A.C. and ter Borg, S.J. (eds) *Biology and Management of Orobanche. Proceedings of the Third International Conference on Orobanche and Related Striga Research*. Royal Tropical Institute, Amsterdam, pp. 505–512.

Oerke, E.C. (2006) Crop losses to pests. *Journal of Agricultural Science* 144, 31–43.

Oerke, E.C. and Dehne, H.W. (2004) Safeguarding production-losses in major crops and the role of crop protection. *Crop Protection* 23, 275–285.

Oerke, E.C., Dehne, H.W., Schonbeck, F. and Webber, A. (1994) *Crop Production and Crop Protection: Estimated Losses in Major Food and Cash Crops*. Elsevier, Amsterdam.

Ogunkoya, A.K., Ukoba, K.O. and Olunlade, B.A. (2011) Development of a low cost solar dryer. *Pacific Journal of Science and Technology* 12, 98–101.

Omoloye, A.A., Odebiyi, J.A., Williams, C.T. and Singh, B.N. (2002) Tolerance indicators and responses of rice cultivars to infestation by the African rice gall midge, *Orseolia oryzivora. Journal of Agricultural Science*, 139, 335–340.

Oswald, A. (2005) Striga control technologies and their dissemination. *Crop Protection* 24, 333–342.

Oswald, A. and Ransom, J.K. (2001) Striga control and improved farm productivity using crop rotation. *Crop Protection* 20, 113–120.

Ou, S.H. (1985) *Rice Diseases*, 2nd edn. Commonwealth Mycological Institute, London.

Parker, C. and Riches, C.R. (1993) *Parasitic Weeds of the World: Biology and Control*, CAB International, Wallingford, UK.

Parsa, S., Morse, S., Bonifacio, A., Chancellor, T.C., Condori, B. *et al.* (2014) Obstacles to integrated pest management adoption in developing countries. *Proceedings of the National Academy of Sciences* 111, 3889–3894.

Pataky, J.K., Raid, R.N., Du Toit, L.J. and Schueneman, T.J. (1998) Disease severity and yield of sweet corn hybrids with resistance to northern leaf blight. *Plant Disease* 82, 57–63.

Pathak, M.D. and Khan, Z.R. (1994) *Insect Pests of Rice*. IRRI, Los Baños, Philippines.

Perkins, J.M. and Pedersen, W.L. (1987) Disease development and yield losses associated with northern leaf blight on corn. *Plant Disease* 71, 940–943.

Perring, T.M., Gruenhagaen, N.M. and Farrar, C.A. (1999) Management of plant viral diseases through chemical control of insect vectors. *Annual Review of Entomology* 44, 457–481.

Pickett, J.A., Woodcock, C.M., Midega, C.A. and Khan, Z.R. (2014) Push–pull farming systems. *Current Opinion in Biotechnology* 26, 125–132.

Pinto, Y.M., Kok, R.A. and Baulcombe, D.C. (1999) Resistance to Rice yellow mottle virus (RYMV) in cultivated African rice varieties containing RYMV transgenes. *Nature Biotechnology* 17, 702–707.

Pratt, R., Gordon, S., Lipps, P., Asea, G., Bigirwa, G. *et al.* (2003) Use of IPM in the control of multiple diseases in maize: strategies for selection of host resistance. *African Crop Science Journal* 11, 189–198.

Ransom, J.K. (1996) Integrated management of *Striga* spp. in the agriculture of sub-Saharan Africa. In: Brown, H., Cussans, G.W., Devine, M.D., Duke, S.O., Fernandez-Quintanilla, C. *et al.* (eds) *Proceedings of the Second International Weed Control Congress.* Department of Weed Control and Pesticide Ecology, Slagelse, Denmark, pp. 623–628.

Rashid, Z., Zaidi, P.H., Vinayan, M.T., Sharma, S.S. and Setty, T.S. (2013) Downy mildew resistance in maize (*Zea mays* L.) across *Peronosclerospora* species in lowland tropical Asia. *Crop Protection* 43, 183–191.

Reddy K.V.S. and Usha, B.Z. (2004) Novel strategies for overcoming pests and diseases in India. New directions for a diverse planet. *Proceedings of the 4th International Crop Science Congress, 26 Sep.–1 Oct. 2004.* Brisbane, Australia. pp. 1–8.

Reddy, T.R., Reddy, P.N., Reddy, R.R. and Reddy, S.S. (2013) Management of Turcicum leaf blight of maize caused by *Exserohilum turcicum* in maize. *International Journal of Scientific and Research Publications* 3, 540–543.

Reynolds, T.W., Waddington, S.R., Anderson, C.L., Chew, A., True, Z. *et al.* (2015) Environmental impacts and constraints associated with the production of major food crops in sub-Saharan Africa and South Asia. *Food Security* 7, 795–822.

Riekert, H.F. and Van den Berg, J. (2003) Evaluation of chemical control measures for termites in maize. *South African Journal of Plant and Soil* 20, 1–5.

Rouland-Lefèvre, C. (2010) Termites as pests of agriculture. In: Roisin, Y., Lo, N. and Bignell, D.E. (eds) *Biology of Termites: A Modern Synthesis.* Springer, Dordrecht, The Netherlands, pp. 499–517.

Rurinda, J., Mapfumo, P., van Wijk, M.T., Mtambanengwe, F., Rufino, M.C. *et al.* (2014) Comparative assessment of maize, finger millet and sorghum for household food security in the face of increasing climatic risk. *European Journal of Agronomy* 55, 29–41.

Salaudeen, M.T. (2014) Relative resistance to Rice yellow mottle virus in rice. *Plant Protection Science* 50, 1–7.

Satapathy, M.K. (1998) Chemical control of insect and nematode vectors of plant viruses. In: Hadidi, A., Khetarpal, R.K. and Koganezawa, H. (eds) *Plant Virus Control.* The American Phytopathological Society, St. Paul, Minnesota, pp. 188–195.

SeedCo. (2010–2011) *Agronomy Manual.* Available at: www.seedco.co.zw (accessed 9 August 2014).

Sekamatte, B.M., Ogenga-Latigo, M. and Russell-Smith, A. (2003) Effects of maize–legume intercrops on termite damage to maize, activity of predatory ants and maize yields in Uganda. *Crop Protection* 22, 87–93.

Séré, Y., Fargette, D., Abo, M.E., Wydra, K., Bimerew, M. *et al.* (2013) Managing the major diseases of rice in Africa. In: Wopereis, M.C.S., Johnson, D.E., Ahmadi, N., Tollens, E. and Jalloh, A. (eds) *Realizing Africa's Rice Promise,* CAB International, Wallingford, UK, pp. 213–228.

Sharma, A., Chen, C.R. and Lan, N.V. (2009) Solar-energy drying systems: a review. *Renewable & Sustainable Energy Reviews* 13, 1185–1210.

Sharma, H.C. (1985) Strategies for pest control in sorghum in India. *International Journal of Pest Management* 31, 167–185.

Sharma, H.C., Agrawal, B.L., Vidyasagar, P., Abraham, C.V. and Nwanze, K.F. (1993) Identification and utilization of resistance to sorghum midge, *Contarinia sorghicola* (Coquillet) in India. *Crop Protection* 12, 343–350.

Shephard, G.S. (2008) Impact of mycotoxins on human health in developing countries. *Food Additives & Contaminants* 25, 146–151.

Shiferaw, B., Prasanna, B.M., Hellin, J. and Bänziger, M. (2011) Crops that feed the world 6. Past successes and future challenges to the role played by maize in global food security. *Food Security* 3, 307–327.

Showemimo, F.A., Kimberg, C.A. and Albi, S.O. (2002) Genotypic response of sorghum cultivars to N-fertilization in the control of *Striga hermonthica*. *Crop Protection* 21, 867–870.

Sileshi, G., Mafongoya, P.L., Kwesiga, F. and Nkunika, P. (2005) Termite damage to maize grown in agroforestry systems, traditional fallows and monoculture on nitrogen-limited soils in eastern Zambia. *Agricultural and Forest Entomology* 7, 61–69.

Singh, D., Singh, A.K. and Kumar, A. (2014) On-farm evaluation of integrated management of rice yellow stem borer (*Scirpophaga incertulas* Walker) in rice-wheat cropping system under low land condition. *Journal of AgriSearch* 1, 40–44.

Singh, S.D. (1995) Downy mildew of pearl millet. *Plant Disease* 79, 545–550.

Smale, M., Byerlee, D. and Jayne, T. (2011) *Maize Revolutions in Sub-Saharan Africa. Policy Research Working Paper 5659*. World Bank, Washington, DC, and Tegemeo Institute, Kenya.

Smyth, S.J. (2017) Genetically modified crops, regulatory delays, and international trade. *Food Energy Security*. DOI:10.1002/fes3.100

Sorho, F., Pinel, A., Traoré, O., Bersoult, A., Ghesquière, A. *et al.* (2005) Durability of natural and transgenic resistances in rice to Rice yellow mottle virus. *European Journal of Plant Pathology* 112, 349–359.

Stathers, T.E., Riwa, W., Mvumi, B.M., Mosha, R., Kitandu, L. *et al.* (2008) Do diatomaceous earths have potential as grain protectants for small-holder farmers in sub-Saharan Africa? The case of Tanzania. *Crop Protection* 27, 44–70.

Storer, N.P., Babcock, J.M., Schlenz, M., Meade, T., Thompson, G.D. *et al.* (2010) Discovery and characterization of field resistance to Bt maize: *Spodoptera frugiperda* (Lepidoptera: Noctuidae) in Puerto Rico. *Journal of Economic Entomology* 103, 1031–1038.

Suh, J.P., Roh, J.H., Cho, Y.C., Han, S.S., Kim, Y.G. *et al.* (2009) The Pi40 gene for durable resistance to rice blast and molecular analysis of Pi40-advanced backcross breeding lines. *Phytopathology* 99, 243–250.

Suleiman, R.A. and Rosentrater, K.A. (2015) Current maize production, postharvest losses and the risk of mycotoxins contamination in Tanzania. *Agricultural and Biosystems Engineering Conference Proceedings and Presentations, Paper Number: 152189434 July 26–29, 2015*. New Orleans, Louisiana.

Sun, X., Cao, Y., Yang, Z., Xu, C., Li, X. *et al.* (2004) *Xa26*, a gene conferring resistance to *Xanthomonas oryzae* pv. *oryzae* in rice, encodes an LRR receptor kinase-like protein. *Plant Journal* 37, 517–527.

Tao, Y.Z., Hardy, A., Drenth, J., Henzell, R.G., Franzmann, B.A. *et al.* (2003) Identifications of two different mechanisms for sorghum midge resistance through QTL mapping. *Theoretical and Applied Genetics* 107, 116–122.

Taylor, D. R. (1989) *Resistance of Upland Rice Varieties to Pale Yellow Mottle Virus Disease in Sierra Leone*. Newsletter, 14:11. IRRI, Los Baños, Philippines.

Tefera, T. (2012) Post-harvest losses in African maize in the face of increasing food shortage. *Food Security* 4, 267–277.

Teka, H.B. (2014) Advance research on *Striga* control: a review. *African Journal of Plant Science* 8, 492–506.

Tsanuo, M.K., Hassanali, A., Hooper, A.M., Khan, Z.R., Kaberia, F. *et al.* (2003) Isoflavanones from the allelopathic aqueous root exudates of *Desmodium uncinatum*. *Phytochemistry* 64, 265–273.

Udomkun, P., Wiredu, A.N., Nagle, M., Bandyopadhyay, R., Müller, J. *et al.* (2017) Mycotoxins in sub-Saharan Africa: Present situation, socio-economic impact, awareness, and outlook. *Food Control* 72, 110–122.

USAID (2008) *Emergency Transboundary Outbreak Pest (ETOP) Situation Update for May, 2008*. Available at: http://pdf.usaid.gov/pdf_docs/PA00J7XN.pdf (accessed 13 March 2017).

USAID (2013) *Emergency Transboundary Outbreak Pest (ETOP) Situation Report for December with a Forecast till Mid-February, 2013*. Available at: http://pdf.usaid.gov/pdf_docs/PA00J7M5.pdf (accessed 13 March 2017).

Verdier, V., Cruz, C.V. and Leach, J.E. (2012) Controlling rice bacterial blight in Africa: needs and prospects. *Journal of Biotechnology* 159, 320–328.

Waage, J.K. and Mumford, J.D. (2008) Agricultural biosecurity. *Philosophical Transactions of the Royal Society of London. Series B, Biological Sciences* 363, 863–876.

Wangai, A., Kinyua, Z.M., Otipa, M.., Miano, D.W., Kasina, J.M. *et al.* (2012) *Maize (Corn) Lethal Necrosis (MLN) Disease*. KARI Information Brochure. Available at: www.fao.org/fileadmin/user_upload/drought/docs/1%20%20Maize%20Lethal%20Necrosis%20KARI.pdf (accessed 13 March 2017).

Warburton, M.L. and Williams, W.P. (2014) Aflatoxin resistance in maize: What have we learned lately? *Advances in Botany* 2014, 352831. DOI:10.1155/2014/352831

Ward, J.M. and Nowell, D.C. (1998) Integrated management practices for the control of maize grey leaf spot. *Integrated Pest Management Reviews* 3, 177–188.

Ward, J.M., Laing, M.D. and Rijkenberg, F.H.J. (1997) Frequency and timing of fungicide applications for the control of gray leaf spot in maize. *Plant Disease* 81, 41–48.

Ward, J.M., Stromberg, E.L., Nowell, D.C. and Nutter Jr, F.W. (1999) Gray leaf spot: a disease of global importance in maize production. *Plant Disease* 83, 884–895.

Webster, C.G., Wylie, S.J. and Jones, M.G. (2004) Diagnosis of plant viral pathogens. *Current Science* 86, 1604–1607.

Webster, R.K. and Gunnell, P.S. (1992) *Compendium of Rice Diseases*. The American Phytopathological Society, St. Paul, Minnesota.

Welz, H.G. and Geiger, H.H. (2000) Genes for resistance to northern corn leaf blight in diverse maize populations. *Plant Breeding* 119, 1–14.

White, D.G. (1999) *Compendium of Maize Diseases*. The American Phytopathological Society, St. Paul, Minnesota.

Williams, J.H., Phillips, T.D., Jolly, P.E., Stiles, J.K., Jolly, C.M. *et al.* (2004) Human aflatoxicosis in developing countries: a review of toxicology, exposure, potential health consequences, and interventions. *American Journal of Clinical Nutrition* 80, 1106–1122.

Williams, R.J. and Singh, S.D. (1981) Control of pearl millet downy mildew by seed treatment with metalaxyl. *Annals of Applied Biology* 97, 263–268.

Williams, W.P. (2006) Breeding for resistance to aflatoxin accumulation in maize. *Mycotoxin Research* 22, 27–32.

Windham, G.L. and Williams, W.P. (2002) Evaluation of corn inbreds and advanced breeding lines for resistance to aflatoxin contamination in the field. *Plant Disease* 86, 232–234.

Wolfe, M.S. (1985) The current status and prospects of multiline cultivars and variety mixtures for disease resistance. *Annual Review of Phytopathology* 23, 251–273.

Wood, T.G. and Cowie, R.H. (1988) Assessment of on-farm losses in cereals in Africa due to soil insects. *Insect Science and Its Application* 9, 709–716.

Worsham, A.D., Moreland, D.E. and Klingman, G.C. (1959) Stimulation of *Striga asiatica* (Witchweed) seed germination by 6-Substituted Purines. *Science* 130, 1654–1656.

Youm, O. and Owusu, E.O. (1998) Assessment of yield loss due to the millet head miner, *Heliocheilus albipunctella* (Lepidoptera: Noctuidae) using a damage rating scale and regression analysis in Niger. *International Journal of Pest Management* 44, 119–121.

Zhu, Y., Chen, H., Fan, J., Wang, Y., Li, Y. *et al.* (2000) Genetic diversity and disease control in rice. *Nature* 406, 718–722.

Zou, J.Q., Li, Y.Y., Zhu, K. and Wang, Y.Q. (2010) Study on inheritance and molecular markers of sorghum resistance to head smut physiological race 3. *Scientia Agricultura Sinica* 43, 713–720.

6 Integrated Pest Management in Tropical Food Legumes

Giuseppe E. Massimino Cocuzza[1],*, Salvatore Bella[2] and Tsedeke Abate[3]

[1]Dipartimento di Agricoltura, Alimentazione e Ambiente, Università degli Studi, Catania, Italy; [2]Council for Agricultural Research and Economic Analysis, Acireale, Italy; [3]International Maize and Wheat Improvement Center, Nairobi, Kenya

6.1 Background

Tropical food legumes, otherwise known as pulses, are grown worldwide, especially in the tropics and subtropics, where they represent one of the main components of the diet and an important source of essential proteins, minerals, calories and antioxidants, ranking them as high-quality health foods (Kushi *et al.*, 1999; Kouris-Blazos and Belsky, 2016). In addition, cultivation of legumes is very advantageous for its sustainability in both developing and more technologically advanced countries (Siddique *et al.*, 2012).

Despite the excellent properties of food legumes, their production is struggling to increase compared to cereal crops. Production should be better exploited, considering the positive socioeconomic effects that would ensue (Foyer *et al.*, 2016). In developing countries, in addition to having an improving effect on soil fertility, the cultivation of legumes can contribute to reducing problems of malnutrition and be an interesting alternative to traditional cash crops (Abate, 2012; Schreinemachers *et al.*, 2014). Table 6.1 depicts the area occupied by some of the major tropical legumes, along with the number of countries that grow them, and the top ten countries. Collectively, these crops occupy a total area of about 87 million ha. Between 1985 and 2007, overall production of legumes in tropical countries of sub-Saharan Africa and south Asia increased at the rate of 6.5% and 2.8% per year, respectively (Abate *et al.*, 2012). However, the growth is mostly due to the expansion of the cultivated areas rather than an increase in productivity (Abate and Orr, 2012).

In tropical and subtropical areas, mainly smallholder farmers grow food legumes, involving more than 700 million people (Abate, 2012). In sub-Saharan and tropical Africa, legumes are grown in small plots and often in association with other crops, in which the use of fertilizers and pesticides is rather low. Consequently, the amount of the harvest may vary greatly over the years, being closely dependent on a number of factors. In southeast Asia, control of the numerous insect and mite pests is mainly carried out by chemicals, which, however, do not always ensure satisfactory results (Srinivasan, 2014). From the intensive and frequently inappropriate use of pesticides emerge several problems, such as

---

* Corresponding author e-mail: cocuzza@unict.it

Table 6.1. Indicators for major grain legumes in the world. (From FAOSTAT, 2017.)

Crop		Area	Total number of	Top ten producing countries
Scientific name	Common name	(000 ha)	countries	(descending order)
Arachis hypogaea	Groundnut	26,206	115	India, China, Nigeria, Sudan, Tanzania, Chad, Senegal, Niger, USA, Indonesia
Phaseolus vulgaris	Common bean	16,015	114	Brazil, Mexico, Tanzania, Kenya, Angola, Mozambique, Uganda, USA, Rwanda, DR Congo
Cicer arietinum	Chickpea	13,046	55	India, Pakistan, Iran, Australia, Turkey, Myanmar, Ethiopia, Tanzania, Mexico, Malawi
Vigna unguiculata	Cowpea	12,278	37	Niger, Nigeria, Burkina Faso, Mozambique, Mali, Kenya, Cameroon, Tanzania, Sudan, DR Congo
Cajanus cajan	Pigeon pea	6290	23	India, Myanmar, Kenya, Tanzania, Malawi, Haiti, Uganda, Dominican Republic, Nepal, DR Congo
Pisum sativum	Field pea	6946	95	Canada, Russia, China, India, Iran, USA, Australia, Ethiopia, Tanzania, Ukraine
Lens culinaris	Lentil	4467	51	India, Canada, Turkey, Nepal, Australia, Iran, USA, Syria, Ethiopia, Bangladesh
Vicia faba	Faba bean	2318	59	China, Ethiopia, Morocco, Australia, Sudan, France, Peru, Tunisia, UK, Italy
Lupinus album	Lupin	716	25	Australia, Poland, Russia, Germany, Chile, Ukraine, Belarus, South Africa, Peru, Spain
Vigna subterranea	Bambara bean	226	5	Niger, Burkina Faso, Cameroon, Mali, DR Congo
Phaseolus vulgaris	String bean	204	15	USA, France, Philippines, Iraq, Mexico, Peru, Turkey, Argentina, Poland, Japan

Note: The figures were calculated by the authors from FAOSTAT (2017), using the 2012–14 averages.

the health risk to farm workers exposed to them, as well as environmental pollution (Hoai *et al.*, 2011; Lamers *et al.*, 2011; Praneetvatakul *et al.*, 2013); for these it is difficult to intervene, because of lack of awareness, especially among smallholder farmers (Schreinemachers *et al.*, 2011; Schreinemachers and Tipraqsa, 2012). Moreover, especially for product intended for export, the demand for products with very low levels of chemical residues is growing, and this forces farmers to adopt alternative control methods, such as integrated pest management (IPM). The IPM production models, especially in emerging countries, should be based on the combined use of agronomic techniques (intercropping, varietal mixtures, optimum sowing date, tolerant or resistant varieties), biological control and insecticide use, trying to enhance the role of botanical pesticides and other locally available material (Abate *et al.*, 2011). In this context, the research has a central role in defining, promoting and disseminating the integrated techniques of production and control of pests. This information activity is particularly necessary in countries where the IPM practices are poorly known (Schreinemachers *et al.*, 2014). In recent years, some governments have adopted policies that encourage the improvement of agricultural production standards (legumes among them) that increase their market value (Abate and Orr, 2012; Praneetvatakul *et al.*, 2013). Among the factors that improve the quality of production, the availability of legumes grown under integrated control practices lends the products a further market value and makes them more easily exportable. As with IPM, the incentive to raise productivity came from growing market demand and the promise of higher cash income (Abate and Orr, 2012).

6.2 The Insects

A wide range of insect pests attack tropical legumes and limit the production of tropical food legumes worldwide. Insect pests of legume plants can be classified into five broad categories, according to the plant growth stage or plant part they primarily attack: seedlings, foliage, flowers, pods and stored seeds. Numerous insect pests attack legume plants during all stages of growth, from seedling to stored product, but only a few of these are recognized as major pests. Examples of these are presented in Table 6.2.

6.2.1 Seedling pests

Bean stem maggots (BSM) or bean flies (Diptera, Agromyzidae) belonging to the genus *Ophiomyia* are considered the most important seedling pests of legume crops in several growing areas of the world, especially in various parts of eastern and southern Africa (Sariah and Makundi, 2007), where they can cause losses in the range of 30–100% (Karel and Matee, 1986). In tropical and subtropical Africa, Asia, and

Australia the most damaging species for beans are *Ophiomyia phaseoli* (Tryon) and *O. centrosematis* (de Meijere), whereas *O. spencerella* (Greathead) is confined in the African continent (Greathead, 1968; Abate and Ampofo, 1996; Shanower *et al.*, 1999; Abate *et al.*, 2000).

O. phaseoli can be a destructive pest for common bean, cowpea, mung bean, black gram, adzuki bean, *Vigna* spp., pea and lima bean. In general, plants are more heavily damaged in the seedling stage than when they are mature. Damage is permanent in plants surviving the insect attack in the seedling stages. In general, the yield losses during the dry season are higher than in the rainy season. The eggs of BSM (60–300 depending on the species) are laid singly on different parts of the seedling, inserted between the epidermis and spongy parenchyma. *O. phaseoli* oviposits mostly on the newly emerging leaves, whereas *O. spencerella* prefers hypocotyl and sometimes the stem, this latter being the favourite site of *O. centrosematis* (Talekar, 1990). On hatching, after 3–10 days, depending on the temperature, the small white maggots of *Ophiomyia* spp. initially burrow in the leaf and then bore into the stem (Talekar, 1990). Clearly visible mines dug by the larvae on

Table 6.2. Most important insect pests of major grain legumes. (Adapted from Abate and Ampofo (1996) and Abate *et al.* (2012).)

	Insect		Important pest on					
Scientific name	Common name	Most important crop stage attacked	Chickpea	Common bean	Cowpea	Groundnut	Pigeon pea	Soybean
Ophiomyia spp.	Bean stem maggots	Seedling	–	+++	±	–	–	±
Aphis craccivora	Cowpea aphid	Vegetative growth	–	–	+++	+++	–	–
Megalurothrips (= *Taeniothrips*) *sjostedti*	Flower thrips	Flowering	–	++	+++	–	–	–
Clavigralla spp.	Pod bugs	Podding	–	–	+++	–	++	–
Helicoverpa armigera	African bollworm	Podding	+++	±	–	–	+++	–
Maruca vitrata (= *testulalis*)	Cowpea pod borer	Podding	–	–	++	–	–	–

– = Not important; ± = minor pest; ++ = moderately important; +++ = highly important.

the stems lead to yellowing of leaves, stunting of plant growth (similar to those caused by moisture stress or root diseases) and even plant mortality.

The agromyzid *Melanagromyza obtusa* Malloch and *M. chalcosoma* Spencer are important pests of pigeon pea in Asia, Australia and eastern-southern Africa, respectively. The cosmopolitan *M. soyae* (Zehntner), over the soybean, attack various legume crops, such as bean, adzuki bean, pea and others (Dempewolf, 2004). The larvae of *Melanagromyza* spp. start their feeding activity on leaves, continue downwards into the stem, and make a tunnel into the pith, reaching the stem–root junction. In case of less severe damage, the root–shoot junction area of the plant appears swollen or with adventitious roots above the bulged area on the stem (Shanower *et al.*, 1999).

In areas where infestations of *Ophiomyia* spp. are recurrent, pest control must be started in the first month after germination. Abate and Ampofo (1996) reported that BSM infestation level is influenced by a combination of different environmental factors and cultural practices, including availability of water, crop variety, varietal mixtures and growth stage of the host plant. Several studies showed that IPM of BSM can be obtained by using resistant cultivars and adopting appropriate cultural practices, such as the sowing time, optimum plant population, mulches with organic material, varietal mixtures, irrigation, soil fertility and intercropping, which represent the most useful cultural methods (Galindo *et al.*, 1983; Mohamed and Teri, 1989; Ampofo, 1993; Letorneau, 1994; Ampofo and Massomo, 1998). In Uganda, the lowest root damage on common bean was recorded using a mixture of resistant traditional varieties and popular susceptible commercial varieties in a proportion of 1:1 (Ssekandi *et al.*, 2016). In an experimental trial carried out in Sri Lanka, mixed cropping with leek showed good results in control of bean fly (Bandara *et al.*, 2009). Another interesting strategy is the cultivation of moth bean, chickpea, lentil and cluster bean on which bean flies lay their eggs that, however, fail to hatch (Talekar and Lee, 1988). Also, early or delayed cultivation can significantly reduce the BSM infestation, as showed in India on pea (Singh *et al.*, 1981) or in Tanzania with bean (Sariah and Makundi, 2007). The control of wild legume plants, such as *Crotolaria* spp., should not be overlooked, considering that they are alternative host plants of bean flies and important sources of infestation (Abate, 1990). Breeding for resistance is actively pursued. Lines of legume food plants tolerant to bean flies have been identified in common bean and cowpea (Ojwang *et al.*, 2009), mung bean (Chiang and Talekar, 1980) and pigeon pea (Shanower *et al.*, 1999) but their diffusion among farmers is rather limited. However, several promising research programmes are carried out for the genetic improvement for tolerance against bean flies (Okwiri Ojwang *et al.*, 2010, 2011). Bean flies have a large number of natural enemies that attack them during nymphal stages and provide a good control. *Opius phaseoli* (Tryon) (Hymenoptera, Braconidae) and *Eucoilidea* sp. (Hymenoptera, Cynipidae) are the major parasitoids of *Ophiomyia phaseoli*, *O. spencerella* and *O. centrosematis*, respectively (Abate and Ampofo, 1996). In Asia, on *Melanagromyza obtusa* (Malloch), the most important parasitoids belong to the genera *Ormyrus* (Hymenoptera, Ormyridae) and *Euderus* (Hymenoptera, Eulophydae), that can reach parasitism rates between 13 and 30% (Shanower *et al.*, 1999). The use of insecticides continues to be the most common control measure practised by farmers. Against adults of *Ophiomyia* spp., the use of chemicals must take place in the early stages of germination or as soil applications during the sowing of the crop. Another method, less practised, is the seed-dressing insecticides (Abate, 1991). The use of broad-spectrum insecticides should be avoided, to protect the natural enemies that, anyway, contribute to limit the agromyzid populations.

Besides BMS, several other insect species can attack bean seedlings, such as *Tanaostigmodes cajaninae* La Salle (Hymenoptera, Tanaostigmatidae), *Agrotis* spp. (Lepidoptera, Noctuidae), *Microtermes* spp. (Isoptera, Termitidae) and *Gryllus bimaculatus* De Geer (Orthoptera, Gryllidae) (Abate and Ampofo, 1996; Shanower *et al.*, 1999). In India, an important stem feeder on mung bean, especially in seedling stage, is the girdle beetle, *Oberia brevis* (Swedenbord) (Coleoptera, Cerambycidae) (Shanower *et al.*, 1999). However, all these insects are considered as minor pests, although they may have occasionally economic significance in some environments.

6.2.2 Flower and foliage pests

Among beetles, the most important bean foliage beetles include *Ootheca bennigseni* Weise, *O. mutabilis* (Schönherr) and *Medythia quaterna* (Fairmaire) (=*Luperodes lineata*) (Coleoptera, Chrysomelidae) (Abate and Ampofo, 1996). *Ootheca* spp. are widely distributed in the Afrotropical region and considered as important pests especially of cowpea and beans (Karel and Rweyemamu, 1984; Wagner, 2010). *O. mutabilis* is prevalent at lower altitudes, while *O. bennigseni* dominates in the medium-altitude zones (1000–1500 m) (Abate and Ampofo, 1996). Their feeding activity on foliage soon after germination can cause yield losses of 27–100% (Singh and Allen 1980; Raheja 1981). Infestation and damage is usually most severe on young seedlings, but sometimes the infestation may persist through the post-flowering stages. Infestation by the early instars may go unnoticed, but the older larvae remove lateral roots and cause wilting and premature senescence in bean plants. Heavy attack to young seedlings may completely destroy a crop (Abate and Ampofo, 1996; Srinivasan, 2014). The adult beetles can provoke severe defoliations on crops (Srinivasan, 2014) and can be effective vectors of various viruses of cowpea and beans (Allen *et al.*, 1981). Damage of *M. quaterna* is caused by the adult, which feeds at the margins of newly opened leaves and the walls of developing pods. *M. quaterna* is also reported to be a vector of cowpea mottle virus (Robertson, 1963; Whitney and Gilmer, 1974; Allen *et al.*, 1981). The use of

chemical insecticides is frequently applied to control foliage pests. However, the use of natural products is very popular in many rural areas of sub-Saharan Africa, mostly due to the high cost or unavailability of chemical insecticides (Kareru *et al.*, 2013). The application of extract of powder from seeds of the neem tree (*Azadirachta indica*) inhibited feeding of *O. bennigseni* (Karel, 1989); an interesting repellent effect was observed by using cow urine and extract of Asteracea *Vernonia lasiopus* O. Hoffm. (Paul *et al.*, 2007). Similarly, a reduction of damage from foliage beetles was recorded in field with the use of various botanical extracts on common bean (Mkenda *et al.*, 2015) and cowpea (Udo and Akpan, 2012). In Tanzania, some breeding lines and local varieties showed interesting tolerance characteristics towards *Ootheca* spp. (Karel and Rweyemamu, 1985).

Normally, the weevils (Coleoptera, Curculionidae) are considered secondary pests, although exceptionally they can cause some loss of production. The bean leafroller *Apoderus humeralis* Olivier, endemic in Madagascar, causes circular holes between the leaf veins and lays single eggs at the tip of leaves, rolling them in the shape of a cigar (Allen *et al.*, 1996). On cowpea and common bean, larvae of *Alcidodes leucogrammus* Erichs bore into the base of the stem producing nematode-like galls on roots, whereas adults feed on leaf margins producing circular holes (Amoako-Atta, 1983). The pea leaf weevil *Sitona lineatus* (L.) occasionally infests legume food plants in Maghreb regions, on which it causes defoliations and reduction of nitrogen root nodules (Blaeser-Diekmann, 1982).

The blister beetles, *Hycleus apicicornis* (Guérin-Ménéville) (=*Coryna apicicornis*), *Hycleus* spp., *Coryna kersteni* Gerstaecker and *Mylabris* spp. (Coleoptera, Meloidae) are widely distributed throughout sub-Saharan Africa and are mainly known as pollen and flower feeders of various legumes (Abate and Ampofo, 1996; Hillocks *et al.*, 2000). Adults are easily recognizable by the showy coloration characterized by red or yellow bands or spots on black elytra. They feed on flowers of various legumes and severe infestations may reduce pod production (Singh and Taylor, 1978). The emission by adults of a substance irritating the human skin is also typical of these species. The larvae undergo hypermetamorphosis, during which each instar looks different. The populations of blister beetles are favoured by intercropping legumes with other crops (e.g. maize, sorghum or millet) or abundance of grasshoppers, whose eggs the larvae of the beetles actively prey on (Lebesa *et al.*, 2011). The use of insecticides is frequent in case of severe infestations by these meloids. However, especially in small agricultural acreage, handpicking larvae and adults may be useful to control low infestations (Abate and Ampofo, 1996). Recently, it has been demonstrated that blue traps are strongly attractive for *H. apicicornis* and their use can find a role in an integrated strategy of control (Lebesa *et al.*, 2011).

Both nymphs and adults of leafhoppers (Homoptera, Cicadellidae), with their feeding bites on foliage, cause conspicuous yellowing to vegetation (known as 'hopper burn') that leads to a quality deterioration. However, normally, leafhoppers are considered minor pests and only particular weather conditions can favour their population build-up to damaging levels. Among leafhoppers, it is worth mentioning, as potentially damaging insects, *Empoasca kerri* Singh-Pruthi (India and Pakistan) on pigeon pea (Yadav *et al.*, 2016) and groundnut (Wightman and Amin, 1988), *E. facialis* Jacobi, *E. dolichi* Paoli (Africa) and *E. kraemeri* Ross and Moore in Latin America on cowpea (Jackai and Daoust, 1989). Their management consists of a continuous visual monitoring or placing yellow sticky traps at random in the field, to detect throughout the year the trend of the leafhopper populations. Normally, leafhoppers are well controlled by natural enemies, so it is advisable to avoid using broad-spectrum pesticides, in order to maximize their performance (Srinivasan, 2014). The cultivation of okra, sesame, sorghum and pearl millet as intercrops in vegetable legume fields reduces the leafhopper damage significantly (Chakravorty and Yadav, 2013).

Thrips (Thysanoptera, Thripidae) attack both foliage and flowers of legumes. Thrips infestation causes premature drop of flowers, reduction of photosynthetic capacity of the plant and pod necrosis, that results in a loss of production (Tang *et al.*, 2015). *Megalurothrips usitatus* (Bagnall), *M. distalis* Karny, *M. sjostedti* (Trybom), *Thrips palmi* Karny, *Caliothrips indicus* Bagnall, *Taeniothrips nigricornis* (Schmutz), *Frankliniella schultzei* (Trybom) and *Sericothrips occipitalis* Hood are potential harmful pests to a wide range of leguminous plants (Taylor, 1969; Chang, 1988; Srinivasan, 2014). Monitoring the populations, adopting good cultural practices and using insecticides only when really necessary may help to preserve the numerous natural enemies and to achieve good thrips control. Populations can be monitored with blue, yellow or white (depending on the species) PVC plate traps coated with sticky substance (Chang, 1990) or checking for the presence of thrips inside the flowers. Legume seedlings grown in protected areas can prevent early infestation. Often, thrips move to the cultivated areas at the beginning of the dry season, when they leave the wild host plants at the end of cycle (Chang, 1990). It could be useful to avoid cultivation in these periods or to control the nearby weeds that share thrips with legumes. Several natural enemies are known and some can be extremely effective, such as the predator *Orius* spp. (Hemiptera, Anthocoridae) (Letourneau and Altieri, 1983; Tamò *et al.*, 1993) or the parasitoids *Ceranisus menes* (Walker) and *C. femoratus* (Gahan) (Hymenoptera, Eulophidae) (Tamò *et al.*, 2012). In India, some genotypes were detected for tolerance or less attractiveness to *M. distalis* and *C. indicus* (Chhabra and Kooner, 1994; Sahu and Shaw, 2005). A reduction of *M. sjostedti* infestations was recorded by bean-maize or cowpea-maize intercropping culture (Kyamanywa and Ampofo, 1988; Kyamanywa *et al.*, 1993).

The groundnut leafminer, *Aproarema modicella* (Deventer) (Lepidoptera, Gelechiidae) is an important pest of groundnut and soybean in south and southeast Asia (Shanower *et al.*, 1993) and in Africa (Buthelezi *et al.*, 2016), where it was introduced from Indo-Asia (Kenis and Cugala, 2006). A research found that a major loss in harvest or a minor swelling of pods derive from attacks to more than 25% of the leaves (Wightman and Amin, 1988). The control of groundnut leafminer can be achieved through the adoption of resistant varieties or agronomic measures (i.e. intercropping groundnut with sorghum, cowpea or millet) and promoting the activity of the numerous natural enemies by minimizing the use of chemicals (Shanower *et al.*, 1993).

The common armyworm, *Spodoptera litura* (Fabr.) (Lepidoptera, Noctuidae), is a polyphagous species, considered to be an important pest in many agricultural legume crops, especially in tropical south and southeast Asia. The larvae of the noctuid feed actively at night, resting in the soil under the plants during the day. With their feeding activity, the larvae may cut the seedlings or young plants at soil level or cause leaf erosions on mature plants, and severe infestations may destroy the crop. Traps activated with sex pheromones can monitor the presence of the noctuid in the field. Nucleopolyhedrovirus or *Bacillus thuringiensis* formulations, alone or in combination with neem, are effective in replacing the chemical pesticides (Lingappa and Hegde, 2001; Sharma *et al.*, 2011). In addition, botanical insecticides, such as neem, showed promising results (Huang *et al.*, 2004; Rajguru and Sharma, 2012). The egg parasitoid *Trichogramma chilonis* Ishii (Hymenoptera, Trichogrammatidae) and the larval parasitoid *Campoletis chlorideae* Uchida (Hymenoptera, Ichneumonidae) contribute to control *S. litura* when released in legume fields at regular intervals (Bajpai *et al.*, 2006).

The two-spotted spider mite, *Tetranychus urticae* Koch (= *T. cinnabarinus*) (Acari, Tetranychidae) is a cosmopolitan and polyphagous species considered a major pest of common bean in several tropical and subtropical countries (Abate, 1988a; Vacante, 2016). *T. urticae* feeds empting the cells mainly on lower leaf surface, which assumes a yellow-brown coloration. Especially on legume plants, the mite develops a dense web that at high density may cover all vegetation. However, serious infestations can be observed only in neglected crops or in particular environmental conditions. It would be advisable not to use acaricide, particularly in small-scale bean production. In snap bean production, high populations tend to curtail the harvest period, so control is often necessary. Several predators of spider mites, such as *Mallada basalis* (Walker), *Phytoseiulus persimilis* Athias-Henriot and *Amblyseius* spp. are well known as effective natural enemies (Srinivasan, 2014). Among insects, *Chrysoperla carnea* Stephens (Neuroptera, Chrysopidae), *Stethorus* spp. (Coleoptera, Coccinellidae), *Oligota* spp. (Coleoptera, Staphylinidae), *Anthrocnodax occidentalis* Felt and *Feltiella minuta* Felt (Diptera, Cecidomyiidae) are also effective to control the two-spotted spider mite. In general, broad-spectrum insecticides should be avoided because of their harmfulness to predators, leading to outbreaks of the mite pest populations (Vacante, 2016).

The legume pod borer, *Maruca vitrata* F. (= *Maruca testulalis*) (Lepidoptera, Crambidae) is a serious pest of food legumes largely present in all tropical and subtropical countries of the world (EPPO, 2016). On cowpea, *Vigna radiata*, *V. mungo*, pigeon pea, *Lablab purpureus*, common bean, *Sesbania cannabina* and *S. grandiflora*, the moth can cause yield losses ranging from 20% to 80% (Karel, 1985; Singh *et al.*, 1990). The moth has light-brown forewings with white markings and a wingspan of 15–28 mm. Early instars are dull white, with irregularly shaped brown or black spots, especially on the dorsal side. The larvae feed on flower buds, pods and sometimes on young tender shoots, peduncles and stems (Srinivasan, 2014). The attack by the early instars on flower buds and flowers may go unnoticed and the damage is evident only when the flowers fall. The larvae attack at the junction between two pods or between a pod and a leaf or stem. They frequently web together flowers, pods and leaves, living in these sheltered sites during the day, coming out during the night to look for fresh flowers and pods to feed on. To complete larval development, each larva may consume about four to six flowers

(Abate and Ampofo, 1996). *M. vitrata* has many effective natural enemies, such as the parasitoids *Phanerotoma syleptae* Zettel, *P. leucobasis* Kriechbaumer, *Therophilus javanus* Bhat and Gupta, *T. marucae* van Achterberg and Long and *Braunsia kriegeri* (Hymenoptera, Braconidae) (Huang *et al.*, 2003; Arodokoun *et al.*, 2006; Muniappan *et al.*, 2012). Actually, the application of chemical pesticides remains the most applied method to manage the pod borer in the main legume production areas of tropical and subtropical countries. The numerous chemical interventions have a rather limited effectiveness due to the very limited time in which the larvae stay on leaves, and are exposed to the pesticides, before entering floral buds, flowers and pods (Schreinemachers *et al.*, 2014). Moreover, it has been documented that the pod borer may develop resistance to some broad-spectrum insecticides (Atachi and Sourokou, 1989; Ekesi, 1999; Ulrichs *et al.*, 2001). Several cowpea varieties tolerant to the pod borer are known and grown in some rural areas or they are reared at research centres (Jackai, 1982; Adekola and Oluleye, 2008). Pheromone traps and lures of *M. vitrata* have been already developed and usefully applied for monitoring populations (Downham *et al.*, 2003; Schlägers *et al.*, 2012). *Maruca vitrata* multiple nucleopolyhedrovirus (MaviM-NPV) has been developed as a biopesticide in Taiwan and Benin, and can successfully flank botanical or chemical pesticides (Sokame *et al.*, 2015; Tamò *et al.*, 2012). In addition, *Bacillus thuringiensis* formulations and the entomopathogenic fungi *Beauveria bassiana* and *Metarhizium anisopliae* can be used to manage the pest (Srinivasan, 2008; Sreekanth and Seshamahalakshmi, 2012). Management strategies against the pod borer are more effective before the larvae enter into the reproductive parts of the plants.

6.2.3 Sap-sucking pests

Aphids (Hemiptera, Aphididae) damage legumes by removal of sap, production of honeydew (on which sooty mould develops) and transmission of viral diseases, capable to seriously compromise production (Katis *et al.*, 2007). On legumes, the latter causes the major economic losses, which can reach 100% of the harvest (Makkouk and Kumari, 2009). The main aphid species that affect legume crops are the black bean aphid, *Aphis fabae* Scopoli, the cowpea aphid, *Aphis craccivora* Koch, and the pea aphid, *Acyrthosiphon pisum* Harris. They frequently form dense colonies on growing vegetation of several legume plants and may transmit a number of viruses, among which are the Bean leafroll virus, the Beet western yellows virus, the Bean yellow mosaic virus, the Pea enation mosaic virus-1, the Pea seed-borne mosaic virus and the Chickpea chlorotic stunt virus (Makkouk and Kumari, 2009). *A. craccivora*, among others, is vector of the Groundnut rosette virus, the most destructive viral disease of sub-Saharan Africa (Naidu *et al.*, 1999). Although aphids have many natural enemies (Muniappan *et al.*, 2012), these latter are of scarce utility in preventing the transmission of viral diseases. Control will be based overall on cultural strategies (such as removing the symptomatic plants, anticipating or delaying the sowing period, minimum tillage and the use of short-cycle and resistant varieties). Chemicals (both synthetic or neem oil) should be limited and applied only when cultural strategies are insufficient and preferably on the early stages of crop growth (El-Hawary and Abd El-Salam, 2008; Makkouk and Kumari, 2009).

The tobacco whitefly (TWF), *Bemisia* spp. gr. *tabaci* (Gennadius) (Hemiptera, Aleyrodidae), is a serious pest for numerous legumes. Both adults and nymphs of this insect suck the plant sap and produce honeydew on which sooty mould grows (reducing the photosynthetic efficiency of the plants), causing a reduction of production. However, the main damage caused by *B. tabaci* species group on legumes is the transmission of several viral diseases, including the dangerous group of Begomovirus (Allen, 1987; Bob *et al.*, 2005; Qazi *et al.*, 2007; Brito *et al.*, 2012). As for other sap-sucking pests, the best strategies to control them are cultural ones, and the same recommended agronomic measures suggested for aphids remain valid. Neem formulations can be applied as a soil drench or foliar application to control whitefly on legume seedlings (Srinivasan, 2014) but, in any case, insecticides are not particularly useful in limiting the spread of viral diseases.

6.2.4 Pod and seed feeders

The major pod-sucking bugs (Hemiptera, Coreidae) are the spiny brown bugs *Clavigralla tomentosicollis* Stål (=*Acanthomia tomentosicollis*), *C. gibbosa* Spinola, *C. scutellaris* (Westwood), *C. schadabi* Dolling, *C hystricodes* Stål and *Anoplocnemis curvipes* (F.) (Abate and Ampofo, 1996; Srinivasan, 2014). *C. tomentosicollis* is mainly present in Africa, whereas *C. gibbosa* and *C. scutellaris* are widely distributed in tropical and subtropical areas. These pests frequently cause severe damage on developing pods and seeds, especially on cowpea, soybean, green and black chickpeas, with yield loss of 20–100% (Srinivasan, 2014). Their nymphs and adults penetrate the pod walls and suck the sap from developing seeds inside. Occasionally, they also feed on stems, leaves and flower buds; *A. curvipes* sometimes sucks the sap out of young shoots and causes them to wilt (Materu, 1970; Abate and Ampofo, 1996). Pod-sucking bugs can be managed only through an integrated strategy based on prevention (early planting, good fertilization and use of tolerant varieties) and a careful use of insecticide, preferring pyrethrin-based or neem oil applied 1–2 times, from about 35 days after planting and once more if necessary, during podding stage (Dugje *et al.*, 2009). In small plots, the bugs can be collected by hand to reduce their population. Several natural enemies are known, such as *Gryon charon* (Nixon), *G. fulviventris* (Crawford), *Protelenomus anoplocnemidis* (Ghesquière) (Hymenoptera,

Scelionidae) and *Ooencyrtus* spp. (Hymenoptera, Encyrtidae) against *C. tomentosicollis* in Africa, *Gryon clavigrallae* Mineo for *C. gibbosa*, *C. scutellaris* in Asia (Bhagwat *et al.*, 1994; Asante *et al.*, 2000; Romeis *et al.*, 2000).

The African bollworm, *Helicoverpa* (=*Heliothis*) *armigera* (Lepidoptera, Noctuidae), is widely distributed in tropical and subtropical areas of the world, where it can cause yield losses to various legume crops (Abate and Negasi, 1981; Shanower *et al.*, 1999). The lima bean pod borer, *Etiella zinckenella* Treitschke (Lepidoptera, Pyralidae), occurs in Asia, sub-Saharan Africa, South and North America, Australia, Europe and the Caribbean. It has been reported on several legume crops, but it causes significant yield losses especially to lentil, peas and soybean (Sharma *et al.*, 2005).

The larval instars of the two moths feed on tender leaves and stems, floral buds and the seeds inside the pods. Population levels are influenced by several factors, including cropping patterns and climatic conditions (Abate, 1988b). As for other insect pests, *H. armigera* damage to intercropped bean is less serious than to sole bean crops. The IPM of *H. armigera* relies on cultural practices (including intercropping, use of varietal mixtures, timely sowing and optimum plant density). Results obtained with some tolerant varieties of chickpea (Kembrekar, 2016), pigeon pea (Sharma, 2016) and bean (Abate and Ampofo, 1996) seem to be promising against *H. armigera*. Sex pheromone traps can be used to monitor the population dynamic of male moths. Broad-spectrum insecticides should be avoided, both for the presence of resistant populations (i.e. pyrethroids, carbamates, organophosphates) (Kembreker, 2016) and for harmful secondary effects that they might have on natural enemies (Nyambo, 1990; Srinivasan, 2014). In any case, it is necessary to minimize and alternate chemical pesticides and use them before larvae penetrate into floral buds or pods (Murray *et al.*, 2005; Srinivasan, 2014). Biopesticides based on *B. thuringiensis*, *Helicoverpa armigera* nucleopolyhedrovirus (HaNPV) and neem can also be used against the moth (Sharma, 2001).

The same cultural practices mentioned for the African bollworm may be adopted also to control *E. zinckenella*. For this piralid, the most important parasitoids reported are *Apanteles hanoii* Tobias and Long, *A. taragamae* Viereck, *Trathala flavo-orbitalis* (Cameron) and *Tropobracon luteus* Cameron (Long and Hoa, 2012). *Trichogrammatoidea bactrae* Nagaraja, and *Trichogramma ostriniae* Pang and Chen were found to parasitize 80% of eggs of *E. zinckenella* (AVRDC, 1992). Biopesticides such as *B. thuringiensis* and neem are highly effective against *E. zinckenella* (Byrappa *et al.*, 2012).

The blue or bean butterfly, *Lampides boeticus* (L.), and *Euchrysops cnejus* (F.) (Lepidoptera, Lycaenidae) occasionally cause serious yield losses to food legumes in Asia, Africa, Europe, Australia and the Pacific (Srinivasan, 2014). In Vietnam, *L. boeticus* is considered a major pest (Schreinemachers *et al.*, 2014). The blue butterfly larvae are attacked by several parasitoids such as *Trichogramma chilotraeae* Nagaraja and Nagarkatti and *T. bactrae* Nagaraja (Hymenoptera, Trichogrammatidae) and *Cotesia specularis* Szepligeti (Hymenoptera, Braconidae). Biopesticides based on *Paecilomyces lilacinus*, *Verticillium lecanii* and neem oil, are superior to *B. thuringiensis* and other entomopathogenic fungi (Arivudainambi and Vijay Chandar, 2009).

6.3 Concluding Remarks

Tropical food legumes are a major source of protein for hundreds of millions of people in Africa, southeast Asia and Latin America. They are also an important source of income, especially as export crops to different parts of the world. Insect pests are among the major biotic constraints hampering the growth of productivity of these crops. Those insects attacking the pods are of particular concern, as they are the major cause of yield losses in tropical food legumes.

Chemical control measures are available for most of the insect pests, but they are not simply applicable to the smallholder farmers' conditions under which most of the tropical grain legumes are produced. An IPM approach that relies mainly cultural measures is the only feasible approach. Crop diversity is an important component of cultural methods and can be achieved through intercropping, strip-cropping or trap-cropping. Crop diversity also encourages natural biological control of insect pests. Research has also developed legume varieties that are resistant or tolerant to many of the major insect pests. The Purdue Improved Cowpea Storage (PICS) is an affordable technology for controlling bruchids on cowpea but it is equally applicable to other bruchid species in other crops under smallholder conditions. Future tropical food legume insect pest efforts need to pay particular attention to scaling available technologies alongside trying to find modern means of pest management techniques.

References

Abate, T. (1988a) *Insect and Mite Pests of Horticultural and Miscellaneous Plants in Ethiopia*. IAR Handbook, No. 1. Institute of Agricultural Research, Addis Ababa.

Abate, T. (1988b) Experiments with trap crops against African bollworm, *Heliothis armigera*, in Ethiopia. *Entomologia Experimentalis et Applicata* 48, 135–140.

Abate, T. (1990) Studies on genetic, cultural and insecticidal controls against the bean fly, *Ophiomyia phaseoli* (Tryon) (Diptera: Agromyzidae), in Ethiopia. PhD thesis. Simon Fraser University, Burnaby, Canada.

Abate, T. (1991) Seed dressing insecticides for bean fly [*Ophiomyia phaseoli* (Tyron) (Diptera: Agromyzidae)] control in Ethiopia. *Tropical Pest Management* 37, 334–337.

Abate, T. (2012) *Four Seasons of Learning and Engaging Smallholder Farmers.* International Crops Research Institute for the Semi-Arid Tropics, Nairobi.

Abate, T. and Ampofo, J.K.O. (1996) Insect pests of beans in Africa: their ecology and management. *Annual Review of Entomology* 41, 45–73.

Abate, T. and Negasi, A. (1981) Chemical control of African bollworm (*Heliothis armigera*) (Hübner) with ultra-low-volume sprays. *Ethiopian Journal of Agriculture Science* 3, 49–55.

Abate, T. and Orr, A. (2012) Research and development for tropical legumes: towards a knowledge based strategy. *Journal of ICRISAT Agricultural Research* 10, 12.

Abate, T., van Huis, A. and Ampofo, J.K.O. (2000) Pest management strategies in traditional agriculture: an African perspective. *Annual Review of Entomology* 45, 631–659.

Abate, T., Shiferaw, B., Gebeyehu, S., Fenta, B., Negash, K., Assefa, K., Eshete, M., Aliye, S. and Hagmann, J. (2011) A systems and partnership approach to agricultural research for development: lessons from Ethiopia. *Outlook on Agriculture* 40 (3), 213–220.

Abate, T., Alene, A.D., Bergvinson, D., Shiferaw, B., Silim, S., Orr, A. and Asfaw, S. (2012) *Tropical Grain Legumes in Africa and South Asia: Knowledge and Opportunities.* International Crops Research Institute for the Semiarid Tropics, Nairobi.

Adekola, O.F. and Oluleye, F. (2008) Induced tolerance of cowpea mutants to *Maruca vitrata* (Fabricius) (Lepidoptera: Pyralidae). *African Journal of Biotechnology* 7, 878–883.

Allen, D.J. (1987) *Principal Diseases of Beans in Africa: Study Guide.* CIAT, Cali, Colombia.

Allen, D.J., Anno-Nyako, F.O., Ochieng, R.S. and Ratinam, M. (1981) Beetle transmission of cowpea mottle and southern bean mosaic viruses in west Africa. *Tropical Agriculture (Trinidad)* 58, 267–274.

Allen, D.J., Ampofo, J.K.O. and Wortman, C.S. (1996) *Revageurs, maladies et carences nutritives du haricot commun en Afrique. Guide pratique.* No. 265. CIAT, Cali, Colombia.

Amoako-Atta, B. (1983) Observations on the pest status of the striped bean weevil *Alcidodes leucogrammus* Erichs on cowpea under intercropping systems in Kenya. *Insect Science Application* 4, 351–356.

Ampofo, J.K.O. (1993) Host plant resistance and cultural strategies for bean stem maggot management. In: *Proceedings of the 2nd Meeting of Pan-African Work Group Bean Entomologists.* CIAT Africa, Workshop Ser. No. 25. Cali, Colombia, pp. 4–73.

Ampofo, J.K.O. and Massomo, S.M. (1998) Some cultural strategies for management of bean stem maggots (Diptera: Agromyzidae) on beans in Tanzania. *African Crop Science Journal* 6, 351–356.

Arivudainambi, S. and Vijay Chandar, A. (2009) Management of pulses blue butterfly, *Lampides boeticus* L. in green gram. *Karnataka Journal of Agricultural Science* 22, 624–625.

Arodokoun, D.Y., Tamò, M., Cloutier, C. and Brodeur, J. (2006) Larval parasitoids occurring on *Maruca vitrata* Fabricius (Lepidoptera: Pyralidae) in Benin, West Africa. *Agriculture, Ecosystems and Environment* 113, 320–325.

Asante, S.K., Jackai, L.E.N. and Tamo, M. (2000) Efficiency of *Gryon fulviventris* (Hymenoptera: Scelionidae) as an egg parasitoid of *Clavigralla tomentosicollis* (Hemiptera: Coreidae) in northern Nigeria. *Environmental Entomology* 29, 815–821.

Atachi, P. and Sourokou B. (1989) Use of Decis and Systoate for the control of *Maruca testulalis* (Geyer) in cowpea. *International Journal of Tropical Insect Science* 10, 373–381.

AVRDC (1992) *Progress Report.* AVRDC Publication (92-374), Asian Vegetable Research and Development Center, Shanhua, Tainan, Taiwan.

Bajpai, N.K., Ballal, C.R., Rao, N.S., Singh, S.P. and Bhaskaran, T.V. (2006) Competitive interaction between two ichneumonid parasitoids of *Spodoptera litura*. *BioControl* 51, 419–438.

Bandara, K.A.N.P., Kumar, V., Ninkovic, V., Ahmed, E., Petterson, J. and Glinwood, R. (2009) Can leek interfere with bean plant-bean fly interaction? Test of ecological pest management in mixed cropping. *Journal of Economic Entomology* 102, 999–1008.

Bhagwat, V.R., Ariëns, S.J.A. and Shanower, T.G. (1994) Effect of host age and intercropping on parasitization of *Clavigralla gibbosa* eggs by *Gryon* sp. *International Chickpea Newsletter* 1, 38–39.

Blaeser-Diekmann, M. (1982) Survey on pests and diseases of faba beans (*Vicia faba*) in Egypt, Morocco and Tunisia. *Fabis Newsletter* 4, 44–45.

Bob, M.A., Odhiambo, B., Kibata, G. and Ong'aro, J. (2005) Whiteflies as pests and vectors of viruses in vegetables and legume mixed cropping systems in Eastern and Southern Africa – Kenya. In: Anderson, P.K. and Morales, F.J. (eds) *Whitefly and Whitefly-borne Viruses in the Tropics:*

Building a Knowledge Base for Global Action. Centro Internacional de Agricultura Tropical (CIAT), Cali, Colombia.

Brito, M., Fernández-Rodríguez, T., Garrido, M.J., Mejías, A., Romano, M. and Marys, E. (2012) First report of cowpea mild mottle carlavirus on yard-long bean (*Vigna unguiculata* subsp. *sesquipedalis*) in Venezuela. *Viruses* 4, 3804–3811.

Buthalezi, N.M., Zaharare, G.E. and Conlong, D.E. (2016) Seasonal monitoring of the flight activity and the incidence of the groundnut leaf miner *Aproaerema modicella* (Lepidoptera: Gelechiidae) at five sites in South Africa. *Austral Entomology.* DOI:10.1111/aen.12249

Byrappa, A.M., Kumar, N.G. and Divya, M. (2012) Impact of biopesticides application on pod borer complex in organically grown field bean ecosystem. *Journal of Biopesticides* 5, 148–160.

Chakravorty, S. and Yadav, D. (2013) Effect of intercrops on Jassid (*Empoasca kerry* Pruthi) infesting green gram, *Vigna radiata* (L.) Wilczeck. *Annals of Plant Protection Sciences* 21, 229–232.

Chang, N.T. (1988) The preference of thrips, *Megalurothrips usitatus* (Bagnall), for three leguminous plants. *Plant Protection Bulletin (Taiwan)* 30, 68–77.

Chang, N.T. (1990) Color preference of thrips (Thysanoptera: Thripidae) in the adzuki bean field. *Plant Protection Bulletin (Taiwan)* 32, 307–316.

Chhabra, K.S. and Kooner, B.S. (1994) Reaction of certain summer mungbean genotypes towards thrips, *Megalurothrips distalis* (Karny). *Indian Journal of Agricultural Research* 28, 251–256.

Chiang, H.S. and Talekar, N.S. (1980) Identification of sources of resistance to beanfly and two other agromyzid flies in soybean and mungbean. *Journal of Economic Entomology* 73, 197–199.

Dempewolf, M. (2004) Arthropods of economic importance – Agromyzidae of the world (CD-ROM). ETI University of Amsterdam, Amsterdam.

Downham, M.C.A., Hall, D.R., Chamberlain, D.J., Cork, A., Farman, D.I., Tamo, M., Dahounto, D., Datinon, B. and Adetonah, S. (2003) Minor components in the sex pheromone of legume pod-borer: *Maruca vitrata* development of an attractive blend. *Journal of Chemical Ecology* 29, 989–1012.

Dugje, I.Y., Omougi, L.O., Ekeleme, F., Kamara, A.Y. and Ajegbe, H. (2009) *Farmers' Guide to Cowpea Production in West Africa.* International Institute of Tropical Agriculture (IITA), Ibadan, Nigeria.

Ekesi, S. (1999) Insecticide resistance in field populations of the legume pod borer, *Maruca vitrata* Fabricius in Nigeria. *International Journal of Pest Management* 45, 57–59.

El-Hawary, F.M. and Abd El-Salam, A.M.E. (2008) Effect of neem and antitranspirant products against *Aphis craccivora* Koch and its biology. *Egyptian Academic Journal of Biological Sciences. A, Entomology* 1, 189–196.

EPPO (2016) *EPPO Global Database.* Available at: https://gd.eppo.int (accessed 25 July 2017).

FAOSTAT (2017) Statistic Division. Available at: www.fao.org/faostat/en/#data (accessed 4 July 2017).

Foyer, C.H., Lam, H.-M. and Considine, M.J. (2016) Neglecting legumes has compromised human health and sustainable food production. *Nature Plants* 2. DOI:10.1038/nplants.2016.112

Galindo, J.J., Abawi, G.S., Thuston, H.D. and Galvez, G. (1983) Effect of mulching on web blight of beans in Costa Rica. *Phytopathology* 73, 610–615.

Greathead, D.J. (1968) A study in East Africa of bean flies (Dipt., Agromyzidae) affecting *Phaseolus vulgaris* and their natural enemies, with the description of new species of *Melanagromyza* Hend. *Bulletin of Entomological Research* 59, 541–561.

Hillocks, R.J., Mwanga, A., Nahdy, M.S. and Subrahma-Nyam, P. (2000) Diseases and pests of pigeonpea in eastern Africa: a review. *International Journal of Pest Management* 46, 7–18.

Hoai, P.M., Sebesvari, Z., Minh, T.B., Viet P.H. and Renaud, F.G. (2011) Pesticide pollution in agricultural areas of Northern Vietnam: case study in Hoang Liet and Minh Dai communes. *Environmental Pollution* 159, 3344–3350.

Huang, C.C., Peng, W.K. and Talekar, N.S. (2003) Parasitoids and other natural enemies of *Maruca vitrata* feeding on *Sesbania cannabina* in Taiwan. *Biological Control* 48, 407–416.

Huang, Z., Shi, P., Dai, J. and Du, J. (2004) Protein metabolism in *Spodoptera litura* (F.) is influenced by the botanical insecticide azadirachtine. *Pesticide Biochemistry and Physiology* 80, 85–93.

Jackai, L. (1982) A field screening technique for resistance of cowpea (*Vigna unguiculata*) to the pod-borer *Maruca testulalis* (Geyer) (Lepidoptera: Pyralidae). *Bulletin of Entomological Research* 72, 145–156.

Jackai, L. and Daoust, R.A. (1989) Insect pests of cowpeas. *Annual Review of Entomology* 31, 95–119.

Karel, A.K. (1985) Yield losses from and control of bean pod borers *Maruca testulalis* (Lepidoptera: Pyralidae) and *Heliothis armigera* (Lepidoptera: Noctuidae). *Journal of Economic Entomology* 77, 761–765.

Karel, A.K. (1989) Response of *Ootheca bennigseni* (Coleoptera: Chrysomelidae) to extracts from neem. *Journal of Economic Entomology* 82, 1799–1803.

Karel, A.K. and Matee, J.J. (1986) Yield losses in common beans following damage by bean fly, *Ophiomyia Phaseoli* Tryon (Diptera: Agromyzidae). *Bean Improvement Cooperative. Annual Report* 29, Washington State University, Pullman, Washington, pp. 115–116.

Karel, A.K. and Rweyemamu, C.L. (1984) Yield losses in field beans following foliar damage by *Ootheca bennigseni* (Coleoptera: Chrysomelidae). *Journal of Economic Entomology* 77, 761–765.

Karel, A.K. and Rweyemamu, C.L. (1985) Resistance to foliar beetle, *Ootheca bennigseni* (Coleoptera: Chrysomelidae) in common beans. *Environmental Entomology* 14, 662–664.

Kareru, P., Rotich, Z.K., and Maina, E.W. (2013) *Use of Botanicals and Safer Insecticides Designed in Controlling Insects: The African Case.* InTech, Winchester, UK.

Katis, N.I., Tsitsipis, J.A., Stevens, M. and Powell, G. (2007) Transmission of plant viruses. In: van Emden, H.F. and Harrington, R. (eds) *Aphid as Crop Pests.* CAB International, Wallingford, UK, pp. 353–390.

Kembrekar, D.N. (2016) Management of legume pod borer, *Helicoverpa armigera* with host plant resistance. *Legume Genomics and Genetics* 7, 1–19.

Kenis, M. and Cugala, D. (2006) Prospects for the biological control of the groundnut leaf miner, *Aproaerema modicella*, in Africa. *CAB Review: Veterinary Science, Nutrition and Natural Resources* 1, 1–9.

Kouris-Blazos, A. and Belsky, R. (2016) Health benefits of legumes and pulses with focus on Australian sweet lupins. *Asia Pacific Journal of Clinical Nutrition* 25, 1–17.

Kushi, L.H., Meyer, K.A. and Jacobs, D.R. Jr. (1999) Cereals, legumes, and chronic disease risk reduction: evidence from epidemiologic studies. *American Journal of Clinical Nutrition* 70, 451–458.

Kyamanywa, S. and Ampofo, J.K.O. (1988) Effect of cowpea/maize mixed cropping on the incident light at the cowpea canopy and flower thrips (Thysanoptera: Thripidae) population density. *Crop Protection* 7, 186–189.

Kyamanywa, S., Baliddawa, C.W. and Ampofo, J.K.O. (1993) Effect of maize plants on colonization of cowpea plants by bean flower thrips *Megalurothrips sjostedti*. *Entomologia Experimentalis et Applicata* 69, 61–68.

Lamers, M., Anyusheva, M., La, N., Nguyen, V.V. and Streck, T. (2011) Pesticide pollution in surface- and groundwater by paddy rice cultivation: a case study from Northern Vietnam. *CLEAN – Soil, Air, Water* 39, 356–361.

Lebesa, N.L., Khan, Z.R., Hassanali, A., Pickett, J.A., Bruce, T.J.A., Skellern, M. and Krüger, K. (2011) Responses of the blister beetle *Hycleus apicicornis* to visual stimuli. *Physiological Entomology* 36, 220–229.

Letourneau, D.K. (1994) Beanfly, management practices, and biological control in Malawian subsistence agriculture. *Agriculture, Ecosystems and Environment* 50, 103–111.

Letourneau, D.K. and Altieri, M.A. (1983) Abundance patterns of a predator, *Orius tristicolor* (Hemiptera: Anthocoridae) and its prey *Frankliniella occidentalis* (Thysanoptera: Thripidae): habitat attraction in polyculture versis monoculture. *Environmental Entomology* 12, 1464–1469.

Lingappa, S. and Hegde, R. (2001) Exploitation of biocontrol potential in the management of insect pests of pulse crops. In: Upadhyay, R.K., Mukerji, K.G. and Chamola, B.P. (eds) *Biocontrol Potential and Its Exploitation in Sustainable Agriculture, Volume 2: Insect Pests.* Kluwer Academic/Plenum Publishers, New York, pp. 321–344.

Long, K.D. and Hoa, D.T. (2012) Notes on parasitoid assemblage reared from larvae of legume pod borers *Maruca vitrata* (Fabricius) and *Etiella zinckenella* Treitschke. *Journal of Biology* 34, 48–58.

Makkouk, K.M. and Kumari, S.G. (2009) Epidemiology and integrated management of persistently transmitted aphid-borne viruses of legume and cereal crops in West Asia and North Africa. *Virus Research* 141, 209–218.

Materu, M.E.A. (1970) Damage caused by *Acanthomia tomentosicollis* Stål and *A. horrida* Germ. (Hemiptera: Coreidae). *East African Agricultural and Forestry Journal* 35, 429–435.

Mkenda, P., Mwanauta, R., Stevenson, P.C., Ndakidemi, P., Mtei, K. and Belmain, S.R. (2015) Extracts from field margin weeds provide economically viable and environmentally benign pest control compared to synthetic pesticides. *PLOS ONE* 10, e0143530.

Mohamed, R.A. and Teri, J.M. (1989) Farmers' strategies of insect pest and disease management in small-scale bean production systems in Mgeta, Tanzania. *Insect Science and its Application* 10, 821–825.

Muniappan, R., Shepard, B.M., Carner, G.R. and Ooi, P.A.C. (2012) *Arthropod Pests of Horticultural Crops in Tropical Asia.* CAB International, Wallingford, UK.

Murray, D.A.H., Lloyd, R.J. and Hopkinson, J.E. (2005) Efficacy of new insecticides for management of *Helicoverpa* spp. (Lepidoptera: Noctuidae) in Australian grain crops. *Australian Journal of Entomology* 44, 62–67.

Naidu, R.A., Kimmins, F.M., Deom, C.M., Subrahmanyam, P., Chiyemekeza, A.J. and van der Merwe, P.J.A. (1999) Groundnut rosette: a virus disease affecting groundnut production in sub-Saharan Africa, *Plant Disease* 83, 700–709.

Nyambo, B.T. (1990) Effect of natural enemies on the cotton bollworm, *Heliothis armigera* (Hübner) (Lepidoptera: Noctuidae) in Western Tanzania. *Tropical Pest Management* 36, 50–58.

Ojwang, P.P.O., Melis, R., Songa, J.M. and Githir, M. (2009) Participatory plant breeding approach for host plant resistance to bean fly in common bean under semi-arid Kenya conditions. *Euphytica* 170, 383–393.

Okwiri Ojwang, P.P., Melis, R., Songa, J.M. and Githiri, M. (2010) Genotipic response of common bean to natural field populations of bean fly under diverse environmental conditions. *Field Crop Research* 117, 139–145.

Okwiri Ojwang, P.P., Melis, R., Githiri, M. and Songa, J.M. (2011) Breeding options for improving common bean for resistance against bean fly: a review of research in eastern and southern Africa. *Euphytica* 179, 363–371.

Paul, U.V., Ampofo, J.K.O., Hilbeck, A. and Edwards, P. (2007) Evaluation of organic control method of the bean beetle, *Ootheca bennigseni,* in East Africa. *New Zealand Plant Protection* 60, 189–198.

Praneetvatakul, S., Schreinemachers, P., Pananurak, P. and Tipraqsa, P. (2013) Pesticides, external costs and policy options for Thai agriculture. *Environmental Science and Policy* 27, 103–113.

Qazi, J., Ilyas, M., Mansoor, S. and Briddou, R.W. (2007) Legume yellow mosaic viruses: genetically isolated begomovirus. *Molecular Plant Pathology* 8, 343–348.

Raheja, A.K. (1981) Status of *Ootheca mutabilis* as a pest of cowpea in northern Nigeria. *Samaru Journal of Agricultural Research* 1, 111–118.

Rajguru, M. and Sharma, A.N. (2012) Comparative efficacy of plant extract alone and in combination with *Bacillus thuringiensis* subsp. *kurstaki* against *Spodoptera litura* Fab. larvae. *Journal of Biopesticides* 5, 81–86.

Robertson, D.G. (1963) Further studies on the host range of cowpea yellow mosaic virus. *Tropical Agriculture* 40, 319–324.

Romeis, J., Shanower, T.G. and Madhuri, K. (2000) Biology and field performance of *Gryon clavigrallae* (Hymenoptera: Scelionidae), an egg parasitoid of *Clavigralla* spp. (Hemiptera: Coreidae) in India. *Bulletin of Entomological Research* 90, 253–263.

Sahu, I.K. and Shaw, C. (2005) Influence of tomato genotypes against *Caliothrips indicus* (Bagnall) with reference to biophysical parameters. *Journal of Soils and Crops* 15, 274–279.

Sariah, J.B. and Makundi, R.H. (2007) Effect of sowing time on infestation of beans (*Phaseolus vulgaris* L.) by two species of the Bean Stem Maggot, *Ophiomyia spencerella* and *Ophiomyia phaseoli* (Diptera: Agromyzidae). *Archives of Phytopathology and Plant Protection* 40, 45–51.

Schläger, S., Ulrichs, C., Srinivasan, R., Beran, F., Bhanu, K.R.M., Mewis, I. and Schreiner, M. (2012) Developing pheromone traps and lures for *Maruca vitrata* in Taiwan. *Gesunde Pflanzen* 64, 183–186.

Schreinemachers, P. and Tipraqsa, P. (2012) Agricultural pesticides and land use intensification in high, middle and low income countries. *Food Policy* 37, 616–626.

Schreinemachers, P., Sringarm, S. and Sirijinda, A. (2011) The role of synthetic pesticides in the intensification of highland agriculture in Thailand. *Crop Protection* 30, 1430–1437.

Schreinemachers, P., Srinivasan, R., Wu, M.-H., Bhattarai, M., Patricio, R., Yule, S., Quang, V.H. and Hop, B.T.H. (2014) Safe and sustainable management of legume pests and disease in Thailand and Vietnam: a situational analysis. *International Journal of Tropical Insect Science* 34, 88–97.

Shanower, T.G., Wightman, J.A. and Gutierrez, A.P. (1993) Biology and control of the groundnut leaf miner, *Aproarema modicella* (Deventer) (Lepidoptera: Gelechiidae). *Crop Protection* 12, 3–10.

Shanower, T.G., Romeis, J. and Minja, E.M. (1999) Insect pests of pigeonpea and their management. *Annual Review of Entomology* 44, 77–96.

Sharma, H.C. (2001) *Cotton Bollworm/Legume Pod Borer,* Helicoverpa armigera *(Hübner) (Noctuidae: Lepidoptera): Biology and Management.* Crop Protection Compendium, CAB International, Wallingford, UK.

Sharma, H.C. (2016) Host plant resistance to insect pests in pigeopea: potential and limitation. *Legume Perspective* 11, 24–29.

Sharma, H.C., Clement, S.L., Ridsdill-Smith, T.J., Ranga Rao, G.V., Bouhssini El, M. *et al.* (2005) Insect pest management in food legumes: the future strategies. In: Kharkwar, M.C. (ed.) *Proceedings of the Fourth International Food-Legumes Research Conference* (IFLRC-IV), 18–22 October, New Delhi.

Sharma, O.P., Bambawale, O.M., Gopali, J.B., Bhagat, S., Yelshetty, S., Singh, S.K., Anand, R. and Singh, O.P. (2011) *Field Guide – Mungbean and Urdbean.* National Centre for Intergrated Pest Management, Indian Council of Agricultural Research, LBS Building, IARI Campus, New Delhi.

Siddique, K.H.M., Johansen, C. and Turner, N.C. (2012) Innovation in agronomy for food legumes. *Agronomy for Sustainable Development* 32, 45–64.

Singh, G., Misra, P.N. and Srivastava, B.K. (1981) Effect of date of sowing pea on the damage caused by *Ophiomyia centrosematis. Indian Journal of Agricultural Sciences* 51, 340–343.

Singh, S.R. and Allen, D.J. (1980) Pests, diseases and protection in cowpeas. In: Summerfield, R.J. and Bunting, A.H. (eds) *Advances in Legume Science.* Royal Botanic Gardens/Ministry of Agriculture, Fisheries and Food, London, pp. 419–443.

Singh, S.R. and Taylor, T.A. (1978) Pests of grain legumes in Nigeria and their control. In: Singh, S.R., van Emden, H.F. and Taylor, T.A. (eds) *Pests of Grain Legumes: Ecology and Control.* Academic Press, London.

Singh, S.R., Jackai, L.E.N., Dos Santos, J.H.R. and Adalla, C.B. (1990) Insect pests of cowpeas. In: Singh, S.R. (ed.) *Insect Pests of Tropical Legumes.* John Wiley & Sons, Chichester, UK (pp. 43–90).

Sokame, B.M., Tounou, A.K., Datinon, B., Dannon, E.A., Agbaton, C., Srinivasan, R., Pittendrigh, B.R. and Tamò, M. (2015) Combined activity of *Maruca vitrata* multi-nucleopolyedrovirus, MaviMNPV, and oil from neem, *Azadirachta indica* Juss and *Jatropha curcas* L., for the control of cowpea pests. *Crop Protection* 72, 150–157.

Sreekanth, M. and Seshamahalakshmi, M. (2012) Studies on relative toxicity of biopesticides to *Helicoverpa armigera* (Hübner) and *Maruca vitrata* (Geyer) on pigeonpea (*Cajanus cajan* L.). *Journal of Biopesticides* 5, 191–195.

Srinivasan, R. (2008) Susceptibility of legume pod borer (LPB), *Maruca vitrata* to δ-endotoxins of *Bacillus thuringiensis* (Bt) in Taiwan. *Journal of Invertebrate Pathology* 97, 79–81.

Srinivasan, R. (2014) *Insect and Mite Pests on Vegetable Legumes: A Field Guide for Identification and Management.* AVRDC – The World Vegetable Center, Shanhua, Tainan, Taiwan. AVRDC Publication No. 14-778.

Ssekandi, W., Mulumba, J.W., Colangelo, P., Nankya, R., Fadda, C., Karungi, J., Otim, M., De Santis, P. and Jarvis, D.I. (2016) The use of common bean (*Phaseolus vulgaris*) traditional varieties and their mixtures with commercial varieties to manage bean fly (*Ophiomyia* spp.) infestations in Uganda. *Journal of Pest Science* 89, 45–57.

Talekar, N.S. (1990) *Agromyzid Flies of Food Legumes in the Tropics.* Wiley Eastern Limited, New Delhi.

Talekar, N.S. and Lee, Y.H. (1988) Biology of *Ophiomyia centrosematis* (Diptera: Agromyzidae), a pest of soybean. *Annals of the Entomological Society of America* 6, 938–942.

Tamò, M., Baumgärtner, J., Delucchi, V. and Herren, H.R. (1993) Assessment of key factors responsible for the pest status of the bean flower thrips *Megalurothrips sjostedti* (Trybom) (Thysanoptera: Thripidae). *Bulletin of Entomological Research* 83, 251–258.

Tamò, M., Srinivasan, R., Dannon, E., Agboton, C., Datinon, B. *et al.* (2012) Biological control: a major component for the long-term cowpea pest management strategy. In: Boukar, O., Coulibaly, O., Fatokun, C., Lopez, K. and Tamò, M. (eds) *Innovative Research along the Cowpea Value Chain.* Proceedings of the Fifth World Cowpea Conference on Improving Livelihoods in the Cowpea Value Chain through Advancement in Science, 27 September–1 October 2010, Saly, Senegal, pp. 249–259.

Tang, L.-D., Yan, K.-L., Fu, B.-L., Wu, J.-H., Liu, K. and Lu, Y.-Y. (2015) The life table parameters of *Megalurothrips usitatus* (Thysanoptera: Thripidae) on four leguminous crops. *Florida Entomologist* 98, 620–625.

Taylor, T.A. (1969) On the population dynamics and flight activity of *Taeniothrips sjostedti* (Trybom) (Thysanoptera: Thripidae) on cowpea. *Bulletin of the Entomological Society of Nigeria* 2, 60–71.

Udo, I.O. and Akpan, E.A. (2012) Evaluation of local spices as biopesticides for the control of *Ootheca mutabilis*, Shalbena and *Clavigralla tomentosicollis* (Stål.) on cultivated cowpea (Vigna unguiculata L.) in Nigeria. *Journal of Agricultural Science* 4, 7–11.

Ulrichs, C., Mewis, I., Schnitzler, W.H. and Burleigh, J.R. (2001) Effectivity of synthetic insecticides against *Maruca vitrata* F. and the parasitoid *Bassus asper* Chou and Sharkey in the Philippines. *Mitteilungen der Deutschen Gesellschaft für allgemeine und angewandte Entomologie* 13, 279–282.

Vacante, V. (2016) *The Handbook of Mites of Economic Plants: Identification, Bio-Ecology and Control*. CAB International, Wallingford, UK.

Wagner, T. (2010) Revision of *Ootheca* Chevrolet, 1837 from tropical Africa – redescription, descriptions of new species and identification key (Coleoptera: Chrysomelidae, Galerucinae). *Zootaxa* 2659, 1–52.

Whitney, W.K. and Gilmer, R.M. (1974) Vectors of cowpea viruses in Nigeria. *Annals of Applied Biology* 77, 17–21.

Wightman, J.A. and Amin, P.W. (1988) Groundnut pests and their control in the semi-arid tropics. *Tropical Pest Management* 34, 218–226.

Xu, R., Kuang, R., Pay, E., Dou, H. and de Snoo, G.R. (2008) Factors contributing to overuse of pesticides in western China. *Environmental Sciences* 5, 235–249.

Yadav, D.K., Sachan, S.K., Singh, G. and Singh, D.V. (2016) Insect pests associated with pigeon pea variety UPAS 120 in western Uttar Pradesh, India. *Plant Archives* 16, 140–142.

Yule, S. and Srinivasan, R. (2013) Evaluation of biopesticides against legume pod borer, *Maruca vitrata* Fabricius (Lepidoptera: Pyralidae), in laboratory and field conditions in Thailand. *Journal of Asia-Pacific Entomology* 16, 357–360.

7 Integrated Pest Management of Root and Tuber Crops in the Tropics

James Legg[1],*, Joshua Okonya[2] and Daniel Coyne[3]

[1]*International Institute of Tropical Agriculture, Dar es Salaam, Tanzania;*
[2]*International Potato Center, Kampala, Uganda;* [3]*International Institute of Tropical Agriculture, Nairobi, Kenya*

7.1 Introduction

7.1.1 Root and tuber crop production in the tropics

All the major root and tuber crops (RTCs) have their origins in the tropical regions of the world. Potato (*Solanum tuberosum* L.) has the greatest current global production (385 million tonnes), and was first cultivated in the Andean zone of present day Peru and Bolivia, although it is now primarily a crop of temperate parts of the world. Cassava (*Manihot esculenta* Crantz – 270 MT) (FAOSTAT, 2016) and sweet potato (*Ipomoea batatas* L. – 104 MT), similarly, have their origins in Latin America, but these are now widely cultivated throughout the tropics. There are many species of yam, belonging to the genus *Dioscorea*, although just a handful are widely cultivated as crop plants – mostly in the tropical regions of Africa, Asia, Latin America, the Caribbean and the Pacific Islands (Coursey, 1967). The most important species are white yam (*Dioscorea cayannensis* subsp. *rotundata* (Poir.) J. Miège), water yam (*D. alata* L.), Chinese yam (*D. polystachya* Turczaninow), air potato (*D. bulbifera* L.), lesser yam (*D. esculenta* Lour.), bitter yam (*D. dumetorum* Kunth)

and cush-cush yam (*D. trifida* L.). The total annual production of cultivated yams is 68 MT, most of which is in west and central Africa. Several other root and tuber crops, notably the edible aroids – *Xanthosoma sagittifolium* (L.) Schott and *Colocasia* spp. – are cultivated in the tropics, but these contribute a small percentage of the overall RTC production. In this chapter, we focus on the major tropical RTCs, which are cassava, sweet potato and yam. As well as being among the most widely grown food crops in the tropics, RTCs are also crops for which production is increasing at some of the highest rates. Of the five major global staples (wheat, rice, maize, potato and cassava), production of cassava is increasing at the fastest rate, at ~70% since the millennium (FAOSTAT, 2016). The importance of cassava seems likely to continue as the crop is predicted to be one of the most suitably adapted to cope with the anticipated environmental impacts of global climate change (Jarvis *et al.*, 2012).

7.1.2 Key features of pests and diseases affecting RTCs in the tropics

Pests and diseases are some of the most important constraints to the production of

---

* Corresponding author e-mail: j.legg@cgiar.org

root and tuber crops in the tropics. Cassava, sweet potato and yam are all affected by each of the main groups of pests and diseases, comprising arthropods (both insects and mites), nematodes, fungi, bacteria, phytoplasmas and viruses. Within these groups of noxious organisms, and for each of the RTCs, there is considerable geographic diversity. Much of this is a consequence of the different histories that the crops have in each of the major tropical continents. Both cassava and sweet potato have been cultivated for thousands of years in the Americas, but are relatively recent introductions to Africa and Asia, while the different species of yam have diverse geographical origins and have not been disseminated to the same extent.

All the major tropical RTCs are primarily propagated using vegetative propagules (stem cuttings for cassava, vines for sweet potato and tubers for yam). This has important consequences for the spread of pests and diseases from one season's crop to the next. Disease-causing organisms are readily spread through infected planting material, and this is commonly the most important means of disease transmission. Farmers usually plant new crops from planting material saved from the previous season. This fact, coupled with increasing trade and movement of people between countries and continents, means that many of the most damaging pest/disease problems in recent decades have been the result of the inadvertent introduction through planting material of alien invasive species. Local spread is also promoted through the increasing intensity of RTC production in the tropics, which means that it is easier for pest organisms to move from one crop to the next. Although climate change may have a positive effect on RTCs, this agronomic benefit may be partially negated by increases in the abundance and severity of pests and diseases. Temperature increase is thought likely to favour arthropod pests, but elevated temperatures can often result in reduced virus titre. Consequently, while raised temperatures generally lead to shortened life cycles and more

rapid population growth, the effects of climate change on pests and diseases are likely to be mixed. Further research will be required to determine the likely climate change impacts on the wide diversity of species and strains of pests and diseases affecting RTCs.

7.1.3 Key features of the farming environment, and the development and application of IPM

Across the greater proportion of the tropics, cassava, sweet potato and yams are grown for subsistence by smallholder farmers. RTCs are often intercropped. Farm sizes are typically small, labour is usually sourced from within the family, hand tools are used for cultivation, there is little or no application of inputs, and produce is consumed directly by the household rather than being sold. This situation has an important influence on the pest/disease occurrence, as well as the nature of tactics that are appropriate for pest/disease management. Small plot size, heterogeneity of the cropping environment and the absence of input use are all factors that are favourable for pest management. In these situations, the relatively high degree of biodiversity (in comparison with large-scale, input intensive monoculture production systems) favours the activity of natural control agents, and there is no opportunity for the emergence of explosive pest outbreaks following the development of resistance to heavy and repeated applications of insecticides. This complex combination of factors, however, also makes it more difficult to develop simple and uniform pest and disease management recommendations. A thorough understanding of the dynamics of each of the main pests and diseases is essential as the basis for the design of appropriate and effective pest management tactics. These data are not available for RTC pests and diseases in many of the regions in which they are most damaging. Where data are available, they are often derived from one or two pests and not sufficiently comprehensive to facilitate the development of integrated management approaches. Integrated Pest Management (IPM) was initially conceived as a way of overcoming the disastrous negative impacts on both pests/diseases as well as the environment that resulted from the uncontrolled and excessive use of organic chemical pesticides (Stern *et al.*, 1959; Carson, 1962). The focus of the approach was to develop pest/disease management systems in which the chemical and biological components functioned in as complementary a manner as possible. More than 50 years after its introduction, IPM continues to be widely advocated and applied in both temperate and tropical cropping systems. The four basic elements of IPM were defined as: the determination of economic thresholds (above which pests/diseases caused economic damage); scouting to monitor pest/disease populations; understanding and augmentation of the natural biological control in the crop; and the application of pesticides which complement rather than harm biological controls only when economic thresholds are exceeded (Naranjo and Ellsworth, 2009). In view of the characteristics of RTC cultivation in the tropics, there are relatively few situations where conditions are appropriate for the application of a typical IPM programme. More often, IPM when applied to RTCs in the tropics, involves the combined application of several pest/disease management tactics that directly apply or promote existing biological control processes while at the same time using complementary approaches such as cultural practices or host-plant resistance. In this chapter, we describe these sorts of approaches for the integrated management of pests and diseases of RTCs. However, as an important characteristic of the RTCs is the vegetative propagation of 'seed' material and the consequent carry-over of pests and diseases in this material, a key facet of IPM in these crops comprises the development, use and maintenance of pest and disease-free (healthy) planting material and sustaining healthy seedling systems.

7.2 Major Pests and Diseases of RTCs and Their Management Using IPM

7.2.1 Cassava

Major pests and diseases of cassava

The greatest diversity of cassava pests and diseases is reported from Latin America. However, relatively less pest/disease damage occurs in the neotropics, in comparison with Africa and Asia, largely because of the long period of coevolution in the Americas between cassava and its pathogens, arthropod herbivores and their natural enemies. Some of the most damaging diseases and pests of cassava in Africa and Asia are recent exotic introductions from Latin America. The most economically important disease of cassava in both Asia and Africa is cassava mosaic disease (CMD), caused by cassava mosaic begomoviruses (CMBs) (Bock and Woods, 1983; Legg and Fauquet, 2004). It might have been anticipated that these viruses would be introduced to Latin America, as major pests and diseases have moved in the opposite direction. However, CMBs have yet to be detected from cassava in Latin America, a fact which is probably due to the failure of genotypes of the whitefly *Bemisia tabaci* (Genn.) to colonize cassava there (Carabali *et al.*, 2005). *B. tabaci* is the only known vector of CMBs. At the global level, the most damaging arthropod pests of cassava are the mealybugs, mites and whiteflies. Whiteflies are also important as vectors of the viruses causing the most damaging viral diseases. In addition to CMD, the most important diseases of cassava are cassava brown streak disease (CBSD), cassava bacterial blight (CBB), cassava frogskin disease (CFSD) and cassava witches' broom (CWB).

Although many species of mealybugs occur on cassava in Latin America, only two are of economic importance. These are *Phenacoccus manihoti* Matt.-Ferr. and *P. herreni* Cox and Williams (Bellotti *et al.*, 1994). Although localized damaging outbreaks have been reported, notably from north-eastern Brazil, *P. manihoti* is best known as the pest that devastated cassava production in Africa, following its inadvertent introduction in the 1970s. Although the papaya mealybug (*Paracoccus marginatus* Williams and Granara de Willink) has recently begun to spread through Africa, it remains of secondary importance there to *P. manihoti*. Eight species of mealybugs have been described attacking cassava in southeast Asia (SEA) (Parsa *et al.*, 2012), and these are considered as a mealybug complex. The relative importance of these species varies between regions.

More than 50 species of herbivorous mite have been reported from cassava in the Americas, but as with mealybugs, the greatest damage inflicted by these pests has been in continents to which they spread as alien invasives (Africa and Asia). The cassava green mite (CGM) *Mononychellus tanajoa* (Bondar) devastated African cassava production following its spread there in the 1970s (Yaninek and Herren, 1988), while in Asia there have been significant economic losses after the introduction of a second green mite species, *M. mcgregori* (Flechtmann and Baker) (Parsa *et al.*, 2015).

Several whitefly species are reported as physical pests of cassava. These include: *Trialeurodes variabilis* (Quaintance), *Aleurotrachelis socialis* Bondar and *Aleurothrixus aepim* Goeldi in Latin America (Bellotti *et al.*, 1994); *Aleurodicus dispersus* Russell in Africa and Asia; and *B. tabaci* in Africa. More important, however, is the role played by *B. tabaci* as vector of CMBs in Africa and Asia and cassava brown streak ipomoviruses (CBSIs) in Africa. Losses caused by whiteflies as direct physical pests are localized and seasonal, and of much less economic significance than the damage caused by the *B. tabaci*-transmitted CMBs and CBSIs (Legg *et al.*, 2014a).

Some of the larger insect pests of cassava are unique to the Americas. The most important of these are the cassava hornworm (*Erinnyis ello* (L.)), which defoliates plants and can cause losses of up to 64% (Arias and Bellotti, 1984), and the cassava burrowing bug (*Cyrtomennis bergi* Froeschner), which attacks root systems causing direct damage, susceptibility to fungal infection and losses of up to 80% (Arias and Bellotti, 1985).

There are relatively few studies on nematodes and the damage they cause to cassava, although the root knot nematodes (*Meloidogyne* spp.) and lesion nematodes (*Pratylenchus* spp.) are considered the principal nematode pests (Bridge *et al.*, 2005). The most economically important are the *Meloidogyne* spp., which have been recorded from all cassava-growing regions and can cause losses as high as 98% (Théberge, 1985). A survey in Uganda revealed most fields to be infected, many with high levels of severity which indicated significant levels of production loss (Coyne and Namaganda, 1996).

Although more virus species (11) have been reported from cassava in Latin America than the other major producer continents, virus damage to cassava is greatest in the regions affected by CMD (Africa and south Asia) and CBSD (Africa) (Legg *et al.*, 2015). These virus diseases have become increasingly important from the end of the 20th century into the early part of the 21st century. CMD was reported for the first time from southeast Asia in 2015 (Wang *et al.*, 2015). CFSD is the most damaging disease associated with virus infection in Latin America, where the aetiology is also thought to be linked to the presence of phytoplasmas (Alvarez *et al.*, 2009). Phytoplasmas also cause widespread damage to cassava in southeast Asia as the causal agent of witches' broom disease (CWB) (Alvarez *et al.*, 2014).

Xanthomonas axonopodis pv. manihoti (Xam) is the gram-negative bacteria that causes cassava bacterial blight (CBB) through most of the cassava-growing regions of the world, although like many of the globally important pests and diseases of cassava, it originates from Latin America, and is an invasive disease in Africa and Asia (Hillocks and Wydra, 2002). Losses attributed to CBB can be locally severe, where conducive environmental conditions lead to rapid infection resulting in necrotic leaf blight, shoot die-back and plant death (Lozano and Booth, 1974).

There are dozens of fungal pathogens that can cause foliar symptoms (such as super-elongation, leaf spots or stem cankers) or root rots. Leaf spots, such as those caused by *Mycosphaerella henningsii* Sivan. or *Passalora manihotis* (Stevens & Solheim) Braun & Crous may cause yield loss under favourable humid conditions, but they are generally of minor importance. Root rots may be severe, particularly where crops are grown in poorly drained or recently opened fields, but these typically cause significant losses only where crops are harvested beyond full maturity.

Management of major pests and diseases of cassava

A wide range of tactics has been deployed in efforts to control pests and diseases of cassava. Since many cassava farmers produce for subsistence, particularly in Africa, there are relatively low levels of investment in control measures. Agro-inputs – including pesticides – are rarely used, and many farmers apply no measures at all to control pests or diseases that cause significant reductions in yield. Failure to apply control measures is particularly common for some of the more cryptic diseases, such as CBSD, and for soil-dwelling pests such as nematodes. Overall, the most widely applied management approaches have been the use of resistant varieties, which are often provided to farmers free of charge, or biological control using natural enemies that are able to spread effectively following initial introduction. Although many management techniques have been applied to cassava pests in different parts of the world, there are few situations where these have been applied together as a classical IPM package, involving the estimation of economic thresholds, scouting, the limited and targeted use of pesticides and the management of natural enemy populations (Naranjo and Ellsworth, 2009). Some of the most important management methods that have been applied to cassava pests and diseases are as follows.

1. Monitoring and surveillance. Region-wide surveillance has been widely used to monitor changes in the status of major pests and diseases, and invasive patterns of spread. For viruses, monitoring surveys have been used to identify new areas of spread for CMBs in southern India (Dutt *et al.*, 2005), and both CMBs and CBSIs in Africa (Legg and Fauquet, 2004; Sseruwagi *et al.*, 2004; Mbanzibwa *et al.*, 2011). Two common objectives of such surveillance work have been the determination of new areas of virus occurrence, and identifying priority areas for control interventions. Extensive surveillance has been a key part of biological control programmes for mite and mealybug invasive pests in Africa (Neuenschwander, 1994, 2001) and Asia (Parsa *et al.*, 2012). Monitoring programmes are typically combined with the application of classical taxonomic or molecular diagnostic methods. Molecular techniques have been widely used in the surveillance of cassava pathogens (Swanson and Harrison, 1994; Bart *et al.*, 2012; Carvajal-Yepes *et al.*, 2014). Molecular markers on mitochondrial DNA have similarly been applied in identifying genotypes of *B. tabaci* occurring on cassava in Africa (Berry *et al.*, 2004), as well as for the more specific task of tracking genotypes putatively associated with cassava virus pandemics (Legg *et al.*, 2002, 2014b). Modelling techniques are being applied to identify agro-ecological zones suitable for potentially invasive pests and diseases (Campo *et al.*, 2011), and they offer great potential for predicting the likely impacts of climate change. Communications and smartphone technology are increasingly being used to monitor the distribution and spread of cassava pests and diseases (IITA, 2011, 2016). Rigorous field-level monitoring or scouting is not a common feature of cassava pest and disease management. These approaches have been used, however, as part of a robust protocol for producing near-virus-free planting material in Tanzania (IITA, 2014). In addition to bi-monthly scouting visits to assess pest/disease incidence and remove any virus-symptomatic plants, the protocol incorporates systematic virus testing for CBSIs, and a 'three adults per shoot-tip' threshold for *B. tabaci*. If the sampled abundance of *B. tabaci* exceeds this threshold, the systemic neonicotinoid insecticide 'imidacloprid' is applied as a soil drench.

2. Host-plant resistance. Breeding to produce varieties that resist disease infection or pest attack has become arguably the most important strategy for managing biotic threats to cassava. The earliest formal cassava breeding work was initiated in the 1930s at the Amani Research Station, in the north-eastern part of present-day Tanzania. Here, classical intra- and inter-specific crosses were made using cultivated cassava and its closest wild relatives, with the aim of introgressing resistance to CMD and CBSD (Jennings, 1957). This pioneering work formed the basis for subsequent conventional breeding that additionally targeted the development of resistance to CBB (Dixon *et al.*, 2003). More recently, molecular markers have been developed to speed up the breeding cycle (Akano *et al.*, 2002; Okogbenin *et al.*, 2007), and transgenic approaches are being used to produce resistance to viruses in both Africa (Taylor *et al.*, 2004; Yadav *et al.*, 2011) and Asia (Ntui *et al.*, 2015). Resistant varieties have been widely disseminated throughout Africa (Manyong *et al.*, 2000) and have been highly effective in controlling the severe CMD pandemic in east and central Africa (Thresh *et al.*, 1994; Legg *et al.*, 2005, 2015). While information is limited, it is recognized that avoiding nematode damage through using resistant material can lead to significant improvements in productivity (Dixon *et al.*, 2003). It is important to assess new cassava germplasm for resistance to prevailing root knot nematode species or geographical strains. Varieties developed and performing well in one location have been shown to be heavily affected by *Meloidogyne* spp. in other regions (Coyne *et al.*, 2004; Coyne *et al.*, 2006).

3. Biological control. Most of the major arthropod pest problems of cassava have been addressed using biological control. The most prominent examples are the classical biological control of cassava mealybug (CM) in Africa (Neuenschwander *et al.*, 2001) and Asia (Graziosi *et al.*, 2016) using the parasitoid wasp – *Anagyrus lopezi* De Santis, and the use of a similar approach to manage cassava green mite (CGM) in Africa (Yaninek *et al.*, 1993). The encyrtid wasp,

Acerophagus papayae Noyes and Schauff, has proved to be highly effective in controlling the papaya mealybug, and has most recently been introduced to India and west Africa for this purpose. *B. tabaci* has been more difficult to manage using natural enemies, since it is considered to be native in areas where it causes damage through transmitting viruses. However, several studies have provided valuable baseline data on the role of the local natural enemy fauna in restricting *B. tabaci* population increase; most attention has focused on aphelinid parasitoid wasps (Otim *et al.*, 2006; Guastella *et al.*, 2014) and coccinellid predators (Asiimwe *et al.*, 2007). Granulosis virus has been effective in controlling the cassava hornworm in Colombia (Bellotti *et al.*, 1990) and entomopathogenic *Neozygites* spp. have supplemented predatory mite (*Typhlodromalus aripo* De Leon) control of cassava green mite in both South America and Africa (Delalibera *et al.*, 2004).

4. Healthy seed and quarantine. Quarantine guidelines have been established by FAO for movements of cassava germplasm between continents (Frison and Feliu, 1991). These require that cassava is moved in virus-indexed tissue culture form. Although plant protection authorities usually permit movements of cassava germplasm in cutting form between neighbouring countries, recent outbreaks of CBSIs in east Africa (Legg *et al.*, 2011) and CMBs in southeast Asia (Wang *et al.*, 2015) have led to countries imposing stricter quarantine controls. However, prior to this development, open quarantine was used in east Africa to facilitate the transfer of resistant varieties between neighbouring countries (Mohamed, 2002). Open quarantine sites were fields that were fenced and closely monitored to prevent pest and disease introductions. They were planted in the receiving country, but located close to the shared border with the donating country. This approach was used with considerable success for moving resistant germplasm from Uganda to the neighbouring countries of Kenya, Tanzania and Rwanda to limit the worst effects of the expanding pandemic of severe CMD in the late 1990s/early 2000s.

As quarantine controls have become more stringent, an increasing focus has been placed on propagation through tissue culture and the establishment of formal seed certification systems to improve the quality of seed health within cassava-growing countries.

5. Cultural methods. A wide variety of cultural methods are of value in controlling cassava pests and diseases. Crop hygiene, and most importantly the destruction of crop trash after harvest, prevents pests such as mealybugs, mites and scale insects, as well as CBB, leaf spot and virus diseases from readily infecting a new crop planted in the same field. Crop rotation is generally beneficial in minimizing levels of re-infestation for similar reasons (Bellotti *et al.*, 1994). There is some evidence that intercropping can reduce the spread of CMBs (Fondong *et al.*, 2002), although the benefits have not been generally considered to be great enough for this practise to be widely recommended for CMD management. Soil fertility management has little net effect on virus diseases, but it has been recommended as an important complement to the biological control of cassava mealybug (Schulthess *et al.*, 1997). Isolation of healthy 'seed' production sites is required as part of new certification guidelines for cassava planting material, since this greatly reduces the likelihood of infection by vector-borne viruses.

6. Chemical control. There is relatively little use of chemical products to control pests and diseases of cassava, although pesticides are used more commonly in Latin America and Asia where there is more commercial cassava production. Insecticides are recommended for the control of the cassava hornworm during the early instar stages (Bellotti *et al.*, 1990). Insecticides and acaricides have limited effectiveness in controlling mealybugs and mites, and these are rarely used both because of this, and since biocontrol is so effective for these pests. Neonicotinoids are increasingly used in an uncontrolled manner against whiteflies and mealybugs in southeast Asian countries (Parsa *et al.*, 2012), although resistance build-up has been demonstrated for tetranychid mites (Sclar *et al.*, 1998). Whiteflies are susceptible to chemical control in Africa. Neonicotinoids have provided effective control of *B. tabaci* in clean 'seed' production systems, but soft-chemistry alternatives are being sought that have a significantly reduced impact on beneficial insects.

7.2.2 Sweet potato

Major pests and diseases of sweet potato

Among the most economically important insect pests of sweet potato (*Ipomoea batatas* (L.) Lam.) in tropical countries are the sweet potato weevil *Cylas formicarius* (Fabricius), the African sweet potato weevil (*C. brunneus* (Olivier), *C. puncticollis* Boheman and *Euscepes postfasciatus* (Fairmaire)), larvae of the sweet potato butterfly (*Acraea acerata* Hew.), the clearwing moth (*Synanthedon* spp.), the sweet potato hornworm (*Agrius convolvuli* L.) and vectors of the viruses that cause sweet potato virus disease, such as the whitefly *B. tabaci* and aphids (Fuglie, 2007). *Cylas formicarius* is globally distributed and a main pest of sweet potato in Asia and North and Central America while *C. puncticollis* and *C. brunneus* are limited to Africa and reported in 24 tropical countries (Okonya *et al.*, 2016a; Musana *et al.*, 2016). Adult *Cylas* spp. feed on sweet potato leaves while larvae feed on stems and roots. Stem damage is thought to be the main reason for yield loss, although damage to the vascular system caused by feeding, larval tunnelling and secondary rots substantially reduces root yields and quality (Sorensen, 2009). Infested roots are bitter and can neither be marketed nor used as animal feed. *Cylas* spp. can cause root yield losses of up to 100% during prolonged dry seasons in warm lowlands. The quality of planting material is also greatly reduced when seed multiplication fields are infested with *Cylas* spp. *Acraea acerata* occurs in 15 tropical African countries and is a major pest in Ethiopia and Rwanda and a minor pest in Uganda and Kenya (Okonya *et al.*, 2016b).

Complete defoliation of sweet potato vines due to feeding by *A. acerata* larvae may lead to a failure of the crop to establish resulting in reduced yields (Smit, 1997). *Acraea acerata* outbreaks are common during the dry season and at such times, 3rd to 5th instar larvae can also feed on the sweet potato vines in case all plants have been defoliated. The extent of root yield losses due to defoliation by *A. acerata* larvae is not well studied but anecdotal evidence from farmers in Uganda, Rwanda and Burundi indicates that they are substantial. During outbreaks, especially in Uganda, Kenya and Tanzania, *A. convolvuli* has been reported to completely defoliate plants, which has a significant effect on root and vine yields. However, there is limited literature on the scale of yield losses for this pest, as well as for *Synanthedon spp.* In farmers' fields, yield losses due to viruses vectored by *B. tabaci* range from 30% to 40% (Gibson and Kreuze, 2015).

The major diseases of sweet potato in the tropics include leaf and stem blights caused by several *Alternaria* spp., storage root rots caused by various bacteria and fungi, and a number of sweet potato virus diseases. Root knot nematodes cause significant damage to sweet potato root systems where conditions are favourable. More than 30 viruses infect sweet potato, with the whitefly-transmitted *Sweet potato chlorotic stunt virus* (SPCSV) and *Sweet potato feathery mottle virus* (SPFMV) being most important (Clark *et al.*, 2012). Infection by SPCSV increases the susceptibility of the plant to infection by other viruses. Other important viruses infecting sweet potato in the tropics include *Sweet potato collusive virus* (SPCV), Badnaviruses, *Sweet potato symptomless virus 1* (SPSMV-1), *Sweet potato mild mottle virus* (SPMMV), *Sweet potato chlorotic fleck virus* (SPCFV), *Sweet potato virus G* (SPVG), *Sweet potato virus Z* (SPVZ) and *Sweet potato virus W* (SPVW) (Fei, 2015). These viruses are transmitted by aphids such as the cotton aphid (*Aphis gossypii* Glover), the green peach aphid (*Myzus persicae* (Sulzer)) and a number of the genotypes of the *B. tabaci* species complex. Accumulation of viruses in planting material often leads to cultivar decline with yield losses ranging from 30% to 40% (Gibson and Kreuze, 2015). Alternaria blight is a problem only at medium to high elevations (>2300 masl) with high humidity or rainfall. This disease causes dark brown or black lesions full of spores on both the shoot and the stem, and eventually kills affected leaves, petioles or stems. Rhizopus soft rot (*Rhizopus stolonifer* (Ehrenberg) Vuillemin) commonly causes postharvest losses. Several nematodes have been associated with sweet potato, but root knot nematodes (*Meloidogyne* spp.) and the reniform nematode (*Rotylenchulus reniformis* Linford and Oliveira) are the most important, causing disfigurement (especially cracking) of tubers and reduced yield (Coyne, 2005; Clark *et al.*, 2013). Millipedes and mole rats also cause substantial root losses in some areas where they occur.

Management of major pests and diseases of sweet potato

There is currently little use of chemical pesticides for pest and disease management in sweet potato, largely since the crop is similar to the other RTCs in being primarily grown on small plots for subsistence. Nevertheless, there is a broad range of measures that have been developed for the management of pests and diseases in this crop, and consequently there is great potential for the more widespread application and development of IPM packages that combine these elements.

IPM components for sweet potato include the following.

1. Monitoring adult pest populations. Regular monitoring of pest populations, for example, by using insect traps (yellow sticky or pheromone) or the occurrence of *A. acerata* larvae on the foliage. *C. formicarius* populations in the field and associated root damage have been reduced by using pheromone-baited traps, while sex pheromones have been artificially synthesized for *C. puncticollis* and *C. brunneus* and their field application has been evaluated (Reddy *et al.*, 2014).

2. Host-plant resistance. Some varieties escape damage from *Cylas* spp. by having deep roots or through producing root latex. No varieties that are truly resistant to *Cylas* weevils have yet been identified, but varieties like 'New Kawogo' are partially tolerant (Anyanga *et al.*, 2013). By contrast, host-plant resistance is a front-line tactic for the control of sweet potato viruses. SPVD can be very effectively managed through the deployment of resistant varieties, such as New Kawogo, Naspot 11 and Tomulabula (Aritua *et al.*, 1999; Valkonen *et al.*, 2015). Resistance against *Meloidogyne* spp. is complicated by multiple species affecting sweet potato and the development of 'resistance breaking populations' of *M. incognita* (Kofoid and White) Chitwood, the main species. However, interest in breeding for resistance against nematodes, as well as other diseases, has intensified in line with the rapid increase in production and interest of sweet potato in the USA.

3. Cultural control. Use of healthy (insect- and virus-free) planting material sourced from reliable vine multipliers or private tissue culture facilities can reduce virus propagation through seed. Crop rotation of periods longer than one year with crops such as onion and garlic has been reported to reduce *C. puncticollis* and *C. brunneus* damage on roots and vines. Proper field sanitation, involving strict roguing and removal of any infested crop roots and vines from the field and its surroundings can prevent re-infestation by *Cylas* species and viruses. Mulching, re-hilling and early harvesting are all cultural control techniques that can be used against *Cylas* species.

4. Biological control. Biopesticides such as *Bacillus thuringiensis* and entomopathogens such as *Beauveria bassiana* have been used to control *Cylas* spp. and *A. acerata*. Naturally occurring parasitoids, such as *Charops* spp., have been reported to cause high mortality of *A. acerata* in sweet potato growing countries in east and west Africa (Okonya and Kroschel, 2013).

5. Chemical control. Contact insecticides have been used in Uganda under some circumstances to control emergency outbreak situations for *A. acerata* and *A. convolvuli.* In commercial sweet potato fields, especially in the United States, *C. formicarius* is mainly controlled using foliar chemical insecticide sprays. *Rhizopus* soft rot may be controlled with chemical fungicides.

7.2.3 Yam

Major pests and diseases of yam

Most information on pests and diseases of yam comes from the yam belt, in west and central Africa, where most yam is produced. The relative abundance and incidence of pests and diseases affecting yam differs depending on geography and species of yam. Proportionately, however, yams are most affected by virus and nematode problems, with particularly high incidences of infection recorded for these two groups (Asala *et al.*, 2012; Kolombia *et al.*, 2015). Viruses in the genus *Badnavirus*, generically referred to as yam badnaviruses (YBVs) tend to be the most frequently encountered, followed by potyviruses, such as *Yam mosaic virus* (YMV) and *Yam mild mosaic virus* (YMMV), and *Cucumber mosaic virus* (CMV) (genus *Cucumovirus*) (Kenyon *et al.*, 2008; Seal *et al.*, 2014). Although more than 30 species have been encountered on yam worldwide, these four viruses are the most economically important, largely due to their occurrence in the African yam belt, although badnaviruses are highly prevalent on yam in the Pacific (Kenyon *et al.*, 2008). Single or mixed infections result in mosaic, chlorosis, mottling and vein banding symptoms in leaves, which may be distorted, crinkled, puckered or stunted, affecting photosynthetic potential. Tubers may be reduced in size and can present brown necrotic spots or become tough and corky. Plants may also be stunted. Mixed infections of viruses can lead to severe disease. This is a common feature of mixed infections of YMV and YMMV.

Of the numerous nematodes associated with yam (Bridge *et al.*, 2005), the lesion nematodes (*Scutellonema bradys* (Steiner & LeHew) Andrassy and *Pratylenchus* spp.)

and root knot nematodes (*Meloidogyne* spp.) are most damaging. They feed on the roots and tubers, reducing crop productivity, but they also cause loss of viability of seed tubers, reducing germination and causing tuber deterioration during storage. Infected tubers are less marketable, although mild infections of lesion nematodes can often go unnoticed, while those infected with root knot nematodes are unsightly due to disfigurement, further affecting marketability. Combined infections are not uncommon. In the African yam belt *S. bradys* is endemic and the causal agent of 'dry rot' disease. *Pratylenchus coffeae* Goodey, however, which is also endemic, is not encountered on yam, despite *P. coffeae* being a major pest of yam elsewhere, such as Central America. This suggests that there are host-specific pathotypes of *P. coffeae*. Where both *S. bradys* and *P. coffeae* jointly infect yam, *P. coffeae* gradually becomes the dominant species. In east Africa, *P. sudanensis* Loof & Yassin infects yam and may present symptoms of dry rot, surface cracking and tuber deterioration, which are similar symptoms to those caused by *P. coffeae* and *S. bradys*. Several *Meloidogyne* species have been recorded as infecting yam. The two most important are *M. incognita* and *M. javanica* (Treub) Chitwood, and to a lesser extent *M. arenaria* (Neal) Chitwood. The highly aggressive species *M. enterolobii* Yang & Eisenback has recently been recovered from infected yam in Nigeria (Kolombia *et al.*, 2016). Over recent decades, it appears that *Meloidogyne* spp. are becoming increasingly prevalent in farmers' fields, resulting in increasing numbers of damaged tubers present in the markets, possibly due to less fallowing and more intensified cropping of yam (Kolombia *et al.*, 2015). Apart from the visible damage caused to tubers, root knot nematodes reduce tuber weight and productivity. During storage, infected tubers desiccate more rapidly, and in a way that is proportionate to the level of infection (Mudiope *et al.*, 2012). Nematode survival reduces over time, but tubers with only mild infection and used for planting relatively soon after harvest, provide inoculum for the next season. Nematode (lesion and root knot) infection may originate from the field, as well as being seed-borne (Bridge *et al.*, 2005).

The most important foliage pathogen of yam is *Colletotrichum gloeosporioides*, the causal agent of anthracnose (Lebot, 2009). Globally widespread, the disease can be highly destructive, particularly on *D. alata*. It can also infect tubers and thereby lead to tuber-borne infections (Green and Simons, 1994). Several pathogens give rise to fungal tuber rots during storage. These include the fungi, *Lasiodiplodia theobromae*, *Penicillium oxalicum*, *Aspergillus niger* and *Fusarium oxysporum* and the bacteria, *Corynebacterium* spp., *Erwinia* spp. and *Serratia* spp. (Emehute *et al.*, 1998; Okigbo and Ikediugwu, 2000). Combined infections can occur, while nematode infection in-field is commonly understood to facilitate or exaggerate fungal or bacterial infection (Bridge *et al.*, 2005). Wounding tuber tissue, such as during harvest, may facilitate entry and infection by rot pathogens (Simons, 1997).

Several invertebrate pests damage yam, either in the field or under storage or both. Mealybugs, belonging to at least three genera (*Planacoccus*, *Geococcus* and *Phenacoccus*) and scales (*Aspidiella hartii* (Cockerell)) may infest tubers in the field and then manifest during storage. Termites (*Microtermes* spp. and *Amitermes* spp.) and yam beetles (*Heteroligus* spp.) both tunnel into tubers while in the field, but may continue to cause damage during storage. Other insect pests such as the yam weevil (*Palaeopus costicollis* Marshall) in Central America and some lepidopteran species, e.g. *Euzopherodes vapidella* Mann and *Dasyses rugosella* Stainton cause damage to stored yam in west Africa (Ashamo and Odeyemi, 2004).

Management of major pests and diseases of yam

The most important pest and disease problems of yam are associated with seed-(tuber-) borne infections of viruses and nematodes. This arises because of the regular exchange and use of farmer-derived planting material, which may often be infected, in addition to a lack of sustainable healthy seed production systems. Improved awareness by farmers of the causes of pest and disease problems as well as the benefits of using healthy seed material, and a better availability and access to healthy planting material would significantly help in overcoming pest and disease-based losses in yam.

1. Monitoring and surveillance. While farmers remain largely ignorant of the biological basis of seed infection, seed degeneration, and the benefits of investing in the use of healthy planting material, progress in this area will remain limited. However, with steady improvements in the development of seedling systems (Aighewi *et al.*, 2015), a systematic approach to monitoring pest and disease dynamics will be essential. Knowledge of the biological basis of seed degeneration is critical for farmers to understand how to select as accurately as possible against disease. Improved farmer awareness of plant health issues and the implications of this for seed systems is pivotal to the implementation of any such strategy. Rapid detection kits for viruses facilitate their detection and enable research or extension staff to monitor their occurrence and incidence both in farmers' fields as well as at seed production sites.

2. Host-plant resistance. The highly heterozygous nature of yam, difficulties of flowering, issues of sterility and the long growth cycle of the crop make breeding difficult. Mechanisms to improve and speed up the breeding process have therefore been a key focus (Petro *et al.*, 2011). A major challenge for yam improvement has been to breed for resistance against the major pests and diseases and introgress resistance genes into locally preferred varieties. Key foci towards this goal have included anthracnose, nematodes and viruses. Despite the heterogeneity of *C. gloeosporioides* populations (Abang *et al.*, 2005), breeders have been able to develop anthracnose-resistant varieties which have been widely distributed to the major yam-producing countries

in west Africa. There is no confirmed evidence for complete resistance against *S. bradys* (Bridge *et al.*, 2005); however, there appears to be greater scope for developing resistance against *Meloidogyne* spp. The task of breeding for nematode resistance is made more difficult because of the diversity of species that attack the crop. An example of this is the recent discovery, in Nigeria, of *M. enterolobii* at relatively high incidence levels (Kolombia *et al.*, 2016). Such information on pest and disease pathogen occurrence and variability is pivotal to establishing effective and durable resistance.

3. Healthy seed systems. Improved seed yam quality and availability can contribute greatly to improved yields and reduced storage losses. Access to and use of healthy planting material, especially of resistant cultivars, is therefore an obvious strategy for reducing the pest and disease losses. Generating healthy seed stocks, regular treatment of tuber seed material, and in-field selection against disease infected plants all help to improve the health of seed stocks. Tuber treatment using hot water or synthetic pesticides (Coyne *et al.*, 2010a,b; Claudius-Cole *et al.*, 2014), has proved effective. Alternatively, generating healthy planting propagules through *in vitro* tissue cultured plantlets, or vine cuttings offers the potential to generate healthy seed tubers that can easily be transported and comply with phytosanitary regulations for the international movement of germplasm (Coyne *et al.*, 2010b; Maroya *et al.*, 2014). Improved seed systems alone are not sufficient to maintain standards, however, and the support of the plant health sector is required. The sustainability of seed systems requires the integration of pest and disease management options in combination with diagnostics support and durable seed certification schemes. Field-based options for preventing re-infection by pests and diseases are relatively limited. Strict roguing of infected plants helps to reduce field spread of viruses, but is rarely applied or enforced.

4. Cultural control. Yam productivity is maximized by combining the use of high-quality planting material of an improved variety, with effective agronomic and pest management practices. Timely weeding is important to reduce weed competition, which is a key challenge for yams in the *D. cayenensis-rotundata* group, which are readily overcome by weeds due to their limited leaf cover. Regular weeding also reduces alternative pest and disease hosts, especially those harbouring viruses. Rotation of land, which has been practised traditionally and is generally beneficial for pest and disease management, is becoming less common as competition for land increases. Cover crops have been assessed for their potential to reduce soil-borne pests and diseases. Those that have multiple effects are more attractive choices, such as *Aeschynomene histrix* Poir. and *Crotolaria juncea* L. These are nitrogen-fixing legumes, which provide a mulch cover and reduce *M. incognita* when intercropped with yam (Claudius-Cole *et al.*, 2014). Intercropping yam, however, is not commonly practised in the yam belt, and so traditional customs may need to be influenced for such practices to be adopted.

5. Biological control. Several studies have assessed the potential effectiveness of biopesticides or microbial antagonists, mainly against fungal pathogens causing tuber rots. Some of the most promising results have been achieved with applications to stored tubers of *Trichoderma viride* (personal communication, S.F. Gray).

6. Chemical control. Pesticides are not commonly used on yam. An exception is the occasional use of broad spectrum herbicides, such as glyphosate or paraquat. The chemical control of yam anthracnose requires regular sequential fungicide application which is not viewed as cost-effective or environmentally sensitive (Onyeka *et al.*, 2006). In addition to this, various chemical treatments have been assessed and used for treatment of seed material for protection against soil- and seed-borne pests and diseases. Some impressive results observed from treating planting setts have led to greater uptake of the use of pesticide 'dips' to reduce losses and produce healthy seed, with significantly extended storage life and viability (Claudius-Cole *et al.*, 2014). For yam producers in sub-Saharan Africa, the limited availability and accessibility of pesticides mean that they will continue to be infrequently used, irrespective of their potential effectiveness.

7.3 Case Studies of Successful IPM for RTCs in the Tropics

7.3.1 Case study 1: IPM to manage the health of cassava planting material in Tanzania

Background

The cassava virus diseases – CMD and CBSD – are the main constraints to cassava production in Africa. While these diseases cause direct yield losses to farmers, they also hinder the free movement of germplasm within a country, as well as regional-level variety dissemination programmes, since the viruses are readily propagated through stem cuttings. Provision of 'clean seed' has therefore been identified as a priority for cassava development work in east Africa. In Tanzania, this has been addressed by establishing a national system for the production of clean cassava seed (=planting material for the example used here). Key components of this approach include the following.

1.   Support for the establishment of clean seed production sites with research institutions (pre-basic seed), private seed companies and farmers (basic seed), commercial farmers (certified seed) and rural farmers/farmer groups (quality declared planting material – QDPM).

2.   The design and implementation of a formal seed certification system for cassava, with certification guidelines appended to the Government of Tanzania Seed Act.

3.   Protocols for field-based management of cassava seed quality and virus testing required for monitoring and regulatory testing.

4.   Strengthening the capacity of researchers, seed inspectors and growers on methods required for managing the health of cassava seed crops.

The IPM approaches employed in this programme have varied depending on the level of seed production (from pre-basic down to QDPM). For high-value pre-basic seed, a more commercial and intensive method has been used, while at the QDPM level, the focus has been on communicating a small number of key messages.

IPM for pre-basic cassava seed production

1. Host-plant resistance. All varieties being multiplied have the best levels of resistance currently available (moderate) for CBSD. The system provides a pipeline for the introduction of higher levels of resistance over time.

2. Site location. Fields are isolated (>300 m from the nearest neighbouring cassava field) and grown at an altitude that is high for the region being served. Whiteflies are less abundant at higher altitudes, and isolation reduces the likelihood of virus spread from neighbouring fields.

3. Source of starter seed material. Planting material is obtained from symptom-free parent material, which is also tested using virus diagnostics for the absence of CBSIs. The system is transitioning to an approach in which starter material will be exclusively obtained from virus-indexed tissue culture material, which is then mass-propagated in an insect-proof screenhouse.

4. Roguing. Any plants expressing symptoms of CMD, CBSD or CM are removed from the field and destroyed.

5. Crop hygiene. Crop debris/trash is removed from the field after all harvest operations to minimize carry-over of CBB and fungal pathogens.

6. Monitoring. Fields are monitored every two months for occurrence of major pests and diseases. During monitoring, assessments are made of the abundance, damage and/or incidence of CBSD, CMD, CM, CGM and CBB.

7. Scouting and thresholds for the control of arthropod pests. If, during monitoring assessments, the following thresholds are exceeded (*B. tabaci* whiteflies > 3 adults per top five leaves; CGM – average severity > 3; CM – incidence greater than 20%), then a pesticide application is immediately made. For both whiteflies and CM this is imidacloprid, while for CGM it is bifenthrin.

8. Certification for quality control. Certification inspections are conducted by the Tanzania Official Seed Certification Institute (TOSCI) during early growth and pre-harvest stages. In addition to checking incidence/damage of all major cassava pests and diseases and ensuring that levels are below prescribed tolerance values, leaves from 200 plants/ha are sampled for diagnostic testing of CBSIs using sensitive real-time PCR techniques (Adams *et al.*, 2013).

IPM for QDPM cassava seed production

At the QDPM level, elements 1, 4 and 8 described above are emphasized. Varieties are the same as those produced by pre-basic sites, there are isolation requirements for seed certification (100 m for certified sites and 50 m for QDPM fields), and planting material is sourced (element 3) from higher-level clean-seed sites. All fields must be certified by TOSCI if the planting material is to be distributed/sold. Certification tolerance levels are higher for lower level sites, since the risk of pest/disease spread declines as distribution of planting material becomes increasingly localized. Training guides have been developed in both English and Kiswahili (the national language in Tanzania) that illustrate the main pests and diseases of cassava, and how to manage them when producing cassava planting material for dissemination/sale. Capacity building for all stakeholders involved in the clean cassava seed system in Tanzania has been a vital element in the successful implementation of the approach.

Although the IPM approach for clean seed production in Tanzania is still being fine-tuned, it has already demonstrated effectiveness as a means to produce and disseminate high-quality cassava planting material. Consequently, similar approaches are being piloted in other parts of east, west and southern Africa.

7.3.2 Case study 2: sweet potato weevil management in Cuba

Background

An example of a successful IPM programme in sweet potato is the management of *C. formicarius* in Cuba. Losses attributed to this pest dropped from 45% to 6% through mass trapping of adult male weevils (Lagnaoui *et al.*, 2000). Farmers were applying up to 14 insecticidal sprays in a single cropping season against *C. formicarius* before 1990. However, Cuba reduced its import of chemical insecticides by 63% in the 1990s due to economic challenges, which led to significant increases in root damage caused by *C. formicarius*. Consequently, root production fell by 50–60% between 1991 and 1992. This dramatic increase in *C. formicarius* damage in the absence of chemical pesticides forced the Cuban government to explore the use of IPM to manage this pest. It is reported that *C. formicarius* damage was reduced to 12% in the pilot site without the use of chemical insecticides. This IPM programme involved the following methods.

1. Clean seed. The use of clean (insect-free) vines dipped in a solution of entomopathogenic fungus (*Beauveria bassiana*) prior to planting. This killed all weevils present in 2–3 days (Lagnaoui *et al.*, 2000). Under the guidance of the national research organization (Instituto Nacional de Investigaciones de Viandas Tropicales), local cooperatives were able to produce large quantities of the clean vines for planting.

2. Biological control. Two species of predatory ants (*Tetramorium guineense* Fabricius and *Pheidole megacephala* (Fabricius)) were used to control weevils. About 100 ant nests per ha were needed to reduce *C. formicarius* damage at harvest from 6–10% to 2.5–3.5%.

3. Mass trapping. The commercially available sex pheromone (Z)-3-dodecen-1-ol (E)-2-butenoate was used in pheromone traps. These successfully attracted adult male weevils, resulting in reduced populations in the field and significantly less root infestation and damage.

4. Crop rotation. Rotating sweet potato with potato or maize over a two-year period led to a reduction in *C. formicarius* damage in the sweet potato (Suris *et al.*, 1995 as cited by Nicholls *et al.*, 2002).

5. Use of irrigation. Irrigation water provided the required moisture in the soil for *B. bassiana* to thrive and prevented the soil from cracking hence preventing the entry/access of *C. formicarius* to sweet potato roots. Covering of soil cracks through re-hilling at the time of weeding has also been very effective for reducing the number of eggs laid in roots by the two African *Cylas* species.

6. Host-plant resistance. Early maturing and deep rooting varieties, such as INIVIT B-88 and Yabu-8, were shown to be less damaged by *C. formicarius* (Lima and Morales, 1992, as cited by Lagnaoui *et al.*, 2000).

7. Field sanitation. This involved destroying crop residues, uprooting volunteer plants and avoiding planting new fields next to old infested fields.

Factors that contributed to the success of this IPM programme

1. Large-scale adoption by most sweet potato farmers of the IPM package because of the effectiveness against three life stages (larva, pupa and adult), its availability and the low cost of *B. bassiana*. The fungus has also proved to be safe to both humans and animals.

2. Adequate commitment by the government for the research effort, which enabled the research team to develop the IPM package and make the programme a success. It is noteworthy that the Cuban government has had a policy on IPM since 1982.

3. Creation of the National Program for the Production of Biological Agents by the Ministry of Agriculture in 1988, which led to the expansion of the rearing laboratory network (Centros de Reproduccion de Entomofagos y Entomopatogenos) from 82 in 1992 to 227 in 1994.

4. Chemical insecticides were not an option, since they were not available. There was therefore no chance that they could interfere with biocontrol agents such as the predatory ants.

5. Weather conditions (moderate to high humidity) were conducive for *B. bassiana* multiplication in the field and subsequent colonization of *C. formicarius*.

6. The presence of a successful public–private partnership that facilitated the identification and mass production of the local *B. bassiana* strains.

7. The commercially available sex pheromones that were effective at attracting *C. formicarius* males also significantly reduced weevil damage. Up to 16 pheromone traps per ha were used.

No single method has proved to be successful for the management of *C. formicarius*, as is the case for *Cylas* spp. in other countries and continents where sweet potato is cultivated. IPM therefore continues to be the most promising approach for the management of these elusive pests. In east Africa, cultural practices such as hilling-up and early harvesting are key IPM components in areas where *Cylas* spp. are particularly damaging, and significant recent effort has been devoted to the identification and deployment of sources of resistance to further enhance the effectiveness of integrated weevil management strategies.

7.3.3 Case study 3: IPM for the development of sustainable yam seed systems

Background

Sustainable seed systems, which consistently supply seed of high quality and which farmers trust, are fundamentally important to improving yam productivity, especially in the African yam belt, where most yam seed material is infected with viruses and plant-parasitic nematodes (Aighewi *et al.*, 2015). Poor-quality seed material is a key cause for the under-performance of yam in the yam belt. Poor seed affects sprouting ability and plant vigour, and gives rise to infected tubers that deteriorate during storage. It may also perpetuate the disease cycle into the following season if the infected tubers are used as planting material. Consequently, the cost of planting materials is high and may constitute up to 50% of the total production cost (Aighewi *et al.*, 2003; Kambaska *et al.*, 2009). This case study is based on the outcomes of a specific project that aimed to improve the health of seed yams.

To improve seed yam systems, the Yam Minisett Technology (YMT) was developed in the 1970s. Farmers rejected YMT for various reasons, including the high labour costs and the high losses of seed pieces (Agbaje and Oyegbami, 2005). In order to address some of these issues, a project entitled 'Evaluation and Promotion of Crop Protection Practices for Clean Seed Yam Production Systems in Central Nigeria' was implemented in Nigeria between 2003 and 2005. This case study summarizes some of the results of this project (see Odu *et al.*, 2016).

Following feedback on the YMT, the project modified and relabelled it the Adapted Yam Minisett Technique (AYMT). The size of the minisetts was increased to 80–100 g. Several potential chemical seed treatments were evaluated, the most effective (a combination of the fungicide mancozeb and the insecticide diazinon) was applied to minisetts for protection from pests and diseases, and setts were planted directly in the field, bypassing pre-planting in nurseries (McNamara *et al.*, 2012; Morse and McNamara, 2015). The AYMT was designed to address the concerns of farmers who had previously rejected YMT. It proved simple and cost-effective, with high germination rates, and quickly attracted farmers' attention, becoming a preferred technique for seed yam production in west Africa (Aighewi *et al.*, 2015). It is now being scaled out and promoted through a regional project entitled 'Yam Improvement for Income and Food Security in West Africa' (www.iita.org/web/yiifswa). The model adopted was a simple method which showcased the new techniques on farmers' trial plots and on-station to improve farmers' awareness. The model was based on an improved understanding of farmer knowledge and practices and on producing appropriate extension materials. Farmers were also encouraged to select seed material from healthy plants, which did not present virus symptoms. In order to create awareness, simple leaflets and posters about the pests and diseases, the damage they cause and a step-by-step procedure to produce healthy seed yams were developed and distributed. These were provided to participating farmers as well as being made available at demonstration field days, which welcomed neighbouring farmers. Field days included a demonstration of how to prepare and treat yams for AYMT; the results and merits were discussed with participating farmers and aired on local radio and television.

The combined fungicide and insecticide treatment of minisetts was highly effective in producing high-quality yam plants when evaluated in on-station trials. In evaluation trials in farmers' fields, plants from treated setts produced more and heavier tubers than the other treatments. The seed yams produced from the pesticide-treated setts were of higher quality and weight, stored better, and had less rotting and deterioration than the yams from untreated setts. Ware yams produced from the healthy tubers (derived from treated setts) were also more and heavier than those obtained from seed derived from other treatments. On-farm, seed production was double that of the farmer treatment (Claudius-Cole *et al.*, 2004).

The project confirmed that healthy seed yams are expensive and in short supply and that the lack of household capital and labour are major constraints to yam production. Prior to this project, two moderately virus-resistant *D. rotundata* varieties (TDr 89/02665 and TDr 98/00804) were multiplied and distributed free of charge to farmers for on-farm testing and subsequently used to produce healthy seed yams. Farmers were willing to produce their own seed yams by adapting technology that had been learned from this project. They were also willing to buy seed yams to make up for the ones they were unable to produce (McNamara *et al.*, 2012; Odu *et al.*, 2016).

Benefits for farmers (Odu et al., 2016)

ADVANTAGES
1. Farmers can use their own seed.
2. AYMT enables farmers to plant 5–10 times more yams than traditional methods.
3. Better and higher yields than when produced through their traditional methods.
4. Tubers produced using AYMT have less pest and disease infection.
5. Offers a more profitable enterprise for farmers leading to improved livelihoods.

DISADVANTAGES
1. The high cost of labour.
2. The technique depends on chemicals which may not always be available.
3. Seed yams derived from AYMT are not resistant to pests and diseases and thus good management is key to realizing all potential benefits.
4. Care is required to prevent pesticide contamination and ensure the safe disposal of excess chemicals and packages.
5. Not suitable for organic farming.*

Key criteria for success identified during the intervention were as follows.

1. Good integration among the stakeholders (farmers, extension workers, NGOs, donor agency, researchers, journalists etc.).
2. Development, testing and distribution of simple extension materials that met farmers' needs.
3. Connecting seed yam growers to credit providers (in this case, GORTA).

---

* Hot water therapy provides an organic alternative for disinfesting/cleansing seed sett material for producing healthy seed, but is more complicated for farmers to manage and also led to high initial losses following treatment (Coyne *et al.*, 2010a).

4.   Connecting seed yam growers and linking the project to a reliable and reputable agro-input supplier with a national network, for consistent and reliable chemical supply.

5.   Participation of dynamic farmers in on-farm trials.

6.   Creating a sense of ownership for the seed yam farmers.

7.   Creating a platform for lobbying policy makers.

7.4 Future Outlook

The tropical root and tuber crops are certain to play a central role in assuring sustainable and nutritious food systems in future years. The crops are expected to perform better than most of the cereal and legume crops under predicted climate change scenarios, and new work on biofortification coupled with the promotion of existing vitamin- and mineral-rich varieties will expand their appeal and value. Tropical RTCs are currently some of the least commercialized of the major food staples. Growing industrialization in the developing world, however, is already driving change in the RTCs sub-sectors and the near future is likely to see an increasing diversification in value chains offering new commercial opportunities for growers. New technology will become more widely used at all levels, including: mechanical equipment for cultivation and processing; tissue culture, aeroponics and other rapid propagation systems for seed; genetic transformation and gene-editing techniques for germplasm improvement; and more sensitive, faster and cheaper field- and laboratory-based tools for diagnostics. All of these changes will have a profound effect on the application of IPM. The sophisticated IPM systems that are now being successfully applied for pest management in parts of the developed world, including the sweet potato example from Cuba described in this chapter, are based on a thorough knowledge of pests/diseases and the interactions with the crop plants that they attack. Current understanding of the wide gamut of pests and diseases affecting RTCs in the tropics is improving, but remains well short of what would be required to build robust knowledge-based IPM strategies. Tackling this deficiency will be a key target for the near future. As commercial demands for increased productivity increase, there is certain to be a temptation for 'quick-fix' solutions, many of which may involve the use of pesticides. In this respect, it will be important to learn lessons from the experience of temperate regions where balanced IPM strategies are now favoured which involve the rational and targeted use of pesticides coupled with the encouragement of natural control processes. An important facet of this is knowing how to calculate the economic benefit of biological control processes, which are frequently undervalued (Naranjo *et al.*, 2015).

The future will bring increasing travel between countries and continents, which in turn will provide greater opportunity for inadvertent spread of pests/diseases. IPM practitioners will therefore need to work closely with quarantine and plant health authorities at country and continental levels to ensure that the likelihood of such spread is minimized. Arguably, the greatest technological transformation in the world over the last decades has been the explosion in information technology capacity. The benefits of this technology are already being felt by researchers, extensionists and farmers working to manage pests and diseases of tropical RTCs, and some examples have been described in this chapter. This ongoing development offers huge promise as a driver for information- and knowledge-sharing. RTCs stakeholders will need to maximize opportunities for applying this technology in building robust IPM strategies if they are to ensure a sustainable and expanding future role for these vitally important tropical crops.

References

Abang, M.M., Fagbola, O., Smalla, K. and Winter, S. (2005) Two genetically distinct populations of *Colletotrichum* causing anthracnose disease of yam (*Dioscorea* spp.). *Journal of Phytopathology* 153, 137–142.

Adams, I.P., Abidrabo, P., Miano, D.W., Alicai, T., Kinyua, Z.M. *et al.* (2013) High throughput real-time RT-PCR assays for specific detection of cassava brown streak disease causal viruses, and their application to testing of planting material. *Plant Pathology* 62, 233–242.

Agbaje, G.O. and Oyegbami, A. (2005) Survey on the adoption of yam minisett technology in south-western Nigeria. *Journal of Food, Agriculture and Environment* 3, 134–137.

Aighewi, B.A., Akoroda, M.O. and Asiedu, R. (2003) Seed yam (*Dioscorea rotundata* Poir.) production, storage and quality in selected yam zones of Nigeria. *African Journal of Root and Tubers* 5, 20–23.

Aighewi, B.A., Asiedu, R., Maroya, N. and Balogun, M. (2015) Improved propagation methods to raise the productivity of yam (*Dioscorea rotundata* Poir.). *Food Security* 7, 823–834.

Akano, A.O., Dixon, A.G.O., Mba, C., Barrera, E. and Fregene, M. (2002) Genetic mapping of a dominant gene conferring resistance to cassava mosaic disease. *Theoretical and Applied Genetics* 105, 521–535.

Alvarez, E., Mejía, J.F., Llano, G., Loke, J., Calari, A., Duduk, B. and Bertaccini, A. (2009) Detection and molecular characterization of Phytoplasma associated with frogskin disease in cassava. *Plant Disease* 93, 1139–1145.

Alvarez, E., Pardo, J.M. and Truke, M.J. (2014) Detection and identification of 'Candidatus Phytoplasma asteris'-related phytoplasma associated with a witches' broom disease of cassava in Cambodia. *Phytopathology* 104 (Suppl. 3), S3.7.

Anyanga, M.O., Muyinza, H., Talwana, H., Hall, D.R., Farman, D.I., Ssemakula, G.N. *et al.* (2013) Resistance to the weevils *Cylas puncticollis* and *Cylas brunneus* conferred by sweetpotato root surface compounds. *Journal of Agricultural and Food Chemistry* 61 (34), 8141–8147.

Arias, B. and Bellotti, A.C. (1984) Pérdidas en rendimiento (dao simulado) causadas por *Errinyis ello* (L.) y niveles críticos de poblaciones en diferentes etapas de desarrollo en tres clones de yuca. *Revista Colombiana de Entomologia* 10, 28–35.

Arias, B. and Bellotti, A.C. (1985) Aspectos ecológicos y de manejo de *Cyrtomenus bergi* Froeschner, chinche de la viruela en el cultivo de la yucca (*Manihot esculenta* Crantz). *Revista Colombiana de Entomologia* 11, 42–46.

Aritua, V., Legg, J.P., Smit, N.E.J.M. and Gibson, R.W. (1999) Effect of local inoculum on the spread of sweet potato virus disease: widespread cultivation of a resistant sweet potato cultivar limits infection of susceptible cultivars. *Plant Pathology* 48, 655–661.

Asala, S., Alegbejo, M.D., Kashina, B., Banwo, O.O., Asiedu, R. and Kumar, P.L. (2012) Distribution and incidence of viruses infecting yam (*Dioscorea* spp.) in Nigeria. *Global Journal of Bioscience and Biotechnology* 1, 163–167.

Ashamo, M.O. and Odeyemi, O.O. (2004) Effect of temperature on the development of the yam moth, *Dasyses rugosella* Stainton (Lepidoptera: Tineidae). *Journal of Stored Products Research* 40, 95–102.

Asiimwe, P., Kyamanywa, S., Gerling, D. and Legg, J.P. (2007) Evaluation of *Serangium* n. sp. (Coleoptera: Coccinellidae) a predator of *Bemisia tabaci* (Homoptera: Aleyrodidae) on cassava. *Journal of Applied Entomology* 131 (2), 76–80.

Bart, R., Cohn, M., Kassen, A., McCallum, E.J., Shybut, M. *et al.* (2012) High-throughput genomic sequencing of cassava bacterial blight strains identifies conserved effectors to target for durable resistance. *Proceedings of the National Academy of Sciences* 109 (32), 13130–13130.

Bellotti, A.C., Arias, B. and Reyes, J.A. (1990) Biological control of the cassava hornworm, *Erinnyis ello* (Lepidoptera: Sphingidae) with emphasis on the hornworm virus. In: Howeler, R.H. (ed.) *Proceedings of the 8th Symposium of the International Society of Tropical Root and Tuber Crops, Bangkok, Thailand, 1988*, pp. 354–362.

Bellotti, A.C., Braun, A.R., Arias, B., Castillo, J.A. and Guerrero, J.M. (1994) Origin and management of neotropical cassava arthropod pests. *African Crop Science Journal* 2, 407–418.

Berry, S.D., Fondong, V., Rey, C., Rogan, D., Fauquet, C.M. and Brown, J.K. (2004) Molecular evidence for five distinct *Bemisia tabaci* (Homoptera: Aleyrodidae) geographic haplotypes associated with cassava in sub-Saharan Africa. *Annals of the Entomological Society of America* 97, 852–859.

Bock, K.R. and Woods, R.D. (1983) The etiology of African cassava mosaic disease. *Plant Disease* 67, 994–995.

Bridge, J., Coyne, D. and Kwoseh, C.K. (2005) Nematode parasites of tropical root and tuber crops. In: Luc, M., Sikora, R. and Bridge, J. (eds) *Plant Parasitic Nematodes in Subtropical and Tropical Agriculture.* 2nd edn. CAB International, Wallingford, UK, pp. 221–258.

Campo, B.V.H., Hyman, G. and Bellotti, A. (2011) Threats to cassava production: known and potential geographic distribution of four key biotic constraints. *Food Security* 3, 329–345.

Carabali, A., Bellotti, A.C., Montoya-Lerma, J. and Cuellar, M.E. (2005) Adaptation of *Bemisia tabaci* biotype B (Gennadius) to cassava, *Manihot esculenta* (Crantz). *Crop Protection* 24, 643–649.

Carson, R. (1962) *Silent Spring.* Houghton Mifflin Company, Boston, Massachusetts.

Carvajal-Yepes, M., Olaya, C., Lozano, I., Cuervo, M., Castaño, M. and Cuellar, W.J. (2014) Unravelling complex viral infections in cassava (*Manihot esculenta* Crantz) from Colombia. *Virus Research* 186, 76–86.

Clark, C.A., Davis, J.A., Abad, J.A., Cuellar, W.J., Fuentes, S. *et al.* (2012) Sweetpotato viruses: 15 years of progress on understanding and managing complex diseases. *Plant Disease* 96, 168–185.

Clark, C.A., Ferrin, D.M., Smith, T.P. and Holmes, G.J. (2013) *Compendium of Sweetpotato Diseases, Pests, and Disorders*, 2nd edn. APS, St. Paul, Minnesota.

Claudius-Cole, A.O., Coyne, D., Kenyon, L., Ayodele, M., McNamara, N. and Morse, S. (2004) Seed yam production systems: assessment of various pre-plant treatments of setts for production of nematode free material. First West and Central African Nematology Meeting, Douala, Cameroon, 8–10 November 2004.

Claudius-Cole, A.O., Fawole, B., Asiedu, R. and Coyne, D.L. (2014) Management of *Meloidogyne incognita* in yam-based cropping systems with cover crops. *Crop Protection* 63, 97–102.

Coursey, D.G. (1967) *Yams – An Account of the Nature, Origins, Cultivation and Utilization of the Useful Members of the Dioscoreaceae.* Tropical Agricultural Series. Longman, Green and Co. Ltd, London.

Coyne, D.L. (2005) Pests, disease and the agro-ecosystem. In: Stathers, T., Namanda, S., Mwanga, R.O.M., Khisa, G. and Kapinga, R. (eds) *Manual for Sweetpotato Integrated Production and Pest Management Farmer Field Schools in Sub-Saharan Africa.* International Potato Centre, Kampala, Uganda, pp. 64–65.

Coyne, D.L. and Namaganda, J. (1996) Plant parasitic nematode incidence on root and tuber crops in Masindi District, Uganda. *African Journal of Root and Tuber Crops* 1, 4–7.

Coyne, D.L., Khizzah, W. and Whyte, J. (2004) Root knot nematode damage to cassava in Kenya. *Roots* 9, 3–5.

Coyne, D.L., Toko, M., Andrade, M., Hanna, R., Sitole, A. *et al.* (2006) *Meloidogyne* spp. and associated galling and damage on cassava in Kenya and Mozambique. *African Plant Protection* 12, 35–36.

Coyne, D.L., Claudius-Cole, A.O., Kenyon, L. and Baimey, H. (2010a) Differential effect of hot water treatment on whole tubers versus cut setts of yam (*Dioscorea* spp.). *Pest Management Science* 6, 385–389.

Coyne, D.L., Claudius-Cole, A.O. and Kikuno, H. (2010b) *Sowing the Seeds of Better Yam.* SP-IPM Technical Innovation Brief 7. IITA, Ibadan, Nigeria.

Delalibera Jr, I., Hajek, A.E. and Humber, R.A. (2004) *Neozygites tanajoae* sp. Nov., a pathogen of the cassava green mite. *Mycologia* 96, 1002–1009.

Dixon, A.G.O., Bandyopadhyay, R., Coyne, D., Ferguson, M., Ferris, R.S.B. *et al.* (2003) Cassava: from a poor farmer's crop to a pacesetter of African rural development. *Chronica Horticulturae* 43, 8–14.

Dutt, N., Briddon, R.W. and Dasgupta, I. (2005) Identification of a second begomovirus, Sri Lankan cassava mosaic virus, causing cassava mosaic disease in India. *Archives of Virology* 150, 2101–2108.

Emehute, J.K.U., Ikotun, T., Nwauzor, E.C. and Nwokocha, H.N. (1998) Crop protection. In: Orkwor, G.C., Asiedu, R. and Ekanayake, I.J. (eds) *Food Yams: Advances in Research.* IITA and NRCRI, Ibadan, Nigeria, pp. 187–214.

FAOSTAT (2016) FAO database. Food and Agriculture Organization of the United Nations, Rome. Available at: http://faostat.fao.org/site/567/default.aspx#ancor (accessed 31 October 2016).

Fei, Z. (2015) Determining the pan-African sweet potato virome: understanding virus diversity, distribution and evolution and their impacts on sweet potato production in Africa. Available at: http://bioinfo.bti.cornell.edu/lab/homepage/research.shtml (accessed 23 February 2016).

Fondong, V., Thresh, J.M. and Zok, S. (2002) Spatial and temporal spread of cassava mosaic virus disease in cassava grown alone and when intercropped with maize and/or cowpea. *Journal of Phytopathology* 150, 365–374.

Frison, E.A. and Feliu, E. (eds) (1991) *FAO/IBPGR Technical Guidelines for the Safe Movement of Cassava Germplasm*. Food and Agriculture Organization of the United Nations, Rome/International Board for Plant Genetic Resources, Rome.

Fuglie, K.O. (2007) Priorities for sweet potato research in developing countries: results of a survey. *HortScience* 42, 1200–1206.

Gibson, R. and Kreuze, J. (2015) Degeneration in sweet potato due to viruses, virus cleaned planting material and reversion: a review. *Plant Pathology* 64, 1–15.

Graziosi, I., Minato, N., Alvarez, E., Ngo, D.T., Hoat, T.X., Aye, T.M. and Wyckhuys, K.A. (2016) Emerging pests and diseases of South-east Asian cassava: a comprehensive evaluation of geographic priorities, management options and research needs. *Pest Management Science* 72, 1071–1089.

Green, K.R. and Simons, S.A. (1994) 'Dead skin' on yams (*Dioscorea alata*) caused by *Colletotrichum gloeosporioides*. *Plant Pathology* 43, 1062–1065.

Guastella, D., Lulah, H., Tajebe, L.S., Cavalieri, V., Evans, G.A., Pedata, P.A., Rapisarda, C. and Legg, J.P. (2014) Survey on whiteflies and their parasitoids in cassava mosaic pandemic areas of Tanzania using morphological and molecular techniques. *Pest Management Science* 71, 383–394.

Hillocks, R.J and Wydra, K. (2002) Bacterial, fungal and nematode diseases. In: Bellotti, A.C., Hillocks, R.J. and Thresh, J.M. (eds) *Cassava: Biology, Production and Utilization*. CAB International, Wallingford, UK, pp. 261–279.

IITA (2011) DEWN: a novel surveillance system. Available at: http://r4dreview.iita.org/index.php/ 2011/04/14/dewn-a-novel-surveillance-system/ (accessed 26 October 2017).

IITA (2014) 5CP Updates: highlights of the progress of the 'New Cassava Varieties and Clean Seed to Combat CBSD and CMD Project': piloting a clean seed system. IITA, Dar es Salaam. Available at: www.iita.org/c/document_library/get_file?uuid=c74b9cd7-07c3-47d4-a460-f15acd908076&groupId=25357 (accessed 31 October 2016).

IITA (2016) *Cassava Disease Surveillance Network*. Available at: http://cassavadiseasenet.org (accessed 31 October 2016).

Jarvis, A., Ramirez-Villegas, J., Campo, B.V.H. and Navarro-Racines, C. (2012) Is cassava the answer to African climate change adaptation? *Tropical Plant Biology* 5, 9–29.

Jennings, D. (1957) Further studies in breeding cassava for virus resistance. *East African Agricultural Journal* 22, 213–219.

Kambaska, K.B., Santilata, S., Trinanth, M. and Debashrita, P. (2009) Response of vine cuttings to rooting in different months in three *Dioscorea* species. *Journal of Natural Sciences* 7, 48–51.

Kenyon, L., Lebas, B.S.M. and Seal, S.E. (2008) Yams (*Dioscorea* spp.) from the South Pacific Islands contain many novel badnaviruses: implications for international movement of yam germplasm. *Archives of Virology* 153, 877–889.

Kolombia, Y.A., Viaene, N., Kumar, L., Bert, W. and Coyne, D. (2015) Survey of the incidence and distribution of damages caused by nematodes on yam (*Dioscorea* spp.) in Nigeria. In: *Book of Abstracts 47th Annual Meeting of ONTA, Varadero, Cuba, 24–29 May 2015*.

Kolombia, Y.A., Kumar, L., Claudius-Cole, A.O., Karssen, G., Viaene, N., Coyne, D.L. and Bert, W. (2016) First report of galls on yam (*Dioscorea* spp.) caused by *Meloidogyne enterolobii* in Nigeria. *Plant Disease*. Available at: http://dx.doi.org/10.1094/PDIS-03-16-0348-PDN (accessed 3 August 2017).

Lagnaoui, A., Cisneros, F., Alcazar, J. and Morales, F. (2000) A sustainable pest management strategy for sweet potato weevil in Cuba: a success story. *Extension Bulletin (ASPAC/FFTC) 2000*, no. 493, pp. 1–7.

Lebot, V. (2009) *Tropical Root and Tuber Crops: Cassava, Sweet Potato, Yams, Aroids*. CAB International, Wallingford, UK.

Legg, J.P. and Fauquet, C.M. (2004) Cassava mosaic geminiviruses in Africa. *Plant Molecular Biology* 56, 585–599.

Legg, J.P., French, R., Rogan, D., Okao-Okuja, G. and Brown, J.K. (2002) A distinct, invasive *Bemisia tabaci* (Gennadius) (Hemiptera: Sternorrhyncha: Aleyrodidae) genotype cluster is associated with the epidemic of severe cassava mosaic virus disease in Uganda. *Molecular Ecology* 11, 1219–1229.

Legg, J.P., Whyte, J., Kapinga, R. and Teri, J. (2005) Special topics on pest and disease management: management of the cassava mosaic disease pandemic in East Africa. In: Anderson, P.K. and Morales, F. (eds) *Whiteflies and Whitefly-borne Viruses in the Tropics: Building a Knowledge Base for Global Action.* Centro Internacional de Agricultura Tropical, Cali, Colombia, pp. 332–338.

Legg, J.P., Jeremiah, S.C., Obiero, H.M., Maruthi, M.N., Ndyetabula, I. *et al.* (2011) Comparing the regional epidemiology of the cassava mosaic and cassava brown streak pandemics in Africa. *Virus Research* 159, 161–170.

Legg, J.P., Shirima, R., Tajebe, L.S., Guastella, D., Simon, B., Nsami, E., Chikoti, P. and Rapisarda, C. (2014a) Biology and management of *Bemisia* whitefly vectors of cassava virus pandemics in Africa. *Pest Management Science* 70, 1446–1453.

Legg, J.P., Sseruwagi, P., Boniface, S., Okao-Okuja, G., Shirima, R. *et al.* (2014b) Spatio-temporal patterns of genetic change amongst populations of cassava *Bemisia tabaci* whiteflies driving virus pandemics in East and Central Africa. *Virus Research* 186, 61–75.

Legg, J.P., Lava Kumar, P., Makeshkumar, T., Ferguson, M., Kanju, E. *et al.* (2015) Cassava virus diseases: biology, epidemiology and management. *Advances in Virus Research* 91, 85–142.

Lozano, J.C. and Booth, R.H. (1974) Diseases of cassava (*Manihot esculenta* Crantz). *PANS* 20, 30–54.

Manyong, V.M., Dixon, A.G.O., Makinde, K.O., Bokanga, M. and Whyte, J. (2000) *The Contribution of IITA-improved Germplasm to Food Security in Sub-Saharan Africa: An Impact Study.* IITA, Ibadan, Nigeria.

Maroya, N., Asiedu, R., Kumar, P.L., Mignouna, D., Lopez-Montes, A. *et al.* (2014) Yam improvement for income and food security in West Africa: effectiveness of a multi-disciplinary and multi-institutional team-work. *Journal of Root Crops* 40, 85–92.

Mbanzibwa, D.R., Tian, Y.P., Tugume, A.K., Patil, B.L., Yadav, J.S. *et al.* (2011) Evolution of cassava brown streak disease-associated viruses. *Journal of General Virology* 92, 974–987.

McNamara, N., Morse, S., Ugbe, U.P., Coyne, D. and Claudius-Cole, A. (2012) Facilitating healthy seed yam entrepreneurship in the Niger River system in Nigeria: the value of 'Research-Into-Use'. *Outlook on Agriculture* 41, 257–263.

Mohamed, R.A. (2002) Role of open quarantine in regional germplasm exchange. In: Legg, J.P. and Hillocks, R.J. (eds) *Cassava Brown Streak Virus Disease: Past, Present, and Future.* Proceedings of an International Workshop, Mombasa, Kenya, 27–30 October. Natural Resources International Limited, Aylesford, UK, pp. 28–36.

Morse, S. and McNamara, N. (2015) The adapted yam minisett technique for producing clean seed yams (*Dioscorea rotundata*): agronomic performance and varietal differences under farmer-managed conditions in Nigeria. *Experimental Agriculture* 51, 467–482.

Mudiope, J., Coyne, D.L., Adipala, E. and Talwana, H.A.L. (2012) Damage to yam (*Dioscorea* spp.) by root-knot nematode (*Meloidogyne* spp.) under field and storage conditions in Uganda. *Nematropica* 42, 137–145.

Musana, P., Okonya, J.S., Mujica, N., Carhuapoma, P. and Kroschel, J. (2016) Sweetpotato weevil, *Cylas brunneus* (Fabricius). In: Kroschel, J., Mujica, N., Carhuapoma, P. and Sporleder, M. (eds) *Pest Distribution and Risk Atlas for Africa. Potential Global and Regional Distribution and Abundance of Agricultural and Horticultural Pests and Associated Biocontrol Agents under Current and Future Climates.* International Potato Center (CIP), Lima, Peru, pp. 64–73.

Naranjo, S.E. and Ellsworth, P.C. (2009) 50 years of the integrated control concept: moving the model and implementation forward in Arizona. *Pest Management Science* 65, 1267–1286.

Naranjo, S.E., Ellsworth, P.C. and Frisvold, G.B. (2015) Economic value of biological control in integrated pest management of managed plant systems. *Annual Review of Entomology* 60, 621–645.

Neuenschwander, P. (1994) Control of the cassava mealybug in Africa: lessons from a biological control project. *African Crop Science Journal* 2, 369–384.

Neuenschwander, P. (2001) Biological control of the cassava mealybug in Africa: a review. *Biological Control* 21, 214–229.

Nicholls, C.I., Pérez, N., Vasquez, L. and Altieri, M.A. (2002) The development and status of biologically based integrated pest management in Cuba. *Integrated Pest Management Reviews* 7, 1–16.

Ntui, V.O., Kong, K., Khan, R.S., Igawa, T., Janavi, G.J., Rabindran, R., Nakamura, I. and Mii, M. (2015) Resistance to *Sri Lankan cassava mosaic virus* (SLCMV) in genetically engineered cassava cv. KU50 through RNA silencing. *PLOS ONE* 10 (4), 1–23.

Odu, B.O., Coyne, D. and Kumar, L. (2016) Adapting a yam seed technique to meet farmers' criteria. In: Andrade-Piedra, J., Bentley, J., Almekinders, C., Jacobsen, K., Walsh, S. and Thiele, G. *RTB Working Paper: Case Studies of Root, Tuber and Banana Seed Systems.* RTB, International Potato Center, Lima.

Okigbo, R.N. and Ikediugwu, F.E.O. (2000) Studies on biological control of postharvest rot of yams (*Dioscorea* spp.) with *Trichoderma viride. Journal of Phytopathology* 148, 351–355.

Okogbenin, E., Porto, M.C.M., Egesi, C., Mba, C., Espinosa, E. *et al.* (2007) Marker-assisted introgression of resistance to cassava mosaic disease into Latin American germplasm for the genetic improvement of cassava in Africa. *Crop Science* 47, 1895–1904.

Okonya, J. and Kroschel, J. (2013) Pest status of *Acraea acerata* Hew. and *Cylas* spp. in sweetpotato (Ipomoea batatas (L.) Lam.) and incidence of natural enemies in the Lake Albert Crescent agroecological zone of Uganda. *International Journal of Insect Science* 5, 41–46.

Okonya, J.S., Mujica, N., Carhuapoma, P. and Kroschel, J. (2016a) Sweetpotato weevil, *Cylas puncticollis* (Boheman 1883). In: Kroschel, J., Mujica, N., Carhuapoma, P. and Sporleder, M. (eds) *Pest Distribution and Risk Atlas for Africa. Potential Global and Regional Distribution and Abundance of Agricultural and Horticultural Pests and Associated Biocontrol Agents under Current and Future Climates.* International Potato Center (CIP), Lima, pp. 54–63.

Okonya, J.S., Mujica, N., Carhuapoma, P. and Kroschel, J. (2016b) Sweetpotato butterfly, *Acraea acerata* (Hewitson 1874). In: Kroschel, J., Mujica, N., Carhuapoma, P. and Sporleder, M. (eds) *Pest Distribution and Risk Atlas for Africa. Potential Global and Regional Distribution and Abundance of Agricultural and Horticultural Pests and Associated Biocontrol Agents under Current and Future Climates.* International Potato Center (CIP), Lima, pp. 74–84.

Onyeka, T.J., Petro, D., Ano, G., Etienne, S. and Rubens, S. (2006) Resistance in water yam (*Dioscorea alata*) cultivars in the French West Indies to anthracnose disease based on tissue culture-derived whole-plant assay. *Plant Pathology* 55, 671–678.

Otim, M., Legg, J., Kyamanywa, S., Polazsek, A. and Gerling, D. (2006) Population dynamics of *Bemisia tabaci* (Homoptera: Aleyrodidae) parasitoids on cassava mosaic disease-resistant and susceptible varieties. *Biocontrol Science and Technology* 16, 205–214.

Parsa, S., Kondo, T. and Winotai, A. (2012) The cassava mealybug (*Phenacoccus manihoti*) in Asia: first records, potential distribution, and an identification key. *PLOS ONE* 7 (10), e47675.

Parsa, S., Hazzi, N.A., Chen, Q., Lu, F., Herrera Campo, B.V., Yaninek, J.S. and Vásquez-Ordóñez, A.A. (2015) Potential geographic distribution of two invasive cassava green mites. *Experimental and Applied Acarology* 65, 195–204.

Petro, D., Onyeka, T.J., Etienne, S. and Rubens, S. (2011) An intraspecific genetic map of water yam (*Dioscorea alata* L.) based on AFLP markers and QTL analysis for anthracnose resistance. *Euphytica* 179, 405–416.

Reddy, G.V., Wu, S., Mendi, R.C. and Miller, R.H. (2014) Efficacy of pheromone trapping of the sweetpotato weevil (Coleoptera: Brentidae): based on dose, septum age, attractive radius, and mass trapping. *Environmental Entomology* 43 (3), 767–773.

Schulthess, F., Neuenschwander, P. and Gounou, S. (1997) Multi-trophic interactions in cassava, *Manihot esculenta*, cropping systems in the subhumid tropics of West Africa. *Agriculture, Ecosystems and Environment* 66, 211–222.

Sclar, D.C., Gerace, D. and Cranshaw, W.S. (1998) Observations of population increases and injury by spider mites (Acari: Tetranychidae) on ornamental plants treated with imidacloprid. *Journal of Economic Entomology* 91, 250–255.

Seal, S., Turaki, A., Muller, E., Kumar, P.L., Kenyon, L. *et al.* (2014) The prevalence of badnaviruses in West African yams (*Dioscorea cayenensis-rotundata*) and evidence of endogenous pararetrovirus sequences in their genomes. *Virus Research* 186, 144–154.

Simons, S.A. (1997) Root and tuber crops. In: Hillocks, R.J. and Waller, J.M. (eds) *Soilborne Diseases of Tropical Crops.* CAB International, Wallingford, UK, pp. 109–149.

Smit, N.E.J.M. (1997) Integrated pest management for sweetpotato in Eastern Africa. PhD thesis. Wageningen University, Wageningen, The Netherlands.

Sorensen, K.A. (2009) Sweetpotato insects: identification, biology and management. In: Loebenstein, G. and Thottappilly, G. (eds) *The Sweetpotato.* Springer, Dordrecht, The Netherlands, pp. 161–188.

Sseruwagi, P., Sserubombwe, W.S., Legg, J.P., Ndunguru, J. and Thresh, J.M. (2004) Methods of surveying the incidence and severity of cassava mosaic disease and whitefly vector populations on cassava in Africa: a review. *Virus Research* 100, 129–142.

Stern, V.M., Smith, R.F., van den Bosch, R. and Hagen, K.S. (1959) The integrated control concept. *Hilgardia* 29, 81–101.

Swanson, M.M. and Harrison, B.D. (1994) Properties, relationships and distribution of cassava mosaic geminiviruses. *Tropical Science* 34, 15–25.

Taylor, N., Chavarriaga, P., Raemakers, K., Siritunga, D. and Zhang, P. (2004) Development and application of transgenic technologies in cassava. *Plant Molecular Biology* 56, 671–688.

Théberge, R.I. (ed.) (1985) *Common African Pests and Diseases of Cassava, Yam, Sweet Potato and Cocoyam*. IITA, Ibadan, Nigeria.

Thresh, J.M., Otim-Nape, G.W. and Jennings, D.L. (1994) Exploiting resistance to *African cassava mosaic virus*. *Aspects of Applied Biology* 39, 51–60.

Valkonen, J.P.T., Kreuze, J.F. and Ndunguru, J. (2015) Disease management, especially viruses in potato and sweetpotato. In: Low, J., Nyongesa, M., Quinn, S. and Parker, M. (eds) *Potato and Sweetpotato in Africa: Transforming the Value Chains for Food and Nutrition Security*. CAB International, Wallingford, UK, pp. 339–349.

Wang, H.L., Cui, X.Y., Wang, X.W. and Liu, S.S. (2015) First report of *Sri Lankan cassava mosaic virus* infecting cassava in Cambodia. *Plant Disease* 100, 1029.

Yadav, J.S., Ogwok, E., Wagaba, H., Patil, B.L., Bagewadi, B., Alicai, T., Gaitan-Solis, E., Taylor, N.J. and Fauquet, C.M. (2011) RNAi-mediated resistance to cassava brown streak Uganda virus in transgenic cassava. *Molecular Plant Pathology* 12, 677–687.

Yaninek, J.S. and Herren, H.R. (1988) Introduction and spread of the cassava green mite *Mononychellus tanajoa* (Bondar) (Acari: Tetranychidae), an exotic pest in Africa and the search for appropriate control methods: a review. *Bulletin of Entomological Research* 78, 1–13.

Yaninek, J.S., Onzo, A. and Ojo, J.B. (1993) Continental-wide releases of neotropical predators against the exotic cassava green mites in Africa. *Experimental and Applied Acarology* 17, 145–160.

8 Integrated Pest Management in Sugarcane Cropping Systems

François-Régis Goebel[1,*] and Amin Nikpay[2]

[1]CIRAD, Unité de Recherche AIDA, Montpellier, France; [2]Department of Plant Protection, Sugarcane and By-products Development Company, Ahwaz, Iran

8.1 Introduction and General Context

Sugarcane (*Saccharum officinarum*) is grown in more than 100 countries worldwide producing a total of 170 million tons of sugar. Brazil is the main producer. It is a strategically important crop, having a profound economic impact on social and governmental issues in many countries around the world (James, 2004). According to statistics from the International Sugar Organization (ISO), sugar consumption per capita (world average) stood at 23.3 kg/year or 63.9 g/day in 2014.

World trade, change in climate conditions, and simplification and intensification of agricultural systems has increased the risks of pest/disease incursions and outbreaks. As with many other tropical crops, sugarcane hosts a considerable quantity of insects and diseases, some of them having an economic impact on sugarcane farmers and industries. For example, the sugarcane industry in Australia is always threatened by neighbouring countries such as Indonesia and Papua New Guinea, which have a much wider range of pests. The response is the adoption in Australia of strong quarantine procedures and biosecurity strategies to avoid such risks (Goebel and Salam, 2011).

The expansion of sugarcane areas for the production of bioenergy and bioproducts other than sugar (biofuels, green electricity, bioplastics, pharmaceutics, panels from bagasse, etc.) can also have an impact on pest communities. In this context, it is vital to overcome all factors that encourage pests and their incursion. As many small-scale farmers in developing countries rely heavily on income generated through sugarcane production, losses from pests and disease can significantly impact these communities, while the incursion of a new pest or disease could have devastating consequences. Despite many years of implementation of pest management strategies, some pests remain difficult to manage and their dynamics are still largely unpredictable, with sometimes dramatic yield reduction (Kiritani, 2006; Gregory *et al.*, 2009). Before implementing any control strategy there is a need to fully understand the impact of agricultural practices and ecological parameters on insect pests.

Stakeholders from the sugarcane industries, including crop protection managers, tend to apply agricultural inputs and control techniques (chemical control, fertilizers, biocontrol, new varieties) without properly analysing the damage levels and economic thresholds of their sugarcane farms or areas

---

* Corresponding author e-mail: regis.goebel@cirad.fr

© CAB International 2017. *Integrated Pest Management in Tropical Regions*
(eds C. Rapisarda and G.E. Massimino Cocuzza)

and the cause of pest pressure and outbreaks. Agricultural practices that are able to increase or reduce pest pressure are insufficiently studied, and it is therefore important to spend time studying these impacts.

Another concern often discussed is the application of new technologies (e.g. GMO) to intensify crop productivity, which may increase the risk of insect resistance and disturb natural enemies and, more globally, ecosystem services. Therefore, as with many other crops, it is important with sugarcane to carefully investigate all parameters that are able to impact its natural enemies.

This chapter provides an overview of the main sugarcane insect pests worldwide, and discusses all aspects of pest management strategies and the impact of agricultural practices in a context of globalization, climate change and loss of biodiversity.

8.2 Sugarcane Pests of Economic Importance

The sugarcane crop (*Saccharum officinarum*) is attacked by a wide range of insect pests all through its plant stages (Williams, 1931; Box, 1953; Williams *et al.*, 1969). A catalogue of all recorded insects associated with the sugarcane crop lists over 1500 species worldwide (Box, 1953; Long and Hensley, 1972) and over 80 diseases. Though the majority of these are minor pests, a few major pests exist and cause significant damage to all parts of the crop (i.e. root, stalks and foliage) (Williams *et al.*, 1969; Hall, 1988). The following is a list of major damaging groups categorized based on the nature of damage.

8.2.1 Leaf feeders

The main insect pest species that feed directly on sugarcane leaves are mainly armyworms (Lepidoptera, Noctuidae) and locusts (Orthoptera: Acrididae). Populations of these pests are unpredictable in nature

and certain species can have intermittent outbreaks (Vreysen *et al.*, 2007). Outbreaks of armyworms, which are night-feeding pests, may occur following the intensive use of mechanical harvesting. This infestation seems also linked to the presence of trash blankets (a refuge for the armyworms) used for weed control and preservation of soil humidity in sugarcane inter-rows (Nikpay, 2016).

The locusts can occasionally damage sugarcane, particularly in Africa, in sub-Saharan regions. In South Africa, the species commonly cited in sugarcane are *Nomadacris septemfasciata* and *Petamella prosternalis* (a grasshopper). Chemical control is generally used to combat these insect pests but biocontrol using parasitoids and entopathogens are in progress.

8.2.2 Sap feeders

This category includes mainly Hemipteran species such as aphids (Aphidoidea), scale insects (Coccoidea), whiteflies (Aleyrodidae), mealybugs (Pseudoccidae), planthoppers (Fulgoroidea) and froghoppers (Cercopoidea), In addition to directly feeding on the plant sap, some species are known disease vectors. For example, the sugarcane aphid, *Melanaphis sacchari* Zethntner transmits two viral diseases of sugarcane, and these are the Sugarcane mosaic virus (SCMV) and the more recently discovered Sugarcane Yellow Leaf virus (McAllister *et al.*, 2008). Other potentially destructive viruses are Fiji disease, transmitted by *Perkinsiella saccharicida* (Homoptera, Delphacidae). However, there are many countries where this insect is present but not the disease. These pest species have a worldwide distribution, hence maintenance of strict quarantine procedures is needed to ensure protection against these major diseases.

Fulmekiola serrata (Homoptera: Thripidae) is not a major pest in African and south Asian countries but in South Africa, its introduction in 2004 and rapid spread took the sugar industry by surprise, particularly in the province of Kwa-Zulu-Natal, where most of the sugarcane is grown. The outbreak of this pest was associated with a severe drought and the South African Sugarcane Research Institute (SASRI) is currently working on control strategies.

Another occasional pest is the sugarcane whitefly, *Neomaskellia andropogonis* Corbett (Hemiptera: Aleyrodidae), which is one of the new emerging pests; damage by this whitefly seems to be expanding in recent years (Nikpay and Goebel, 2016; Nikpay, 2017).

The sugarcane yellow mite, *Oligonychus sacchari* (McGregor) (Prostigmata: Tetranychidae), is an occasional pest of sugarcane in Iranian sugarcane fields. Infestations generally occur during late May–early August, and the lower leaves of sugarcane are usually colonized first. However, prolonged heavy infestations are accompanied by extensive damage to the middle and upper leaves of young plants, reducing plant growth (Singh *et al.*, 2003; Nikpay *et al.*, 2013).

8.2.3 Stalk feeder: moth borers (Lepidoptera), key pests of sugarcane

The sugarcane crop is attacked by a wide range of stalk feeders. These can be loosely classified, depending on the time of infestation and the part of the stalk they feed on, as: top feeders, stem feeders and shoot feeders, sometimes causing dead-hearts. The main group of pests in this category is the moth borers, which are by far the most damaging sugarcane pests in all cane-growing countries, except for Australia and Fiji (Sallam, 2006).

There are about 50 species of moth borers worldwide, belonging to the genera *Chilo*, *Eldana*, *Sesamia*, *Diatraea*, *Scirpophaga*, *Elasmopalpus*, *Eoreuma*, *Telchin*, *Tetramoera* and *Acigona* that attack sugarcane (Long and Hensley, 1972; Kfir *et al.*, 2002), many of which are polyphagous species and can attack other graminaceous crops such as maize, rice, millet and sorghum, as well as several species of wild grasses (Kfir *et al.*, 2002). The larval stage of these species bores into the stalk and causes

significant losses in biomass and sugar contents (Goebel and Way, 2009) (Fig. 8.1). Moth borers are difficult to control because of their cryptic biology; hence biological control and varietal resistance are key components of their management.

Besides this main group of stalk borers, there are a few coleopteran species belonging to the family Cerambycidae and Curculionidae. Larvae of the longhorn beetle *Dorysthenes buqueti* burrow in cane stubble and attack the base of the stalk. Plants ultimately die and crops fail to ratoon properly in cases of severe infestation (Sommartya *et al.*, 2007). This insect is widely distributed in southeast Asia and is a major pest of sugarcane in Thailand. It can be controlled by entomopathogens. Also worth mentioning are weevil borers *Metamasius hemipterus* and *Rhabdoscelus obscurus* that are present in many sugarcane-producing countries and attack not only sugarcane but also palm trees.

Lastly, it is important to mention the rats (*Sigmodon hispidus*, *Rattus* sp.) as pests of economic importance of sugarcane in some countries in Central America, particularly in Nicaragua (Goebel, 2014). Rats feed directly on cane stalks, particularly when stalks are lodging. The rats eat through the rind of the cane stalk and feed on the soft and juicy internodes. The effect to the cane stalk is like ringing a tree. After this attack, the stalk above the chewed portion usually dies, and sometimes the lower portion too. Cane stalks with less damage may continue to grow but will produce less sugar. In Nicaragua, control of rats is done by applying a product called Brodifacoum, a powerful anticoagulant raticide.

8.2.4 Root feeders: white grubs (Scarab beetles)

These are underground pests that usually attack the root system, which dries the plants out and makes them vulnerable to tipping (Fig. 8.2). They belong to the families Dysnatinae, Rutelinae and Melolonthinae and most damaging genera are *Hoplochelus*, *Dermolepida*, *Lepidotia*,

Fig. 8.1. Stemborer *C. sacchariphagus* damaging sugarcane stalk (a, b) and the adult (c) (Photo: F.-R. Goebel).

Heteronychus, Adoretus and *Anomala*. Soil applications of chemical granules and entopathogenic fungus can usually control them (Allsopp, 2010).

Other pests include the sugarcane root spittlebug *Mahanarva Fimbriolata* (Stål) and *Aeneolamia varia* (Hemiptera: Cercopidae), which also damage leaves. Termites (Isoptera) can also be devastating in dry regions of Africa and Asia. The species that mostly cause damage in Africa belong to the genera *Macrotermes*.

8.2.5 Natural enemies of insect pests: parasitoids and predators

In the sugarcane agrosystem, there is also a myriad of beneficial insects that play a major role in suppression of pests. Tiny wasps such as *Cotesia flavipes* (Hymenoptera: Braconidae) and *Trichogramma* spp. (Hymenoptera: Trichogrammatidae) are among the most effective parasitoids of stemborers (Conlong and Goebel, 2006; Goebel *et al.*, 2010). Sugarcane fields are also home to a wide range of generalist predators such as spiders, ants and many others that play a major role in the regulation of pest populations (Bonhof *et al.*, 1997; Goebel *et al.*, 1999a). Besides beneficial insects, bats are also known to be good predators of sugarcane pests and can be employed directly in field by settling bat houses next to the fields of farms. However, knowledge of these natural enemies is still needed to improve their use as part of the whole ecological system. Natural enemies to control sugarcane pests should not be applied or released if agricultural practices and other techniques have a negative interaction with them (i.e. chemical treatments, cane burning, etc.).

8.3 How to Manage Insect Pests

Various examples in Africa, Asia, South America, Australia and Réunion Island have shown that the management of insect pests in sugarcane is a time-consuming task,

Fig. 8.2. The grey back canegrub *Dermolepida albohirtum* (adult larva) (Photo: F.-R. Goebel).

which requires the understanding of various factors from field to landscape scale. There are external factors such as climatic conditions that cannot be managed. However, a number of agronomic and ecological factors involving crop husbandry in general and natural resource management can be managed to reduce pest pressure. Managing a cropping system, applying new techniques, or disturbing an ecosystem for land grabbing to settle new sugarcane areas can lead to pest problems and outbreaks (Goebel and Sallam, 2011).

8.3.1 Agricultural practices and pest management

General considerations and recommendations

Cultural control is considered the first step of defence against pests including stalk borers, and includes techniques such as destruction of crop residues, crop rotation, manipulation of planting dates, early harvesting, collecting dead shoots, decreased fertilizer use, field monitoring, avoiding 'stand-over' cane (cane growing longer than the recommended time) whenever possible; further measures include cutting the cane at or below ground level so as to prevent larvae in the stools re-infesting ratoon crops; after cutting, removing all residue stalk and leaf material; covering exposed residue of cane with soil (this kills eggs and young larvae in the stalk stumps); and avoiding the use of broad-spectrum or persistent insecticides (Kfir *et al.*, 2002; Leslie, 2004). In general, all agronomic practices that reinforce the ability of plants to resist insect infestation are to be encouraged; however, if some practices such as fertilizer and water supply are not properly applied in sugarcane fields (i.e. over-application of nitrogen, water mismanagement that leads to plant stresses), this can dramatically increase pest infestation and therefore yield losses (Goebel and Sallam, 2011).

Removing dry cane leaves from the stem or pre-trashing also suppress stalk borer numbers by reducing the number of eggs already in the field, and reduces the preferred, dry-leaf oviposition sites (Leslie, 2004). Also, removing the leaves slows down the penetration time for larvae, thus exposing them to natural enemies. In the USA, to reduce the number of overwintering larvae, stubble in fallow fields should be ploughed out as quickly as possible (LSU AgCenter, 2010).

Planting stem borer-free sugarcane seed pieces is also an elementary recommended management tactic to reduce overwintering populations of stalk borer (Leslie, 2004; Beuzelin, 2011). In the context of new plantations, pieces of sugarcane stalk 'seed cane' are placed horizontally into the soil, and they may contain eggs or larvae of stalk borers, which could re-infest neighbouring crops. These can be killed by immersing the seed cane in water at 50°C for two hours, or dipping it in pesticide mixtures (Carnegie, 1981).

Planting and harvesting dates cause various sugarcane phenological conditions potentially influencing stem borer population dynamics (Beuzelin *et al.*, 2011). For example, the practice of 'carry-over' of cane more than 12 months which is conducted in the south of Durban (South Africa) is favourable to borer development and damage. This was shown with *E. saccharina*, where damage starts to increase markedly after 12 months and thereafter (Goebel *et al.*, 2005). The loss of sugar is often critical at this stage and carry-over cane poses a problem for the industry.

With standard sugarcane management practices, early planting typically provides a better root establishment and higher yields. Viator *et al.* (2005) determined the effect of three planting dates (August, September and October) on the yield of five sugarcane cultivars in Louisiana. Plant cane sugar yields for cultivar LCP 85-384 were not affected by planting date, while for HoCP 85-845 and CP 70-321, sugar yields were higher for the August planting. In a recent study, Beuzelin (2011) investigated the effects of four planting dates on phenological characteristics and the stalk borer *Diatraea saccharalis* infestations. He found that number of *D. saccharalis*-related deadhearts showed that early August planting dates had important *D. saccharalis* infestations and the potential to host major overwintering populations. Later planting dates are likely to minimize overwintering populations.

Regular irrigation has a positive effect in reducing borer infestations in the field. Reay-Jones *et al.* (2005) showed that irrigation significantly reduced the occurrence of bored internodes by 2.5-fold and moth borer exit holes/stalk by 2.5-fold in two commercial sugarcane cultivars. Drought stress can affect sugarcane and may increase the susceptibility of varieties to stalk borer damage. Under drought stress, sugarcane plants have higher levels of several free amino acids and more dry leaves (Reay-Jones *et al.*, 2005), which enhances plant suitability for oviposition and larval development (Showler and Castro, 2010). Protein and free amino acid content changes in drought-stress cultivars and finally these plants will be more attractive for feeding, growth, oviposition and reproduction (White, 1984).

In addition to drought, salinity can affect plant growth and vigour, which also affect stalk borer damage. Reay-Jones *et al.* (2003) reported that high soil salinity, a stress factor that also enhances free amino acid accumulations in plants, increases Mexican rice borer infestations in sugarcane.

Despite positive effects of agricultural practices if well applied, there are cases of mismanagement of these practices and techniques that will normally improve sugarcane yield and productivity. Some of them can lead to dramatic change in pest pressure (Fig. 8.3), as shown in the following examples.

The impact of nitrogen on borer damage

Studies have shown that sugarcane production is positively correlated with the availability of nutrients, especially nitrogen, and this source of nutrient is used in the form of fertigation which is mixed with water (Atkinson and Nuss, 1989). Application of nitrogen fertilizers promote higher cane productivity but have resulted in elevated pests and diseases outbreaks. Nitrogen fertilizers mostly influence fecundity, longevity and damage caused by arthropod pests (Scriber, 1984). This phenomenon is more obvious in the case of stalk borers, as clearly documented by several researchers (Atkinson and Nuss, 1989; Goebel *et al.*, 2005; Pandey, 2014; Mahlanza *et al.*, 2014). In South Africa, many commercial farmers and agricultural managers from sugar estates apply too much fertilizer; this is particularly true for nitrogen (more than 150 kg/ ha). Goebel *et al.* (2005) showed that, by increasing the level of nitrogen fertilizers, the percentage of stalk damage increased for both small-scale and commercial growers (Table 8.1). In a general survey, these authors found that, whatever the sugarcane region, there was a significant difference of 50 kg/ha of nitrogen applied between the two types of farms, with an average of 134 kg/ha for commercial growers versus 81 for small growers (Goebel *et al.*, 2005). The elevated damage was significant in commercial growers because they applied more

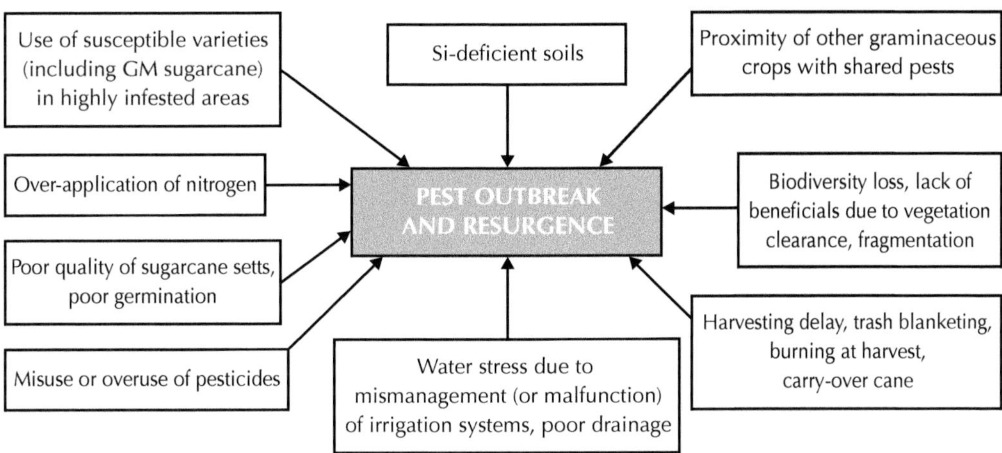

Fig. 8.3. Main management practices and environmental constraints likely to change pest pressure in sugarcane agroecosystems (Goebel and Sallam, 2011).

Table 8.1. Infestation levels and nitrogen categories at Sezela and Felixton mill areas (Goebel *et al.*, 2005).

Classes	Nitrogen class (kg/ha)	% Stalk damaged (mean±SE)
1	0–50	7.3±1.7 b
2	51–75	11.3±1.7 b
3	76–100	8.7±0.9 b
4	101–125	14.6±2.2 b
5	126–150	23.9±1.9 a
6	>150	26.1±5.2 a
CV%		78.4
F		2.3
P		0.0444

Mean within a column followed by the same letters (a, b, c) are not significantly different = P < 0.05, Ranking test of Student-Newmans-Keuls Test (General Linear Model, SAS, SAS Institute).

than 150 kg/ha nitrogen in their fields. A positive correlation between nitrogen input and pest infestation levels was found with the borer *E. saccharina* and proved to be another factor leading to lower pest prevalence in small-scale farms (Goebel *et al.*, 2005). A critical threshold of 100 kg N per ha was found to agree with results obtained in Cuba for *D. saccharalis* (Lopez *et al.*, 1983).

Mahlanza *et al.*, (2014) reported that nitrogen content of resistant and semi-resistant varieties were lower than susceptible varieties and this higher nitrogen level in susceptible varieties makes them prone to infestation of *E. saccharina*. In India, Pandey (2014) studied the effect of nitrogen level on damage by sugarcane borer *Chilo auricilius* Dudgeon. The author indicated that infestation of stalk borer (percentage of stalk damaged) increased significantly with the increase of nitrogen level in treated plots. High levels of nitrogen may result in softening of the plant and can render certain varieties susceptible to a high level of infestation by stalk borers (Goebel *et al.*, 2005). In Brazil, Pannuti *et al.* (2015) tested effects of different nitrogen levels on consumption of cane fragments by *Diatraea saccharalis*. The results showed that the use of nitrogen fertilization favours an increased incidence and consumption of stalk borer. Similar results were found on the same pest recently in Argentina by Salvatore *et al.* (unpublished data), particularly when using high dose of nitrogen (more than 100 kg/ha).

In conclusion, there is a trade-off between fertilizer inputs to increase productivity and associated increased losses to pests.

Silicon in soil: What impact on insect populations?

Through three examples with the Lepidopteran stem borers *Eldana saccharina* in South Africa, *Sesamia* spp. in Iran and *Diatraea saccharalis* in USA, it is shown how this factor is important in pest management.

One relatively new approach to manage stalk borers and pests in general is the application of silicon as a nutritional soil and foliar amendment. This approach is classified as a nutritional integrated pest management (IPM), as it involves improving crop resistance by improving crop vigour (Keeping *et al.*, 2013; Reynolds *et al.*, 2009). Among agricultural crops, sugarcane and rice are classified as silicon-accumulating plants, which can use silicon more efficiently than can other crops (Djamin and Pathak, 1967; Ma, 2004; Ma and Yamaji, 2015). Silicon is absorbed by plants in the form of monosilicic acid $(Si(OH)_4)$, the most common form of Si in the soil solution at a pH below 9 (Jones and Handreck, 1967). After uptake and transport from roots to vegetative shoots, silicic acid become concentrated due to water loss or physiological processes, and finally is concentrated as silica gel (Ma and Yamaji, 2006). Silicon decreases transpiration rate and conductance from stomata of leaves. Beneficial effects of silicon are usually more apparent in Si-accumulating plants under various abiotic (such as salt stress, drought stress, heavy metal toxicity) and biotic stresses (including pests and diseases caused by both fungi and bacteria). There is an evident link between silicon and reduced water stress and between water stress and increased susceptibility to insect attack, at least by borers (Kvedaras *et al.*, 2007b,c) (Fig. 8.4).

Approximately 60% of soils in the South African sugar industry are sandy, acidic soils, typically silicon-deficient (Meyer and Keeping, 2005). There are several publications on the effects of silicon on sugarcane pests, including chewing and sucking arthropod pests (Kvedaras *et al.*, 2007a,b; Keeping *et al.*, 2009, 2013, 2014; Korndorfer *et al.*, 2011; White and White,

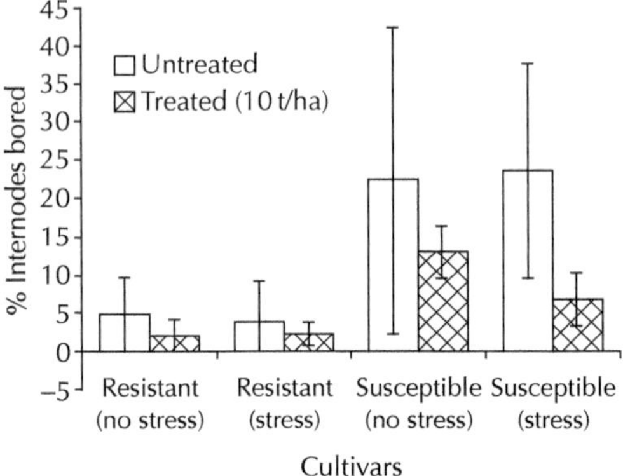

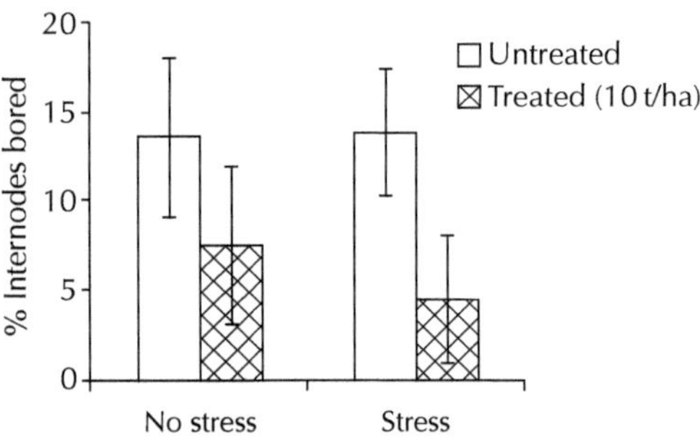

Fig. 8.4. Percentage of internodes bored (grouped cultivars in pot trial). The infestation is 1.8-fold in non-stressed cane; 3.1-fold in stressed cane (Kvedaras *et al.*, 2007b).

2014; Nikpay and Soleyman Nejadian, 2014; Nikpay *et al.*, 2015). Silicon has both direct and indirect effects on stalk borer larvae. Direct effects include reduced larval weight, reduced stalk damage, reduced internodes bored, reduced length of borer tunnel and, finally, reduction in moth borer exit holes. Indirect effects are described as delay in insect development and growth, resulting in retardation in crop penetration and, finally, exposure to adverse climatic conditions (Reynolds *et al.*, 2009). In South Africa, where the most silicon trials are performed, Keeping *et al.* (2013) found that application of calcium silicate could reduce percentage of stalk damage, percentage of internodes bored, length of the borer tunnel and number of *E. saccharina* borer per 100 stalks.

In Iran, Nikpay *et al.* (2015) applied calcium silicate as a soil amendment under field conditions. Before harvest, the results showed that silicon application reduced *Sesamia* spp. damage on three sugarcane varieties. The authors indicated that silicon fertilization reduced percentage of stalk damage, percentage of internodes bored, percentage of moth borer exit holes, length of borer tunnel and number of live *Sesamia* per 100 stalks.

In the United States, White and White (2014) performed pot trials and applied silicon as a soil fertilizer. The authors found reduction in percentage of internodes bored in both susceptible and resistant varieties.

Silicon can also promote the production of volatile defence compounds by damaged plants. These chemical signals 'recruit' predators and parasitoids (Gurr and Kvedaras, 2010). Si-accumulating plants translocate silicic acid throughout their tissues: when attacked, they produce systemic stress signals (e.g. salicylic acid; jasmonic acid) that are key to induced plant defences. For example, jasmonic acid is relevant to biological control of arthropods because this compound is known to trigger production of herbivore-induced plant volatiles by the attacked plant (Kvedaras *et al.*, 2010). These authors also found that application of silicon could enhance predator attraction to treated plants. In their study, they stated that Si-treated plants with a pest infestation were more attractive to natural enemies than Si-untreated plants with a pest infestation. Therefore, it was concluded that Si treatment could enhance biological control in the field. In a new study, Nikpay *et al.* (2017) found that foliar application of silicon on sugarcane can increase the level of parasitism versus untreated sugarcane varieties.

The effect of silicon was also clear on mandibular wear of *E. saccharina* during the experiments in pot trials (Kvedaras *et al.*, 2009) (Fig. 8.5). In fact, silicon particles are distributed along the stalk and were found to act like a mechanical barrier with an abrasive function. This effect on mandibles hampered the ability of larva to penetrate into the stalk and feed on pith, and the chance of being parasitized was greater.

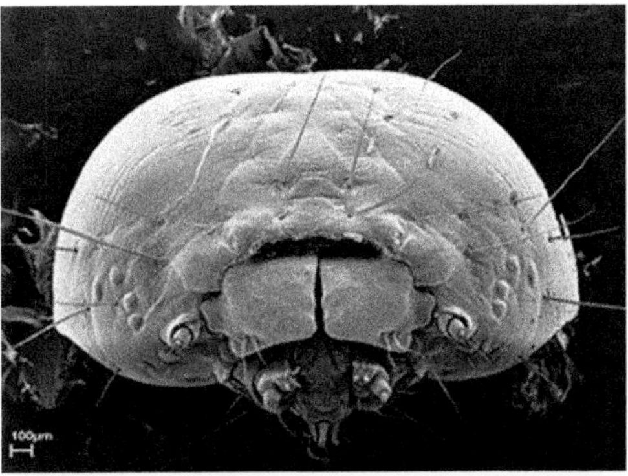

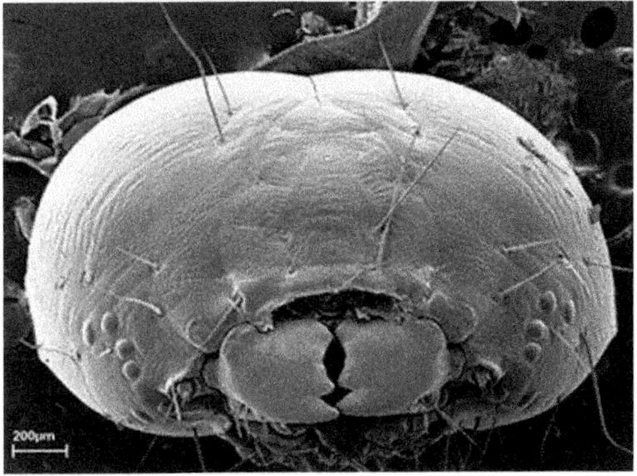

Fig. 8.5. Head capsules of *E. saccharina* (Scanning Electron Microscopy); above: state of larval mandibles without silicon; below: with silicon treatment (the mandibles are severely worn) (Kvedaras *et al.*, 2009).

The impact of cane burning in large sugarcane areas

Biodiversity is severely affected by the burning practice before harvest, which disturbs the entire biological equilibrium (fields and hedges). In Réunion, the banning of cane burning in highly infested areas has reduced borer damage by 50%. Some surveys have proved that borer larvae can survive in the internodes (Goebel *et al.*, 2010). On the basis of these results, and for environmental considerations (pollution by the ashes), the island stopped this practice at the beginning of the 2000s. Many producing countries have also decided to stop burning at harvest and implemented green harvesting. This has also been pushed by the growing demand for cane trashes to use as field blanketing, or for energy use and bioplastics. However, in some countries in Africa, Asia and South America, cane burning is still employed and has a devastating effect on biodiversity (Fig 8.6).

Use of varietal resistance to sugarcane pests and GM sugarcane varieties

Use of conventional (non-GM) pest- or disease-resistant varieties is a major component of IPM in sugarcane ecosystems, but if recommendations are not followed, the risk of elevated pest pressure will remain (Hensley *et al.*, 1977; Meagher *et al.*, 1996; Keeping, 2006). Usually, in sugarcane-breeding programmes there is no particular research activity for borer resistance as such (incorporation of specific genes/traits) but when new varieties are set to be commercialized they are generally tested for their susceptibility to borers. In South Africa, a borer evaluation system is used (ratings from 0 to 9) against *Eldana saccharina*, allowing the ranking of varieties from

susceptible, moderately susceptible and intermediate to resistant (Keeping, 2006). Different statistical analysis using pest data on different varieties collected in different sugarcane areas in South Africa by SASRI (Goebel *et al.*, 2005) concluded that in the most infested areas, the percentage of stalk damaged was significantly different between the varieties surveyed (n = 100); The PCA confirmed three distinct groups (all data mixed), susceptible, intermediate and resistant varieties (Fig. 8.7). Ranking generally follows SASRI resistance ratings.

Again, it is important to apply recommendations from sugarcane research institutes, particularly in highly infested areas ('hot spots'). The choice of varieties on a farm can really change the pest status and pressure. For example, in the context of a high demand for biomass to produce energy and other bioproducts, new 'high fibre' varieties containing a higher proportion of cellulose, hemicellulose and lignin are being developed in different breeding

Fig. 8.6.   Cane burning in a big sugar estate in Sudan, northeast Africa (Photo: F.-R. Goebel).

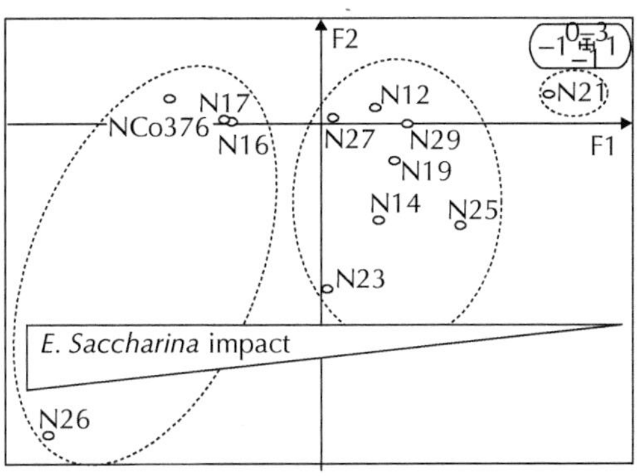

Fig. 8.7. From the factorial drawing F1/F2, the dots corresponding to different samples were grouped per variety. Only the centre of all dots corresponding to each variety was represented. Axis 1 shows a decreasing impact of *E. saccharina* from the negative to the positive value.

centres. However, it is not yet known what impact these varieties will have on pest infestation, particularly stem borers. Similarly, 'high sucrose' varieties also currently under development, are likely to influence pest dynamics because these types of sweet varieties are preferentially attacked by stem borers. Thus there is a need to directly consider the pest consequences of any new varieties and whether pest resistance can be a key component of variety development.

The use of genetically modified (GM) sugarcane has been identified as a future strategy for the expansion of sustainable sugarcane production (Smeets *et al.*, 2008). Beyond all associated risks (public concerns about GM-based sugar products, environment and biodiversity impacts, capacity for gene flow from GM seed cane toward wild species), it is important to consider here the likely development of pest resistance to transgenic crops representing a significant threat to the large-scale adoption of these cultivars (Tabashnik *et al.*, 2008). In particular, the direct and indirect impacts that GM varieties may have on the dynamics of the pests and their associated natural enemies will need to be understood. This is especially important for existing successful biological control strategies, where volatiles emitted by the host plant following a pest

attack play an important role. Research on new traits to be incorporated into sugarcane varieties against pests, particularly stem borer resistance, is not yet complete (Craveiro *et al.*, 2010). As transgenic sugarcane varieties have not yet been commercialized, we need to learn from other industries where GM varieties are under production (e.g. maize, canola, cotton and soybean crops).

8.3.2 Biocontrol and conservation biological control

Classical biological control, which is the introduction and establishment of exotic natural enemies against introduced pest species, is a well-known technology and an essential component in sugarcane pest management. It has been developed for more than 50 years in most sugarcane-producing countries, particularly for controlling stem borers which are difficult to reach once the larvae is inside the sugarcane stalk. In addition, sugarcane is a very dense crop which sometimes reduces the efficacy of the pesticide treatments when air sprayed. In this context, sugarcane is much less treated with chemicals for insect control than other cropping systems such as cotton or horticulture. Biocontrol in sugarcane includes the use of parasitoids and entomopathogenic fungi. While parasitoids are mainly used on Lepidoptera, entomopathogens are rather employed for the control of coleopteran and heteropteran species.

Releases of Hymenoptera parasitoids to control Lepidopteran stemborers

Many research institutions have developed biocontrol using parasitoids via augmentative or inundative releases in the sugarcane fields, such as *Trichogramma* spp., *Cotesia* spp., *Lyxophaga* spp., *Telenomus* spp., *Tetrastichus* spp. and others, with success stories but also several failures. For example, the failure of borer control using *Trichogramma* spp. in the 1960s and 1970s was partly due to lack of research on parasitoids themselves (species, bionomics and efficacy), but also lack of quality control of mass production (Goebel *et al.*, 2010). During this period, biological programmes often introduced exotic parasites and released them without evaluating (in some cases) their impact on pests (Goebel *et al.*, 2010). All these facts have led to a negative image of biocontrol with *Trichogramma* spp. and other parasitoids and loss of interest for this strategy (Goebel *et al.*, 2002).

From our own experience in Réunion, we have learned that a biocontrol programme needs proper research following strict protocols and requires constant technical improvement. In this French overseas department, biocontrol of *Chilo sacchariphagus*, the sugarcane spotted stem borer, using *Trichogramma chilonis* has been constantly improved by spending more time on research and development. The first step was to choose a *T. chilonis* strain with optimal performances (Reay-Jones *et al.*, 2006).

After more than ten years of laboratory studies and field experiments (biology, natural parasitism, ecology, time and rates of field releases, mass production, etc.) the strategy adopted in Réunion was to release 100,000 *T. chilonis* per ha and per week at the beginning of the crop growth (between 1 and 4 months). To ensure a good coverage by *Trichogramma*, which is not a mobile insect, it was decided to set up 100 release points per ha. This technique reduced damage by 50%, with an economic gain estimated between €600 and €1400/ha (Goebel *et al.*, 2010). This result was obtained thanks to a strong partnership between three French organizations – INRA (Institut National de la Recherche Agronomique), CIRAD (Centre de coopération internationale en recherche agronomique pour le développement) and FDGDON (Fédération Départementale des Groupements de Défense contre les Organismes nuisibles-Réunion).

In many other countries, such as Indonesia (Java) and India, this parasitoid is used as the main component of biocontrol strategy. Indonesia is still producing millions of *Trichogramma* in association with the sugar factories, while India has seen small farmers taking over the production and release of *Trichogramma* wasps in their own fields.

Another good example of biological control using parasitoids is Brazil, which has succeeded in controlling *Diatraea saccharalis* using the combination of two parasitoids: *Cotesia flavipes*, a larval parasitoid and *Trichogramma galloi* parasitizing eggs (Botelho *et al.*, 1999). This example is noteworthy because by using key parasitoids in concert, control of stem borer populations is optimal. Some countries also use pupal parasitoids such as *Tetrastichus howardii* (Hymenoptera: Eulophidae) or *Xanthopimpla stemmator* (Hymenoptera, Ichneumonidae). Biocontrol using parasitoids will continue in most sugarcane-producing countries, but in the meantime research and development activities should continue to improve biocontrol in all its components: quality control, cost reduction, conditioning, packaging, efficacy, economic feasibility and adoption by growers (Goebel *et al.*, 2010).

Biocontrol and predation by ants

In implementing such biocontrol strategies, predation of borer eggs parasitized by *Trichogramma* spp. should not be neglected. In Réunion, the importance of predation of *C. sacchariphagus* eggs by ants *Pheidole megacephala* and *Solenopsis geminata* has been reported as an essential component of the natural control of this pest (Goebel *et al.*, 1999b), while ant predation is better known on other stem borer species (Teran, 1980; Bonhof *et al.*, 1997). In this particular context, biological control using field releases of *Trichogramma* spp. should be planned accordingly and focused on younger canes at the period of moth oviposition and where ant predation is still low. Ant colonies tend to build up rapidly particularly when the cane fields become dense, generally between 6 and 10 months, and natural predation of *C. sacchariphagus* is significant, making the use of *T. chilonis* redundant or wasteful. In Réunion, to decrease this negative impact, new dispensers with tiny holes to prevent ants from penetrating and feeding on parasitized eggs were tested with the help of a private company Biotop (Goebel *et al.*, 2010).

Biocontrol of white grubs and other pests using entomopathogenic fungi

In 1973, in Réunion Island (French Overseas Department), the white grub *Hoplochelus marginalis* Fairmaire was accidentally introduced from Madagascar and within ten years became a threat to the whole sugar industry (Jeuffrault *et al.*, 2004). This pest found optimum field conditions for rapid population growth assisted by a lack of predators, parasitoids and entomopathogenic fungi that ensure a natural control in Madagascar. After years of chemical control through the 1980s, an effective fungus *Beauveria hoplochelii* (Ascomycota: Hypocreales), previously described as *B. brongniartii*, was discovered in Madagascar on another species of *Hoplochelus* and introduced into Réunion Island where it successfully controlled the pest in most sugarcane areas (Jeuffrault *et al.*, 2004; Robène-Soustrade *et al.*, 2015).

In Australia, *Dermolepida albohirtum* (Coleoptera, Melonthinae), the greyback cane beetle, is mainly controlled by insecticides applied in the soil but in north Queensland, the green muscardine *Metarhizium anisopliae* has given some good results and is proposed as a biocontrol strategy in this region. This fungus is also widely used to control the longhorn beetle *Dorysthenes buqueti*, a major pest of sugarcane in Thailand (Suasa-ard *et al.*, 2008).

In Brazil and other Central American countries, the spittle bug *Mahanarva fimbriolata* and *Aeneolamia* spp. are controlled mainly using the entomopathogenic fungus *Metarhizium anisopliae*. This fungus is also effective on scarab pests. In Nicaragua at San Antonio Mill, *Aeneolamia* spp. (Homoptera, Cercopidae) is one of the main insect pests, causing significant losses by feeding on roots (nymphs) and foliage (adults). After results with *Metarhizium anisopliae* as an effective agent, San Antonio has built a well-equipped laboratory near the mill to mass-produce the fungus. The product name is Metarhisa-WP and it is applied via aerial treatment in 2 or 3 applications depending on the infestation level by spittlebugs in the field. The fungus is produced on maize and formulated in bags (conidia/spores: $2.5–5 \times 10^{12}$ per ha) and mixed with water before being applied at the rate of 1 kg/ha, with a good efficacy (Goebel, 2014).

Nematodes are also used as biocontrol agents in sugarcane. In Colombia, *Steinernema* sp. and *Heterorhabditis bacteriophora* have been tested with success on the nymphs of *Aeneolamia varia*, a pest detected in sugarcane crops in 2007 in the Cauca Valley (Moreno Salguero *et al.*, 2012).

8.3.3 Chemical control of insect pests: not so used in sugarcane fields

Pesticides are used in sugarcane to control some pests such as white grubs, armyworms and sap feeders. However, a much bigger use of chemicals is for weed management with herbicides. As discussed earlier, stem borer control remains difficult due to the internal development of larvae in the stalk and the density of the crop. Eggs and young larvae before penetrating into the stalks can be easily killed by insecticides such as pyrethrins, but the applications have to be precisely scheduled to coincide with this period. Countries such as South Africa, USA and Australia continue to apply insecticides to control major pests but this has led to significant environmental issues. In Australia, the proximity of sugarcane areas to the Great Barrier Reef World Heritage site has increased runoff pollution risk and led to strict regulations for the use of pesticides and fertilizers (Lewis *et al.*, 2009). In developing countries (particularly in Africa and South America) agrochemicals such as the herbicide glyphosate are used as ripeners to increase sugar content before harvest. These chemicals are applied over large areas using small planes or helicopters and the effect on natural vegetation around sugarcane fields, pests and natural enemies (particularly those that are more exposed such as spiders) is not known. However, application of insecticides induces insecticide resistance, outbreaks of secondary pests and destruction of natural enemy communities. Regulations and science need to underpin sugarcane industries managed by overseas or international private companies and this will be a challenging task to drive them to use more environmentally friendly methods for weed and pest management.

8.3.4 Restoring natural vegetation and planting companion plants in and around sugarcane fields

It is well known that agricultural intensification and large-scale monocultures such as sugarcane lead to considerable losses in habitat and biodiversity at multiple spatial and temporal scales (Tilman *et al.*, 2002). Changes to a simpler landscape structure and reduction of native vegetation/trees alter movements of pests and natural enemies and increases infestation levels and the likelihood of pest outbreaks in agriculture. In surveys conducted in South Africa, small-scale sugarcane farms (<2 ha) had 2–3 times lower damage levels of *E. saccharina* than in larger commercial farms/monocultures (Goebel *et al.*, 2005). In small farms, a high diversification of crops interspersed with mixed marginal and natural vegetation is generally observed and pests are therefore naturally suppressed. Such landscapes are better at supporting natural enemy diversity.

It is therefore necessary to plan and carry out pest control on a broader scale than that of the field: the landscape, in other words, a spatial extent, whether natural or modified by humans, which presents a visual or functional identity. At this level, it is possible to take advantage of the biodiversity present in the landscape components: natural vegetation, cultivated plants, trees, forest corridors, pastures, watercourses, etc. At this level, it is also possible to coordinate the actions of farmers and other stakeholders (state services, development agencies, etc.), or even to develop the landscapes in order to maximize the use of companion plants. In South Africa, the sugarcane borer *Eldana saccharina* invaded sugarcane monocultures when this crop was introduced in the last century. Only the use of insecticides and resistant varieties succeeded in limiting the damage caused by this pest. Biological control was problematic, as the parasitoids of the sugarcane borer did not 'follow' it when it colonized sugarcane from *Cyperus papyrus* (Cyperacae) and wild grasses. Research has identified the companion plants to use or introduce around fields in order to stimulate the natural control of the sugarcane borer: wild plants such as *Cyperus, Erianthus, Pennisetum* or *Desmodium*, and cultivated plants such as maize or sorghum, which attract parasitoids, and trap or repel pests and their natural enemies (Fig. 8.8) (Conlong and Rutherford, 2009). These companion plants can be used to develop a push–pull system (stimulo-deterrent diversion) as an interesting component of pest management.

As shown by this research, developing systems to minimize infestations and to encourage natural pest enemies implies taking into account the interactions between insects and all the components of the landscape: natural vegetation, its location, its characteristics; the fields, their size and the plants grown there; fallow land; corridors; and of course the strategies adopted by farmers and other stakeholders. Knowledge of these elements is therefore essential. It will be critical to understand the impacts of landscape diversity on maximizing natural pest control potential, particularly from parasitoids, at different scales while also maximizing production.

8.4 Conclusion: Toward an Ecologically Based Pest Management (EBPM) at the Landscape Scale

Pest status can change rapidly in the context of simplification and intensification of sugarcane agrosystems. Conditions to avoid high infestation should be observed at any time, including conservation biological

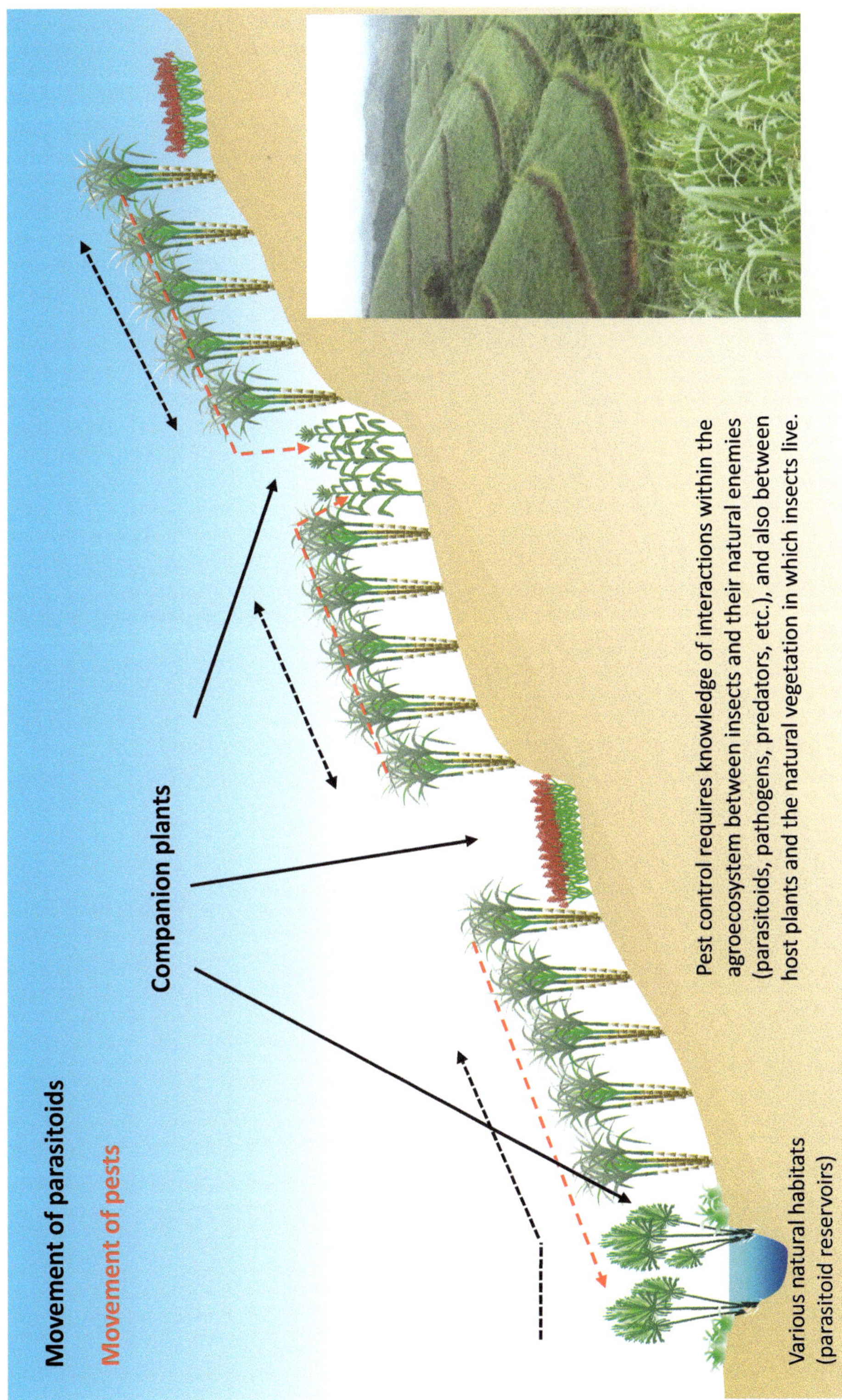

Fig. 8.8. Taking into account landscape components and companion plants for biological control of *Eldana saccharina*, a sugarcane pest in South Africa. (From Conlong and Rutherford, 2009.)

control. All agricultural practices that favour borer infestation should be assessed, revised and adjusted (e.g. nitrogen). For example, plant species diversity in sugarcane agrosystems should be encouraged for sustainable management of crop pests (e.g. push–pull, plant traps). There is also a need for habitat networks that connect farms with surrounding ecosystems (as corridors) that allow natural enemies and other beneficial biota to circulate into fields (Andow, 1991). In sugarcane, like many other cropping systems, the major challenge in implementing ecologically based pest management (EBPM) is to find strategies to overcome the ecological limits imposed by monocultures (Altieri and Nicholls, 1999). There is also a need to change the scale of IPM from conventional methods at the field level to a more global approach, at the landscape level, that is called Area Wide integrated Pest Management (AW-IPM) (Vreysen et al., 2007). This relatively new approach is certainly the way to go mainly because fields and farms are part of an ecosystem where insects have no boundaries and live in different habitats (crops, natural vegetation, weeds and forests). This approach has been successfully implemented in South Africa to control E. saccharina (Webster et al., 2009). Such an approach can be sustained with the use of satellite imagery from remote sensing and Geographic Information Systems (GIS) to generate maps and apprehend the effect of landscape structure on population dynamics of pests and natural enemies via spatial statistics.

Soil fertility is also an important component to associate with an ecologically based pest management system. There are positive interactions between soils and pests that, once identified, can provide guidelines for optimizing the total agroecosystem function (Fig. 8.9).

Despite the potential links between soil fertility and crop protection, the evolution of integrated pest management (IPM) and integrated soil fertility management (ISFM) have proceeded separately. The integrity of the agroecosystem relies on synergies of

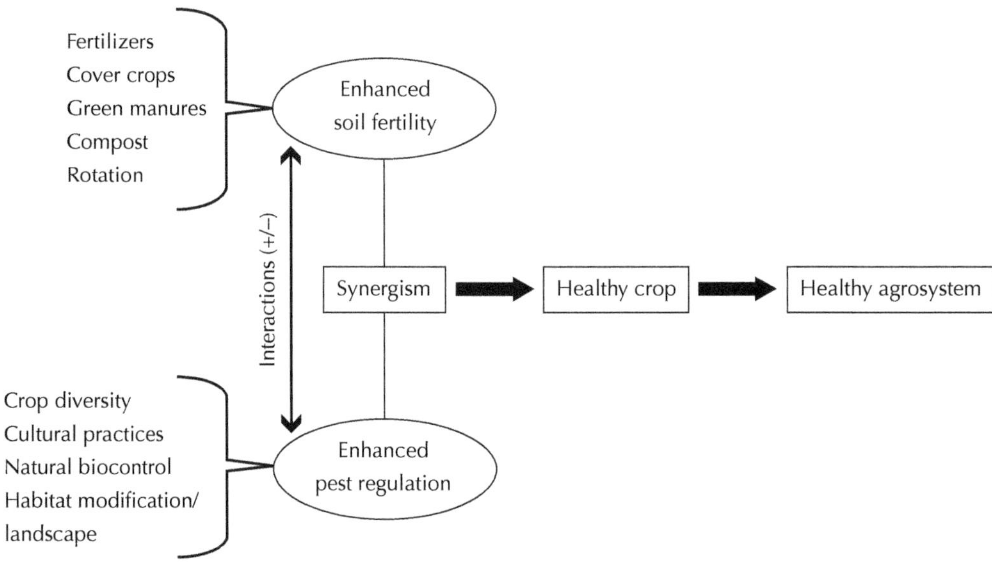

Fig. 8.9. Optimizing soil–pest interactions: a key pathway to achieving sugarcane agroecosystem health. (From Altieri and Nicholls, 2003.)

plant diversity and the continuing function of the soil microbial community, and its relationship with organic matter. Most pest management methods used by farmers can be considered soil fertility management strategies, and vice versa. Increasingly, new research is showing that the ability of a crop plant to resist or tolerate insect pests and diseases is tied to optimal physical, chemical and mainly biological properties of soils.

To illustrate all this, Mark D. Hunter (2002: 159) argues:

> As insect ecologists, we are obligated to understand the processes that influence the abundance, richness and diversity of insects in fragmented landscapes. As pest managers, we need to know how the architecture of landscapes (and crop husbandry) influences pest population dynamics and their interactions with natural enemies and agents of control. As conservation biologists, we must develop strategies to maintain focal insect species, faunal diversity and the trophic interactions that drive key ecosystem processes.

More than ever, there is a need to stimulate multidisciplinary research projects involving agronomists, entomologists, ecologists, managers and farmers to develop this EBPM approach in sugarcane agroecosystems.

References

Allsopp, P. (2010) Integrated management of sugarcane whitegrubs in Australia: an evolving success. *Annual Review of Entomology* 55, 329–349.

Altieri, M.A. and Nicholls, C.I. (1999) Biodiversity, ecosystem function and insect pest management in agricultural systems. In: Collins, W.W. and Qualset, C.O. (eds) *Biodiversity in Agroecosystems*. CRC Press, Boca Raton, Florida.

Altieri, M.A. and Nicholls, C.I. (2003) Soil fertility management and insect pests: harmonizing soil and plant health in agroecosystems. *Soil & Tillage Research* 72, 203–211.

Andow, D.A. (1991) Vegetational diversity and arthropod population response. *Annual Review of Entomology* 36, 561–586.

Atkinson, P.R. and Nuss, K.J. (1989) Associations between host-plant nitrogen and infestation of the sugarcane borer, *Eldana saccharina* Walker (Lepidoptera: Pyralidae). *Bulletin of Entomological Research* 79, 489–506.

Beuzelin, J.M. (2011) Agroecological factors impacting stem borer (Lepidoptera: Crambidae) dynamics in Gulf coast sugarcane and rice. PhD thesis. Louisiana State University and Agricultural and Mechanical College.

Beuzelin, J.M., Meszaros, A., Akbar, W. and Reagan, T.E. (2011) Sugarcane planting date impact on fall and spring sugarcane borer (Lepidoptera: Crambidae) infestations. *Florida Entomologist* 94, 242–252.

Bonhof, M.J., Overholt, W.A., Van Huis, A. and Polaszek, A. (1997) Natural enemies of cereal stemborers in East Africa. *Insect Science and its Application* 17, 19–35.

Botelho, P.S.M., Parra, J.R.P., Das Chagas Neto, J.F. and Oliveira, C.P.B. (1999) Association of the egg parasitoid *Trichogramma galloi* Zuchi (Hymenoptera: Trichogrammatidae) with the larval parasitoid *Cotesia flavipes* (Cam.) (Hymenoptera: Braconidae) to control the sugarcane borer *Diatraea saccharalis* (Fabr.) (Lepidoptera: Crambidae). *Annals of the Entomological Society of Brazil* 28, 491–496.

Box, H.E. (1953) *List of Sugar-cane Insects: A Synonymic Catalogue of Sugar-cane Insects and Mites of the World, and Their Insect Parasites and Predators*. Commonwealth Institute of Entomology, London.

Carnegie, A.J.M. (1981) Combatting *Eldana saccharina* Walker, a progress report. *Proceedings of the South African Sugar Technologists Association* 55, 116–119.

Conlong, D.E. and Goebel, F.R. (2006) *Trichogramma bournieri* Pintureau et Babault (Hymenoptera: Trichogrammatidae) and *Chilo sacchariphagus* Bojer (Lepidoptera: Crambidae) in sugarcane in Mozambique: a new association. *Annales de la Société Entomologique de France* 42, 417–422.

Conlong, D.E. and Rutherford, R.S. (2009) Conventional and new biological and habitat interventions for integrated pest management systems: review and case studies using *Eldana saccharina* Walker (Lepidoptera: Pyralidae). In: Peshin, R. and Dhawan, A.K. (eds) *Integrated Pest Management: Innovation-Development Process.* Springer Science + Business Media, Dordrecht, The Netherlands, pp. 241–261.

Craveiro, K.I., Gomes Júnior, J.E., Silva, M.C., Macedo, L.L., Lucena, W.A., Silva, M.S. *et al.* (2010) Variant Cry1Ia toxins generated by DNA shuffling are active against sugarcane giant borer. *Journal of Biotechnology* 145, 215–221.

Djamin, A. and Pathak, M.D. (1967) Role of silica in resistance to Asiatic rice borer, *Chilo suppressalis* Walker, in rice varieties. *Journal of Economic Entomology* 60, 347–351.

Goebel, F.R. (2014) Main observations and recommendations from a visit to San Antonio (2–9 October), Nicaragua Sugar Estates Limited (NSEL). Consultancy report. CIRAD, Montpellier, France.

Goebel, F.R. and Sallam, N. (2011) New pest threats for sugarcane in the new bioeconomy and how to manage them. *Current Opinion in Environmental Sustainability* 3, 81–89.

Goebel, F.R. and Way, M. (2009) Crop losses due to two sugarcane stemborers in Reunion and South Africa. *Sugar Cane International* 27 (3), 107–111.

Goebel, F.R., Tabone, E., Do Thi Khanh, H., Roux, E., Marquier, M. and Frandon, J. (2010) Biocontrol of *Chilo sacchariphagus* (Lepidoptera: crambidae) a key pest of sugarcane: lessons from the past and future prospects. *Sugar Cane International* 28, 128–132.

Goebel, R., Fernandez, E., Begue, L.J. and Alauzet, C. (1999a) Prédation par *Pheidole megacephala* (*Fabricius*) (Hym.: *Formicidae*) des oeufs de *Chilo sacchariphagus* (Bojer) (Lep.: *Pyralidae*), foreur de la canne à sucre à l'île de la Réunion. *Annales de la Société Entomologique de France* 35, 440–442.

Goebel, R., Fernandez, E., Tiber, R. and Alauzet, C. (1999b) Dégâts et pertes de rendement sur la canne à sucre dus au foreur *Chilo sacchariphagus* (Bojer) à l'île de la Réunion (Lep.: Pyralidae). *Annales de la Société Entomologique de France* 35, 476–481.

Goebel, R., Soula, B. and Tabone, E. (2002) Biological control of the sugarcane stemborer *Chilo sacchariphagus* by the use of *Trichogramma chilonis*. *IOBC Egg Parasitoid News* 14, 36–37.

Goebel, R., Way, M.J. and Gossard, C. (2005) The status of *Eldana saccharina* (Lepidoptera: Pyralidae) in the South African sugar industry based on regular survey data. *Proceedings of the South African Sugar Technologists Association* 79, 337–346.

Gregory, P.J., Johnson, S.N., Newton, A.C. and Ingram, J.S.I. (2009) Integrating pests and pathogens into climate change/food security debate. *Journal of Experimental Botany* 60, 2827–2838.

Gurr, G. and Kvedaras, O.L. (2010) Synergizing biological control: scope for sterile insect technique, induced plant defences and cultural techniques to enhance natural enemy impact. *Biocontrol* 52 (3), 198–207.

Hall, D. (1988) Insects and mites associated with sugarcane in Florida. *Florida Entomologist* 71, 138–150.

Hensley, S.D., Fanguy, H.P. and Giamalva, M.J. (1977) The role of varietal resistance in control of the sugarcane borer, *Diatraea saccharalis* (F.) in Louisiana. *Proceedings of the International Society of Sugar Cane Technologists* 16, 517–552.

Hunter, M.D. (2002) Landscape structure, habitat fragmentation, and the ecology of insects. *Agriculture and Forestry Entomology* 4 (3), 159–166.

James, G. (2004) *Sugarcane*, 2nd edn. Blackwell Science, Oxford, UK.

Jeuffrault, E., Rolet, A., Reynaud, B., Manikom, R., Georger, S., Taye, T., Chiroleu, F., Fouillaud, M. and Vercambre, B. (2004) Vingt ans de lutte contre le ver blanc de la canne à sucre à la Réunion: Un succès, mais il reste des questionnements scientifiques pour confirmer la durabilité de la lutte biologique. *Phytoma – La Défense des Végétaux* 2004 (573), 16–19.

Jones, L.H.P. and Handreck, K.A. (1967) Silica in soils, plants, and animals. *Advances in Agronomy* 19, 107–149.

Keeping, M.G. (2006) Screening of South African sugarcane cultivars for resistance to the stalk borer *Eldana saccharina* Walker (Lepidoptera: Pyralidae). *African Entomology* 14, 277–288.

Keeping, M.G., Kvedaras, O.L. and Bruton, A.G. (2009) Epidermal silicon in sugarcane: cultivar differences and role in resistance to sugarcane borer *Eldana saccharina*. *Environmental and Experimental Botany* 66, 54–60.

Keeping, M.G., Meyer, J.H. and Sewpersad, C. (2013) Soil silicon amendments increase resistance of sugarcane to stalkborer *Eldana saccharina Walker* (Lepidoptera: Pyralidae) under field conditions. *Plant and Soil* 363, 297–318.

Keeping, M.G., Miles, N. and Sewpersad, C. (2014) Silicon reduces impact of plant nitrogen in promoting stalk borer (*Eldana saccharina*) but not sugarcane thrips (*Fulmekiola serrata*) infestations in sugarcane. *Frontiers in Plant Science* 5, 289–300.

Kfir, R., Overholt, W.A., Khan, Z.R. and Polaszek, A. (2002) Biology and management of economically important lepidopteran cereal stem borers in Africa. *Annual Review of Entomology* 47, 701–731.

Kiritani, K. (2006) Predicting impacts of global warming on population dynamics and distribution of arthropods in Japan. *Population Ecology* 48, 5–12.

Korndorfer, A.P., Grisoto, E. and Vendramim, J.D. (2011) Induction of insect plant resistance to the spittlebug *Mahanarva fimbriolata* Stal (Hemiptera: Cercopidae) in sugarcane by silicon application. *Neotropical Entomology* 40, 387–392.

Kvedaras, O.L., Keeping, M.G., Goebel, F.R. and Byrne, M. (2007a) *Eldana saccharina* Walker (Lepidoptera: Pyralidae) larval performance and stalk damage in sugarcane: influence of plant silicon and feeding site. *International Journal of Pest Management* 53 (3), 183–194.

Kvedaras, O.L, Keeping, M.G., Goebel, F.R. and Byrne, M. (2007b) Water stress augments silicon-mediated resistance of susceptible sugarcane cultivars synergy in resistance of sugarcane cultivars to the Stalk Borer, *Eldana saccharina* Walker (Lepidoptera: Pyralidae). *Bulletin of Entomological Research* 97 (2), 175–183.

Kvedaras, O.L., Keeping, M.G., Goebel, F.R. and Byrne, M.J. (2007c) Silicon and water stress synergy in resistance of sugarcane cultivars to the stalk borer, *Eldana saccharina* (Lepidoptera: Pyralidae). *Sugar Cane International* 25, 3–6.

Kvedaras, O.L., Byrne, M.J., Coombs, N.E. and Keeping, M.G. (2009) Influence of plant silicon and sugarcane cultivar on mandibular wear in larval *Eldana saccharina* Walker (Lepidoptera: Pyralidae). *Agricultural and Forest Entomology* 11, 301–306.

Kvedaras, O., An, M., Choi, Y.S. and Gurr, G.M. (2010) Silicon enhances natural enemy attraction and biological control through induced plant defences. *Bulletin of Entomological Research* 100, 367–371.

Leslie, G. (2004) Pests of sugarcane. In: James, G. (ed.) *Sugarcane*, 2nd edn. Blackwell Science, Oxford, UK, pp. 78–100.

Lewis, S.E., Brodie, J.E., Bainbridge, Z.T. Rohde, K.W., Davis, A.M., Masters, B.L., Maughan, M., Devlin, M., Mueller, J.F. and Schaffelke, B. (2009) Herbicides: a new threat to the Great Barrier Reef. *Environmental Pollution* 157, 2470–2484.

Long, W.H. and Hensley, S.D. (1972) Insect pests of sugarcane. *Annual Review of Entomology* 17, 149–176.

Lopez, E., Fernandez, C. and Lopez, O. (1983) Effect of nitrogen fertilization on *Diatraea saccharalis* (Fbr.) incidence on sugarcane. *Proceedings of the International Society of Sugar Cane Technologists* 18, 910–914.

LSU AgCenter (2010) Louisiana recommendations for control of sugarcane insects. In: *Louisiana Insect Pest Management Guide 2010*. Louisiana State University Agricultural Center Publication 1838, Baton Rouge, Louisiana, pp. 130–131.

Ma, J.F. (2004) Role of silicon in enhancing the resistance of plants to biotic and abiotic stresses. *Soil Science and Plant Nutrition* 50, 11–18.

Ma, J.F. and Yamaji, N. (2006) Silicon uptake and accumulation in higher plants. *Trends in Plant Science* 11 (8), 392–397.

Ma, J.F. and Yamaji, N. (2015) A cooperative system of silicon transport in plants. *Trends in Plant Science* 20, 435–442.

Mahlanza, T., Rutherford, R.S., Snyman, S.J. and Watt, M.P. (2014) *Eldana saccharina* (Lepidoptera: Pyralidae) resistance in sugarcane (*Saccharum* sp.): effects of *Fusarium* spp., stalk rind, fibre and nitrogen content. *African Entomology* 22 (4), 810–822.

McAllister, C.D., Hoy, J.W. and Reagan, T.E. (2008) Temporal increase and spatial distribution of Sugarcane yellow leaf and infestations of the aphid vector, *Melanaphis sacchari*. *Plant Disease* 92, 607–615.

Meagher, R.L., Irvine, J.E., Breene, R.G., Pfannenstiel, R.S. and Gallo-Meagher, M. (1996) Resistance mechanisms of sugarcane to Mexican rice borer (Lepidoptera: Pyralidae). *Journal of Economic Entomology* 89, 536–543.

Moreno Salguero, C.A., Bustillo Pardey, A.E., López Núñez, J.C., Castro Valderrama, U. and Ramírez Sánchez, G.D. (2012) Virulencia de nematodos entomopatógenos para el control del salivazo

Aeneolamia varia (Hemiptera: Cercopidae) en caña de azúcar. *Revista Colombiana de Entomología* 38, 260–265.

Nikpay, A. (2016) Effects of burning trashes on leaf feeders *Mythimna* spp. (Lepidoptera: Noctuidae) damage in sugarcane fields. *Proceedings of the Australian Society of Sugarcane Technologists* 38, 16.

Nikpay, A. (2017) Damage assessment of sugarcane whitefly *Neomaskellia andropogonis* Corbett and population dynamics on seven commercial varieties in south west of Iran. *Sugar Tech* 19, 198–205.

Nikpay, A. and Goebel, F.R. (2016) Major sugarcane pests and their management in Iran. *Proceedings of the International Society of Sugar Cane Technologists* 29, 103–108.

Nikpay, A. and Soleyman-Nejadian, E. (2014) Field applications of silicon-based fertilizers against sugarcane yellow mite *Oligonychus sacchari*. *Sugar Tech* 16, 319–324.

Nikpay, A., Arbabi, M. and Sharafizadeh, P. (2013) Integrated management of *Oligonychus sacchari* (Prostigmata: Tetranychidae) in sugarcane commercial fields. In: *2nd Global Conference on Entomology*, Kuching, Sarawak, Malaysia, p. 332.

Nikpay, A., Soleyman-Nejadian, E., Goldasteh, S. and Farazmand, H. (2015) Response of sugarcane and sugarcane stalk borers *Sesamia* spp. (Lepidoptera: Noctuidae) to calcium silicate fertilization. *Neotropical Entomology* 44, 498–503.

Nickpay, A., Soleyman-Nejadian, E., Goldasteh, S. and Farazmand, H. (2017) Efficacy of silicon formulations on sugarcane stalk borers, quality characteristics and parasitism rate on five commercial varieties. Proceedings of the National Academy of Sciences, India, Section B: Biological Sciences, 87, pp. 289–297.

Pandey, S.K. (2014) Effect of nitrogen levels on the incidence of stalk borer (*Chilo auricilius* Dudgeon) in sugarcane varieties. *Agricultural Science Digest* 34 (2), 134–136.

Pannuti, L.E.R., Baldin, E.L.L., Gava, G.J.C., Silva, J.P.G.F., Souza, E.S. and Kolln, O.T. (2015) Interaction between N-fertilizer and water availability on borer-rot complex in sugarcane. *Bragantia, Campinas* 74 (1), 75–83.

Reay-Jones, F.P.F., Way, M.O., Sétamou, M., Legendre, B.L. and Reagan, T.E. (2003) Resistance to the Mexican rice borer (Lepidoptera: Crambidae) among Louisiana and Texas sugarcane cultivars. *Journal of Economic Entomology* 96, 1929–1934.

Reay-Jones, F.P.F., Reagan, T.E., Way, M.O. and Legendre, B.L. (2005) Concepts of areawide pest management of the Mexican rice borer (Lepidoptera: Crambidae). *Sugar Cane International* 23 (3), 20–24.

Reay-Jones, F.P.F., Rochat, J., Goebel, R. and Tabone, E. (2006) Functional response of *Trichogramma chilonis* to *Galleria mellonella* and *Chilo sacchariphagus* eggs. *Entomologia Experimentalis et Applicata* 118, 229–236.

Reynolds, O.L., Keeping, M.G. and Meyer, J.H. (2009) Silicon-augmented resistance of plants to herbivorous insects: a review. *Annals of Applied Biology* 155, 171–186.

Robène-Soustrade, I., Jouen, E., Pastou, D., Payet-Hoarau, M., Goble, T. *et al.* (2015) Description and phylogenetic placement of *Beauveria hoplocheli* sp. nov. used in the biological control of the sugarcane white grub, *Hoplochelus marginalis*, on Reunion Island. *Mycologia* 107 (6), 1221–1232.

Sallam, M.N. (2006) A review of sugarcane stemborers and their natural enemies in Asia and Indian Ocean Islands: an Australian perspective. *Annales de la Société Entomologique de France* 42, 263–283.

Scriber, J.M. (1984) Host-plant suitability. In: Bell, W.J. and Cardé, R.T. (eds) *Chemical Ecology of Insects*. Chapman & Hall, London, pp. 159–200.

Showler, A.T. and Castro, B.A. (2010) Influence of drought stress on Mexican rice borer (Lepidoptera: Crambidae) oviposition preference in sugarcane. *Crop Protection* 29, 415–421.

Singh, M., Jadaun, V.C., Singh, S.R., Singh, A., Lal, K. and Singh, S.B. (2003) Chemical control of sugarcane yellow mite (*Oligonychus sacchari* Hirst). *Sugar Tech* 5, 77–78.

Smeets, E., Junginger, A., Faaij, A., Walter, P., Dolzan, P. and Turkenburg, W. (2008) The sustainability of Brazilian ethanol – an assessment of the possibilities of certified production. *Biomass and Bioenergy* 32, 781–813.

Sommartya, P., Suasa-ard, W. and Puntongcum, A. (2007) Natural enemies of sugarcane longhorn stem borer, *Dorysthenes buqueti* Guerin (Coleoptera: Cerambycidae), in Thailand. *Proceedings of the International Society of Sugar Cane Technologists* 26, 858–862.

Suasa-ard, W., Sommartya, P., Buchatian, P., Puntongcum, A. and Chiangsin, R. (2008) Effect of *Metarhizium anisopliae* on infection of sugarcane stems borer, *Dorysthenes buqueti* Guerin (Coleoptera: Cerambycidae) in laboratory. *Proceedings of the 46th Kasetsart University Annual Conference*, Kasetsart, Thailand, pp. 155–160.

Tabashnik, B.E., Gassmann, A.J., Crowder, D.W. and Carrière, Y. (2008) Insect resistance to Bt crops: evidence versus theory. *Nature Biotechnology* 26, 199–202. DOI:10.1038/nbt1382

Teran, F.O. (1980) Natural control of *Diatraea saccharalis* (Fabr. 1794) eggs in sugarcane fields of São Paulo. *Proceedings of the International Society of Sugar Cane Technologists* 17, 1704–1714.

Tilman, D., Cassman, K.G., Matson, P.A., Naylor, R. and Polasky, S. (2002) Agricultural sustainability and intensive production practices. *Nature* 418, 671–677.

Viator, R.P., Richard Jr, E.P., Garrison, D.D., Dufrene Jr, E.O. and Tew, T.L. (2005) Sugarcane cultivar yield response to planting date. *Journal of the American Society of Sugar Cane Technologists* 25, 78–87.

Vreysen, M.J.B., Robinson, A.S. and Hendrichs, J. (2007) *Area-wide Control of Insect Pests: From Research to Field Implementation*. Springer, Dordrecht, The Netherlands.

Webster, T.M., Brenchley, P.G. and Conlong, D.E. (2009) Progress in the area-wide integrated pest management plan for *Eldana saccharina* Walker (Lepidoptera: Pyralidae) in the Midlands North region of KwaZulu-Natal. *Proceedings of the South African Sugar Technologists Association* 82, 471–485.

White, T.C.R. (1984) The abundance of invertebrate herbivores in relation to the availability of nitrogen in stressed food plants. *Oecologia* 63, 90–103.

White, W.H. and White, P.M. (2014) Sugarcane borer resistance in sugarcane as affected by silicon applications in potting medium. *Journal of the American Society of Sugar Cane Technologists* 33, 38–54.

Williams, F.X. (1931) *The Insects and Other Invertebrates of Hawaiian Sugarcane Fields*. Hawaiian Sugar Planter's Association and Advertiser Publishing Company, Honolulu, Hawaii.

Williams, J.R., Metcalfe, J.R., Mungomery, R.W. and Mathes, R. (eds) (1969) *Pests of Sugar Cane*. Elsevier, Amsterdam.

9 Integrated Pest Management in Cotton

Shoil M. Greenberg[1] and Megha N. Parajulee[2,*]

[1]*Kika de la Garza Subtropical Agricultural Research Center, Weslaco, USA (retired);*
[2]*Texas A&M University System, Lubbock, USA*

9.1 What Is IPM?

IPM stands for Integrated Pest Management. The concept of IPM has been defined in various ways. Some examples follow.

1. IPM is a sustainable approach to managing pests by combining biological, cultural, physical and chemical tools in a way that minimizes economic, health and environmental risks (ND IPM Homepage, Texas Pest Management Association – https://www.ag.ndsu.edu/ndipm/documents/ipm-definition; http://www.tpma.org/) (FAO, 2004; Greenberg *et al.*, 2012b).

2. IPM is a knowledge-based decision-making process that anticipates limits and eliminates or prevents pest problems, ideally before they have become established. IPM typically combines several strategies to achieve long-term solutions. IPM programmes include education, proper waste management, structural repair, maintenance of biological and mechanical control techniques, and pesticide application when necessary (http://pestcontrolcanada.com/integrated-pest-management-ipm/).

3. IPM is a pest management strategy that focuses on long-term prevention or suppression of pest problems through a combination of techniques such as: monitoring for pest presence and establishing treatment threshold levels; using non-chemical practices to make the habitat less conducive to pest development; improving sanitation; and employing mechanical and physical controls. Pesticides that pose the least possible hazard and are effective in a manner that minimizes risk to people, property and the environment are used only after careful monitoring indicates they are needed, according to established guidelines and treatment thresholds (California Department of Pesticide Regulation – http://apps.cdpr.ca.gov/schoolipm/overview/definition_ipm.cfm).

4. IPM employs approaches, methods and disciplines to minimize environmental impact and risks, and optimize benefits (Greenberg *et al.*, 2012b).

These four definitions of IPM clearly suggest that IPM is an approach of pest management where multiple available and compatible tactics and tools are used, simultaneously and/or sequentially, to suppress the pest population below economically damaging levels.

An expansion of the IPM concept is the approach of Integrated Crop Management (ICM), which includes other agricultural decision-making tasks such as fertilizer and soil water management (Greenberg *et al.*, 2012b). An ICM programme would include an IPM component to deal with pest management decisions and address remaining issues applicable to the total crop production

---

* Corresponding author e-mail: m-parajulee@tamu.edu

process (Ohio Pest Management & Survey Program). Thus IPM is a system of pest management decisions based on ecological, economic and sociological values. Pest management practices may be classified according to the approach or method used to deal with a pest problem. In terms of approach, pest management practices may be designed to prevent, suppress or eradicate problems. IPM approaches and methods are used to minimize environmental contamination and risk from harmful organisms, and to optimize benefits (Greenberg *et al.*, 2012b). It is a systems approach to pest management that utilizes decision-making procedures based on either quantitative or qualitative observations of the pest problem and the related host or habitat (Ohio Pest Management & Survey Program).

The US Environmental Protection Agency provides a useful set of IPM principles, as below.

1. *Acceptable pest levels*: emphasis on pest situation under control, but not the total removal of pests. Pest management depends on action thresholds, and the control measures are applied only if the pest densities exceed those thresholds.

2. *Preventive cultural practices*: pest populations can be prevented from reaching or exceeding the action thresholds via pest tolerant crop cultivar selection, plant quarantine, cultural techniques and plant sanitation.

3. *Monitoring*: regular observation of the pest situation (pest scouting) is the cornerstone of IPM.

4. *Mechanical controls*: are the first option to consider, if a pest should reach an unacceptable level (using traps, vacuuming, and tillage to disrupt breeding).

5. *Biological controls*: natural biological processes and materials can provide pest suppression, with minimal environmental impact, and often at low cost.

6. *Chemical controls*: synthetic pesticides are generally only used as required to suppress the population below threshold, and often only at specific times in a pest life cycle, with a preference for pesticide groups that are derived from plants or naturally occurring substances (Bennett *et al.*, 2005; Greenberg *et al.*, 2012b).

9.2 Cotton: An Important Crop Worldwide

Cotton, grown around the world in more than 100 countries, is the most important fibre crop, accounting for 40% of the world's fibre market and generating high levels of income and employment. Its production mainly provides fibre to the textile industry, but it also contributes to food and feed. About 45% of the harvested crop is composed of the seed, which is crushed to separate its three components – oil, meal and hulls. Cottonseed oil is a common component of many food items, used primarily as a cooking oil, shortening and salad dressing. The meal and hulls are used as feed for livestock, poultry and fish, and as fertilizer. Present world production is about 25.5 million tons of seed cotton from 34.8 million ha. Cotton is a major export revenue source in the United States.

The world's land area dedicated to cotton production has not significantly changed during the last 5 years, with an average of 30.0 million ha annually. Average yield of cotton fibre ranges from 740 to 795 kg per ha. India, China, United States, Pakistan and Brazil are the main cotton producing countries in the world, with 5.7, 4.8, 2.8, 1.5 and 1.3 million metric tonnes of lint produced (https://www.statista.com/statistics/263055/cotton-production-worldwide-by-top-countries/), respectively. These five countries account for over 75% of the global cotton production. Altogether, either on-farm or in associated industries such as transportation, storage and ginning, around 350 million people are estimated to be involved in cotton production worldwide.

In the US, cotton production covers an area of around 11.0 million acres (4.5 million ha), being realized in more than 18,600 farms. In 2014, mean yield has been 829.0 kg/ha. Around $5.7 billion in cash are generated by this crop for farmers and $27.6 billion by its industry in revenues for various productive segments. Total economic cotton production activity is estimated at approximately $100 billion. US cotton sold overseas has recently averaged $7 billion per year. The US commonly supplies 12 million bales or more of the world cotton exports, accounting for over 30% of the total world export market. The US also exports more than 3.5 million bale equivalents of cotton textile products annually. Cottonseed is another important component of cotton production, with an average of 5.2 million tons annually, yet being often overlooked. It is used both as cottonseed and cottonseed meal in feed for livestock, dairy cattle and poultry (more than 6 billion pounds) or cottonseed oil for food products ranging from margarine to salad dressing (approximately 90 million gallons).

Cotton is propagated by seed, in low-density (15,000 to 40,000 plants/ha) or high-density (100,000 to 160,000 plants/ha) population. Medium- and heavy-textured, deep well drained, fertile clayey, alluvial, chernozem and laterite soils with good water holding capacity are the best soils for cotton cultivation, though this crop may well adapt on a wide range of soil types (but acidic or dense subsoils limit root penetration). Optimum pH ranges from 7.0 to 8.0, though down to 5.5 can be tolerated. Cotton growing period is about 150 to 180 days, depending on temperature and variety. On average, 50 to 85 days are required from planting to first bud formation, then 25 to 30 days for flower formation and additionally 50 to 60 days from flower opening to boll maturity. Climate, especially temperature, highly influences development of cotton, which is very sensitive to frost (requiring a minimum of 200 frost-free days) and whose vegetative growth is favoured by cool nights and low daytime temperatures, which reduce production of fruiting branches. With the exception of day-neutral varieties, cotton is usually a short-day plant, with temperature influencing however the effect of day length on flowering. Optimum temperatures are: 18 to 30°C for germination (min. 14°C, max. 40°C); around 30°C for early vegetative growth (min. 20°C); 12 to 27°C (night) and 20 to 40°C (daytime) for proper bud formation and flowering; 27 to

32°C for boll development and maturation (yields reduced above 38°C). Wind and rain greatly influence cotton production: young seedlings are seriously affected by strong and/or cold winds; fibre is blown away from opened bolls and dirtied with soil particles; pollination is impaired and fibre quality reduced by continuous rains occurring during flowering and boll opening.

Though cotton can be grown under limited water supply, irrigation is needed for an intensive production, up to a requirement of more than 20,000 litres/kg of cotton; this is why cotton production was concentrated during the past in large river basins, such as the Indus River in Pakistan, the Murray–Darling Basin in Australia and the Rio Grande in the USA and Mexico. Under different climatic conditions and varying length of growing season (150–210 days), with an average daily evapotranspiration rate of 4 to 8 mm/day, if high yields are required, cotton seasonal water requirements are 350 to 900 mm/ha. In many countries, modern techniques for water supply have proved to be applicable and economically viable. In different agro-ecological situations, drip irrigation allows 15 to 30% increment in yields, while reducing water (30 to 45%) and energy (20%) use, as well as labour (15%) and plant protection chemicals (5%) inputs. Compared to conventional furrow irrigation, it also improves lint quality. Drip irrigation systems may be also used to apply nutrients (fertigation), allowing an efficient reaction to plants' fertility needs. Also, the use of daily growing rates or leaf water potential measurements enable efficient use of water, fertilizer and energy inputs.

9.3 Pests Damaging Cotton Production

Cotton production is especially threatened by insect attacks and by weed competition during the early stages of development (Greenberg et al., 2012b). Pathogens may be harmful in some areas and years, but are considered to be only of minor importance. Only recently have viruses reached pest status in south Asia and some states of the United States. The estimates of the potential losses of arthropod pests and weeds averaged worldwide at 37 and 36%, respectively. Pathogens and viruses contribute about 9% to total potential loss. The proportional contribution of crop protection in overall cotton production varies from 0.37 in west Africa to 0.65 in Australia where the intensity in cotton production is very high. It is estimated that 29% of attainable production is lost to crop pests (Oerke, 2006). Insect pests are the principal cause of yield losses in cotton. Estimates indicate that the yield losses due to insects would amount to about 15% of world annual production.

In the United States, the economic arthropod pests of cotton are boll weevil, *Anthonomus grandis grandis* Boheman (current distribution limited to small southern region of Texas); cotton bollworm, *Heliothis zea* (Boddie); tobacco budworm, *Heliothis virescens* (Fabricius); cotton fleahopper, *Pseudatomoscelis seriatus* (Reuter); beet armyworm, *Spodoptera exigua* (Hübner); cabbage looper, *Trichoplusia ni* (Hübner); cotton aphid, *Aphis gossypii* Glover; cotton leaf perforator, *Bucculatrix thurberiella* Busck; black cutworm, *Agrotis ipsilon* (Hufnagel); fall armyworm, *Spodoptera frugiperda* (Smith); pink bollworm, *Pectinophora gossypiella* (Saunders); whitefly, *Bemisia* spp. gr. *tabaci* (Gennadius); spider mite, *Tetranychus* spp.; thrips, *Thrips* spp.; cotton leafminer, *Stigmella gossyppi* (Forbes & Leonard); plant bug, *Creontiades signatus* (Distant); yellowstriped armyworm, *Spodoptera ornithogalli* (Guenée); Texas leafcutter ant, *Atta texana* Buckley; and lubber grasshopper, *Brachystola magna* (Girard) (Greenberg *et al.*, 2010; 2012b).

In the US, arthropod pests alone reduce overall cotton yield by over $400 million (mean value for 2005–2007), in Texas – $99.3 million (Table 9.1; Williams, 2006–2008). However, four or five dominant species constitute 90% of the arthropod pest problem in a given cotton-producing region that require management interventions. Control is needed only when a pest population reaches an economic threshold (Table 9.2) or treatment level at which further increases would result

Table 9.1. US and Texas cotton arthropod losses for 2006 (Williams, 2007).

	US		Texas	
Arthropods	Rank by % loss	Bales lost	Rank by % loss	Bales lost
Bollworm/Budworm	1	229,186	2	78,826
Lygus bugs	2	171,478	6	10,314
Thrips	3	145,040	3	65,062
Fleahopper	4	119,745	1	108,057
Aphids	5	80,418	4	61,162
Stink bugs	6	68,823	5	13,186
Spider mites	7	60,720	10	2,917
Whitefly	8	14,817	8	3,926
Fall armyworm	9	12,071	7	5,404
Boll weevil	10	3,190	9	3,190
Beet armyworm	11	1,104	12	229
Cutworms	12	1,100	0	0
Saltmarsh caterpillars	13	237	0	0
Pink bollworm	14	232	13	28
Loopers	15	131	0	0
Grasshopper	16	144	0	0
Green mirids	17	0	11	685
Total losses (bales*)		908,436		352,985

*One bale of lint = 200 kg

Table 9.2. Economic thresholds of major cotton arthropods in Texas cotton (Norman and Sparks, 2003; Castro *et al.*, 2007).

Insects	Season	Economic threshold
Boll weevil	All	40 overwintered boll weevils per acre; 15–20% damage squares from squaring to peak bloom
Thrips	Germination to 3–4 true leaves	The average number of thrips counted per plant is equal to the number of true leaves at the time of inspection
Fleahoppers (FH)	All	1st–3rd weeks of squaring – 15–25 nymphs and adults per 100 terminals. After 1st bloom – treatment is rarely justified
Aphids	All	≥50 aphids per leaf
Whiteflies	All	When ≥40% of the 5th node leaves infested with 3 or more adults
Plant Bugs (*Creontiades* spp.)	First 4–5 weeks of fruiting	15–25 bugs per 100 sweeps
Spider mites	All	When 50% of the plants show noticeable reddened leaf damage
Bollworm	Before bloom After boll formation	≥30% of the green squares examined are worm-damaged and small larvae are present 10 worms ≤¼-inch in length per 100 plants and 10% damaged fruit for non-*Bt* cottons; or 10 worms >¼-inch in length per 100 plants with 5% damaged fruit
Beet armyworm	All	When leaf feeding and small larvae counts exceed 16–24 larvae per 100 plants and at least 10% of plants examined are infested; when feeding on squares, blooms or bolls the threshold needs to be 8–12 larvae larger than ¼-inch per 100 plants
Fall armyworm	Pre-flower After first flower	30% of the green squares are damaged 15–25% small larvae are present per 100 plant terminals and 10–15% of squares or bolls are worm damaged

in excessive yield or quality losses. This level is one of the important indexes in IPM for optimizing control and minimizing risk from arthropods (Greenberg *et al.*, 2012b).

Pest suppression activities are initiated when pest populations reach treatment thresholds. Treatment thresholds are when a pest population does not exceed size of insect pests that require treatment, so pest population size does not exceed the economic injury level (EIL). Economic injury levels are the point at which the value of the crop lost exceeds the cost of control (Greenberg *et al.*, 2010, 2012b).

Knowledge of growth stages is important to the proper timing of scouting procedures and treatments (Table 9.3).

9.4 Sampling Insect Populations

Field observations (scouting) are a vital component of cotton insect management. It is necessary that fields are inspected at least once and preferably twice a week in order to detect the species present, their stages and densities, and pest damage to the plants (Greenberg *et al.*, 2012b). Scouting should also include monitoring plant growth, fruiting profile, weeds, diseases, beneficial insect activity, and the effects of implemented pest suppression practices. The number of samples taken is based upon the field (plot) size and the variability between samples. There are different sampling methods: visual observation of plants (generally range from 25 to 100 plants, and preferably examine 5 consecutive plants in 10–20 representative locations within a field); sweep net (5 sweeps per sample, at least 20 samples per field); beat bucket (3–5 plants per bucket, sample, at least 20 samples per field); drop cloth (the standard length – three feet long (=0.91 m), used if row spacing is 30 inches (=0.76 m) or wider, a minimum of 4–6 drop cloth samples should

Table 9.3. Cotton development calendar (number of total days) and heat units (Texas Lower Rio Grande Valley, IPM Annual Report, 2006–2008).

Developmental time periods	Number of days		Accumulated heat units, DD60	
	Avg.	Range	Avg.	Range
Planting to emergence	7	5–10	109	59–159
Emergence of:				
First true leaf	8	7–9	166	127–205
Six true leaves	25	23–27	463	321–608
Pinhead square	29	27–30	517	378–663
1/3 grown square	43	35–48	752	508–996
Square initiation to bloom	23	20–25	924	719–1129
Bloom to peak bloom	18	14–21	1280	977–1582
Full grown boll	23	20–25	1383	1091–1674
Open boll	47	40–55	1939	1857–2021

DD60 heat unit measurements are in Fahrenheit (F°). Conversions from degrees Fahrenheit to degrees Centigrade (C°): $C° = (F° - 32) * 5/9$.

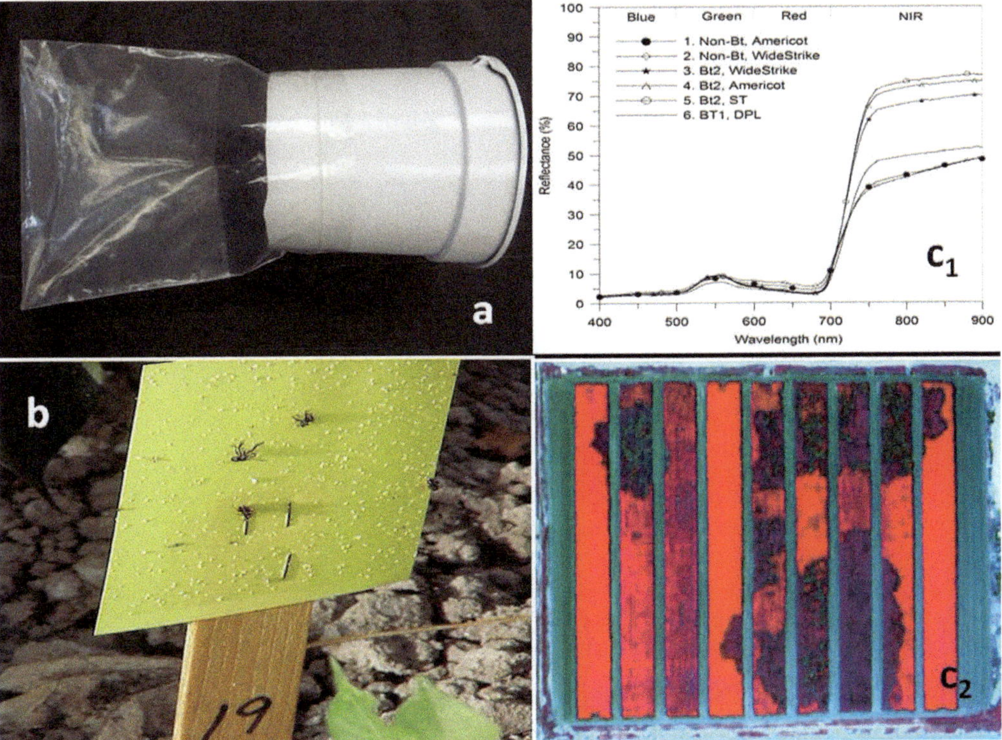

Fig. 9.1. Examples of the sampling methods: (a) modified beat bucket method; (b) yellow-coloured sticky traps; (c_1) and (c_2) remote sensing technology.

be taking per field); coloured sticky traps; and pheromone traps. Some of the common sampling methods are shown in Fig. 9.1 (Greenberg *et al.*, 2012b). Methods of identification and sampling of cotton pests and beneficials, and techniques for collecting them are published in many sources, especially in the Extension entomology literature (Steyskal *et al.*, 1986; Bohmfalk *et al.*, 2002; Norman and Sparks, 2003; Greenberg *et al.*, 2005b). The cost of controlling insects is one of the significant inputs of the crop production budget, annually averaging from $70 to over $100 per acre (https://ipmdata. ipmcenters.org/documents/pmsps/ MidsouthCottonPMSP.pdf).

9.5 Economic and Environmental Impacts of Synthetic Chemicals for Insect Control

Synthetic chemicals continue to be the primary tool for insect control (Greenberg *et al.*, 2012a). Approximately 3,000,000,000 kg of pesticides is applied each year with a purchase price of nearly $40 billion per year worldwide (https:// www.ncbi.nlm.nih.gov/pmc/articles/ PMC2946087/). In the US, approximately 500 million kg of more than 600 different pesticide types are applied annually at a cost of $10 billion (Pimentel and Greiner, 1997). In general, each dollar invested in pesticide control returns approximately $4 in benefit (Pimentel, 1997). While conventionally grown cotton requires significant amount of insecticides for effective and economic production, an extensive array of efficient insecticide chemistries is available for use in cotton production (Table 9.4; Tomilin, 2003; Greenberg *et al.*, 2012a).

Each year cotton producers around the world use nearly $2.6 billion worth of pesticides, equalling more than 10% of the world's pesticides and nearly 25% of the world's insecticides (Perschau and Sanfilippo, 2007). Recent statistics suggested that 150.0 million pounds (74.1 million kg) of insecticides was used annually on agricultural crops in the United States. Over half of this amount was applied to cotton fields, corresponding roughly to 6.0 lb (2.96 kg) of active ingredient per acre or 14.8 lb (7.3 kg) of active ingredient per ha

(Gianessi and Reigner, 2006). Although, pesticides are generally profitable in agriculture, their use does not always decrease crop losses. For example, despite the more than 10-fold increase in insecticide (organochlorides, organophosphates, and carbamates) use in the US from 1945 to 1990, total crop losses from insect damage have nearly doubled from 7% to 13% (Pimentel *et al.*, 1991; Greenberg *et al.*, 2012a). This rise in crop losses to insect pests is, in part, caused by changes in agricultural practices (Greenberg *et al.*, 2012a). Most benefits of pesticides are based on the direct crop returns. Such assessment does not include the indirect environmental and economic costs associated with the recommended applications of pesticides in crops. An investment of about $10 billion in pesticides control each year saves approximately $40 billion in US crops, based on direct costs and benefits. However, the indirect costs of pesticide use to the environment and public health needs to be balanced against these benefits. Based on the available data, the environmental and public health costs of recommended pesticide use total more than $9 billion each year (Table 9.5) (Pimentel, 2005).

Users of pesticides pay directly only about $3 billion, whereas the additional $6 billion is accounted for by indirect costs of pesticides including pesticide resistance and destruction of natural enemies and negative impact on human health (Greenberg *et al.*, 2012a; Pimental and Burgess, 2014).

9.6 Changes in Insect Control of Texas Cotton IPM

During recent years there have been significant changes in Texas cotton IPM, and this system continues to evolve rapidly. These changes are occurring because of three major factors: improving boll weevil eradication; using new and more target-specific insecticides; and implementation

Table 9.4. Insecticides recommended for use on cotton in the USA.

Class	Common name	Brand name	Recommended target pests
OP	Acephate (0.5–1.0)*	Orthene® 90S (generics)	Thrips, cutworms, *Greontiadis* plant bugs, fleahoppers, cutworm, fall armyworm
OP	Dicrotophos (0.25–0.5)	Bidrin	Thrips, plant bugs, fleahoppers, stink bugs, aphids, boll weevil
OP	Dimethoate (0.11–0.22)	Dimethoate® (generics)	Thrips, fleahopper and *Greontiadis* plant bugs
OP	Malathion (0.61–0.92)	Fufanon® ULV9.9	Boll weevil
OP	Methamidophos (0.7–2.2)	Monitor®	Thrips, plant bugs, fleahoppers, whiteflies
C	Oxamyl (0.25)	Vydate® 2L	Boll weevil, plant bugs, fleahoppers
C	Methomyl (0.45)	Lannate®2.4LV	Aphids, beet armyworm, fall armyworm, fleahoppers
C	Thiodicarb (0.6–0.9)	Larvin®3.2	Bollworm, beet armyworm, fall armyworm, tobacco budworm, loopers
CN	Imidacloprid (0.05)	Provado®1.6F	Plant bugs, fleahoppers, aphids, whiteflies
CN	Acetamiprid (0.025–0.05)	Intruder®70WP	Aphids, whiteflies, fleahoppers
CN	Thiamethoxam (0.03–0.06)	Centric® 40WG	Plant bugs, aphids, whiteflies, fleahoppers
IGR	Methoxyfenozide (0.06–0.16)	Intrepid®2F	Beet armyworm, fall armyworm, loopers
OC	Dicofol (0.75–1.5)	Kelthane® MF	Spider mites
P	Bifenthrin (0.37)	Capture® or Discipline®	Bollworms, fall armyworm, aphids, plant bugs
P	Cyfluthrin (0.01–0.06)	Baythroid® 2E	Cutworm, stink bugs, bollworms, boll weevil, whiteflies
P	Cyhalothrin (0.01–0.04)	Karate®-Z	Cutworm, stink bugs, bollworms, boll weevil
P	Deltamelthrin (0.04–0.2)	Decis	Cutworm, stink bugs, bollworms, whiteflies, thrips
	Spiromesifen (0.094–0.25)	Oberon® 2SC	Whiteflies, spider mites
	Other Chemicals:		
	Plant Growth Regulation	Ethephon® (Prep®) Mepiquart® Clorade®	Modified plant growth
	Defoliants	Def®, Dropp®, Ginstar®	For early harvest

*In parentheses – rate active ingredient pound/acre
OP – organophosphate; C – carbamate; CN – Chloro-nicotinyl; IGR – insect growth regulator; OC – organochlorine; P – pyrethroid

Table 9.5. Total estimated environmental and social costs from pesticides in the US.

Costs	Amount (billion $/year)
Public health impacts	1140
Domestic animal deaths and contaminations	30
Losses due to the destruction of natural enemies	520
Cost of pesticides resistance	1500
Honey bee and pollination losses	334
Crop losses	1391
Fishery losses	100
Bird losses	2160
Groundwater contamination	2000
Government regulations to prevent damage	470
Total	9645

of transgenic Bt-cotton and other approaches to managing pests by combining biological, cultural and physical tools (Greenberg *et al.*, 2012b; Pimentel and Burgess, 2014).

A decrease in the availability due to high expenses of registering, producing and purchasing, as well as the adverse effects of pesticides on non-target organisms, particularly humans and the environment, are resulting in a reduced use of conventional pesticides as therapeutic measures to suppress pest populations. Thus, alternative pest control techniques are receiving increased attention, but unanswered questions remain plus a greater effort is needed to move more of these techniques towards acceptance or economic feasibility. The natural occurrences of beneficial populations are often too low to maintain pests at an acceptable level particularly in cases of widespread monocultures. As a result, augmentative biological control is being recognized as an important alternative to chemical control. They also often impact non-target systems. In addition, arthropods often develop resistance to specific pesticide chemistries rendering them ineffective over time.

Based on our studies, we demonstrated effects of abiotic and biotic factors, plant phenology and physiology, insect ecology, behaviour and reproductive potential on interrelationships between cotton, an economically important crop in Texas, and boll weevil, *A. grandis grandis*, beet armyworm, *S. exigua*, and sweet potato whitefly, *B. tabaci* MEAM1 (= *ex* biotype B), examples of some of the key insect pests on cotton, and how those can be useful for the development of IPM strategies. All these findings improved our capability to: (i) develop environmentally safe and efficient strategies focused on long-term prevention of pests or their damage; (ii) predict changes in insect populations relative to crop phenology and initial population density; (iii) develop methods to predict yield losses; and (iv) establish the optimum timing of insecticide applications. These four management guidelines are the cornerstones of any successful IPM strategies.

9.7 Boll Weevil

Within regions where they occur, the boll weevil (BW) is a destructive key pest on cotton. The US National Cotton Council estimated that the boll weevil cost the US cotton industry $300 million annually in the late 1990s. In 1995 (first attempt of BW eradication within Texas), farmers in LRGV (Lower Rio Grande Valley) region lost 13.5 million kg of cotton lint worth $150 million. In addition to 15% yield loss, extensive ULV malathion spraying to combat BW, mostly by plane, led to massive secondary pest (beet armyworm (BAW)) outbreaks and area wide natural enemy disruption (Summy *et al.*, 1996; Greenberg *et al.*, 2001a). In 2006, the second attempt of LRGV boll weevil eradication was initiated and outbreaks of whiteflies brought high economic losses via reduction of lint quality and yields (losses were more than 500 lb/ac, SMG unpublished data); again, due to extensive widespread use of synthetic insecticides disrupting natural enemy populations. Following the success of US boll weevil eradication programmes in all areas except the LRGV region of Texas, the US Cotton Belt has decreased overall insect-induced damage to cotton crop and cotton yield losses. Boll weevils continue to be an issue in the cotton-producing regions of Mexico and South America, where multiple applications of synthetic insecticides are required to manage this pest (Greenberg *et al.*, 2012b).

9.7.1 Boll weevil feeding and reproduction as a function of cotton square availability

The dependence of cotton square availability on boll weevil feeding and reproduction was determined by providing different numbers of cotton squares (flower buds) to individual weevils in the laboratory (Greenberg *et al.*, 2003a). Both males and females puncture squares by feeding. In addition, females oviposit in squares, leaving behind a sealed puncture. Squares were replaced daily. The number of lifetime punctures produced by boll weevil females and males increased with square availability (Greenberg *et al.*, 2003a). The total number of punctures (feeding and oviposition) caused by boll weevil females was 2.7-fold higher than that caused by males (feeding only) (Greenberg *et al.*, 2004c). Our results suggested that boll weevil populations may be regulated in part by a density-dependent mechanism based on the availability of squares of suitable size for oviposition (Greenberg *et al.*, 2007b). When resources are in short supply, competition between individuals within a population may reduce reproductive rate and survival. Fecundity was significantly higher (232.9–283.6 sealed punctures per female) in the 10:1, 15:1, and 20:1 (squares:female) treatments than in the 1:1 treatment (83.2). Survival of weevil progeny to adulthood was about twofold higher in the 10:1, 15:1, and 20:1 treatments than in the 1:1 and 5:1 treatments (Greenberg *et al.*, 2003a). Estimates of the intrinsic rate of increase (r_m) and of the population growth index (GI) indicate that availability of oviposition sites become progressively more limiting as the ratio of squares to

females decreases below 10:1. Boll weevil populations maintained at a square:female ratio of 10:1 would increase about 100-fold each generation (R_0), a rate significantly higher than that exhibited under 5:1 (32.6-fold) or 1:1 (21.5-fold) square:female ratios. When the numbers of cotton squares were increased from 1 to 10, 15 or 20 per female, the total number of progeny produced per female increased 3.2–4.8-fold (Greenberg *et al.*, 2003a).

Summary conclusions included:

1. Females laid fewer eggs per lifetime when provided with five or fewer squares per day. Although the percentage of squares infested with boll weevil larvae was about 2.0-fold greater in the 1:1 treatment than in the 10:1 treatment, there were fewer total punctures in the 1:1 and 5:1 treatments than in the 10:1 treatment. This suggests a reduction in oviposition and feeding behaviour associated with an increase in square damage. These results suggest that females may regulate the number of eggs deposited in squares in response to the number of squares available.

2. The percentage of oviposited eggs that hatched, indicated by the presence of larvae, was lowest in the 5:1 and 1:1 treatments. It is possible that the higher number of punctures in these treatments increased the likelihood of direct damage to eggs already oviposited, or negatively altered the microenvironment of oviposited eggs, perhaps through faster desiccation or decay of the heavily damaged square. Cannibalism or intraspecific competition among multiple larvae in a square may also have limited survival (Greenberg *et al.*, 2004c, 2006).

Our data indicate that the daily number of undamaged squares to which boll weevil females have access affects the number of sealed punctures (and presumably the number of eggs laid), the number of feeding punctures, survival of progeny to adulthood, and the ultimate number of adult progeny produced per lifetime. The threshold at which squares become limiting for these parameters appears to lie between 5 and 10 squares per female per day. The life table statistics generated from these data will help improve our capacity to predict square loss and changes in boll weevil populations in the field, given the initial density of suitable squares for oviposition and a corresponding initial population density of weevils (Greenberg *et al.*, 2003a, 2004c).

9.7.2 Influence of different cotton fruit phenotypic stages on reproductive development and survival of the boll weevil

Boll weevil feeding and reproductive potentials and survivals were determined on 10 different cotton fruit phenological stages. Females did not oviposit in pinhead squares. The fewest eggs were oviposited in bolls >30 mm in diameter (0.1 egg/boll). The highest number of eggs was recorded in squares sizes of 5–6 mm (7.3 eggs/square) and 7–8 mm (8.1 eggs/square). Boll weevil survival to adulthood was highest on squares measuring 7–8 or 9–10 mm in size (58.6–59.7%). No survival occurred in matchhead squares or bolls >30 mm. Duration of development was longest on boll sizes of 15–20 and 20–30 mm (18.2–18.8 d). The growth index (percentage immature survival divided by immature development time) of female boll weevils was 2.8-fold higher in 7–8 or 9–10 mm diameter squares than in 20–30 mm diameter bolls.

Summary: This information will improve our ability to develop methods to predict fruit losses and changes in boll weevil populations in the field, given a starting density of fruit suitable for oviposition, and a corresponding initial population density of weevils (Greenberg *et al.*, 2004c).

9.7.3 Size-dependent feeding and reproduction by boll weevil females

The relationship between the size of boll weevil females, measured in terms of pupal weight, and their feeding and oviposition activities was determined for females of 5 different weight categories: <5.0 mg, 5.1–10.0 mg, 10.1–15.0 mg, 15.1–20.0 mg and >20.1 mg. Egg deposition, the number of feeding punctures and puncture ratios (eggs/total punctures) were significantly affected by female weight. The fewest number of each were recorded for females with weights ≤5.0 mg, and the highest with weights >20.0 mg. Boll weevil females weighing >10.0 mg produced progeny with significantly higher survival to adulthood (66.1–73.6%) than those weighing ≤10.0 mg (38.5–44.2%). They also produced a significantly higher percentage of female progeny (58.3–62.5%) than females with pupal weights ≤10.0 mg (28.0–40.6%). The total number of lifetime oviposition punctures produced by females with pupal weights >10.0 mg were 16-fold higher than for females weighing ≤5.0 mg. The population growth indices for females having pupal weights >10 mg averaged 1.8-fold higher than those with pupal weights ≤10.0 mg. Life table calculations indicated that boll weevil populations with pupal weights of >10.0 mg will increase an average 8.4-fold higher each generation (R_0) than will females with weights ≤5.0 mg.

Summary: Cultural practices that result in the production of small adults may be used to impact the overall weevil populations (Greenberg *et al.*, 2004c, 2005c, 2007b, 2009a).

9.7.4 Temperature-dependent development and reproduction of the boll weevil

Effects of temperature on development, survival and fecundity of boll weevil were assessed at 10, 11, 12, 15, 20, 25, 30, 35, 45 and 46°C; 65% relative humidity, and a photoperiod of 13:11 (L:D, h). The mortality of boll weevil immature stages was 100% at 12°C and decreased to 36.4% as the temperature increased to 25°C. When the temperature increased from 30°C to 45°C, the mortality of weevils again increased from 50.1% to 100%. From 15°C to 35°C, the boll weevil preimaginal development rate was linearly related to temperature. The average developmental time of total boll weevil immature life stages decreased 3.6-fold and the preovipositional period decreased 3.3-fold when the temperature was increased from 15°C to 30°C. The lower threshold for development was estimated to be 10.9°C, 6.6°C, 7.0°C and 9.0°C for eggs, larvae, pupae, and total immature stages, respectively, with total thermal time requirement to complete immature stages of 281.8 DD (degree day) (15°C) and 247.8 DD (35°C). At 11°C and 46°C, female weevils did not oviposit. Longevity of adult females decreased 4.6-fold with increasing temperatures from 15°C to 35°C. Fecundity increased with increasing temperatures up to 30°C and significantly decreased thereafter.

Summary: These findings will be useful in creating a temperature-based degree-day model for predicting the occurrence of key life stages of BW in the field. An accurate predictor of a pest's development can be important in determining sampling protocols, timing insecticide applications, or implementing an IPM control strategy targeting susceptible life stages (Greenberg *et al.*, 2005d).

9.7.5 Circadian rhythms of feeding, oviposition and emergence of the boll weevil

Circadian rhythms of feeding, oviposition, and emergence of boll weevil adults were determined at five different photophases (24, 14, 12, 10 and 0 hours) and a constant 27°C temperature, 65% RH in the laboratory. Cotton squares exposed to boll weevil females in Petri dishes were removed and examined for feeding and oviposition punctures every 4 hours during daylight (0700–1900 h) and after 12 h of darkness (1900–0700 h) over eight consecutive days. Cohorts of randomly selected egg-punctured squares were sampled from oviposition females at 0700, 1100, 1500 and 1900 during 24 hours and under different photophase treatments, and maintained in Petri dishes at 27±1°C, 65% RH. Dishes were observed twice daily (1900 and 0700 h) for adults emerging during the day or night periods. Periodicity of oviposition was not affected by the length of the photophase. The boll weevil has round-the-clock circadian rhythm of oviposition, with a daytime preference. We observed that 82.4–86.0% of the boll weevil eggs were deposited between 0700 and 1900 h, and 14.0–17.6% between 1900 and 0700 h during a 24-h period. Feeding activity of boll weevil females under photoperiods of 24:0 h (complete light) and 0:24 h (complete darkness) did not significantly change between 0700–1900 h versus 1900–0700 h, while the daily cycle of light and darkness in other photoperiods significantly increased the feeding punctures from 0700–1900 compared with 1900–0700 h. The periodicity of emergence depended significantly on the time of oviposition and the length of the photophase.

Summary: Investigation of boll weevil circadian rhythm provides a better understanding of boll weevil ecology and reveals potential weak links for improving control technologies targeting their reproductive strategies (Greenberg *et al.*, 2006).

9.7.6 Cotton stalk destruction

In the subtropical Lower Rio Grande Valley of Texas, cotton regrows and produces fruit from undestroyed stalks throughout the winter, and in the spring period weevils from such locations become a serious threat to the next cropping season due to early colonization. The success of the boll weevil eradication programme in the LRGV, although BW has not been completely eradicated from this region, depended on thorough stalk destruction following harvest (Greenberg *et al.*, 2007b). However, adverse weather conditions and conservation tillage often impeded immediate and complete stalk destruction using typical tool implements, and alternative stalk control methods were needed. We also demonstrated that the herbicide, 2, 4-D-dimethyl ammonium (brand name Savage®), applied twice, immediately and 2 weeks after cotton harvest, was 100% effective in killing stalks, regardless of whether they were shredded or standing, or whether harvest was by stripper or picker.

Summary: Development of efficacious chemical stalk destruction methodology together with physical/cultural methods greatly improved boll weevil management in the LRGV of Texas, as well as other locations where year-round cotton growth and reproduction presented a severe challenge for boll weevil management (Greenberg *et al.*, 2010). Chemical-based stalk destruction offers a valuable tool for use in a variety of situations where mechanical stalk destruction is undesirable, inadequate or ineffective. This tool should greatly aid the establishment and maintenance of a host-free period for boll weevil management in tropical and subtropical environments. The use of 2, 4-D Amine for cotton stalk destruction will effectively eliminate food and reproductive opportunities for overwintering boll weevils and significantly facilitate boll weevil management. In the spring, boll weevils from such locations did not become a serious threat (farmers can eliminate 1–2 chemical applications) (Greenberg *et al.*, 2007a).

9.7.7 Effects of planting dates on boll weevils and cotton fruit in the subtropics

The effects of planting dates 2–3 weeks apart on boll weevil, field-level populations, and feeding and oviposition damage to cotton squares and bolls were studied in LRGV of Texas. Squares were 44–56% more abundant in some later planted treatments than in the earlier planted, but mean cumulative numbers of oviposition- and feeding damaged squares were 2.7–4.8-fold greater in some later planted treatments (Showler *et al.*, 2005). Increased square production in later planted cotton was offset by boll weevil infestation that occurred when squares are most vulnerable and contribute the most toward the pest's reproduction. Early planting avoided boll weevil population build-ups in the field when large squares were abundant (Showler *et al.*, 2005). Insecticide sprays in the earliest planted treatment of each year, based on the 10% damaged squares threshold, were >33% and >43%

fewer than in corresponding middle and late planting treatments, respectively. Delayed planting, relative to the onset of favourable cotton growing weather, at the field levels were more cost-effective when planting in optimal middle or late, than early (mean insecticide treatment cost \$/ha (recommended rate for cyfluthrin by tractor)): early, 5 March 2002, \$121.12/ha; middle, 19 March 2002, \$195.06/ha; and late, 2 April 2002, \$239.08/ha. Lint yields in this study did not differ significantly between the treatments (Showler *et al.*, 2005).

9.7.8 Effects of insecticides and defoliants for control of overwintering boll weevils

Pre-harvest application of the insecticides Karate® or Guthion® at half-rate with the cotton defoliant Def® (synergistic effects) allowed growers to attain the benefits of reduction in late-season pest populations before boll weevils, aphids and whiteflies disperse to overwintering habitats; more appropriate timing and rate of insecticides at the time of defoliation; and reduction in pesticide application costs and insecticide input into the environment. Those provided savings of at least \$6.5 million in the LRGV of Texas (Greenberg *et al.*, 2004b, 2012b).

9.7.9 Reproductive potential of overwintering, F_1, and F_2 female boll weevils in LRGV of Texas

During the cotton-free period, female boll weevils without access to cotton resorb their unlaid eggs and enter reproductive diapause (Greenberg *et al.*, 2007b). However, when they were provided daily with greenhouse grown cotton squares, commencement of oviposition began after 7, 15 or 20 d, depending on when they were captured. Females captured later in the winter fed longer before laying eggs than those captured in the early fall, suggesting that it may take females longer to terminate diapause the longer the time period they have been dormant (Greenberg *et al.*,

2007b). The rate of feeding by females was significantly less during the winter months, and this may have affected the rate of diet-mediated termination of dormancy (Greenberg *et al.*, 2007b). Females of the first and second generations after the overwintering generation produced a significantly higher percentage of progeny surviving to adulthood and a higher proportion of these progeny were females. Offspring development time from overwintering female parents was significantly longer than that from first and second generations under the same laboratory conditions. The total number of lifetime eggs produced by females of the second generation during the cotton growing season was ~9.9-fold higher than for overwintering females and 1.5-fold higher than for first generation females. Life table calculations indicated that the population of second generation boll weevils increased an average of 1.5-fold higher each generation than for females of the first generation and 22.6-fold higher than for overwintering females. Our data showed variation in boll weevil survival, development, and reproductive potential among the overwintering and first- and second-generation females, suggesting inherent seasonal fluctuations in these parameters.

Knowledge of the rate of increase of a population from one generation to the next is basic to an understanding of the degree of control that is needed to hold insects to non-economic levels (Greenberg *et al.*, 2007b).

9.7.10 Overwintering boll weevils

The boll weevil overwinters in the southern United States in the adult stage. At the onset of the cotton-free period, females begin resorbing their unlaid eggs and entering reproductive diapause. Diapause functions to help the boll weevil survive periods of food shortage while permitting activity during extended periods of relatively mild climatic conditions (Greenberg *et al.*, 2009a). Survival of overwintering boll weevils is a critical determinant in the severity of infestation in the subsequent cotton season and

is a key element in ongoing area-wide suppression programmes. Controlling overwintering boll weevil populations in a subtropical environment will be an important factor for the success of the Boll Weevil Eradication Program. The boll weevils are polyphagous pollen feeders and actively feed on pollen from a diverse assemblage of plant species (Greenberg *et al.*, 2007b). Boll weevil ingestion and digestion of pollen may be key survival mechanisms in fall and winter when cotton is not available. We began identifying pollen spectra (pollen fingerprinting) of a large variety of plant species. Because of the diversity of plant species in the LRGV of Texas, pollen fingerprinting can be used to help characterize boll weevil dispersal. Understanding boll weevil dispersal before and after cotton production is paramount in boll weevil management and eradication (Greenberg *et al.*, 2012b). Transfer of pollen feeding overwintering boll weevil females to cotton fruits restored their reproductive maturity and potential (Greenberg *et al.*, 2007c).

9.7.11 Effects of burial and soil condition on postharvest mortality of boll weevil in fallen cotton fruit

During hot weather immediately after summer harvest operations in the LRGV of Texas, burial of infested fruit in conventionally tilled field plots permitted significantly greater survival of weevils than in no-tillage plots (Greenberg *et al.*, 2004a). Burial of infested squares protected developing weevils from heat and desiccation that cause high mortality on the soil surface during and after harvest in midsummer and late summer. A laboratory assay showed that burial of infested squares resulted in significantly greater weevil mortality in wet than in dry sandy or clay soils (Greenberg *et al.*, 2004a). Significantly fewer weevils rose to the soil surface after burial of infested bolls during winter compared with bolls settling on the soil surface, a likely result of wetting by winter rainfall. A combination of leaving infested fruit exposed to heat before the onset of cooler winter temperatures and burial by tillage when temperatures begin to cool might be an important tactic for reducing populations of boll weevils that overwinter in cotton fields (Greenberg *et al.*, 2004a).

9.7.12 Effects of conventional vs conservation tillage systems on population dynamics of boll weevil

The conventional tillage cotton treatment had a greater water stress, causing plants to shed squares and bolls. Cotton plants in the conventional tillage treatment allocated more resources into vegetative growth while the conservation tillage cotton responded by fruiting at a higher rate (Greenberg *et al.*, 2003c). At 110 days after planting, the conservation tillage cotton had an average height of 42.4 cm per plant versus 63.0 cm in conventional tillage, and the number of leaves per plant was 32.4 versus 51.7, while fruit numbers were 13.0 versus 7.1, respectively (Greenberg *et al.*, 2003b). Increased plant height and numbers of leaves in the conventional tillage provided significantly more light interception and shading of the soil surface. In the conservation tillage cotton, 60.2% of the incoming sunlight reached the soil surface, while the conventional tillage had only 36.2%. Soil temperatures between the rows in conservation tillage cotton were 8–11°C higher than in conventional tillage and significantly influenced boll mortality in infested squares shed from plants. The number of boll weevils per plant was 2.3–3.4-fold higher in conventional tillage compared with the conservation tillage. Trap counts of boll weevil populations followed a similar trend with 1.6–2.8-fold more weevils in conventional tillage compared to conservation tillage. The mortality of boll weevils in fallen, naturally infested squares, and in cohorts of laboratory-infested squares collected from the middle of the rows was 1.5–1.8-fold higher in the conservation tillage field than in conventional. The percentage of punctured squares by boll weevils during the growing season averaged 2.1-fold higher in conventional than in conservation tillage fields (Greenberg *et al.*, 2003b,c, 2010).

9.7.13 Other activities for improved insecticide efficiencies against boll weevils

Cotton IPM in LRGV of Texas was also improved by the following management practices:

1. reducing the rate of the primary insecticide (ULV malathion) without reducing efficacy of the programme, for example, the ULV malathion rate was reduced from 16-oz/ac to 12-oz/ac with oil as an adjuvant (Greenberg *et al.*, 2012a);
2. terminating insecticide treatments as soon as crop maturity allowed; and
3. improving pesticide application techniques (e.g., correct nozzle placement, nozzle type, and nozzle pressure) (Leonard *et al.*, 2006; Lopez *et al.*, 2008).

An alternative of chemical control could possibly be the propagation and augmented releases of beneficial insects (example *Catolaccus grandis* (Burks), parasitoid of boll weevil larvae (Table 9.6) (Summy *et al.*, 1994; Greenberg *et al.*, 1998, 2012b)).

These findings will significantly reduce the use of insecticides against boll weevils

Table 9.6. Parasitism of boll weevils by *Catolaccus grandis*.

Texas sites	Date releases	Percentage of parasitism
Monte Alto	28 April 1993	80.0
	5 May 1993	52.8
	12 May 1993	76.4
	19 May 1993	78.3
	26 May 1993	74.9
	24 June 1993	85.2
Weslaco	24 May 1994	83.3
	2 June 1994	69.2
	6 June 1994	62.5
	16 June 1994	50.0

Boll weevil used: 3rd instar larva and pupa; Parasitoid density: Monte Alto – 1000 and Weslaco – 500 females/acre/week.

and improve our capability to develop environmentally safe and efficient strategies for controlling this economically important pest in tropical and subtropical regions of the world similar to LRGV of Texas.

9.8 Changes in the Sucking Bug Complex – Stink Bugs, Plant Bugs and Cotton Fleahoppers

The cotton sucking bug pests belonging to the suborder Heteroptera have been elevated in pest status within the cotton-growing regions of the United States over the past decade. Some of the most notable heteropterans are the plant bugs, including the tarnished plant bug, *Lygus lineolaris* (Palisot de Beauvois); western tarnished plant bug, *Lygus hesperus* Knight; the stink bug complex (Pentatomidae); and the cotton fleahopper *Pseudatomoscelis seriatus* (Reuter). This transition from traditionally being considered secondary pests and now elevated to primary pest status has also coincidentally followed the functional eradication of the boll weevil from the southeastern and southern US cotton belt regions (Grefenstette and El-Lissy, 2008; Greenberg *et al.*, 2012b).

Other reasons often mentioned for increases in bugs infesting cotton across the cotton belt with the progression of eradication is the adoption of varieties containing the *Bt* endotoxins that were being released in conjunction with boll weevil eradication efforts (Greenberg *et al.*, 2012b). Over time, the numbers of boll weevils were reduced, coinciding with a reduction in the number of ULV malathion applications within a season, which may have coincidently been suppressing the plant bugs. Also, the adoption of *Bt* cotton varieties significantly reduced lepidopteran pests that reduced pesticide applications simultaneously with BW eradication. There were three key factors: (i) the progress of boll weevil eradication and the reduction of ULV malathion; (ii) the adoption of variants of cotton varieties with *Bt* toxins; and (iii) the use of target specific insecticides for worm control are most often cited as the reasons for changes

in the shift from lepidopteran management to sucking bug pests attacking cotton (Layton, 2000; Greene and Capps, 2003; Vitale *et al.*, 2011).

The LRGV region is continuing to eradicating the boll weevil from the southern regions including the South Texas Rio Grande Valley and the Winter Garden area south and west of San Antonio, near Uvalde, TX. However, from each cotton-producing region across the US, starting from the southeast to the west, and down to south Texas, the bug problems vary by species, and by severity of the economic threat. For example, the tarnished plant bug *Lygus lineolaris* (Palisot de Beauvois) has increased in pest status in the southern and mid-south cotton regions following boll weevil eradication (Layton, 2000) to the point that it has developed insecticide resistance to a wide variety of insecticides and poses as a significant pest in the southeastern cotton growing areas of the Mississippi Delta (Wrona *et al.*, 1996). Not all bug complexes have increased or are related to boll weevil eradication. California and Arizona have had perennial problems with *Lygus hesperus* Knight and *Lygus elisus* Van Duzee (Heteroptera: Miridae) in lucerne and cotton before eradication of the boll weevil from these two states (Leigh *et al.*, 1985) and after the boll weevil was declared eradicated from the cotton-producing regions of these two states (Zink and Rosenheim, 2005; Greenberg *et al.*, 2012b). Cotton is damaged by tarnished plant bugs as they feed on cotton squares (flower buds) with the most significant impact when fruit abscises or drops to the ground (Tugwell *et al.*, 1976), although they cause significant damage to maturing bolls resulting in yield loss as well as reduced lint fibre quality (Parajulee and Shrestha, 2014). Further to the west in Arizona and California is the western tarnished plant bug which seems to be similar in their feeding injury to cotton (Leigh *et al.*, 1996).

For the last few years, the verde plant bug, *Creontiades signatus* (Distant), has been reported to infest cotton grown in the Lower Rio Grande Valley and the Lower Coastal Bend regions of south Texas, causing injury to developing lint and seed inside cotton bolls (Armstrong *et al.*, 2010). The bug has increased in pest status since the initiation of the second attempt to eradicate the boll weevil in LRGV beginning in late 2005, and during much of the progress of eradication of the Upper and Lower Coastal Bend producing areas from about 1999 to present (MNP, personal observation). Feeding injury from the verde plant bug is similar to *Lygus* bugs, but it has thus far been considered a late season pest injuring and causing abscission in bolls <315 heat units (DD) from anthesis (Brewer *et al.*, 2012). Molecular and taxonomic work identified *C. signatus* as being native up to several miles inland from the Gulf Coast of the US and Mexico (Coleman *et al.*, 2008). Reasons for increases in the densities of this new plant bug pest of south Texas, as it increases as a pest of cotton, can only be speculated. Some known factors in support of increased densities moving into and threatening cotton would be the significant increase in the acres of soybean, *Glycine max* (L.) Merr. planted in the Lower Rio Grande Valley. This mirid plant bug can reproduce on soybean and within the seedhead of grain sorghum, *Sorghum bicolor* (L.) Moesch and several weedy species also serve as reproductive hosts. Cotton may not be the most preferred host of this insect, but it survives on the cotton plant and has a preference for oviposition on the petioles of cotton leaves similar to other *Lygus* species (Armstrong *et al.*, 2009). The stink bug complex of cotton can be varied, and complex but the most frequent species encountered are the southern green stink bug, *Nezara viridula* (L.), the green stink bug, *Acrosternum hilare* (Say), and the brown stink bug, *Euschistus servus* (Say) (Hemiptera: Pentatomidae) (Greenberg *et al.*, 2012b). These three species are considered the primary targets for a significant number of insecticide applications applied to cotton (Williams, 2008), most notably in the midsouth and southern cotton regions, and has also been associated with the elevated pest status following boll weevil eradication (Greene *et al.*, 1999; Turnipseed *et al.*, 2004; Willrich *et al.*, 2004). However, in the

subtropical region of Texas, the diversity of species seems to be broader from central Texas down to the Lower Gulf Coast region. The Lower Gulf Coast region south of Corpus Christi experiences the rice stink bug, *Oebalus pugnax* (F.), and in the LRGV and far west Texas the Conchuela stink bug, *Chlorochroa ligata* (Say) is more prevalent (Muegge, 2002). Stink bugs of all species and localities are noted for being more injurious to small and medium sized cotton bolls, causing more injury from lacerating thicker boll tissues resulting in greater injury to the tissues, seed and lint (Greene *et al.*, 1999; Musser *et al.*, 2009).

The most consistent, early season, truebug pest of cotton in the state of Texas is the cotton fleahopper, which prefers feeding on small developing fruits (squares) in the upper terminal of plants (Stewart and Sterling, 1989). If injured enough, the small squares will abscise from the plant. However, the cotton plant is noted for compensation, and if management practices are instigated, or populations decrease for biotic or abiotic reasons, no economic loss may be seen when compared to cotton that has not sustained injury, and in fact some gain from overcompensation may be realized (Sterling, 1984; Barman and Parajulee, 2013). The length of the growing season is often associated with compensatory gain because of the delayed fruit set. The historical relationship of the severity of cotton fleahopper infestations in cotton with that of the progress of boll weevil eradication in the state of Texas is difficult to make, as severe outbreaks have been noted before, during, and in some cases after an area has been functionally eradicated. The Texas High Plains region was declared functionally eradicated in 2003. Cotton fleahopper populations are now as much of a threat as they were before eradication. In south Texas, cotton fleahoppers are still considered a significant pest, and boll weevil eradication has not been fully realized (Greenberg *et al.*, 2012b).

Alongside the more recent changes in the pest status of heteropteran insects of cotton has been the realization of their feeding injury and its association with boll rot.

The feeding on cotton squares and bolls is made significantly more important because the wounds create an entry site for bacterial and fungal pathogens to enter and invade the interior of the forming fruit, making it difficult, if not impossible, to tell the difference between pathogen-infected and uninfected. Environmental conditions in the cotton field, mostly in the form of temperature, humidity and moisture can prevent or promote the growth of the boll-rotting pathogens (Greenberg *et al.*, 2012b). Economic thresholds established for most sucking pests are generally based on direct feeding injury, and do not include boll rot as a yield-limiting factor. Square and boll rot may promote the delayed abscission of cotton fruit from the production of ethylene produced by the rotting and degradation of fruit tissues (Duffy and Powell, 1979). Cotton bolls are not expected to sustain damage from cotton fleahoppers, due to the fact that the mouthparts, more specifically the stylets, are not long enough to penetrate the exterior carpel wall of the boll. Boll-rot pathogens have, however, been associated with direct transmission, as with the common plant pathogen and cottonseed-rotting bacterium *Pantoea ananatis* (Bell *et al.*, 2006, 2010). The stink bugs and green plant bugs possess stylets that are long and broad enough to cause physical damage from insertion and laceration of the tissue, injection of digestive enzymes, and the ingestion of the enzymatic soup of sap and cytoplasm. This subsequently causes loss of boll, lint and seed tissues, and provides an entry for pathogens that collectively may cause boll rot (Medrano *et al.*, 2009). If the cotton fruit, including bolls, do not abscise, the quality and quantity of lint will be reduced (Greenberg *et al.*, 2012b).

(Vitale *et al.*, 2011; Greenberg *et al.*, 2012b). Host-plant resistances to the cotton fleahopper and plant bugs have been studied extensively during the last four decades. The three main sources of host-plant resistance identified were associated with relatively high gossypol levels (Lukefahr, 1975), smooth-leaf genotypes (Lukefahr *et al.*, 1970), and production of nectar. No active cotton breeding programmes have continued with any forms of resistance since Lidell *et al.* (1986) screened for glabrous, pilose, and nactariless traits. Many of these same traits were screened for *Lygus* bugs in cotton (Tingey and Pillimer, 1977; Gannaway, 1994; Jenkins and Wilson, 1996). No information is available for host-plant resistance to stink bugs in cotton. Treatment thresholds for insecticide applications for these bugs have been provided in several Extension-based publications that list the bug pests and insecticides used for their control (Wrona *et al.*, 1996). Little research based on economic injury levels (EIL) has been provided for the green plant bug (*Creontiades dilutes*), which has, thus far, been considered a late season pest in cotton. Late-season injury levels for the green plant bug based on boll damage parameters, such as boll size (diameter) and age from tagged white-blooms, has been reported by Armstrong *et al.* (2010). Early season infestations occurring during the pre-bloom period have not been observed in south Texas. Economic thresholds could improve if the dynamics of confounding factors, such as the relationship of boll rot and injury levels based on bug pest densities are studied. The overwintering biology and ecology of plant bugs and stink bugs, and the means to monitor movement into the agricultural crops would be of significant use for their management (Greenberg *et al.*, 2012b).

9.8.1 Improving management options for the integrated approach to control bug pests

The plant bugs as a group have, in the past, been targeted for the discovery of host-plant resistance traits that could be integrated into traditional cotton-breeding programmes

9.9 Lepidopteran Insect Pests in Cotton

In the US, bollworm, *Helicoverpa zea* (Boddie), tobacco budworm (TBW), *Heliothis virescens* (F.), fall armyworm (FAW),

Spodoptera frugiperda (Smith), beet armyworm (BAW) and *Spodoptera exigua* (Hübner) are the major insect pests of cotton (Greenberg *et al.*, 2010; Williams, 2011). In the LRGV region of Texas, the most prevalent noctuids captured in the pheromone traps (average per year during 2007–2009) are BAW (63.8%), FAW (26.5%), bollworm (7.9%), and TBW (1.8%). In the US, cotton losses from the heliothine complex were 49.3 million kg ($78.9 million and 28.0% from total insect losses); in Texas – 18.5 m kg, $29.6 m and 49.6%; in LRGV of Texas – 1.0 m kg, $1.6 m, and 63.4%, respectively (Williams 2010). During the last two decades, the BAW has become an increasingly destructive secondary pest in US cotton. FAW is a destructive migratory pest of many crops in the western hemisphere, where it seems to be common and widespread (Greenberg *et al.*, 2012a).

9.9.1 Beet armyworm
(Lepidoptera: Noctuidae)

Beet armyworms attack more than 90 plant species in at least 18 families throughout North America, many of which are crop plants. In 1998 alone, 2.1 m ha of cotton were infested with BAW and losses were $19.2 m in the United States. Epidemic outbreaks in cotton have cost as much as $916.4 per ha in yield reduction, and the cost of insecticides targeted at this pest were as much as $108.7 per application per ha (Williams, 1999; Greenberg *et al.*, 2001b).

Feeding and life history of beet armyworm on different host plants

Consumption rates, development times, and life table parameters of the beet armyworm were determined in the LRGV of Texas on cultivated crops (cabbage (*Brassica*), cotton (*Gossypium*), and pepper (*Piper, Capsicum*)) and widely distributed weeds (pigweed, *Amaranthus* spp., and wild sunflower, *Helianthus*) (Greenberg *et al.*, 2001b). The feeding index (pupal weight divided by total weight of leaf tissue consumed) was highest on pigweed,

followed by cotton, pepper, sunflower and cabbage. On all host plants, significant relationships were found between the amount of leaf tissue consumed and resulting pupal weight. Likewise, significant relationships were found between pupal weight and subsequent adult fecundity on all host plants. The highest percentage of female progeny was recorded in beet armyworms reared on pigweed (62.2%); the lowest for larvae reared on cabbage (43.6%). Duration of the larval stage was shortest on pigweed (12.4 d) and longest on pepper (18.0 d) plants. Larval survival was highest on pigweed (94.4%) and lowest on cabbage (67.1%). The population growth indices for females reared on pigweed averaged 2.1-fold higher than those reared on cabbage. Life table calculations indicated that beet armyworm populations reared on pigweed will increase an average 4.3-fold higher each generation (R_0) than will females reared on cabbage. For an overall assessment of the effects of different host plants on beet armyworm, we focused on three key statistics: feeding index; net reproductive rate, R_0; and growth index (percentage immature survival divided by immature development time). Using these measures, beet armyworm performance was best on pigweed, worst on cabbage, and intermediate on cotton, pepper and sunflower (Greenberg *et al.*, 2001b).

Pigweed and wild sunflower are abundant weeds throughout the LRGV of Texas, as well as in many other areas of the Cotton Belt, and may serve as reservoir hosts for infestation of cultivated crops. The net reproductive rate is highest for beet armyworms that developed on pigweed. If weed host availability becomes limiting, oviposition in crops such as cotton, pepper, and cabbage may occur at levels leading to outbreaks, even if these are not the most preferred or most nutritious hosts.

Summary: Populations of beet armyworms may build in a cultivated or uncultivated host and then spread to other hosts, because crop phenologies change through the season. Understanding such patterns of host utilization and host sequencing will contribute to development of tactics involving manipulation of wild and cultivated hosts for incorporation into area wide management strategies (Greenberg *et al.*, 2001b).

Host plant preference for oviposition

Beet armyworm oviposition preferences were determined on five host plants: cabbage, cotton, bell pepper, pigweed and wild sunflower, in no-choice, two-choice and five-choice tests. Tests were conducted in the laboratory, greenhouse and field cages (Greenberg *et al.*, 2002a). Oviposition preferences were compared on the basis of two measurements: (i) the proportion of eggs laid on the plants to total that were deposited; and (ii) the oviposition preference index defined as ((number of eggs laid on the plant) − (number of eggs laid on the cage)) * 100 / total number of eggs laid. The proportion of eggs laid on the plants to total that were deposited was highest for pigweed (0.612) and lowest for cabbage (0.184). Beet armyworm females were significantly deterred from laying eggs on cabbage and sunflower, while pigweed and cotton elicited a positive oviposition preference. Pepper tended to be neutral or slightly unattractive. Apparent interactions among plant species in choice tests produced measurable shifts in oviposition preference. Most notably, female response to pepper was enhanced in the presence of cotton or pigweed. Egg masses laid on the plants contained a significantly higher number of eggs than those laid on the surface of the cage, except in the case of cabbage leaves (Greenberg *et al.*, 2002a).

Summary: Knowledge of hierarchies of host plant oviposition preference by beet armyworm females will be useful in understanding the population dynamics of this important agricultural pest, for developing effective beet armyworm management strategies, especially through monitoring and prediction of its outbreaks in cotton. It also will be useful in designing cultural management strategies, which may include trap cropping, crop rotations, and weed control. Wild and cultivated plant-hosts of beet armyworms serve as important reservoirs stimulating or deterring egg deposition. Isolation and identification of these

components may provide new tools for management of beet armyworm through oviposition deterrence or bait trapping using oviposition stimulants (Greenberg *et al.*, 2002a).

9.9.2 Controlling lepidopteran pests via transgenic technology

The use of transgenically modified cotton that expresses an insecticidal protein derived from *Bacillus thuringiensis* Berlinger has revolutionized global agriculture (Head *et al.*, 2005; Greenberg *et al.*, 2012a). Transgenic Bt cotton, Bollgard® (Monsanto Co., St Louis, MO) encoding the Cry1Ac insect toxin protein was introduced in the United States in 1996 (Layton, 1997); in 2002, Bollgard® II with stacked Cry1Ac and Cry2Ab endotoxins was launched, with increased efficiency against several lepidopteran species (Sherrick *et al.*, 2002). Dow AgroSciences, LLC (Indianapolis, IN) introduced their pyramided-gene technology onto the market in 2004 as WideStrike® which produced two *Bt* endotoxins, Cry1Ac and Cry1Fa (Adamczyk and Gore, 2004). VipCot is a transgenic cotton technology, the active *Bt* toxin is Vip 3A, which is an exotoxin produced during vegetative stages of *Bt* growth (Mascarenhas *et al.*, 2003). In the first year of commercial availability in the United States, Bollgard® cotton was planted on 850,000 ha or 15% of the total cotton area, while by 2007, about 2.9 million ha or 65.8% of US cotton land area was planted with transgenic *Bt* cotton. However, adoption of *Bt* cotton varied greatly across growing regions in the US, and other countries, depending on the availability of suitable varieties and more importantly depending on the particular combination of pest control problems. Bollgard® cotton varieties have been rapidly accepted by farmers in areas where budworm-bollworm complex is the primary pest problem, particularly when resistance to chemical pesticides is high. There are many reasons which can affect changes in the degree of expression of stacked endotoxins. Individual lepidopteran species vary in their susceptibility to *Bt* proteins (Luttrell and Mink, 1999), and efficacy can be affected by protein expression levels in different plant structures (Adamczyk *et al.*, 2008) and among different varieties (Adamczyk and Gore, 2004). Differences in susceptibility can also occur based on the geographic location of populations (Luttrell *et al.*, 1999).

Microbial insecticides are environmentally friendly and highly selective. Transgenic plants reduce the need for conventional insecticides, providing benefits for human health, and the environment. For example, in US cotton, the average number of insecticide applications used against tobacco budworm-cotton bollworm complex decreased from 5.6 in 1990–1995 to 0.63 in 2005–2009 (Williams, 2010). Carpenter and Ginanessi (2001) estimated that the average annual reduction in use of pesticides on cotton in the US has been approximately 1000 tons of active ingredients. Traxler *et al.* (2003) estimated that the economic benefits from the introduction of *Bt* cotton fluctuated from year to year but averaged $215 million. Adoption of transgenic *Bt*-cotton is described in Table 9.7. *Bt* types, traits and varieties used in LRGV of Texas for 2005–2010 are shown in Table 9.8.

During the 2005–2007 seasons (Greenberg and Adamczyk, 2010; Greenberg *et al.*, 2012a), the average percentage of leaf damage on non-*Bt* trait varieties was 1.5-fold greater than on Bollgard® varieties. Leaf

Table 9.7 *Bt* cotton production area (Williams, 2005–2007).

Year	*Bt* cotton (ha)	% *Bt* cotton from total planted	Hectares *Bt* sprayed	Average number applications
US				
2005	2,994,086	51.8	1,234,855	0.54
2006	3,439,604	57.2	1,603,722	0.59
2007	2,877,114	65.8	895,232	0.50
Texas				
2005	546,898	22.6	75,061	0.78
2006	669,891	27.2	37,823	0.44
2007	929,654	47.5	22,657	0.44

damage was 3.6-fold less on Bollgard® II and WideStrike®-trait varieties than on non-*Bt* cotton, and 2.4-fold less than on Bollgard®-trait varieties. The same trend was observed for the proportion of consumed leaves. On non-*Bt* cotton varieties, the index was 1.6-fold greater than on Bollgard® varieties and 2.4-fold greater than on Bollgard® II and WideStrike® varieties. The proportion of consumed leaves on Bollgard® was 1.5-fold greater than on Bollgard® II or WideStrike® cotton. The differences of leaf damage between varieties containing dual *Bt* endotoxins (Bollgard® II and WideStrike®) during the cotton-growing seasons were not significant ($t = 0.440$; $P = 0.668$), except at the end of the season (110 days of cotton age). The damage to WideStrike® cotton (Phy 485 WRF) was 1.4-fold greater than to the Bollgard® II variety (ST 4357 BG2RF). The seasonal average of fruit on the plant damaged (attributed to bollworm 88.5% and, to a lesser extent, beet armyworm) on non-*Bt* cotton (15.2%) was about 4.6-fold greater than on WideStrike® (3.3%), 3.8-fold greater than Bollgard® II (4.0%), and 1.7-fold greater than Bollgard® (9.0%). Damage by noctuids on abscised cotton fruit was 39.0% for non-*Bt*, 28.5% for Bollgard®, 12.6% for Bollgard® II and 8.5% for WideStrike® cottons. In non-*Bt* cotton, live larvae were 6.2-fold greater than on WideStrike®, 4.5-fold greater than on Bollgard® II and only 1.7-fold greater than on Bollgard®. Live larvae in fallen fruit were 92.6% bollworm and 7.4% beet armyworm (Greenberg and Adamczyk, 2010; Greenberg *et al.*, 2012).

It has been well demonstrated that development and deployment of transgenic plants with insecticidal genes will lead to: (i) reduced quantities of insecticides applied for lepidopteran control; (ii) reduced pesticide exposure to farmers, farm labourers and non-target organisms; (iii) increased activity of natural enemies; (iv) reduced levels of pesticide residues on food crops; and (v) a safer environment.

An important advantage of transgenetically modified *Bt* crops over chemical control of pests is that it prevents insects from damaging plants because the *Bt* insecticidal toxicity is present in the plant throughout the growing season. *Bt* crops also benefit from decreased dependence on weather conditions that affect the timing and effectiveness of insecticide applications (Greenberg *et al.*, 2012a). Chemical pest control is less effective and will generally result in crop losses in the presence of insect infestations. Farmers typically apply chemical controls only after noticing the presence of pests on the cotton plants, by which time some damage has already occurred. As a result, *Bt* varieties have superior yield performance over a wide range of growing conditions (Fernandes-Cornejo and McBride, 2000).

Field level studies of the performance of *Bt* cotton have been completed in Mexico, Argentina, South Africa, China, India, and in the US (Greenberg *et al.*, 2012a; Table 9.9).

Falck-Zepeda *et al.* (2000) evaluated that the average annual distribution of benefits from the introduction of *Bt* cotton to be 45% to US farmers, 36% to germplasm suppliers, and 19% to cotton consumers. In the US, the number of pesticide applications used against lepidopteran pests in conventional cotton has fallen from 4.3 in 1995 to 0.7 in *Bt* cotton in 2009 (Williams, 1999, 2011). Carpenter and Ginanessi (2001) estimated that average annual reductions in the use of pesticides on cotton in the US has been approximately 1000 tons of active ingredients, which can provide significant economic and environmental benefits, thus potential for increase in yield, profit, revenue, convenience for system, potential reduction in spray equipment, improved insect control, reduction in natural target organisms exposure, improved water quality, and reduction in energy use.

Summary: Transgenic cotton has reduced the need for conventional insecticides used against lepidopteran pests on average in the US by about 59.4%, Texas 74.7%, and the LRGV of Texas 60.3%; the average number of pesticide applications in conventional cotton has fallen from 4.3 in 1995 to 2.1 in the USA, 1.8 in Texas, and 2.5 in LRGV for 2009–2011, with benefits to human health and the environment. The revenue differences between *Bt* and conventional cotton for the past four years in LRGV of Texas was $214.3/h and profit about $94.9/ha. Compared to non-Bt cotton, *Bt* cotton increased lint yield (kg/ha) by 150.4 lb/ha and reduced pest control cost by $38.9/ha.

Table 9.8. *Bt* cottons used in LRGV of Texas (Greenberg and Adamczyk, 2010).

Bt type	*Bt* trait	Variety	*Bt* endotoxins	Owner of *Bt* trait	Owner of variety
None	Non-*Bt*	DPL 5415RR	None	None	Delta & Pineland
Single	Bollgard®	NuCotn 33B	Cry1Ac	Monsanto	Delta & Pineland (Monsanto)
Dual	Bollgard® II	DPL424 BGII/RR	Cry1Ac+Cry2Ab	Monsanto	Delta & Pineland
Dual	WideStrike®	Phy485 WRF	Cry1Ac+Cry2F	Dow Agroscience	Dow Agroscience

Table 9.9. Performance differences between *Bt* and conventional cotton.

	Argentina	China	India	Mexico	South Africa	US	Texas (US)	LRGV Texas
Bt – conventional								
Lint yield (kg/ha)	+531.0	+523.0	+699.0	+170.0	+91.0	+181.0	+118.2	+150.4
Insecticide use (%)	−47.0	−65.0	−41.0	−77.0	−33.0	−59.4	−74.7	−60.3
Chemical sprays (number)	−2.4	n/a	−3.0	−2.2	n/a	−2.1	−1.8	−2.5
Revenue ($/ha)	+121.0	+130.0	n/a	+248.0	n/a	+257.9	+168.4	+214.3
Pest control ($/ha)	−34.8	−230.0	−30.0	−157.0	+4.5	−16.2	−9.6	−38.9
Profit ($/ha)	+22.9	+470.0	+135.0	+335.0	+29.0	+140.9	+77.8	+94.9

9.9.3 Parasitism of numerous lepidopteran species by *Trichogramma*

While biological control is not considered to be a primary component of IPM in cotton agroecosystems in the United States, mass rearing and augmentative release of *Trichogramma* spp., an egg parasitoid of numerous lepidopteran pest species, has shown some positive results in our study at LRGV of Texas (Table 9.10).

Encarsia, are among the most important natural enemies of the *Bemisia* spp. worldwide, as well as in southern region of the United States (Cock, 1993). The effective implementation of biological agents for whitefly management depends on our knowledge of biological relationships between host and parasitoids. A comprehensive IPM programme should be the main strategy of crop protection against whiteflies (Greenberg *et al.*, 2008).

9.10 Whiteflies (Homoptera: Aleyrodidae)

The sweet potato whitefly, *Bemisia* spp. gr. *tabaci* (Gennadius), especially its genotype MEAM1 (= *ex* biotype B), is a polyphagous pest species in the tropics and subtropics worldwide. The hosts of the *B. tabaci* complex include more than 500 plant species representing 74 plant families, many of which are of economic importance (Greathead, 1986). The whitefly causes damage directly through feeding and excretion of honeydew, and indirectly by plant virus transmission (Greenberg *et al.*, 2008). Whitefly is a significant economic pest of cotton worldwide. In 2007, whitefly-induced cotton loss in the US was estimated at 14,817 bales (Williams, 2008). Management of *B. tabaci* currently depends primarily on the application of insecticides, but the possibility of biological control of *B. tabaci* exists, which reduces pesticide risks on non-target organisms. Parasitic wasps of two insect genera, *Eretmocerus* and

9.10.1 Stage of *Bemisia tabaci* for parasitoid effectiveness

Eretmocerus mundus (Hymenoptera: Aphelinidae) parasitizes all nymphal host instars. The highest percentage of parasitism occurs on second instars and the least on the fourth (red-eyed nymph) instars (Greenberg *et al.*, 2008). When *E. mundus* parasitizes the early instars of *B. tabaci*, the host insect, it continues to feed, grow and develop. However, the parasitized late instar (third and fourth instars) whitefly nymphs cease their development.

The highest percentage of nymphs parasitized by *Encarsia pergandiella* (Hymenoptera: Aphelinidae) occurs on third instars and the least on the first instars. The second, third and 'early' fourth instars suffer the highest proportion of total host mortality, resulting in the highest level of parasitoid survival. Parasitized 1st, 2nd or 3rd instar whitefly nymphs continue to develop, whereas the parasitized fourth instars cease their development.

Table 9.10. Effectiveness of *Trichogramma* (Greenberg *et al.*, 1998) in reducing the number of lepidopteran pests via parasitism.

| | Percentage of Parasitized or Desiccated Eggs | | | | | |
| | Parasitized eggs | | Desiccated eggs | | Total mortality | |
Treatment	BAW*	bollworm	BAW	bollworm	BAW	bollworm
T. pretiosum	44.8±4.3	90.3±1.7	24.9±2.1	5.5±1.2	69.7±5.6	95.8±0.5
T. minutum	51.6±3.7	88.9±1.6	29.3±2.2	6.5±1.6	80.9±3.3	95.4±1.6
Control	0	0	5.8±1.2	4.4±1.7		

*Beet armyworm

In no-choice experiments with both parasitoids together, we found that *E. mundus* successfully parasitized the younger host instars, while *E. pergandiella* parasitized a greater proportion of the older instars. Similar results were observed when parasitoids were provided a choice of two instars in six different paired combinations, but the trends were not as pronounced as in the no-choice tests (Greenberg *et al.*, 2008). When all four instars were provided simultaneously, the numbers of first, second and third instars parasitized by *E. mundus* did not differ significantly from each other (range 10.3–16.4%), but all were significantly higher than parasitism on fourth instar nymphs (2.1%). The highest percentage parasitization by *E. pergandiella* was of third instars (17.2%), and the lowest was of the first instars (2.8%). These data will be useful in designing rearing protocols, in helping make decisions in augmentative releases, and in developing predictive models for specific crop production scenarios (Greenberg *et al.*, 2008).

9.10.2 Effects of host–parasitoid ratios on parasitoid effectiveness

B. tabaci (second instars) and Eretmocerus mundus

A laboratory functional response study (Greenberg *et al.*, 2008) estimated that *E. mundus* could reduce *B. tabaci* populations by up to 95%. A linear regression, fitting a Type I functional response curve, estimated the parasitoid attack rate (*a*) of 0.18 per day and a handling time (t) of 12 min. Successful emergence of *E. mundus* was not significantly different at the different parasitoid ratios (75–90%). Approximately 60% of parasitoid progeny were females when >10 host nymphs were provided per parasitoid female compared to 40% female offspring produced when <5 host nymphs were provided per parasitoid. Nymphal mortality decreased from 78.3 to 18.8% with increasing *E. mundus* densities from 1 to 15 per 100 second instar hosts. Conversely, the numbers of host nymphs killed and female

parasitoid progeny per female decreased from 18.8 to 5.2 and from 14.3 to 0.7, respectively. The emergence rate, sex ratio and longevity of progeny were significantly greater at the lowest parasitoid density, while developmental time was significantly shorter. These data suggest that the searching efficiency of *E. mundus* decreases with increasing parasitoid density.

B. tabaci (third instars) and Encarsia pergandiella

The total nymphal mortality decreased from 91.3% to 19.1% when host–parasitoid ratio increased from 1:1 to 1:50, with an attack rate of 0.27 per day and estimated handling time of 72 min. When parasitoid densities increased from 1 to 15 females, the number of host nymphs killed and female parasitoid progeny per female were decreased 3.2- and 27.1-fold, respectively. Emergence rate, development time, and body lengths were significantly higher in lower host–parasitoid ratios compared to that in higher host–parasitoid ratios. Based on these data, we propose a generalized index of efficacy (GIE) to compare effects of host–parasitoid ratios. This is calculated by multiplying the proportion of parasitized nymphs, emergence of parasitoids, and the number of female progeny per parasitoid–host treatment. The most efficient ratio was 1 parasitoid female *E. mundus* per 10 second-instar *B. tabaci*, or 1 female *E. pergandiella* to 10 third-instar *B. tabaci*. The results provided important information on parasitoids for possible mass propagation and augmentation against *B. tabaci* (Jones and Greenberg, 1999; Jones *et al.*, 1999; Greenberg *et al.*, 2000a, 2001a, 2002b).

9.10.3 Temperature-dependent life history of *Eretmocerus eremicus* on *B. tabaci*

Maximum oviposition rate for *B. tabaci* is 12 eggs per 48-h period at 24°C. The temperature threshold for oviposition is 12.5°C. Developmental thresholds were 10.3°C and 8.7°C for *B. tabaci* and *E. eremicus* on *B. tabaci*, respectively. Mean degree-day requirements for egg to adult development for *B. tabaci* and *E. eremicus* on *B. tabaci* were 319.7 and 314.4, respectively. The temperature-dependent life history parameters can be useful in designing mass rearing protocols, in helping to make decisions in augmentative releases, and in development of predictive models (Greenberg *et al.*, 2000b).

9.10.4 Effectiveness of parasitoids on whitefly suppression

We have estimated that native parasitoids can control whiteflies in organic cotton (95–100%), sustainable agriculture cotton (80–90%), *Bt* cotton (59–60%), conventional cotton (25–30%), and under production scenarios where frequent insecticide applications on a calendar basis are necessary (0–5%) (Greenberg *et al.*, 2012a). Therefore, it is demonstrated that the whitefly biological control via parasitoids can be used in all cotton-production settings.

9.11 Thrips (Thysanoptera: Thripidae)

Thrips is one of the key insect pests on cotton. The cotton yield losses from thrips during 2007 in all of the cotton belt areas in the US was 145,040 bales (ranked 3rd by percentage losses as compared with other insects), with 84.7% of the total planted cotton acreage infested. The cotton yield reduction from thrips for this same period of time in Texas was 65,062 bales (ranked 3rd by percentage losses), and in the LRGV of Texas – 1563 bales (ranked 6th by percentage losses) (Williams, 2007; Greenberg *et al.*, 2009b). This estimate of cotton losses from thrips in the LRGV is an underestimation. First, the cotton in the LRGV was 100% infested with thrips in four of the last five years. Second, the subtropical climate contributed to growing different agricultural crops year-round and some of those hosts were good sources of thrips inundating the cotton crop. For example, onion is a major vegetable crop in south Texas (Liu and Chu,

2004) and cotton usually germinates at about the same time that onions are harvested (Sparks *et al.*, 1998; Greenberg *et al.*, 2009b). Management decisions about distribution and density of thrips during the cotton-growing season should be based on actual field observations. Sampling is a vital component of cotton insect control. Sampling must be done to estimate insect pest populations, pest damage or beneficial insects. Scouting for thrips is difficult because thrips are nearly microscopic and are cryptic feeders. The most widely accepted method of examination for thrips in Texas cotton is visual examination *in situ*.

9.11.1 Species composition

At locations where onion fields were within 0.5 km of cotton fields, the predominant species in LRGV are western flower thrips, *Frankliniella occidentalis* (Pergande) (61.7%) and onion thrips (27.2%). Observations from cotton fields located at least 50 km from commercial onion-growing areas contained western flower thrips (68.5%), followed by bean thrips, *Caliothrips fasciatus* (Pergande) (29.2%) as the two predominant species. At both situations, young seedling cotton plants, cotyledon stage to 4–5 true leaf stage, were more susceptible to onion thrips damage. Western flower thrips and bean thrips were found more dominant on 5–6 true leaf stage and older cotton. Thrips generally begin colonizing in young seedling cotton as early as at cotyledon stage, increase in numbers gradually, peaking in abundance when plants attain 4–5 true leaf stage. Thrips populations generally decline when cotton plants reach 5–6 true leaf stage, at which time the plants simply become tolerant to thrips injury.

9.11.2 Population dynamics

Thrips damage to early-season cotton results in significant leaf area reduction, delayed maturity and retarded plant growth. Thrips feed on the plant terminal, disrupting normal plant growth and causing silvery, crumpled, crinkled, cupped and distorted leaves. Cotton plant responses to thrips feeding include pre-bloom square loss, reduced leaf area, poor root development, delayed crop maturity and decreased lint yield. While the presence of thrips in cotton is observed throughout the growing season, the cotton plant is generally vulnerable to yield loss during the pre-square stage (5–7 true leaves). Visual examination, beat bucket method, and sticky blue card-traps can all be used to measure population dynamics of thrips in cotton. However, the thrips sampling in cotton during the most susceptible stage (presquaring stage) is primarily done via visual sampling. As soon as cotton germinates, thrips are attracted to cotyledon cotton from non-cotton hosts. During the early infestation period, thrips may be more injurious to young cotton under cool, wet conditions compared to that under warmer environmental conditions.

9.11.3 Predator populations

Several species of predators are found on cotton plants that can feed and suppress thrips populations, including minute pirate bug, *Orius tristicolor* (White), bigeyed bug, *Geocoris punctipes* (Say), assassin bug, *Sinea* spp., convergent lady beetle, *Hippodamia convergens* (Guerin-Meneville), green lacewing, *Chrysoperla rufilabris* (Burmeister), and several species of spiders and predacious mites. In Texas cotton, *O. tristicolor* is the most abundant predator species (60–70%) during early stage of cotton, but predators are generally not a major factor regulating thrips population.

9.11.4 Chemical control for thrips

Acephate (Orthene®) is the product of choice for foliar applications to manage thrips in Texas, unless the cotton producer opts for systemic seed-coated insecticides. Seed treatments with systemic insecticides, including thiamethoxam and imidacloprid, have shown significant thrips suppression in Texas conditions. These seed-applied insecticides provide about three weeks of protection from thrips injury in cotton, but heavy thrips pressure may require additional foliar applications. Other thrips-control insecticide products include seed-applied aldicarb, foliar-applied dimethoate, dicrotophos and spinetoram.

9.11.5 Cultural control

Thrips are early season pests of seedling cotton which move to this crop from a multitude of wild hosts, wheat, onions and other vegetables (Greenberg *et al.*, 2010). Thrips are particularly a problem under cool, wet conditions when plant growth is slowed. Therefore, adjusting the planting date to avoid the heavy influx of thrips from surrounding habitat is one of the best cultural management options for thrips management (Sansone and Minzenmayer, 2005; Slosser *et al.*, 2005). All *et al.* (1992) and Leonard (1995) showed that population densities of thrips are lower in conservation tillage than in conventional cotton fields. Cotton plants in conventional tillage plots developed more rapidly and may have been more attractive to migrating adult thrips (Greenberg *et al.*, 2010). DeSpain *et al.* (1992) reported that there was no effect of tillage system on thrips numbers.

9.12 Cotton aphid (Homoptera: Aphididae)

The cotton aphid, *Aphis gossypii* Glover, is the primary aphid species occurring in cotton and is the only aphid species within the cotton aphid complex which is of economic importance. This species attacks plants by removing the phloem sap from leaves, stems, terminals and fruit bracts (Bagwell and Baldwin, 2005). Aphids are normally found on the undersides of leaves, but they can be present on branches, stems and fruits. Heavy infestations cause smaller leaves to crinkle and cup downward. Older, larger leaves gradually turn yellow and wilt. In extreme cases, aphids can cause cotton to shed foliage and fruits (Bagwell and Baldwin, 2005). Aphids will sometimes infest the tender young foliage in the main stem terminal, causing it to appear warped or twisted. These terminal infestations can be observed easily while walking across the field. Aphids on the undersides of large leaves, however, are not noticeable unless the leaf or entire plant is bent over. Another symptom of a high aphid population is visible honeydew, a clear sticky substance secreted by aphids. Although the aphids are on the underneath surface of the leaves, the honeydew will drip down onto the upper surface of lower leaves, giving them a varnished, shiny appearance which is easily observed. Aphids are not always evenly distributed. Sometimes they are concentrated in certain areas such as field margins or hot spots within the field. If the infested area is large enough, it may require spot treatment to prevent aphids from uniformly spreading across the field. Conversely, these isolated hot spots may be as small as one or two heavily infested plants. There is no well-defined economic threshold for aphids, but the chemical treatment is generally triggered when the density of 50 aphids per leaf persists for two weeks. The decision to treat is based on several considerations, including the stage of crop development, weather and the presence or absence of natural control factors such as beneficial insects or the fungal disease associated with aphids (Bagwell and Baldwin, 2005).

Only females are found in cotton throughout the growing season. Reproduction occurs without mating. A new generation can develop approximately every 7 days at temperatures around 26.7°C, resulting in rapid population build-up, especially when low numbers of predators are present. The cotton aphid will be larger and darker during cooler weather and will be small and yellow in summer. A single female can produce up to 60 offspring during her three- to four-week lifespan. Excessive nitrogen fertilizer rates may favour heavy aphid infestations because there are indications that aphids develop and reproduce better on plants with high nitrogen levels (Parajulee *et al.*, 2010). Adult females can be wingless, or winged forms can develop in response to crowding or poor food quality. Heavy rainfall may reduce aphid numbers by washing them off plants (Bagwell and Baldwin, 2005).

9.12.1 Natural control

Several biological control agents are available for cotton aphid suppression in the field. Primary among these is the pathogenic fungus *Neozygites fresenii* (Nowakowski), which rapidly decreases aphid populations during mid-season (Bagwell and Baldwin, 2005). Aphids may show some resurgence during late season, after this disease has occurred. Spores of this fungus overwinter in the soil, which may result in the reappearance of the aphid fungal disease the following crop year. Disease symptoms can be recognized by the presence of brownish, fuzzy dead aphids on the undersides of leaves. A tiny parasitic hymenopteran wasp, *Lysiphlebus testaceipes* (Cresson), may also help to suppress aphid populations. Parasitism is evidenced by the appearance of mummified aphid bodies on leaf undersides, particularly in late season. As the parasitoid develops within the aphid body, it causes the aphid to swell and eventually turn a papery light brown when the parasite is ready to emerge. Parasites emerge by chewing through the mummy cuticle, leaving a large hole that is easily visible. Predators such as lady beetles, lacewings and syrphid fly larvae also can help control aphids. Their beneficial actions are important in early-season cotton to help prevent rapid aphid build-up (Bagwell and Baldwin, 2005). Shrestha and Parajulee (2013) reported a 2–3-week lag between cotton aphid and predator density peaks in an irrigated cotton study, wherein naturally occurring predator population, via predator conservation approach, was attributed to delaying the cotton aphid population build-up in cotton.

9.12.2 Chemical control

While cotton aphids are well suppressed by natural enemies in Texas cotton, cotton aphids require significant use of insecticide control method for its management in other cotton states. However, several insecticides are available for cotton aphid management (see UC IPM, 2015 for a detailed insecticide list used in California cotton aphid management programme). Insecticides must be chosen with respect to resistance management. Cotton aphids in several states have developed resistance to all major insecticide classes, including the carbamate, organochlorine, organophosphate and pyrethroid classes (Bagwell and Baldwin, 2005). Neonicotinoid class of insecticides are the most reasonable option at the present time for cotton aphid management in the United States, but insecticides must be integrated with other non-insecticide options when possible. To delay the onset of resistance, avoid unnecessary insecticide applications during early season, because this will reduce beneficial insects as well as selecting for further resistance. The use of an in-furrow systemic insecticide at planting may be considered (Morgan, 2002).

9.13 Useful Tools in Cotton IPM that Minimize Insect Pest Risks and Enhance Natural Pest Suppression

9.13.1 Some biorational and botanical insecticides that are available for cotton IPM (Tables 9.11 and 9.12)

Biorational pesticides have shown some promising potential to be used in cotton IPM against several insect pests. For example, our work suggested that three commercial neem-based insecticides, Agroneem®, Ecozin® and Neemix®, displayed oviposition deterrence in beet armyworm (Greenberg *et al.* 2005a). Neem-based insecticides also deterred feeding by beet armyworm larvae. Agroneem®, Ecozin® and Neemix® caused 78, 77 and 72% beet armyworm egg mortality after direct contact with these insecticides, respectively, while the non-treated control egg mortality was only 7.4%. Survival of beet armyworm larvae fed for 7 days on cotton leaves treated with neem-based insecticides was reduced to 33, 60 and 61% for Ecozin®, Agroneem® and Neemix®, respectively, compared with 93% in the non-treated control (Greenberg *et al.* 2005a). Neem-based insecticides could also control other lepidopteran pests (Saxena and Rembold, 1984; Isman, 1999; Ma *et al.*, 2000).

Table 9.12. Registered and available botanical insecticides (Isman, 1999).

Common name	Producer	Azadirachtin (%)	Target insects
Neemix®	W.R. Grace & Co., Columbia, MD	0.25	Noctuids, aphids
Neemix® 4.5	Certis USA, LLC	4.5	Noctuids, aphids, whiteflies, thrips
Ecozin® EC	Amvac, USA, CA	3.0	Noctuids, whiteflies
Agroneem®	AgroLogistic Systems, Inc., CA	0.15	Noctuids

Table 9.11. Registered and available biorational pesticides (Copping, 2003).

Country	Product name	Based on	Target insects
USA	DiPel® DF or ES, Condor®, Javelin® WG	*Bacillus thuringiensis*	Noctuids
USA	Mycotrol®	*Beauveria bassiana*	Sucking insects
USA	Naturalis®	*Beauveria bassiana*	Sucking insects
USA	BioBlast®	*Metarhizium anisopliae*	Thrips, mites, Coleoptera
USA-Europe	PFR-97TM	*Paecilomyces fumosoroseus*	Whiteflies, thrips
USA	SpinTor® (spinosad)	*Saccharopolyspora spinosa*	Noctuids, thrips

9.13.2 Microbial biopesticides

Effectiveness of some biopesticides based on *B. bassiana* and *M. anisoplia* against sucking insects is not significantly different from synthetic insecticides used (Table 9.13). *B. thuringiensis* showed satisfactory results against lepidopterans (Table 9.14).

9.13.3 Role of natural enemy complex during boll weevil eradication

Naturally occurring beneficial insects (Table 9.15) were attributed to have provided 10–15% suppression of cotton pests in conventional cotton during active boll weevil eradication in LRGV. It is expected that the role of naturally occurring biological control agents will increase significantly as frequency of insecticide applications decline. In the Texas High Plains, insecticide application (frequency, amount and total insecticide load) has declined by 70% during the current years, compared to that during the boll weevil years.

9.13.4 Cultural control and IPM system

Among important alternatives to insecticides in cotton are cultural control techniques. Different tillage systems are one of the important cultural tools (Greenberg *et al.*, 2012b). Conservation tillage has found some acceptance among growers because it reduces soil erosion, conserves soil moisture, and substantially lowers cost of field operations compared to conventionally tilled systems. In the Texas LRGV, 30% of the cotton acreage is under conservation tillage (Greenberg *et al.*, 2010). Water availability for irrigation has become a major concern for most Texas production regions. In this case, conservation tillage can be a good tool for improving soil moisture status. Our results demonstrated that different tillage practices had indirect potentially

Table 9.14. Effectiveness of biorational pesticides against lepidopteran pests (Greenberg *et al.*, 2012a).

Insect larvae	Pesticides	Mortality (%)
Fall armyworm	SpinTor® (spinosad), 12–150 g a.i. per ha	72.3±1.6
Complex (Fall and beet armyworms, bollworm)	spinosad, 1st spray; DiPel® (2nd spray 100–300 g a.i. per ha)	76.2±3.8
Beet armyworm	DiPel®	65.3±3.6
Bollworm	spinosad	71.3±5.8
Bollworm	DiPel®	61.3±2.1

Table 9.13. Effects of different biorational and botanical pesticides on sucking insects (Greenberg *et al.*, 2012a).

		Mortality (%)			
		Bemisia tabaci			
Pesticides	Rate	Younger	Older	*Aphis gossypii*	*Thrips* spp.
B. bassiana	2 g/l	98.8±0.6a	97.6±1.4a	96.4±2.1a	90.4±1.8a
M. anisoplia	5 g/l	90.4±4.8a	91.4±3.1a	91.6±3.6a	98.6±0.8a
Neemix®	41.3 g/l	41.6±10.4b	26.0±6.7d	72.1±9.7b	51.4±4.2c
Azadirect®	32.3 g/l	68.0±10.2b	64.6±2.5c	90.4±6.5a	46.7±1.8c
QRD®	1.3 g/gal	82.1±5.5a	80.3±5.5a	92.4±2.7a	69.1±7.7b
Insecticides:					
Fulfil®	0.4 g/l	–	–	100a	–
Oberon®	0.2 g/l	98.9±0.8a	95.9±3.3a	–	–
Control (H₂O)		6.2±2.0c	1.8±0.8e	4.6±2.0c	1.4±0.9d

Table 9.15. Common beneficial insects occurring in LRGV cotton (based on Extension Entomologist of LRGV of Texas and authors' observations).

Beneficial insects	Target insects
Minute pirate bug, *Orius tristicolor* (White)	Aphids, thrips, whiteflies, mites, and moth eggs and small larvae
Bigeyed bug, *Geocoris uliginosus* (Say)	Mites, whiteflies, thrips, plant bug *Creontiades*, fleahoppers and moth eggs
Lady beetles, *Hippodamia convergens* (Guerin-Meneville)	Aphids, moth eggs and small larvae
Green lacewings, *Chrysopa rufilabris* (Burmeister)	Immatures feed on aphids, spider mites, whiteflies
Syrphid fly larva	Aphids
Spider, *Hibana futilis* (Banks)	Fleahoppers, *Pseudomatoscelis seritatus* (Reuter), plant bug, *Creontiades signatus* (Distant)
Encarsia pergandiella Howard	Parasites on whitefly nymphs
Trichogramma spp.	Egg parasite
Bracon spp.	Larva parasite mostly on lepidopteran pests

positive or negative effects on pest and beneficial populations in cotton. The effects are influenced by both abiotic and biotic factors, which can be created or manipulated by conventional (CV) and conservation (CS) tillage systems. Tillage operations modify soil habitats where some insect pests and beneficial insects reside during at least part of their life cycles (Greenberg *et al.*, 2010). These modifications can alter survival and development of both soil- and foliage-inhabiting insects. Conventional tillage in dryland cotton had greater water stress causing plants to shed squares and bolls while the conservation tillage cotton under the same dryland environment responded by fruiting at a higher rate (Greenberg *et al.*, 2003b,c). Adjustment of planting dates to avoid early cooler planting temperatures and escape thrips moving from drying wheat to cotton is an excellent example of cultural control tactic in cotton IPM. Skip-row planting, cultivar selection, proper irrigation timing, targeted tillage practices to conserve natural enemies, habitat diversification, and crop mix in the agricultural landscape all complement ecologically intensive IPM in cotton.

9.14 Cotton Diseases and Nematodes

9.14.1 Seedling disease complex

Seedling diseases are caused by a complex of soil fungi which may occur separately or in combinations. These fungi include *Pythium* spp., *Fusarium* spp., *Rhizoctonia solani* and *Thielaviopsis basicola*. Symptoms include decay of the seed before germination, decay of the seedling before emergence, girdling of the emerged seedling at or near the soil surface, and rotting of root tips. Crop rotation, quality of the seed, timely planting, and the use of fungicides such as Captan®, Maxim®, Nu-Flow® ND, Nu-flow® M, Vitavax® and Baytan® can reduce losses to seedling diseases and are registered for commercial seed and soil treatments (Allen *et al.*, 2010).

9.14.2 Root rot

Root Rot (fungus – *Phymatotrichum omnivorum*) typically starts to appear in early summer. It causes rapid wilting, followed by death of the plants within a few days. Leaves shrivel, turn brown and die, but they remain attached to the plant. The disease kills plants in circular areas ranging from a few square metres to a hectare or more in size. Dead plants will remain standing in the field but can be easily pulled from the soil. Soil amendments may be applied to alter the growing environment in the root zone. In particular, increase in organic matter and modifying soil pH by using chelated element sulfur and inorganic trace elements (Zn and Fe) is recommended. In addition, winter cover of *Brassicae* plants are used for cultural control via disease suppression, whereas fumigating infested planting holes will usually only delay the onset of disease in non-infested plants. Applying sulfur in trenches 10–15 cm wide and 1.5 m deep around the outside of the drip line of infested plants has shown to prevent the spread of root rot (Matocha *et al.*, 2008, 2009).

9.14.3 Boll rot

Boll rot is common in very heavy cotton growth. If excessive vegetative growth has occurred, one may encounter boll-rot problems. Reducing some of the leaf tissues with the selective use of defoliants may be a practical answer. Effective weed and insect management will decrease incidence of boll rot (Allen *et al.*, 2010).

9.14.4 Nematodes

Rotylenchus reniformis (Linford and Oliveira) is a major problem confronting cotton production in the Lower Rio Grande Valley of Texas. Root-knot nematode, *Meloidogyne incognita* (Kofoid and White) Chitwood, is widespread in cotton of Texas in sandy or sandy loam soils. Larvae feed on plant roots causing swellings (galls). Control practices for nematodes include crop rotation and chemical control with nematicides or soil fumigants (Robinson *et al.*, 2008).

9.15 Weed Control

In the LRGV of Texas, the primary winter and spring weeds in cotton are common purslane (*Potulaca oleracea* L.), pigweed (*Amaranthus palmeri* Wats.), wild sunflower (*Helianthus annuus* L.), and Johnson grass (*Sorghum halepense* (L.) Person). Controls are by use of conventional tillage systems, winter cover crops, and selective herbicides. Moran and Greenberg (2008) showed that black oat (*Avena strigose* Schreb.) and hairy vetch (*Vicia villosa* Roth) suppressed winter weeds to the same extent or more than did winter tillage in no-cover plots. In the spring, soil incorporated black oats cover was slightly more beneficial to cotton than incorporated hairy vetch, but neither cover controlled spring weeds. Two years of winter cover cropping did not obviate the need for cultivation and hand-weeding for sustainable spring weed management in cotton (Greenberg *et al.*, 2012b).

9.16 IPM Models

The model of a pyramid can be used to highlight the method that growers might productively use to construct their pest management programmes. There are different models of IPM pyramids, but all models essentially deliver the same IPM value. In Fig. 9.2, the foundation of a sound pest and disease management programme in an annual cropping system begins with cultural practices that alter the farm landscape to promote crop health. These include crop rotations that limit the availability of host material used by plant pathogens, judicious use of tillage to disrupt pest and pathogen life cycles, burying weeds, and preparing seed beds of optimal moisture and bulk density. Management of soil fertility and moisture can also limit plant diseases by minimizing plant stress. Environmentally mediated pest controls can be regulated in terms of temperature, light, moisture and soil composition. However, the design of such systems cannot wholly eliminate pest problems (Greenberg *et al.*, 2012b). The

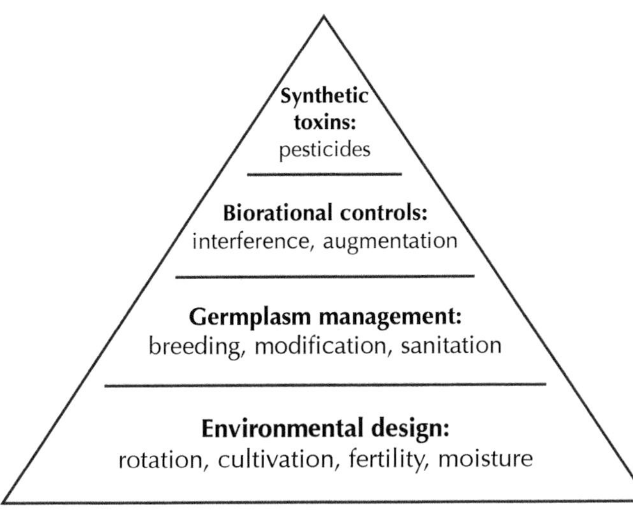

Fig. 9.2. Generalized model for an integrated pest management programme (McSpadden Gardener and Fravel, 2002).

second layer of defence against pests consists of the quality of crop germplasm. Newer technologies that directly incorporate genes into crop genomes, commonly referred to as genetic modification or genetic engineering, are bringing new traits into crop germplasm. The most widely distributed are the different insecticidal proteins derived from *Bacillus thuringiensis*. Upon these two layers, growers can further reduce pest pressure by considering both biological and chemical inputs (McSpadden Gardener and Fravel, 2002; Greenberg *et al.*, 2012b).

High yields of agricultural crops can only be obtained if there is sufficient control of pests. In the mid-20th century, development of chemical pesticides provided an effective answer, but pests became resistant and by killing naturally occurring beneficial species, resurgence of pest populations occurred. The devastating outbreaks of tobacco budworm (*Heliothis virescens*) in the LRGV of Texas during the late 1960s and early 1970s (and the similar outbreaks of *Heliothis armigera* in Australia during the same period) demonstrated conclusively that unilateral reliance on pesticides for insect control was not sustainable and could lead to economic calamities (Greenberg *et al.*, 2012a). This led to the concept of integrated pest management utilizing a range of control tactics in a harmonious way (Fig. 9.3). The diagram shows the

different aspects of IPM – avoidance of pest, then surveillance, and finally, if necessary, control using a bio- or chemical pesticide. In Texas, IPM implies integration of approaches and methods into a pest management system, which takes into consideration that environmental impacts and economic risks have been minimized (Greenberg *et al.*, 2012b).

IPM models presented in Figs 9.4 and 9.5 are successfully used in managing row crop pests in Texas, particularly in the Texas High Plains. Our cotton IPM approach emphasizes the need to maximize the number of pest management tactics integrated and no single pest control method should be used by itself. Chemical controls are used only when needed (in relation to economic thresholds). It is important to optimize their application. Nozzles need to be selected to optimize the droplet sizes, so that the pesticides can be distributed where the pests are located with minimal spray drift. Monitoring (sampling) of the pest is constantly needed. Mere presence of a pest is not a reason to justify action for control.

IPM has largely been expanded to Integrated Cropping Systems Management (ICSM) in Texas and many other cotton-growing states in the United States. It is expected that a holistic cropping systems approach that encompasses all agricultural

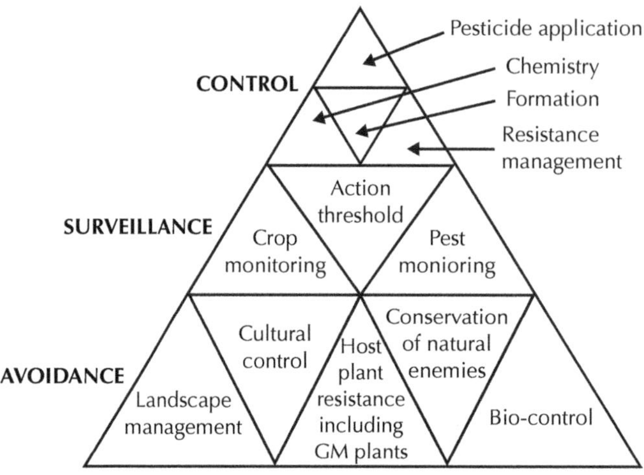

Fig. 9.3. IPM model used in managing *Bemisia tabaci* in Arizona, USA (Naranjo *et al.*, 2001).

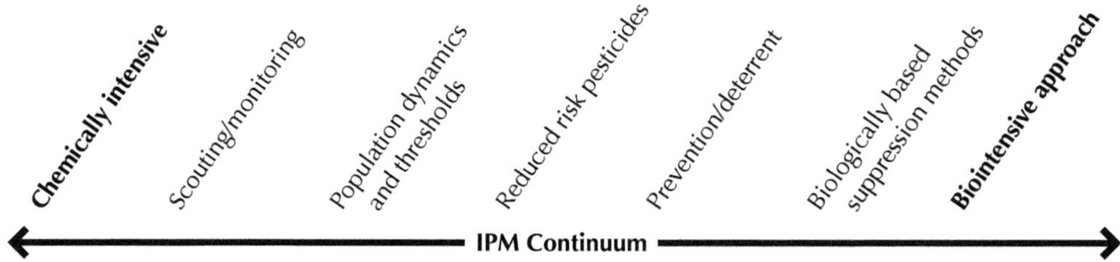

Fig. 9.4. Linearized IPM continuum used in Texas cotton pest management programme.

input management and economics of production system, research-extension-producer partnerships, and ecological considerations is the future of a sustainable agriculture production, regardless of the crop in question (Fig. 9.5). This will include increased use of reduced-risk pesticides and genetically engineered crops. Recent surveys of both conventional and organic growers indicate an interest in using bio-control products (Van Arsdall and Frantz, 2001). The future success of the biological control industry will depend on innovative business management, product marketing, Extension education, and research (Mathre *et al.*, 1999; Greenberg *et al.*, 2012b). These will contribute substantially to making the 21st century the age of biotechnology by the development of innovative biocontrol strategies. For those who are committed to practising effective IPM, it will be necessary to develop (i) a workable definition that incorporates the key components of IPM, and (ii) a set of performance standards to permit a quantitative assessment of IPM implementation in the field.

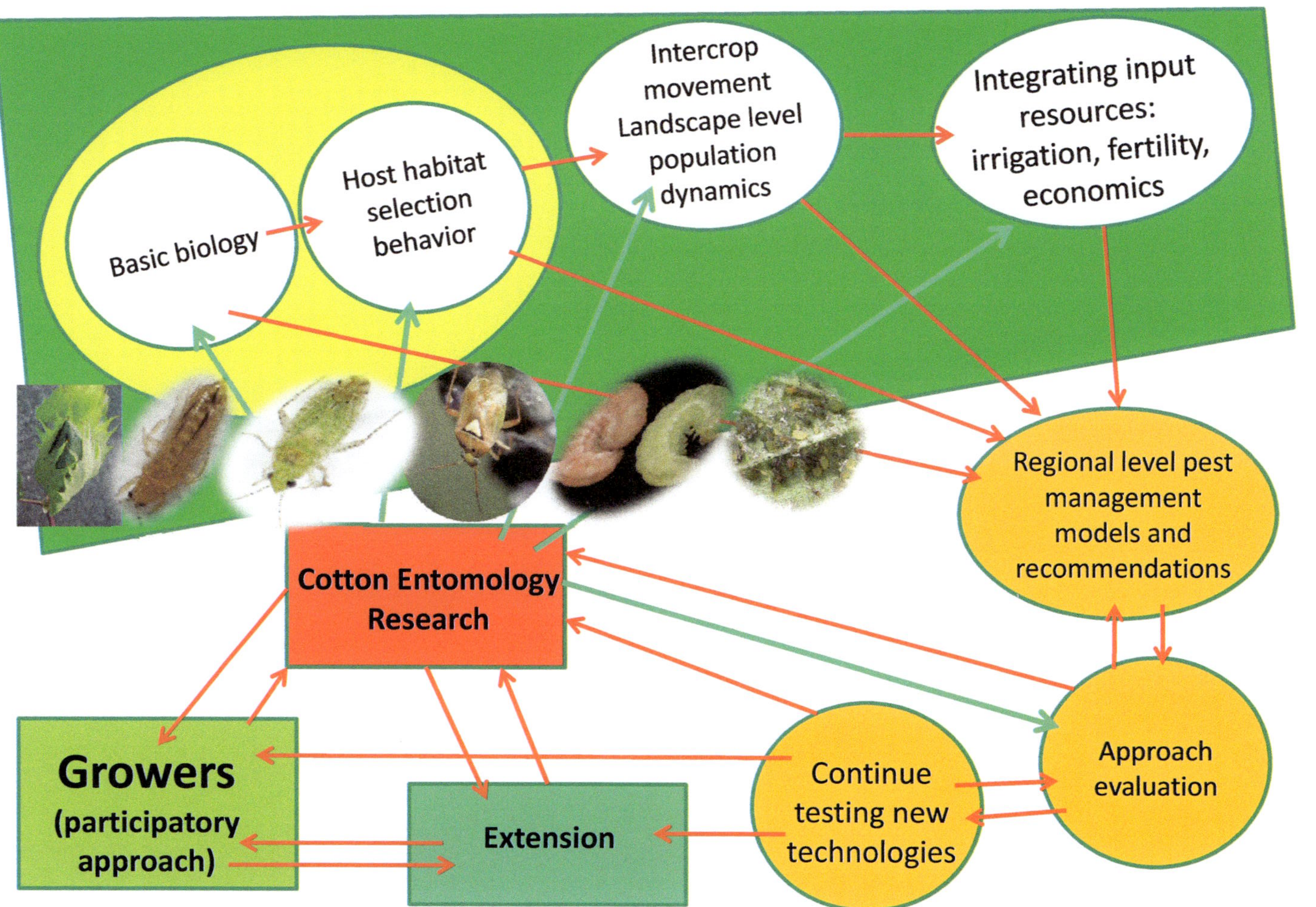

Fig. 9.5. Ecologically intensive clientele-oriented IPM model used in the Texas High Plains (MNP, unpublished data).

References

Adamczyk Jr, J.J. and Gore, J. (2004) Laboratory and field performance of cotton containing Cry1Ac, Cry1F, and both Cry1Ac and Cry1F (WideStrike) against beet armyworm and fall armyworm larvae (Lepidoptera: Noctuidae). *Florida Entomologist* 87, 427–432.

Adamczyk Jr, J.J., Greenberg, S., Armstrong, S., Mullins, W.J., Braxton, L.B., Lassiter, R.B. and Siebert, M.W. (2008) Evaluation Bollgard II and WideStrike technologies against beet and fall armyworms. *Proceedings Beltwide Cotton Conferences*. CD-ROM. National Cotton Council, Memphis, Tennessee.

All, J.N., Tanner, B.H. and Roberts, P.M. (1992) Influence of no-tillage practices on tobacco thrips infestations in cotton. *Proceedings Southern Conservation Tillage Conference.* Jackson and Milan, Tennessee, pp. 77–78.

Allen, S.J., Scheikowski, L.J., Gambley, C., Sharman, M. and Maas, S. (2010) Integrated disease management for cotton. *Cotton Pest Management Guide 2010.* CRDC and CottonInfo, Narrabri, Australia, pp. 115–121.

Armstrong, J.S., Adamczyk, J.J. and Coleman, R.J. (2009) Determining the relationship between boll age and green plant bug feeding injury to South Texas cotton. *Proceedings Beltwide Cotton Conferences*, San Antonio, Texas, pp. 717–720.

Armstrong, J.S., Coleman, R.J. and Duggan, B.L. (2010) Actual and simulated injury of *Creontiades signatus* Distant (Hemiptera: Miridae) feeding on cotton bolls. *Journal of Entomological Science* 45, 170–177.

Bagwell, R.D. and Baldwin, J.L. (2005) *Aphids on Cotton.* Louisiana Cooperative Extension Service. Available at: www.lsuagcenter.com/NR/rdonlyres/CA621AD0-89B2-471F-87ED-91CFBFC3097A/11536/pub2455aphids2.pdf (accessed 3 August 2017).

Barman, A.K. and Parajulee, M.N. (2013) Compensation of *Lygus hesperus* induced pre-flower fruit loss in cotton. *Journal of Economic Entomology* 106, 1209–1217.

Bell, A.A., Lopez, J.D. and Medrano, E.G. (2006) Frequency and identification of cottonseed-rotting bacteria from cotton fleahoppers. *Proceedings of the Beltwide Cotton Conferences.* National Cotton Council, Memphis, Tennessee (CD-ROM).

Bell, A.A., Medrano, E.G. and Lopez, J.D. (2010) Transmission and importance of *Pantoea ananatis* during feeding on cotton buds (*Gossypium hirsutum* L.) by cotton fleahoppers (*Pseudatomoscelis seriatus* Reuter). Proceedings of World Cotton Research Conference 4, Lubbock, Texas, September 2007 (CD-ROM).

Bennett, G.W., Owens, J.M. and Corrigan, R.M. (2005) *Truman's Scientific Guide to Pest Management Operations*, 6th edn. Purdue University/Questex, West Lafayette, Indiana, pp. 10–12.

Bohmfalk, G.T., Frisbie, R.E., Sterling, W.L., Metzer, R.B. and Knutson, A.E. (2002) *Identification, Biology, and Sampling of Cotton Insects.* Texas A&M University System, College Station, Texas.

Brewer, M.J., Armstrong, J.S., Medrano, E.G. and Esquivel, J.F. (2012) Association of verde plant bug, *Creontiades signatus* (Hemiptera: Miridae), with cotton boll rot. *The Journal of Cotton Science* 16, 144–151.

Carpenter, J.E. and Ginanessi, L.P. (2001) *Agricultural Biotechnology: Updated Benefits Estimates.* National Center for Food and Agricultural Policy, Washington, DC.

Castro, B.A., Cattaneo, M. and Sansone, C.G. (2007) *Managing Cotton Insects in the Lower Rio Grande Valley.* Texas Agricultural Extension Service, Texas A&M University System, College Station, Texas.

Cock, M.J.W. (1993) Bemisia tabaci: *An Update 1986–1992 (on the Cotton Whitefly with an Annotated Bibliography).* CAB International, Silwood Park, Ascot, UK.

Coleman, R.J., Hereward, H.P., De Barro, P.J., Frohlich, D.R., Adamczyk Jr, J.J. and Goolsby, J. (2008) Molecular comparison of *Creontiades* plant bugs from south Texas and Australia. *Southwest Entomologist* 33, 111–117.

Copping, L.G. (2003) The Biopesticide Manual. British Crop Protection Council, Alton, UK.

DeSpain, R.R., Benedict, J.H., Landivar, J.A., Eddleman, B.R., Goynes, S.W., Ring, D.R., Parker, R.D. and Treacy, M.F. (1992) Cropping systems and insect management. *1992 Proceedings Beltwide Cotton Conferences*, National Cotton Council, Memphis, Tennessee, pp. 811–812.

Duffey, J.E. and Powell, R.D. (1979) Microbial induced ethylene synthesis as a possible factor of square abscission and stunting in cotton infested by cotton fleahopper. *Annals of the Entomological Society of America* 72, 599–601.

Falck-Zepada, J.B., Traxler, G. and Nelson, R.G. (2000) Surplus distribution from the introduction of a biotechnology innovation. *American Journal of Agricultural Economics* 82, 360–369.

FAO (2004) Economic impacts of transgenic crops. In: *The State of Food and Agriculture*. FAO, Rome. Available at: http://www.fao.org/docrep/006/Y5160E/y5160e09.htm#P1_30 (accessed 21 June 2017).

Fernandez-Cornejo, J. and McBride, W.D. (2000) *Genetically Engineered Crops for Pest Management in U.S. Agriculture: Farm Level Effects*. Agricultural Economic Report n. 786, Economic Research Service, US Department of Agriculture, Washington, DC.

Gannaway, J.R. (1994) Breeding for insect resistance. In: Matthews, G.A. and Tunstall, J.P. (eds) *Insect Pests of Cotton*. CAB International, Wallingford, UK, pp. 431–453.

Gianessi, I. and Reigner, N. (2006) *Pesticides Use in U.S. Crop Production: Insecticides & Other Pesticides*. CropLife Foundation, Washington, DC.

Greathead, A.N. (1986) Host plants. In: Cock, M.J.W. (ed.) Bemisia tabaci – *A Literature Survey on the Cotton Whitefly with an Annotated Bibliography*. CAB International Institute of Biological Control, Ascot, UK, pp. 17–25.

Greenberg, S.M. and Adamczyk, J.J. (2010) Effectiveness of transgenic Bt cottons against noctuids in the Lower Rio Grande Valley of Texas. *Southwestern Entomologist* 35, 539–549.

Greenberg, S.M., Summy, K.R., Roulston, J.R. and Nordlund, D.A. (1998) Parasitism of beet armyworm by *T. pretiosum* and *T. minutum* under laboratory and field conditions. *Southwestern Entomologist* 23 (2), 183–188.

Greenberg, S.M., Jones, W.A. and Legaspi Jr, B.C. (2000a) Interaction between *Encarsia pergandiella* (Hymenoptera: Aphelinidae) and its host *Bemisia argentifolii* (Homoptera: Aleyrodidae): effects of parasitoid densities and host-parasitoid ratios. *Subtropical Plant Science* 52, 36–41.

Greenberg, S.M., Legaspi Jr, B.C., Jones, W.A. and Enkegaard, A. (2000b) Temperature dependent life history of *Eretmocerus eremicus* (Hymenoptera: Aphelinidae) on two whitefly hosts (Homoptera: Aleyrodidae). *Environmental Entomology* 29 (4), 851–860.

Greenberg, S.M., Legaspi Jr, B.C. and Jones, W.A. (2001a) Comparison of functional response and mutual interference between two aphelinid parasitoids of *Bemisia argentifolii* (Homoptera: Aleyrodidae). *Journal Entomological Science* 36 (1), 1–8.

Greenberg, S.M., Sappington, T.W., Legaspi, B.C., Liu, T.-X. and Setamou, M. (2001b) Feeding and life history of *Spodoptera exigua* (Lepidoptera: Noctuidae) on different host plants. *Annals Entomological Society America* 94, 566–575.

Greenberg, S.M., Sappington, T.W., Setamou, M. and Liu, T.-X. (2002a) Beet armyworm (Lepidoptera: Noctuidae) host plant preferences for oviposition. *Environmental Entomology* 31 (1), 142–148.

Greenberg, S.M., Jones, W.A. and Liu, T.-X. (2002b) Interaction among two species of *Eretmocerus* (Hymenoptera: Aphelinidae), two species of whiteflies (Homoptera: Aleyrodidae), and tomato. *Environmental Entomology* 31(2), 397–402.

Greenberg, S.M., Sappington, T.W., Spurgeon, D.W. and Setamou, M. (2003a) Boll weevil (Coleoptera: Curculionidae) feeding and reproduction as functions of cotton square availability. *Environmental Entomology* 32 (3), 698–704.

Greenberg, S.M., Smart, J.R., Bradford, J.M., Sappington, T.W., Norman, J.W. and Coleman, R. (2003b) Effects of conventional vs. conservation tillage systems on population dynamics of boll weevil (Coleoptera: Curculionidae) in dryland cotton. *Subtropical Plant Science* 55, 32–39.

Greenberg, S.M., Smart, J.R., Bradford, J.M., Sappington, T.W., Norman, J.W. and Coleman, R.J. (2003c) Effects of different tillage systems in dryland cotton on population dynamics of boll weevil (Coleoptera: Curculionidae). *Subtropical Plant Science* 55, 32–39.

Greenberg, S.M., Showler, A.T., Sappington, T.W. and Bradford, J.M. (2004a) Effects of burial and soil condition on post-harvest mortality of boll weevils (Coleoptera: Curculionidae) in senesced cotton squares and bolls. *Journal of Economic Entomology* 97 (2), 409–413.

Greenberg, S.M., Sappington, T.W., Elzen, G.W., Norman, J.W. and Sparks, A.N. (2004b) Effects of insecticides and defoliants applied alone and in combination for control of overwintering boll weevil (*Anthonomus grandis*, Coleoptera: Curculionidae) – laboratory and field studies. *Pest Management Science* 60, 849–859.

Greenberg, S.M., Sappington, T.W., Setamou, M. and Coleman, R.J. (2004c) Influence of different cotton fruit sizes on boll weevil (Coleoptera: Curculionidae) oviposition and survival to adulthood. *Environmental Entomology* 33 (2), 443–449.

Greenberg, S.M., Showler, A.T. and Liu, T.-X. (2005a) Effects of neem-based insecticides on beet army-worm (Lepidoptera: Noctuidae). *Insect Science* 12, 17–23. CD-ROM.

Greenberg, S.M., Yang, C. and Everitt, J.H. (2005b) Evaluation effectiveness of some agricultural operations on cotton by using remote sensing technology. In: *Proceedings of the 20th Workshop of Remote Sensing Technology* (CD-ROM). Weslaco, Texas, 4–6 October.

Greenberg, S.M., Spurgeon, D.W., Sappington, T.W. and Setamou, M. (2005c) Size-dependent feeding and reproduction by boll weevil (Coleoptera: Curculionidae). *Journal of Economic Entomology* 98 (3), 749–756.

Greenberg, S.M., Setamou, M., Sappington, T.W., Liu, T.-X., Coleman, R.J. and Armstrong, J.S. (2005d) Temperature-dependent development and reproduction of the boll weevil (Coleoptera: Curculionidae). *Insect Science* 12 (6), 449–459.

Greenberg, S.M., Armstrong, J.S., Setamou, M., Sappington, T.W., Coleman, R.J. and Liu, T.-X. (2006) Circadian rhythms of feeding, oviposition, and emergence of the boll weevil (Coleoptera: Curculionidae). *Insect Science* 13 (6), 461–467.

Greenberg, S.M., Sparks, A.N., Norman, J.W., Coleman, R.J., Bradford, J.M., Yang, C., Sappington, T.W. and Showler, A. (2007a) Chemical cotton stalk destruction for maintenance of host-free periods for the control of overwintering boll weevil in tropical and subtropical climates. *Pest Management Science* 63, 372–380.

Greenberg, S.M., Sappington, T.W., Setamou, M., Armstrong, J.S., Coleman, R.J. and Liu, T.-X. (2007b) Reproductive potential of overwintering, F1, and F2 female boll weevils (Coleoptera: Curculionidae) in the Lower Rio Grande of Texas. *Environmental Entomology* 36 (2), 256–262.

Greenberg, S.M., Jones, G.D., Eischen, F., Coleman, R.J., Adamczyk, J.J., Liu, T.-X. and Setamou, M. (2007c) Survival of boll weevil (Coleoptera: Curculionidae) adults after feeding on pollens from various sources. *Insect Science* 14, 503–510.

Greenberg, S.M., Jones, W.A. and Liu, T.-X. (2008) *Bemisia tabaci* (Homoptera: Aleyrodidae) instar effects on rate of parasitism by *Eretmocerus mundus* and *Encarsia pergandiella* (Hymenoptera: Aphelinidae). *Entomological Science* 11, 97–103.

Greenberg, S.M., Jones, G.D., Adamczyk, J.J., Eischen, F., Armstrong, J.S., Coleman, R.J., Sétamou, M. and Liu, T.-X. (2009a) Reproductive potential of field-collected overwintering boll weevils (Coleoptera: Curculionidae) fed on pollen in the laboratory. *Insect Science* 16, 321–327.

Greenberg, S.M., Liu, T.-X. and Adamczyk, J.J. (2009b) Thrips (Thysanoptera: Thripidae) on cotton in the Lower Rio Grande Valley of Texas: species, composition, seasonal population dynamics, damage, and control. *Southwestern Entomologist* 34 (4), 417–430.

Greenberg, S.M., Bradford, J.M., Adamczyk, J.J., Smart, J.R. and Liu, T.-X. (2010) Insects population trends in different tillage systems of cotton in south Texas. *Subtropical Plant Science* 62, 1–17.

Greenberg, S.M., Alejandro, J. and Sétanou, M. (2012a) Economic and environmental impact transgenically modified cotton comparative with synthetic chemicals for insect control. *Journal of Agricultural Science and Technology* B2, 750–757.

Greenberg, S.M., Adamczyk, J.J. and Armstrong, J.S. (2012b) Principles and practices of integrated pest management on cotton in the Lower Rio Grande Valley of Texas. In: Larramendy, M.L. and Soloneski, S. (eds) *Integrated Pest Management and Pest Control – Current and Future Tactics*. InTech, Rijeka, Croatia, pp. 3–34.

Greene, J.K. and Capps, C.D. (2003) Efficacy of selected insecticides for control of stink bugs, 2003. *Summaries of Arkansas Cotton Research,* Research Series 521. Arkansas Agricultural Experiment Station, Division of Agriculture, University of Arkansas, Fayetteville, Arkansas, pp. 227–232.

Greene, J.K., Turnipseed, S.G., Sullivan, M.J. and Herzog, G.A. (1999) Boll damage by southern green stink bug (Hemiptera: Pentatomidae) and tarnished plant bug (Hemiptera: Miridae) caged on transgenic *Bacillus thuringiensis* cotton. *Journal of Economic Entomology* 92 (4), 941–944.

Grefenstette, W.J. and El-Lissy, O. (2008) *Boll Weevil Eradication – 2008*. Available at: www.aphis.usda.gov/plant_health/plant_pest_info/cotton_pests/downloads/bwep-update-spring2008.pps#1

Head, G., Moar, M., Eubanks, M., Freeman, B., Ruberson, J., Hagerty, A. and Turnipseed, S. (2005) A multiyear, large-scale comparison of arthropod populations on commercially managed Bt and non-Bt cotton fields. *Environmental Entomology* 34, 1257–1266.

Isman, M.B. (1999) Neem and related natural products. In: Hall, F.R. and Menn, J.J. (eds) *Biopesticides: Use and Delivery*, Humana Press, Totowa, New Jersey, pp. 139–154.

Jenkins, J.N. and Wilson, F.D. (1996) Host plant resistance. In: King, E.C., Phillips, J.R. and Coleman, R.J. (eds) *Cotton Insects and Mites: Characterization and Management.* Memphis, Tennessee, pp. 563–597.

Jones, W.A. and Greenberg, S.M. (1999) Host instar suitability of *Bemisia argentifolii* (Homoptera: Aleyrodidae) for the parasitoid *Encarsia pergandiella* (Hymenoptera: Aphelinidae). *Journal of Agricultural and Urban Entomology* 16, 49–57.

Jones, W.A., Greenberg, S.M. and Legaspi, B. (1999) The effect of varying *Bemisia argentifolii* and *Eretmocerus mundus* ratios on parasitism. *Biocontrol* 44 (1), 13–28.

Layton, M.B. (1997) *Insect Scouting and Management in Bt-transgenic Cotton.* Mississippi Cooperative Extension Service Publication 2108. Mississippi State University, Oktibbeha County, Mississippi.

Layton, M.B. (2000) Biology and damage of the tarnished plant bug, *Lygus lineolaris*, in cotton. *Southwestern Entomologist* Suppl. 23, 7–20.

Leigh, T.F., Hyer, A.H., Benedict, J.H. and Wynholds, P.F. (1985) Observed population increase, Nymphal weight gain, and oviposition nonpreference as indicators of Lygus Hesperus Knight (Heteroptera: Miridae) resistance in glandless cotton. *Journal of Economic Entomology* 78, 1109–1113.

Leigh, T.F., Roach, S.H. and Watson, T.F. (1996) Biology and ecology of important insect and mite pests of cotton. In: King, E.C., Phillips, J.R. and Coleman, R.J. (eds) *Cotton Insects and Mites: Characterization and Management. Number Three.* The Cotton Foundation Reference Book Series. Cotton Foundation Publishers, Cordova, Tennessee.

Leonard, B.R. (1995) Insect pest management in conservation tillage systems: a mid south perspective. *Proceedings Beltwide Cotton Conferences*, National Cotton Council of America, Memphis, Tennessee, pp. 74–78.

Leonard, B.R., Core, J., Temple, J. and Price, P.P. (2006) Insecticide efficacy against cotton insect pests using air induction and hollow cone nozzles, 957961. *Proceedings Beltwide Cotton Production Conferences*, National Cotton Council of America, Memphis, Tennessee.

Lidell, M.C., Niles, G.A. and Walker, J.K. (1986) Response of nectariless cotton genotypes to cotton fleahopper (Heteroptera: Miridae) infestation. *Journal of Economic Entomology* 79, 1372–1376.

Liu, T.-X. and Chu, C.-C. (2004) Comparison of absolute estimates of *Thrips tabaci* (Thysanoptera: Thripidae) with field visual counting and sticky traps in onion field in south Texas. *Southwestern Entomologist* 29 (2), 83–89.

Lopez, J.D., Fritz, B.K., Latheef, M.A., Lan, Y., Martin, D.E. and Hoffmann, W.C. (2008) Evaluation of toxicity of selected insecticides against thrips on cotton in laboratory bioassays. *Journal of Cotton Science* 12, 188–194.

Lukefahr, M.J. (1975) Fleahoppers vs leaf-hoppers as pests of glabrous cottons. *Proceedings of the Beltwide Cotton Production Research Conference.* National Cotton Council of America, Memphis, Tennessee.

Lukefahr, M.J., Cowan Jr, C.B. and Houghtaling, J.E. (1970) Field evaluations of improved cotton strains resistant to the cotton fleahopper. *Journal of Economic Entomology* 63, 1101–1103.

Luttrell, R.G. and Mink, J.S. (1999) Damage to cotton fruiting structures by the fall armyworm, *Spodoptera frugiperda* (Lepidoptera: Noctuidae). *Journal of Cotton Science* 3, 35–44.

Luttrell, R.G., Wan, L. and Knighten, K. (1999) Variation in susceptibility of noctuid (Lepidoptera) larvae attacking cotton and soybean to purified endotoxin protein and commercial formulations of *Bacillus thuringiensis*. *Journal of Economic Entomology* 92, 21–32.

Ma, D.-L., Gordh, G. and Zalucki, M.P. (2000) Biological effects of azadirachtin on *Helicoverpa armigera* (Hubner) (Lepidoptera: Noctuidae) fed on cotton and artificial diet. *Australian Journal of Entomology* 39, 301–304.

Mascarenhas, V.J., Shotkoski, F. and Boykin, L. (2003) Field performance of Vip Cotton against various lepidopteran cotton pests in the U.S. *Proceedings Beltwide Cotton Production Conferences*, National Cotton Council of America, Memphis, Tennessee, pp. 1316–1322.

Mathre, D.E., Cook, R.J. and Callan, N.W. (1999) From discovery to use: traversing the world of commercializing biocontrol agents for plant disease control. *Plant Disease* 83, 972–983.

Matocha, J.E., Greenberg, S.M., Bradford, J.M. and Wilborn, J.R. (2008) Biofumigation and soil amendment effects on cotton root rot suppression. *Proceedings Beltwide Cotton Production Research Conferences*, National Cotton Council, Memphis, Tennessee. CD-ROM.

Matocha, J.E., Greenberg, S.M., Bradford, J.M., Yang, C., Wilborn, J.R. and Nichols, R. (2009) Controlled release fungicides, soil amendments, and biofumigation effects on cotton root rot suppression. *Proceedings Beltwide Cotton Production Research Conferences*. National Cotton Council, Memphis, Tennessee. CD-ROM.

McSpadden Gardener, B.B. and Fravel, D.R. (2002) Biological control of plant pathogens: Research, commercialization, and applications in the USA. *Plant Health Progress*. DOI:10.1094/PHP-2002-0510-01-RV

Medrano, E.G., Esquivel, J.F., Bell, A.A., Greene, J., Roberts, P. *et al.* (2009) Potential for *Nezara virdulaa* (Hemiptera: Pentatomidae) to transmit bacterial and fungal pathogens into cotton bolls. *Current Microbiology* 59, 405–412.

Moran, P.J. and Greenberg, S.M. (2008) Winter cover crop and vinegar for early-season weed control in sustainable cotton. *Journal of Sustainable Agriculture* 32 (3), 483–506.

Morgan, E.R. (2002) *Crop Profile for Cotton in Mississippi*. Mississippi State University Extension Service, Mississippi State University, Oktibbeha County, Mississippi. Available at: https://ipmdata.ipmcenters.org/documents/cropprofiles/MScotton.pdf (accessed 3 August 2017).

Muegge, M. (2002) *Far West Texas Cotton Insects Update* 1(3). Texas AgriLife Extension Service, Ft. Stockton, Texas.

Musser, F.R., Knighten, K.S. and Reed, J.T. (2009) Comparison of damage from tarnished plant bug (Hemiptera: Miridae) and southern green stink bug (Hemiptera: Pentatomidae) adults and nymphs. *MidSouth Entomologist* 2, 1–9.

Naranjo, S.E., Ellsworth, P.C., Chu, C.C. and Henneberry, T.J. (2001) Conservation of predatory arthropods in cotton: role of action thresholds for *Bemisia tabaci*. *Journal of Economic Entomology* 95, 682–691.

Norman, J.W. and Sparks, A.N. (2003) *Managing Cotton Insects in the Lower Rio Grande Valley*. Texas Agricultural Extension Service, Texas A&M University System, College State, Texas.

Oerke, E.C. (2006) Crop losses to pests. *Journal of Agricultural Science* 144, 31–43.

Parajulee, M.N. and Shrestha, R.B. (2014) Metapopulation approach for landscape level management of western tarnished plant bug, *Lygus hesperus,* in Texas (Hemiptera, Miridae). *Plant Protection Journal (Acta Phytophylacica Sinica)* 41, 761–768.

Parajulee, M.N., Carroll, S.C., Shrestha, R.B., Kesey, R.J., Nesmith, D.M. and Bordovsky, J.P. (2010) Effect of nitrogen fertility on agronomic parameters and arthropod activity in drip irrigated cotton. *Proceedings Beltwide Cotton Conferences*, National Cotton Council, Memphis, Tennessee, pp. 1163–1169.

Perschau, A. and Sanfilippo, D. (2007) Promoting organic cotton: from field to fashion. *Pesticide Action Network*, November, 16.

Pimentel, D. (1997) Pest management in agriculture. In: Pimental, D. (ed.) *Techniques for Reducing Pesticide Use: Environmental and Economic Benefits*. John Wiley & Sons, Chichester, UK, pp. 1–11.

Pimentel, D. (2005) Environmental and economic costs of the application of pesticides primarily in the United States. *Environment, Development, and Sustainability* 7, 229–252.

Pimentel, D. and Burgess, M. (2014) Environmental and economic costs of the application of pesticides primarily in the United States. In: Pimentel, D. and Peshin, R. (eds) *Integrated Pest Management,* Springer Science + Business Media, Dordrecht, The Netherlands, pp. 48–69.

Pimentel, D. and Greiner, A. (1997) Environmental and socio-economic costs of pesticide use. In: Pimental, D. (ed.) *Techniques for Reducing Pesticide Use: Environmental and Economic Benefits*. John Wiley & Sons, Chichester, UK, pp. 51–78.

Pimentel, D., McLaughlin, L., Zepp, A., Latikan, B., Kraus, T. and Kleinman, P. (1991) Environmental and economic impacts of reducing U.S. agricultural pesticides use. In: Pimental, D. (ed.) *Handbook and Pest Management in Agriculture*. CRC Press, Boca Raton, Florida, pp. 679–718.

Robinson, A.F., Westphal, A., Overstreet, C., Padgett, G.B., Greenberg, S.M., Wheeler, T.A., Stetina, S.R. (2008) Detection of suppressiveness against *Rotylenchulus reniformis* in soil from cotton (*Gossypium hirsutum*) fields in Texas and Louisiana. *Journal of Nematology* 40 (1), 35–38.

Sansone, C.G. and Minzenmayer, R.R. (2005) The impact of different tillage practices on arthropods in cotton in the southern rolling plains of Texas. *Proceedings Beltwide Cotton Conferences*, National Cotton Council, Memphis, Tennessee, pp. 1164–1170.

Saxena, K.N. and Rembold, H. (1984) Orientation and ovipositional responses of *Heliothis armigera* to certain neem constituents. *Proceedings of the 2nd International Neem Conference*, Rauischholzhausen, Germany, pp. 199–210.

Sherrick, S.S., Pitts, D., Voth, R. and Mullins, W. (2002) Bollgard II performance in the Southeast. *Proceedings Beltwide Cotton Production Conferences*, National Cotton Council of America, Memphis, Tennessee, pp. 1034–1049.

Showler, A., Greenberg, S.M., Scott, A.W. and Robinson, J.R.C. (2005) Effects of planting dates on boll weevil (Coleoptera: Curculionidae) and cotton fruit in the Lower Rio Grande Valley of Texas. *Journal of Economic Entomology* 98 (3), 796–804.

Shrestha, R.B. and Parajulee, M.N. (2013) Potential cotton aphid, *Aphis gossypii,* population suppression by arthropod predators in upland cotton. *Insect Science* 20, 778–788.

Slosser, J.E., Cole, C.L., Boring III, E.P., Parajulee, M.N. and Idol, G.B. (2005) Thrips species associated with cotton in the northern Texas Rolling Plains. *Southwestern Entomologist* 30, 1–7.

Sparks, A.N., Anciso, J., Riley, D.J. and Chambers, C.C. (1998) Insecticidal control of thrips on onions in south Texas: insecticide selection and application methodology. *Subtropical Plant Science* 50, 58–62.

Sterling, W.L. (1984) Action and inaction levels in pest management. *Texas Agricultural Experiment Station Bulletin*, B-1480, College Station, Texas.

Stewart, S.D. and Sterling, W.L. (1989) Causes and temporal patterns of cotton fruit abscission. *Journal of Economic Entomology* 82, 954–959.

Steyskal, G.C., Murphy, W.L. and Hoover, E.M. (1986) *Insects and Mites: Techniques for Collection and Preservation*, Publication No. 1443, U.S. Dept. of Agriculture, Agricultural Research Service, Washington, DC.

Summy, K.R., Morales-Ramos, J.A., King, E.G., Wolfenbarger, D.A., Coleman, R.J. and Greenberg, S.M. (1994) Integration of boll weevil parasite augmentation into the short-season cotton production system of the Low Rio Grande Valley. *Proceedings Beltwide Cotton Conferences.* National Cotton Council, Memphis, Tennessee, pp. 953–964.

Summy, K.R., Raulston, J.R., Spurgeon, D. and Vargas, J. (1996) An analysis of the beet armyworm outbreak on cotton in the Lower Rio Grande Valley of Texas during the 1995 production season. *Proceedings of the Beltwide Cotton Conferences.* National Cotton Council, Memphis, Tennessee, pp. 837–842.

Texas Lower Rio Grande Valley, IPM Annual Report, 2006–2008. Texas AgriLife Extension Service District 12, Weslaco, Texas.

Tingey, W.M and Pillimer, E.A. (1977) Lygus bugs: crop resistance and physiological nature of feeding injury. *Bulletin of the Entomological Society of America* 23, 277–287.

Tomilin, C.D.S. (ed.) (2003) *The Pesticide Manual*, 13th edn. British Crop Production Council (BCPC), Alton, UK.

Traxler, G., Godoy-Avila, S., Falck-Zepeda, J. and Espinoza-Arellano, J. de J. (2003) Transgenic cotton in Mexico: a case study of the Comarca Lagunera. In: Kalaitzandonakes, N. (ed.) *The Economic and Environmental Impacts of Agbiotech*. Springer Science, New York, pp. 183–202.

Tugwell, P., Young, S.C., Dumas, B.A. and Phillips, J.R. (1976) *Plant Bugs in Cotton: Importance of Infestation Time, Types of Cotton Injury, and Significance of Wild Hosts Near Cotton*. University of Arkansas Agricultural Experiment Station, Report Series 227. Arkansas Agricultural Experiment Station, Division of Agriculture, University of Arkansas, Fayetteville, Arkansas.

Turnipseed, S., Sullivan, M. and Khalilian, A. (2004) Optional management tactics for the sucking bug complex in advanced B.t. cotton. In: *Proceedings of the Beltwide Cotton Conference*, National Cotton Council of America, Memphis, Tennessee, pp. 1534–1537.

UC IPM (2015) *How to Manage Pests: UC Pest Management Guidelines*. Available at: http://ipm.ucanr.edu/PMG/r114300111.html (accessed 6 November 2017).

Van Arsdall, R.T. and Frantz, C. (2001) *Potential Role of Farmer Cooperatives in Reducing Pest Risk: Final Report*. National Council of Farmer Cooperatives, US-EPA, Pesticide Environmental Stewardship Program, Washington DC, pp. 195–112.

Vitale, J., Ouattarra, M. and Vognan, G. (2011) Enhancing sustainability of cotton production systems in West Africa: a summary of empirical evidence from Burkina Faso. *Sustainability* 3, 1136–1169.

Williams, M.R. (1999, 2005–2007, 2010–2011) Cotton insect losses. *Proceedings of the Beltwide Cotton Production Conferences.* National Cotton Council of America, Memphis, Tennessee. CD-ROM.

Willrich, M.M., Leonard, B.R., Gable, R.H. and LaMotte, L.R. (2004) Boll injury and yield losses in cotton associated with brown stink bug (Heteroptera: Pentatomidae) during flowering. *Journal of Economic Entomology* 97, 1928–1934.

Wrona, A.F., Carter, F., Diehl, J., Ellsworth, K.H., Harder, D. *et al.* (1996) Cotton and resistant insects. *Cotton Physiology Today* 7, 13–19.
Zink, A.W. and Rosenheim, J. (2005) State-dependent feeding behavior by western tarnished plant bugs infuences flower bud abscission in cotton. *Entomologia Experimentalis et Applicata* 117, 235–242.

10 Integrated Pest Management in Tropical Vegetable Crops

Luko Hilje[1],*, Edison R. Sujii[2] and Urbano Nava-Camberos[3]

[1]*Tropical Agricultural Research and Higher Education Center, Turrialba, Costa Rica;*
[2]*Empresa Brasileira de Pesquisa Agropecuária, Brasilia, Brazil;* [3]*Facultad de Agricultura y Zootecnia, Universidad Juárez del Estado de Durango, México*

10.1 Introduction

New World tropics do not exactly coincide with the neotropics, as the latter correspond to one of the seven biogeographical regions of the Earth, and include portions of Mexico and South America located outside the boundaries delimited by the latitudinal imaginary lines of Cancer (23.5°N) and Capricorn (23.5°S). Concerning climatic conditions, in the actual tropics, temperature, rainfall and air humidity are normally much higher than in subtropical or temperate areas, whereas photoperiod varies only slightly throughout the year, all of which have a strong influence on the biology and ecology of organisms.

Likewise, lacking the mathematically predictable four seasons observed elsewhere on the planet, which are determined by the Earth's annual translation movement around the sun, in the tropics it is rainfall – often haphazard – that is the driving factor for seasonality, giving rise to only two seasons (dry and rainy). In addition, whereas mainly in the Pacific watershed of tropical America there is a sharp contrast between both seasons, in vast zones of the Atlantic watershed the normal pattern is one of almost permanent rainfall all year long, with short and rather unpredictable dry periods.

Moreover, because of geographic, altitudinal and local topographic features, the American tropics are very complex and heterogeneous in climatic and ecological terms. Both in Caribbean islands and in the mainland, natural ecosystems extend from sea level to great heights, reaching up to 6960 m in the Andes mountains. Such ecosystems are part of very contrasting life zones *sensu* Holdridge (1978), that include dry forests, thorn woodlands, natural pure stands of pine and oak, broadleaf humid forests, deserts and paramos. As a matter of fact, some wild plant species domesticated by man arose and evolved in some of these life zones, as is the case with maize, beans, potato, tomato, cucurbits, cassava, pineapple, rubber, cacao and other fruit trees (León, 1987). In summary, the American tropics harbour a unique plethora of genes, organisms and ecosystems that makes them, along with the Old World tropics, the planet's richest regions in biodiversity.

Regarding agriculture, in several of the aforementioned natural habitats or life zones, land has been cleared by man throughout history, in order to plant both native and exotic crops, sometimes over

---

* Corresponding author e-mail: luko@ice.co.cr

large extensions devoid of their original vegetation. Nowadays, agricultural lands range from sea level to very high altitudes, in the Andes mountains, although they are mainly concentrated below 2000 m. In the case of vegetables, the focus of this chapter, some of them can be grown between 0 and 3800 m, depending on each crop and its particular cultivars or varieties.

Within the altitudinal ranges where vegetables are grown, temperature is quite stable throughout the year and rainfall provides enough moisture for promoting insect development, reproduction and dispersal, making insect pests a continuous threat to crops all year round; the only exceptions are highly seasonal areas, plus deserts and the semi-arid areas, like Brazilian caatinga, but in such cases, water has been supplied by artificial irrigation. These situations force growers and specialists to invest large efforts and resources to deal with pests in general, and insects in particular, on a permanent basis.

The objective of this chapter is to provide an overview of the identity and relative importance of insects that affect vegetable production in the American tropics, as well as an appraisal of the advances, achievements, shortcomings, threats and opportunities, in regards to ways of dealing with them, from an agroecological or integrated pest management (IPM) perspective. Even though there are several definitions of what a vegetable is, we classify a vegetable as a herbaceous annual crop that provides some edible structure (shoots, leaves, inflorescences, pods, seeds, fruits, roots, tubers or bulbs) for human consumption, either fresh or cooked. Accordingly, annual plants that are sources of fodder, grains and hard and starchy tubers (like cassava, taro, tannia and yams) do not qualify here as vegetables.

For the purposes of this chapter, IPM is considered as a strategy consisting of a combination of two or more compatible methods (tactics) to keep insect pest populations at levels of non-economic importance, while minimizing harm to people and the environment. In short, it is a paradigm based upon three key concepts: *prevention*, *coexistence with pests*, and *ecological and*

economic sustainability, the latter meaning that any tactics to be implemented in farmers' fields ought to be amenable with environmental protection and public health, while being cost-effective, in order for them to be attractive for and adoptable by farmers (Hilje *et al.*, 2003).

10.2 Vegetables and Their Pests in Tropical Environments

Since colonial times, export crops such as sugarcane and cotton, followed by coffee and bananas later on – all of them planted over huge extensions – have driven the economy of the majority of American countries located in the tropics, a picture that pretty much holds until today. Therefore, in addition to receiving the bulk of attention from policy makers and banks, farmers involved in their production have created commodity groups and even funded their own research and extension centres. In contrast, in spite of their value as food crops, vegetables have not counted upon such an institutional support.

In the American tropics, agriculture is represented by a vast gamut of contrasting cropping systems, which range from small plots of polycultures to very extensive area-wide monocultures, of hundreds to thousands of hectares (Fig. 10.1); for instance, in countries like Mexico, Peru, Venezuela or Brazil, some single vegetable farms, owned by either individuals or companies, can reach as much as 1200 ha. In the first case, which are labour-intensive, farmers rely on low external inputs, whereas the latter are input-intensive (certified seeds, including hybrids and transgenic varieties, plus synthetic fertilizers, pesticides and growth regulators, as well as mechanization, including precision-agriculture techniques), closely resembling production systems in developed countries.

Small and medium-size farmers comprise a wide spectrum, from indigenous communities who plant small patches of staple crops surrounded by primary forest, to commercially oriented growers, either as individuals or in cooperatives, who plant

Fig. 10.1. Contrasting vegetable cropping systems in the American tropics: a mosaic of vegetables (potato, cabbage, onion, carrot, etc.) in the Irazú volcano foothills, in Costa Rica (a); a conventional tomato monoculture habilitated with drip irrigation, in a desert coastal area of Peru (b); a large onion plot, planted along with tomato and other vegetables, in a farm owned by a cooperative, in Güines, Cuba (c); a broccoli plot surrounded by pastures with forest patches and windbreaks in Machachi, in the highlands of Ecuador (d). Photos by Luko Hilje.

several crops in different spatial and temporal schemes throughout the year (Hilje *et al.*, 2003).

Many of these farmers are illiterate, whereas some of them do not easily communicate in Spanish or Portuguese, the predominant languages, which hinders technology transfer. In contrast, wealthy farmers or transnational companies involved in export crop production (bananas, sugarcane, melons, watermelons, pineapple and soybean) have ready access to new information and technologies by the internet and other means, as well as through representatives of agrochemical companies.

In regard to vegetables, they can also be found in a wide range of cropping systems, from home backyards to large monocultures, produced either for local consumption (cabbage, carrot, cauliflower, celery, cucumber, aubergine, garlic, green bean, hot pepper, lettuce, okra, onion, pumpkin, squash, sugar beet and sweet potato) or as cash crops for export (asparagus, sweet pepper, broccoli, melon, potato, snow peas, tomato and watermelon).

Moreover, they can be grown along a wide altitudinal range. In general, agricultural lands extend from sea level to about 3800 m, although they are mainly concentrated below 1200 m, with bananas, rice, melons, watermelon, cacao and cotton usually planted below 300 m, and sugarcane and pineapple below 900 m; coffee extends its range between 500 m and 1200 m, whereas maize and beans are planted from 0 m to 2500 m. In the case of vegetables, some of them can be grown between 0 m and 3800 m, depending on the crop (Table 10.1).

Vegetables currently planted in tropical America include a mixture of both native and exotic species. In the first case,

Table 10.1. Names, family, geographical origin, and planting altitudes for the main vegetable crops grown in tropical America.

Common name	Scientific name	Family	Probable centre of origin*	Altitude (m)**
Leaves, stems and flowers				
Asparagus	*Asparagus officinalis*	Liliaceae	Europe and Asia	0–2000
Broccoli	*Brassica oleracea* var. *italica*	Brassicaceae	Mediterranean Basin, SW Europe and England	0–2000
Cabbage	*Brassica oleracea* var. *capitata*	Brassicaceae	Mediterranean Basin, SW Europe and England	800–2800
Cauliflower	*Brassica oleracea* var. *botrytis*	Brassicaceae	Mediterranean Basin, SW Europe and England	1800–2900
Celery	*Apium graveolens*	Apiaceae	Europe and Asia	1000–2500
Lettuce	*Lactuca sativa*	Asteraceae	Middle East	800–2500
Fruits, pods and seeds				
Cucumber	*Cucumis sativus*	Cucurbitaceae	India	0–1300
Aubergine	*Solanum melongena*	Solanaceae	India	0–900
Green bean	*Phaseolus vulgaris*	Fabaceae	Mexico or Peru	0–2800
Hot pepper	*Capsicum* spp. (5 species)	Solanaceae	Peru or Mexico	0–2700
Melon	*Cucumis melo*	Cucurbitaceae	Iran	0–1100
Okra	*Abelmoschus esculentus*	Malvaceae	Tropical Africa	0–1500
Pea	*Pisum sativum*	Fabaceae	Central Asia or Middle East	0–2200
Pumpkin	*Cucurbita moschata*	Cucurbitaceae	Mexico or Peru	0–2400
Squash or marrow	*Cucurbita pepo*	Cucurbitaceae	Mexico	0–2600
Strawberry	*Fragaria x ananassa* (hybrid)	Rosaceae	Temperate America	800–2500
Sweetcorn, maize	*Zea mays*	Poaceae	Mexico	0–3300
Sweet pepper	*Capsicum annuum*	Solanaceae	Peru or Mexico	0–1500
Tomato	*Solanum lycopersicum*	Solanaceae	Peru and Ecuador	0–1400
Watermelon	*Citrullus lanatus*	Cucurbitaceae	South Africa	0–400
Roots, bulbs and tubers				
Carrot	*Daucus carota*	Apiaceae	Europe, Asia and Africa	600–3000
Garlic	*Allium sativum*	Liliaceae	Asia	600–1800
Onion	*Allium cepa*	Liliaceae	Iran and Pakistan	0–2800
Potato	*Solanum tuberosum*	Solanaceae	Andes Mountains	1500–3000
Sugar beet	*Beta vulgaris*	Amaranthaceae	Europe	600–3000
Sweet potato	*Ipomoea batatas*	Convolvulaceae	Tropical America	0–2750

*Purseglove (1974, 1975) and other sources; **Ruiz *et al.* (2013b), plus other literature sources, including vegetable specialists in some countries.

herbivorous insects that have coevolved with them in their centres of origin (Table 10.1) are the natural candidates for becoming their primary or secondary pests, depending on a number of agroecological factors.

For instance, in Central America – a quite small and narrow region, although rich in microclimates – 1800 insect species are reported as pests of 84 crops, both perennial and annual, including vegetables (Coto *et al.*, 1995). For the latter, numbers amount to at least 182 insect species (Saunders *et al.*, 1998), a figure that has probably changed only slightly in the last 20 years; they distribute as follows: Coleoptera (64), Lepidoptera (53), Hemiptera (26), Homoptera (20), Diptera (8), Thysanoptera (6), Orthoptera (3) and Hymenoptera (2).

This is a clear indication of the high levels of insect biodiversity and endemism occurring in the tropics, which is also valid for other plant-associated organisms in this realm. Even though endemism is an asset from a biological viewpoint, it may become kind of a burden for growers and entomologists, because null or scanty information on the biology, ecology, and management approaches of undescribed or poorly known insects strongly impedes their appropriate management.

Likewise, field experiences from the authors and colleagues, as well as from the available literature, overwhelmingly show that for any crop in the tropics, regardless of how simplified it is in structural and floristic terms, there is an amazing food web of herbivorous and carnivorous insect species (parasitoids and predators) associated with it.

The complexity of tropical agroecosystems was appropriately depicted by Hart (1985) based upon a systems approach, as a result of his extensive field experiences with small farmers in Central America. His chart (Fig. 10.2), which is still valid as a conceptual framework, considers a number of key components plus their interactions, as well as the limits, inputs and outputs of a particular agricultural system, including the management or decision-making plan applied by the farmer.

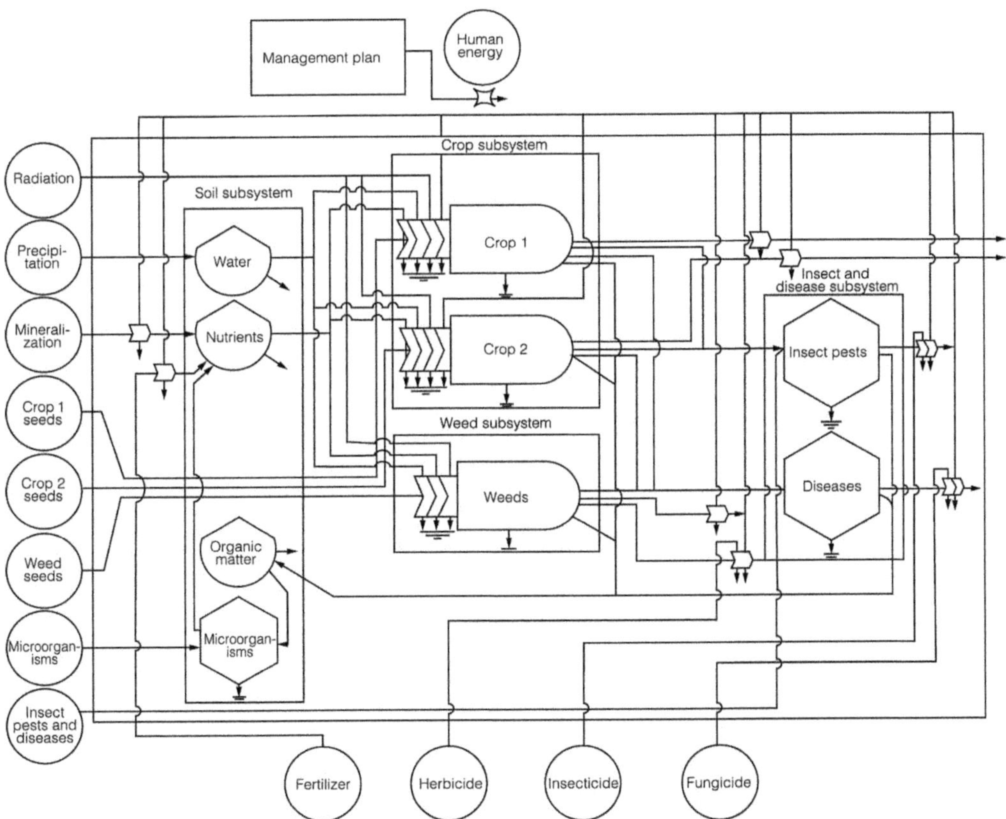

Fig. 10.2. Flow of materials, energy and information through an agroecosystem which includes a subsystem with two crops, as depicted by Hart (1985).

From a plant protection viewpoint, it can readily be seen that in the real world, no farm is isolated from its surroundings, and that for any particular crop it is possible to recognize subsystems for weeds – direct competitors to the crop – as well as for insect pests and diseases – both actual injurious agents – which either as seeds, individuals, spores or cysts can immigrate from crops, fallow fields or feral vegetation present in the neighbouring landscape.

A short digression: landscape management has become a key issue in ecology in the last 25–30 years, thanks to conservation biologists concerned about how fragmented landscapes pose a risk, mainly to some bird species and top carnivore mammals that need very large home ranges; thus, connectivity through biological corridors have been suggested – and even implemented in some cases – as means to favour their reproduction and survival. However, since the emergence of IPM as a paradigm, area-wide approaches have been a major issue for dealing with insect pests, due to their high mobility and adaptability, as well as to the spatial and temporal crop assemblages within a given area.

Nevertheless, despite the fact that managing insect pests from a regional or landscape perspective would be desirable and should be promoted, it is quite difficult to accomplish, especially because of socioeconomic factors and constraints (land tenure patterns, farmer organization, education and awareness, etc.). In fact, practices like crop residue disposal, crop-free periods, planting dates and weed removal, are sound ways to reduce both insect populations or the inoculum of noxious microorganisms (viruses, mycoplasmae, bacteria, etc.) associated with them. This is why some of these practices are included in the national legislation related to plant protection, although law enforcement is not easy to accomplish, due to a number of factors, ranging from logistical to political ones.

A classical example of the adoption of crop-free periods comes from the Dominican Republic, where the tomato yellow leaf curl virus (TYLCV), vectored by the whitefly *Bemisia tabaci*, devastated the tomato industry in the Azua valley and other areas, from 1992 to 1994 (Álvarez and Abud-Antún, 1995; Villar *et al.*, 1998). In contrast to 8805 ha harvested, with an average yield of 21.6 t/ha, in 1989, figures decreased to 3729 ha and 11.3 t/ha in 1993. Therefore, regulatory measures were issued, aimed at banning cultivation of whitefly hosts three months before the main tomato growing season; sorghum was deployed as a rotational crop. This practice, along with planting of tolerant hybrids and rational insecticide use, allowed the tomato industry to bounce back over the coming years; by 1997, harvested area reached 8940 ha and yields amounted to 30.4 t/ha.

An analogous procedure was implemented in northeast Brazil in 1992 to deal with the tomato pinworm (*Tuta absoluta*), a key pest demanding up to 36 sprayings per season (Benvenga *et al.*, 2007), as well as in recent years to manage *B. tabaci* as a vector of viruses affecting beans and tomato in the central-west region of the country (Michereff and Nagata-Inoue, 2015).

Now, coming back to Hart's scheme, when the insect pest subsystem is taken apart and dissected (Fig. 10.3), other elements become obvious, such as natural enemies (parasitoids, predators and entomopathogens) of both primary and secondary pests. Moreover, even though it is not shown there, for every herbivorous insect species in any given environment, there is always an assemblage of associated natural enemies. These have evolved along with their host or prey species, and developed well-defined functional roles (niches), in a way that they complement each other for finally regulating herbivorous populations to a rather low level, which reaches a steady-state equilibrium.

In addition, both parasitoids and predators may act at different trophic levels, as primary, secondary or tertiary predators – mainly vertebrate animals in the last two categories – or as primary or secondary parasitoids (hyperparasitoids). For instance, in the case of the whitefly *B. tabaci*, there are at least 15 and eight *Encarsia* species (Aphelinidae) that are primary parasitoids of their nymphs in Central America and

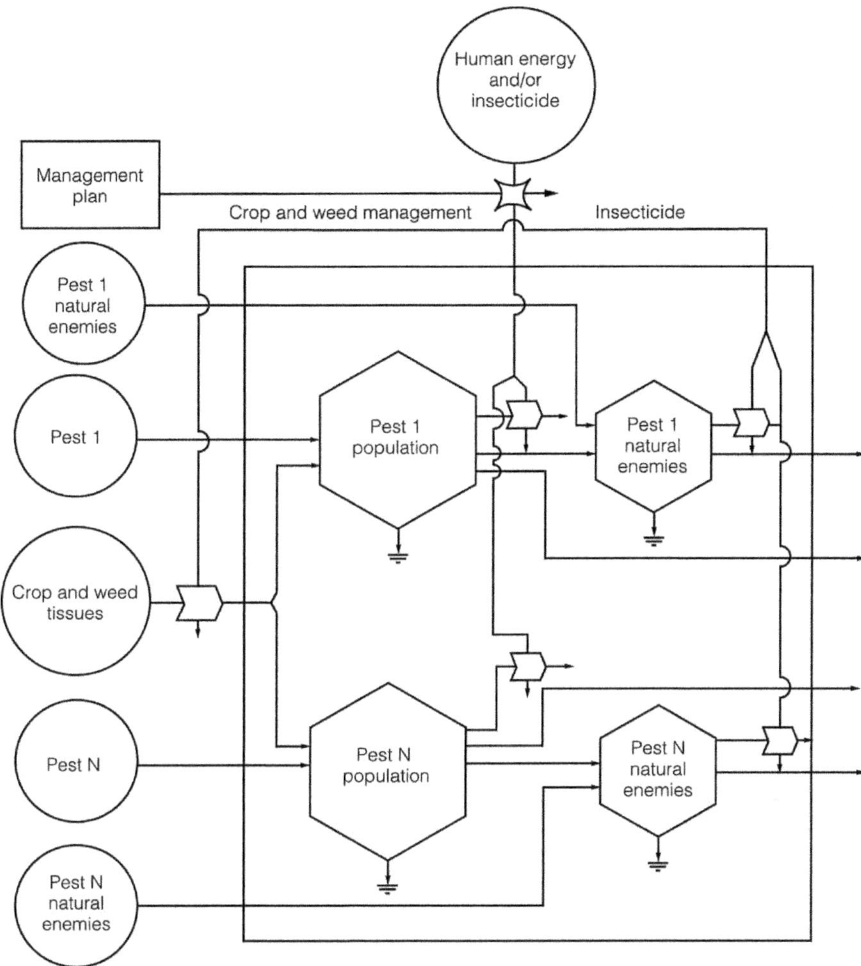

Fig. 10.3. Flow of materials and energy in a pest system acting as a subsystem within an agroecosystem, as depicted by Hart (1985).

central Brazil, respectively, whereas *Signiphora aleyrodis* is a hyperparasitoid of *Encarsia formosa* and *E. hispida* (Cave, 1996; Oliveira *et al.*, 2003).

Moreover, from a landscape perspective, large patches of wild vegetation, along with living fences, hedgerows and windbreaks, may act as perching sites and corridors for different species of parasitoids and predators, thus providing connectivity between farms and favouring their action as pest population regulators.

All these intricate and fragile food webs may be resilient to a certain extent, but when external harmful factors appear over and over and intensify, as occurs with insecticides, population disturbances (secondary pest outbreaks, pest resurgences, etc.) may occur. This explains the so-called insecticide treadmill, which was quite well documented for both cotton and bananas in Central America in the past, and that has been taking place in a silent way – not published in formal journals, but witnessed by farmers and extension agents over the years – in tropical crops.

In other words, disturbances resulting from insecticide overuse have been instrumental in revealing how much complexity underlies tropical agroecosystems

– vegetables are not as simple as they appear to be.

10.3 Native and Exotic Insect Pests

Among insects that feed on vegetables in tropical America, not all of them are equally important on economic grounds, so that they can be classified as primary or key pests (more abundant and harmful, persistent and ubiquitous), as well as secondary or tertiary pests (if they are either less abundant, less harmful, or restricted in time or space).

A compilation of the 70 key pests of vegetables in tropical environments reveals that they belong to only six orders, ranked as follows: Lepidoptera (42%), Homoptera (19%), Coleoptera (17%), Hemiptera (9%), Diptera (7%) and Thysanoptera (4%) (Table 10.2). The overwhelming majority of them are generalist feeders, including plants from different species and families in their diets. Examples of very generalist pests with different types of mouthparts and/or feeding habits are the beet armyworm (*Spodoptera exigua*), the green peach aphid (*Myzus persicae*), the western flower thrips (*Frankliniella occidentalis*) and the vegetable leafminer (*Liriomyza sativae*), which feed on 15, 14, 12 and 11 vegetable crops, respectively, plus a number of other crops and wild plants.

Moreover, they have wide geographic distributions within the American tropics. The majority (52%) of them are present in the four regions analysed, whereas 19, 12 and 17% are in three, two or only one of such regions, respectively. In addition, the 70 pest species may be either native, exotic or cosmopolitan, with native ones representing 43% of them, followed by exotic (19%) and cosmopolitan ones (7%) (Table 10.2). The latter are those species of uncertain origin and present in tropical, subtropical and even temperate regions of the world; as expected, of the seven cosmopolitan vegetable pests inhabiting the American tropics, all are present in the four regions, the only exception being *Delia antiqua*, which is not in the Caribbean so far.

In the case of exotic species, it is widely accepted that they reach pest status because when a crop is introduced and planted in a new country or region, some of their associated insect herbivores are introduced alongside them, but leaving behind their natural enemies. Therefore, if the latter succeed in adapting to the new physical and agronomic environment, and also lack their natural enemies, conditions are very suitable for their survival, reproduction and establishment, reaching pest status. Good examples are the diamondback moth (*Plutella xylostella*), probably of European origin, which reached America decades or centuries ago, as well as the potato psyllid (*Bactericera cockerelli*), which arrived in Mexico from the US in the mid-1950s, and has continued expanding southwards in recent years.

Conversely, in the case of native species, a number of factors or circumstances solely or in combination have favoured previously innocuous insects to become pests. They include the development and deployment of varieties of a given crop that are more susceptible than landraces; a switch in host preference, from a taxonomically related wild species to a crop planted at a commercial scale, as a result of insect species recruitment in response to planted area, according to the concept of species-area (MacArthur and Wilson, 1967), borrowed from island biogeography theory; and natural enemy decimation due to insecticide overuse, giving rise to secondary pest outbreaks, pest resurgences, etc.

But, also, native insect species can adapt their feeding habits and developmental traits and shift from native wild plants to exotic crops, and are thus able to colonize the latter and become primary, secondary or tertiary pests. For instance, all over the tropics, most insect species affecting cacao and sugarcane are native to each country or region where these crops have been introduced (Strong, 1974; Strong *et al.*, 1977), which is probably valid for other crops, both perennial and annual, including vegetables, but data are badly lacking to substantiate this on scientific grounds.

In contrast to perennial (coffee, bananas, cacao and oil palm) or semi-perennial

Table 10.2. Main insect pests of vegetables in tropical America (in alphabetical order), including their origin, geographical distribution and hosts. (From Argüello *et al.* (2007), CATIE (1990a,b, 1993), Kranz *et al.* (1982), Saunders *et al.* (1998), Schmutterer (1990) and Núñez (2005). In addition to the authors' experience, information was also gathered from specialists in different countries.)

Common name	Scientific name and family	Origin*	Distribution	Vegetable hosts**
Coleoptera				
Pepper weevil	*Anthonomus eugenii* (Curculionidae)	N	M, CA	Hot pepper, sweet pepper
Leaf beetle	*Cerotoma arcuata* (Chrysomelidae)	N	SA	Green bean, sweetcorn, potato, sweet pepper
Potato wireworm	*Conoderus* sp. (Elateridae)		SA, CA	Potato, sweet potato
Sweet potato weevil	*Cylas formicarius* (Curculionidae)	N	M, CI, SA	Sweet potato
Banded cucumber beetle	*Diabrotica balteata* (Chrysomelidae)	N	M, CA, CI, SA	Cabbage, cauliflower, cucumber, melon, onion, okra, potato, sweetcorn, sweet pepper, sweet potato, tomato, watermelon
Leaf beetle	*Diabrotica bicolor* (Chrysomelidae)	N	SA	Hot pepper, sweet pepper
Leaf beetle	*Diabrotica speciosa* (Chrysomelidae)	N	M, CA, SA	Lettuce, green bean, cabbage, cauliflower, cucumber, melon, okra, onion, potato, pumpkin, squash, sweetcorn, sweet potato, tomato, watermelon
Spotted cucumber beetle	*Diabrotica undecimpunctata howardii* (Chrysomelidae)	N	M, CA	Asparagus, cabbage, aubergine, green bean, melon, onion, pea, pumpkin, squash, sugar beet, sweetcorn, sweet potato, watermelon
Flea beetle	*Epitrix cucumeris* (Chrysomelidae)	N	M, CA, CI, SA	Cabbage, cucumber, aubergine, garlic, lettuce, onion, potato, pumpkin, sweet pepper, tomato
Brown flea beetle	*Epitrix fasciata* (Chrysomelidae)	N	M, CA, CI, SA	Aubergine, potato, tomato
Sugarcane weevil	*Exophthalmus quadrivittatus* (Curculionidae)	N	CI	Strawberry
Tortoise beetle	*Metriona flavolineata* (Chrysomelidae)	N	CI	Sweet potato
Whitegrubs	*Phyllophaga* spp. (Scarabaeidae)	N	M, CA, CI, SA	Cabbage, green bean, hot pepper, potato, strawberry, sweetcorn, sweet potato
Andean potato weevil	*Premnotrypes vorax* (Curculionidae)	N	SA	Potato
American flea beetle	*Systena basalis* (Chrysomelidae)	N	CI	Potato, sweet potato, tomato
Diptera				
Onion fly	*Delia antiqua* (Anthomyiidae)	C	M, CA	Garlic, onion
Vegetable leafminer	*Liriomyza huidobrensis* (Agromyzidae)	N	CA, SA	Garlic, lettuce, pea, potato, sugar beet

continued

Table 10.2. *continued.*

Common name	Scientific name and family	Origin*	Distribution	Vegetable hosts**
Vegetable leafminer	*Liriomyza sativae* (Agromyzidae)	N	M, CA, CI, SA	Celery, cucumber, aubergine, lettuce, pea, potato, squash, sugar beet, sweet pepper, tomato, watermelon
Serpentine leafminer	*Liriomyza trifolii* (Agromyzidae)	N	M, CA, CI, SA	Carrot, celery, cucumber, aubergine, lettuce, melon, okra, onion, pea, potato, squash, sugar beet, sweet pepper, tomato
Sweet pepper fly	*Neosilba certa* (Lonchaeidae)	N	CA, SA	Aubergine, hot pepper, sweet pepper
Hemiptera				
Whitefly	*Aleurotrachelus trachoides* (Aleyrodidae)	N	CA, CI, SA	Aubergine, sweet pepper, sweet potato, tomato
Sweet potato whitefly	*Bemisia tabaci* (Aleyrodidae)	E	M, CA, CI, SA	Asparagus, sweet pepper, broccoli, cabbage, cauliflower, cucumber, aubergine, hot pepper, lettuce, melon, okra, potato, pumpkin, squash, sweet potato, tomato, watermelon
Aubergine lace bug	*Corythaica cyathicollis* (Tingidae)	N	CA, CI	Aubergine, potato, tomato
Tomato mirid	*Cyrtopeltis tenuis* (Miridae)	E	CI	Tomato
Green stink bug	*Nezara viridula* (Pentatomidae)	E	M, CA, CI, SA	Cabbage, cucumber, aubergine, green bean, hot pepper, melon, okra, pea, potato, sweetcorn, sweet pepper, sweet potato, tomato, watermelon
Greenhouse whitefly	*Trialeurodes vaporariorum* (Aleyrodidae)	E	M, CA, CI, SA	Cabbage, cucumber, aubergine, green bean, hot pepper, lettuce, melon, okra, pea, potato, squash, strawberry, sweet pepper, tomato, watermelon
Homoptera				
Groundnut aphid	*Aphis craccivora* (Aphididae)	C	M, CA, CI, SA	Celery, okra, pea, sugar beet, tomato
Bean aphid	*Aphis fabae* (Aphididae)	E	M, CA	Green bean, pea, sugar beet, tomato
Cotton aphid	*Aphis gossypii* (Aphididae)	C	M, CA, CI, SA	Asparagus, carrot, celery, aubergine, lettuce, melon, okra, pea, potato, pumpkin, sweet pepper, tomato, watermelon
Foxglove aphid	*Aulacorthum solani* (Aphididae)	E	CA, CI, SA	Aubergine, lettuce, pea, potato
Potato psyllid	*Bactericera cockerelli* (Triozidae)	E	M, CA, CI	Hot pepper, potato, sweet pepper, tomato
Cabbage aphid	*Brevicoryne brassicae* (Aphididae)	E	M, CA, CI, SA	Broccoli, cabbage, cauliflower
Corn leafhopper	*Dalbulus maidis* (Cicadellidae)	N	M, CA, CI, SA	Sweetcorn

continued

Table 10.2. *continued.*

Common name	Scientific name and family	Origin*	Distribution	Vegetable hosts**
Potato leafhopper	*Empoasca fabae* (Cicadellidae)	N?	M, CA, CI, SA	Celery, aubergine, green bean, hot pepper, melon, potato, sweetcorn, tomato
Bean leafhopper	*Empoasca kraemeri* (Cicadellidae)	N?	M, CA, CI, SA	Pea, green bean, potato, sweet potato
Mustard aphid	*Lipaphis erysimi* (Aphididae)	E	M, CA, CI, SA	Broccoli, cabbage, cauliflower, celery, cucumber, onion, pea, potato
Potato aphid	*Macrosiphum euphorbiae* (Aphididae)	E	M, CA, CI, SA	Aubergine, lettuce, pea, potato, sugar beet, sweet potato, tomato
Green peach aphid	*Myzus persicae* (Aphididae)	C	M, CA, CI, SA	Asparagus, cabbage, celery, aubergine, green bean, hot pepper, lettuce, melon, onion, pea, potato, sweet pepper, tomato, watermelon
Green corn aphid	*Rhopalosiphum maidis* (Aphididae)	E	M, CA, CI, SA	Sweetcorn, green beans, potato, squash
Rice root aphid	*Rhopalosiphum rufiabdominalis* (Aphididae)	C	M, CA, CI, SA	Aubergine, potato, tomato
Brown ambrosia aphid	*Uroleucon ambrosiae* (Aphididae)	E	M, CA, CI, SA	Lettuce
Lepidoptera				
Black cutworm	*Agrotis ipsilon* (Noctuidae)	C	M, CA, CB, SA	Asparagus, celery, lettuce, onion, potato, sugar beet, strawberry, sweet potato, tomato
Cabbageworm	*Ascia monuste* (Pieridae)	N	M, CA, CI, SA	Broccoli, cabbage, cauliflower, lettuce
Soybean looper	*Chrysodeixis includens* (Noctuidae)	N	M, CA, CI, SA	Asparagus, broccoli, cabbage, carrot, celery, cucumber, aubergine, pea, potato, sweet potato, tomato
Cabbage heart worm	*Copitarsia decolora* (Noctuidae)	N	M, CA, SA	Asparagus, cabbage, cauliflower, garlic, onion, potato, strawberry, sugar beet, sweet pepper
Melon worm	*Diaphania hyalinata* (Crambidae)	N?	M, CA, CI, SA	Cucumber, melon, pumpkin, squash
Pickleworm	*Diaphania nitidalis* (Crambidae)	N?	M, CA, CI, SA	Cucumber, melon, squash
Lesser cornstalk borer	*Elasmopalpus lignosellus* (Pyralidae)	N?	M, CA, CI, SA	Asparagus, cabbage, melon, pea, potato, sugar beet, sweet pepper, tomato
Granulate cutworm	*Feltia subterranea* (Noctuidae)	N	M, CA, CI, SA	Broccoli, carrot, cauliflower, celery, aubergine, lettuce, onion, pea, potato, strawberry, sweet pepper, sweet potato, tomato, watermelon
Old World bollworm	*Helicoverpa armigera* (Noctuidae)	E	SA	Tomato
Tomato fruitworm	*Helicoverpa zea* (Noctuidae)	N	M, CA, CI, SA	Asparagus, celery, aubergine, lettuce, okra, onion, sweetcorn, sweet pepper, tomato
Tobacco bollworm	*Heliothis virescens* (Noctuidae)	N	M, CA, CI, SA	Asparagus, lettuce, pea, sweet pepper, tomato

continued

Table 10.2. *continued.*

Common name	Scientific name and family	Origin*	Distribution	Vegetable hosts**
Cabbage budworm moth	*Hellula phidilealis* (Crambidae)	N	M, CA, Cl, SA	Broccoli, cabbage, cauliflower
Tomato pinworm	*Keiferia lycopersicella* (Gelechiidae)	N	M, CA, Cl	Aubergine, potato, tomato
Cabbage moth	*Leptophobia aripa* (Pieridae)	N	M, CA, SA	Broccoli, cabbage, cauliflower
Tobacco hornworm	*Manduca sexta* (Sphingidae)	N	M, CA, Cl, SA	Aubergine, potato, sweet pepper, tomato
Sweet potato moth	*Microthyris anormalis* (Crambidae)	N	Cl, SA	Sweet potato
Fruit borer	*Neoleucinodes elegantalis* (Crambidae)	N	CA, Cl, SA	Aubergine, sweet pepper, tomato
Potato tubermoth	*Phthorimaea operculella* (Gelechiidae)	N	M, CA, Cl, SA	Aubergine, potato, tomato
Imported cabbageworm	*Pieris rapae* (Pieridae)	E	M	Broccoli, cabbage, cauliflower
Diamondback moth	*Plutella xylostella* (Plutellidae)	E	M, CA, Cl, SA	Broccoli, cabbage, cauliflower
Armyworm	*Spodoptera albula* (Noctuidae)	N	CA, Cl, SA	Cabbage, melon, tomato
Southern armyworm	*Spodoptera eridania* (Noctuidae)	N	M, CA, Cl, SA	Cabbage, carrot, aubergine, okra, onion, potato, sugar beet, sweet pepper, sweet potato, tomato, watermelon
Beet armyworm	*Spodoptera exigua* (Noctuidae)	E	M, CA, Cl, SA	Asparagus, broccoli, cabbage, celery, cauliflower, aubergine, garlic, lettuce, melon, onion, pea, potato, sugar beet, sweet pepper, sweet potato, tomato
Fall armyworm	*Spodoptera frugiperda* (Noctuidae)	N	M, CA, Cl, SA	Asparagus, lettuce, melon, onion, strawberry, sweetcorn, sugar beet
Stripped armyworm	*Spodoptera latifascia* (Noctuidae)	N	M, CA, Cl, SA	Asparagus, cabbage, carrot, cucumber, lettuce, melon, onion, pea, potato, sugar beet, sweet pepper, sweet potato, tomato
Stripped armyworm	*Spodoptera ornithogalli* (Noctuidae)	N	M, CA, Cl, SA	Asparagus, cabbage, carrot, cucumber, lettuce, melon, onion, pea, potato, sugar beet, sweet pepper, sweet potato, tomato
Armyworm	*Spodoptera sunia* (Noctuidae)	N	CA, Cl, SA	Asparagus, melon, onion, pea, potato, sugar beet, tomato, watermelon
Guatemalan potato moth	*Tecia solanivora* (Gelechiidae)	N	M, CA, SA	Potato

continued

Table 10.2. *continued.*

Common name	Scientific name and family	Origin*	Distribution	Vegetable hosts**
Cabbage looper	*Trichoplusia ni* (Noctuidae)	C	M, CA, CI, SA	Broccoli, cabbage, cauliflower, celery, cucumber, lettuce, melon, pea, potato, squash, sweet pepper, sweet potato, tomato, watermelon
Sweet potato moth	*Trichotaphe melissia* (Gelechiidae)	N	CI	Sweet potato
Tomato borer	*Tuta absoluta* (Gelechiidae)	N	SA	Aubergine, potato, sweet pepper, tomato
Thysanoptera				
Western flower thrips	*Frankliniella occidentalis* (Thripidae)	E	M, CA, SA	Cabbage, carrot, cucumber, aubergine, lettuce, melon, onion, pea, sugar beet, strawberry, sweet pepper, tomato
Melon thrips	*Thrips palmi* (Thripidae)	E	CI, SA	Cucumber, green bean, aubergine, melon, pea, potato, pumpkin, squash, sweet pepper, onion, tomato, watermelon
Onion thrips	*Thrips tabaci* (Thripidae)	E	M, CA, CI, SA	Asparagus, cabbage, carrot, celery, cucumber, aubergine, garlic, green bean, onion, pea, tomato

*Native (N) or exotic (E) to tropical America. Cosmopolitan (C) are species with a wide distribution and unknown origin.
**For some pest species, in addition to vegetables listed in Table 10.1, other crops, as well as some wild plants, also act as feeding and/or reproductive hosts.
Abbreviations: M (Tropical Mexico); CA (Central America); CI (Caribbean islands); SA (South America). In the case of CA, CI and SA, we refer to its presence in at least one country in each of these regions.

(sugarcane and pineapple) crops, which occupy land for long periods and in many cases share the same region or landscape with vegetables, the latter represent an ephemeral resource for herbivorous insects. Then, since the cropping season of vegetables ranges from 3 to 5 months, for insects to successfully take advantage of such a short-lived resource, their reproductive and developmental traits have to fit this particular condition.

Therefore, in response to the agroecological vacuum represented by a newly planted vegetable field, insects endowed with favourable biological characteristics, such as high fecundity and a short lifespan, which in turn determine the intrinsic rate of increase, can arrive, establish, readily increase their population and outcompete other herbivorous species, thus reaching pest status; the intrinsic rate (r_m) is the maximum rate of increase of a population, per individual. Even though these population parameters strongly depend on temperature, relative humidity and host quality, a few examples are in place, as follows.

For instance, at 25°C and other suitable conditions, for *L. sativae* such values correspond to 100–130 eggs, 16 days and $r_m = 90$ (Araujo *et al.*, 2013); for *Diabrotica speciosa*, 6500 eggs, 35 days and $r_m = 83$ (Ávila and Parra, 2002); for *T. absoluta*, 150 eggs, 19 days and $r_m = 85$ (Thomazini *et al.*, 2001); for *P. xylostella*, 200 eggs, 17 days and $r_m = 90$ (Boiça Junior *et al.*, 2011); for *Diaphania nitidalis*, 250–360 eggs, 28 days and $r_m = 120$–180 (Pratissoli *et al.*, 2007); and for *Spodoptera frugiperda*, 1100 eggs, 32 days and $r_m = 10$–35 (Santos *et al.*, 2004).

Nevertheless, for any given insect pest, a population crash is likely to occur once the crop vanishes, and survival is at serious risk. Therefore, under such circumstances, they can rely upon various particular adaptations or take advantage of some farmers' practices.

In the case of specialized pests (monophagous or oligophagous) of staple crops, despite expecting harsh conditions, some cultural practices can benefit them. For instance, in the Irazú volcano foothills, in Costa Rica, farmers plant potato, cabbage and other vegetables all year round in a staggered way, as a means to lessen economic uncertainty, thus ensuring good yields and market prices, to compensate for lower prices during other times of the year. Therefore, pests like the potato tubermoths *Tecia solanivora* and *Phthorimaea operculella*, as well as the diamondback moth (*P. xylostella*), can search over the landscape to locate either potato or cabbage (plus broccoli and cauliflower) fields at any time during the year. In other words, despite fields belonging to different farmers, as a whole they represent a functional monoculture, from a landscape perspective.

A slightly different pattern occurs with export crops like melon, watermelon and some other cucurbits. Since their commercialization depends on temporal 'windows' in the international markets, during the dry season they are planted in a staggered sequence, every two weeks, in irrigated fields. Therefore, oligophagous or polyphagous pests like the melon worm (*Diaphania hyalinata*), the whitefly *B. tabaci*, and the aphid *Aphis gossypii* – just to mention three of them – can reach high populations and remain active in the field throughout the dry season; once this cornucopia is over, they move to other crops or wild plants during the rainy season.

On the other hand, for polyphagous or generalist vegetable insect pests, the typical short cropping season of any crop does not pose so much of a risk. Despite being certainly forced to search for alternate hosts once they emigrate from a vanishing vegetable crop, they search for other crops in the surroundings, some of which may be interspersed among fields with perennial or semi-perennial crops within a given landscape. In a way, from an insect's perspective,

such crop 'mosaics' are functional monocultures on a regional scale.

For instance, the fruitworm *Helicoverpa zea* feeds on vegetables in the families Solanaceae (sweet pepper, aubergine and tomato), Apiaceae (celery), Asteraceae (lettuce), Liliaceae (asparagus and onion) and Malvaceae (okra), but also on other crops such as cotton (Malvaceae), soybean and cowpea (Fabaceae), maize and sorghum (Poaceae). If a combination of any these crops is simultaneously planted in a landscape, *H. zea* will prefer those on which it usually performs better, but continuous food supply will allow this insect to prosper.

Moreover, in extratropical regions – since *H. zea* is present all the way from Canada to Argentina – in addition to many wild hosts and several ornamental plants, it has also been reported on a wide range of crops (Capinera, 2014), including lucerne, clover, lima bean, pea, snap bean and vetch (Fabaceae), cucumber, melon, pumpkin, squash and watermelon (Cucurbitaceae), potato, sweet potato and tobacco (Solanaceae), sunflower and artichoke (Asteraceae), maize, rice, wheat, oat and millet (Poaceae), cabbage and collard (Brassicaceae), pear, peach, plum, raspberry and strawberry (Rosaceae), grape (Vitaceae), avocado (Lauraceae), spinach (Amaranthaceae) and flax (Linaceae). Therefore, *H. zea* could also thrive on some of these crops if grown in the tropics, thus increasing the likelihood to survive and reproduce on a continuous basis.

Another good example of a polyphagous and cosmopolitan species is the whitefly *B. tabaci*, whose adults are key virus vectors (mainly of begomoviruses, but also carlaviruses, luteoviruses, nepoviruses, potyviruses and closteroviruses). In the Caribbean watershed of Costa Rica, they can move from small plots of tomatoes to small patches of sweet pepper, cucumber or beans – and even wild plants – depending on the availability of each of these crops; aside from the fact that adults can fly up to 7 km (Byrne and Blackmer, 1996), they can feed on about 25 crops in the American tropics and some 500 hosts, both cultivated and wild, worldwide (Greathead, 1986).

Therefore, viruses can remain in wild reservoirs – which sometimes are symptomless – and, thanks to *B. tabaci*'s polyphagous habits it can be disseminated between crops all year round.

In summary, the key attribute for a successful vegetable insect pest is to be an opportunistic species, that is, an r-strategist, as proposed in his seminal paper by Pianka (1970), or close to it, as the insect could rapidly maximize its rate of increase, in turn monopolizing the short-lived crop resources available. Moreover, polyphagy and high mobility are complementary and very desirable adaptive traits, in order to enable the insect to readily locate alternative hosts as the crop being affected fades away; in fact, both traits are common among insects that feed on vegetables.

A fourth characteristic of high adaptive value is diapause, which allows an interruption in an insect's life cycle in case of food shortages – it occurs in subtropical and temperate regions, due to climatic harshness and lack of food during autumn and winter. However, in the tropics, even though under certain circumstances, reproductive diapause has been shown to occur in some insect species, adults remain active in the field (Janzen, 1983). In other words, in tropical environments insect diapause is rather unusual.

10.4 Farmers' Response to Insect Threat

Insect pest control in tropical agroecosystems, including vegetables, is a relentless task for farmers and extension agents, as well as for entomologists. On the one hand, insects commonly remain active all year round, showing overlapping generations. But, also, pest management scenarios are dynamic and ever-changing, not only due to unpredicted upsurges of already established pests, but also to invasive species that are able to evade quarantine procedures and facilities. For instance, by just taking a glimpse at literature concerning tomato production in tropical America (CATIE, 1990b), it can be noticed that current key pests, like whiteflies and the potato psyllid, were either absent or economically insignificant about 25 years ago.

Therefore, farmers face uncertainty on a daily basis, not only because of the pests themselves, but also due to factors associated with them, like unfavourable weather conditions, low market prices, stringent banking loans, etc. In order to reduce risk, farmers spray insecticides on a calendar basis – usually every single week – as a 'preventative' approach. Within this context, this is certainly a misleading term, as actual preventative methods (tolerant varieties, cultural practices, biological control, etc.) are those aimed at creating conditions well in advance for insect pests not to express their damaging potential, while not causing undesirable disturbances; in addition, hopefully, such methods should be long-lasting, as well as cost-effective in the long run.

However, in contrast to other cropping systems, there are two factors that strongly promote insecticide overuse for vegetable production: their short growing season and the high economic return obtained from produce. Both of these factors merge in such a way that prompts farmers to over-apply insecticides – at shorter frequencies, as well as at doses higher than recommended – in some cases, disregarding the high price of the chemicals, as the invested money can soon be recovered. For instance, in a survey conducted in 1998 in Costa Rica (Hilje, unpublished), a farmer said that he had made 25 Confidor (imidacloprid) applications within a year against the whitefly *B. tabaci*, in his 1.5 ha planted with an undetermined sweet pepper variety; he argued that profits were quite high, so that he could afford to buy such expensive insecticide.

Unfortunately, a current overview on vegetable cropping systems all over the American tropics reveals that insecticides – as well as other pesticides – are employed in a unilateral, indiscriminate and excessive way; for the majority of vegetables, a short interview with any farmer is enough to confirm this pattern. Their use is *unilateral*, as farmers seldom consider pest control tactics other than insecticides, because of their perceived advantages (high efficacy, high profitability and ready availability); *indiscriminate*, because with a few exceptions, most insecticides are not specific, killing both insect pests and beneficial animals (natural enemies, pollinators and vertebrates); and *excessive*, since they are generally applied at doses and frequencies higher than recommended, and on a calendar basis, regardless of insect pest density or crop damage levels.

In consequence, insecticide overuse poses serious health risks to farmers themselves by labour exposure, as well as to consumers, since many vegetables are eaten fresh. Moreover, water, soil, wildlife, marine ecosystems – mainly mangroves and coral reefs – and other components of the environment can be negatively impacted by insecticides, although scientific data to substantiate such effects are scanty in the American tropics.

Now, from a strictly economic perspective, insecticide overuse immediately increases production costs, which are transferred to consumers later on, while fostering resistance development in insects. Under the spraying regimes typical of vegetable production in the American tropics, it is expected that intensive selection pressure, in combination with short life cycles, suitable climatic conditions and food availability will lead insect pests to evolve resistance. Field observations strongly suggest that this is a common phenomenon, but studies on a continuous basis are lacking to scientifically substantiate its existence.

Finally, farmers who export vegetables to US and European markets are confronted with a paradox. In many cases, they sign contracts with export companies, committing themselves to apply certain technology 'packages'. Since harvesting dates are strongly dependent on market 'windows', farmers are required to apply insecticides on a strict calendar basis, which increases the likelihood of exceeding maximum residue levels set by regulatory agencies in the importing country. But, at the same time, once in the food stores, produce is subject to rigorous scrutiny by consumers, who are willing to buy only unblemished goods.

Therefore, farmers are faced with the dilemma of exporting 'perfect' products, while complying with stringent insecticide tolerances, which is not always possible, giving rise to detentions and rejections of shipments at the ports of entry.

In summary, regardless of their economic and social status, vegetable farmers in the American tropics are subject to insect pest threat on a permanent basis, and the conventional way of dealing with them has been shown to cause a gamut of undesirable effects, both for society as a whole and for farmers themselves, especially in regard to their health, by exposure to insecticides. Therefore, when talking to them, almost all express their willingness to adopt other methods, provided that alternatives are as practical and cost-effective as insecticides. This is the real challenge for IPM practitioners.

10.5 Capacity-building and Institutionalization of IPM

Since its coining by University of California researchers more than half a century ago, the term 'integrated pest management', as well as its practices, have gained acceptance and support worldwide (Kogan and Bajwa, 1999). Initially, IPM grew in credibility mainly thanks to the United Nations Food and Agriculture Organization (FAO). For tropical America, in response to critical economic and environmental situations associated with cotton production in the 1960s and 1970s, FAO promoted successful approaches, in collaboration with scientists from the University of California, to deal with insect pests in both Peru (Herrera, 2010) and Nicaragua (Daxl, 1989).

Thereafter, several individuals and institutions became aware of the importance of IPM as a component of sustainable cropping systems, and fostered both formal training and field activities associated with it. Since then, local universities in almost every Latin American and Caribbean country started to include IPM topics in courses related to plant protection (entomology, plant pathology and weed science) or created specific courses focused on IPM. In addition, national ministries, along with agricultural research institutes, established research and extension programmes on the matter, and sent abroad personnel for graduate training, as did also universities. Today, national institutions in almost every country, in collaboration with international programmes developed by centres belonging to the CGIAR Consortium, like CIMMYT (Mexico), CIAT (Colombia) and CIP (Peru), as well as FAO's Regional Office (Chile), continue to promote IPM at different levels.

It is worth mentioning the key role of the Tropical Agricultural Research and Higher Education Center (CATIE), based in Costa Rica, in promoting IPM in the American tropics. Due to its strategic position and geographical scope, the US-based Consortium for International Crop Protection (CICP) envisioned that CATIE could be the leader for launching an ambitious regional IPM project focused mainly on vegetables, in Central America. Therefore, from 1984 to 1995 a project was implemented to develop technical assistance, graduate education, training and publication activities, pest diagnosis and identification services, validation of IPM alternatives for vegetables in several countries, and the establishment of a Central American Plant Protection Network, involving universities and ministries.

CATIE's contributions were truly seminal in legitimizing IPM as a feasible alternative for crop protection in Central America, but their coverage extended far beyond, through their publications, like the journal *Manejo Integrado de Plagas* (80 issues, from 1986 to 2007) and some 25 books, including IPM guidelines for tomato, sweet pepper, cabbage and maize. Moreover, a multiplicative effect has been accomplished, thanks to more than 100 MSc graduates, who have acted to the present day – some of them from key institutional positions – as IPM promoters in their respective Latin American and Caribbean countries.

In summary, IPM has achieved its institutionalization all over Latin America and the Caribbean, thanks to joint efforts of international cooperation and funding agencies, plus sustained efforts from local universities, national ministries and other agricultural research institutes.

10.6 Advances and Achievements in IPM

It is impossible to summarize in just a few pages what has been accomplished so far, in regards to practical aspects of IPM for vegetable insect pests in the American tropics. Nevertheless, at least a general overview is in place, to highlight both advances and achievements, as well as some shortcomings, threats and opportunities.

In order to better understand the contents of this section, it has been divided and organized in accordance to major areas (management tools, resources, tactics and approaches) related to IPM, as they apply to the American tropics.

10.6.1 Diagnosis and pest identification

Obviously, the first requisite for a correct design and development of an IPM programme for vegetables is to exactly determine the real cause of a given plant injury, as well as the identity of its causal agent; once this is accomplished, it is possible to define how to tackle it, in accordance to prior experiences or information available in the scientific literature.

Fortunately, in the case of insects affecting vegetables, in almost all Latin American and Caribbean countries there are qualified taxonomists at either local universities or ministries, who are able to identify both immature and adult stages. To a great extent, this is possible because of the existence of several resources, such as comprehensive and profusely illustrated books or handbooks on agricultural insect pests. In addition, today's professional entomologists, students, extension agents and even farmers, count upon a wide gamut of multimedia and internet resources, some of which can be accessed through smartphones. Also, it is becoming common practice that any such individuals are able to take a shot at identifying a plant health problem and immediately send it for diagnosis and identification.

In line with this, it is worth mentioning Plantwise, a global initiative launched by CABI in 2011. This programme relies on a network of national institutions in more than 30 countries, including many Latin American and Caribbean ones, through what is called plant clinics, in charge of adequately trained local scientists, in order to provide farmers with services of pest diagnosis and management recommendations. In addition, taking advantage of internet resources, there is an online knowledge bank, that is periodically updated by both local scientists and specialists, in response to novel plant health problems arising from real-world situations, so that it is very dynamic, practical and fast.

However, a major constraint prevails, which is the lack of both infrastructure and expertise in molecular methods for diagnosing insect-borne microorganisms (viruses, mycoplasmas, bacteria, etc.). Centres that provide some of these services are CINVESTAV and CIMMYT (Mexico), CIBCM (Costa Rica), EAP (Honduras), CENSA (Cuba), CIAT (Colombia), IVIC (Venezuela), CIP (Peru) and EMBRAPA (Brazil).

Moreover, the worldwide crisis elicited by the whitefly *B. tabaci* in the last 25 years has warned entomologists of the need to consider an insect species not as single genetic entity, but as a set of races or biotypes, which are morphologically identical to one another but show differences in host preference, reproductive traits, etc., including unusual plant damage types, like physiological disorders (squash silverleaf, uneven ripening of tomato, white streaking in cole crops, stem blanching of lettuce, etc.).

Along the same lines, given the high levels of diversity existing in the tropics, it would be expected to find many sibling insect species, which may cause different types of damage in comparison to the supposed species to which they belong. Hopefully, they could be recognized by means of DNA bar-coding methods, for which international cooperation would be desirable; once in place, DNA barcode datasets can help in readily identifying vegetable insect pests without the need of a taxonomist.

10.6.2 Decision-making criteria

Coexistence with pests, one of the three key concepts of the IPM paradigm – as it was previously stated – is meaningful if, and only if, farmers refrain from overspraying on a calendar basis and allow a given herbivorous insect population to remain in the field, provided that such density does not exceed a certain point. This is why criteria like economic injury levels (EIL) and economic thresholds (ET) jointly have been one of the cornerstones of IPM (Stern *et al.*, 1959).

On conceptual grounds, IPM plans should start well before planting, even during the phase of selection of vegetables to be planted, and farmers must be aware of the potential for both population growth and degree of damage inflicted by key pests, as well as some secondary ones that may eventually become troublesome. Therefore, once an insect pest problem is properly characterized, which includes its taxonomic identification, as well as a general risk assessment of its damage potential under existing agroecological conditions (crop variety, routine agronomic practices, climate, natural enemies, etc.), thresholds for phenological stages, during which crop susceptibility is higher, should be in place.

In addition, the usefulness of thresholds for deciding when to start controlling an insect pest also depends on the availability of different tactics, so that cultural practices as well as behavioural and biological control methods may be used either for containment or mitigation, whereas bioinsecticides or conventional insecticides ought to be reserved as a last resource. For each method, its logistic feasibility, cost and foreseen effectiveness in reducing insect pest damage has to be appraised, at least on general grounds, aimed at obtaining some sort of cost–benefit assessment – plus field worker and consumer safety and environmental considerations – which in turns depends on the expected price of produce in the market.

Nevertheless, despite the obvious relevance of EILs and ETs as decision-making criteria, both their determination and implementation have been quite limited in the American tropics and elsewhere. This is so because ETs are not fixed but dynamic, and vary depending on the variables that compose the general model of the EIL, so that they are directly affected by the cost of management actions (inputs and labour), and indirectly by the commercial value of produce; in addition, they may vary in accordance to the level of pest infestation, the response of the crop to damage, and the efficiency of management tactics.

In this regard, Nava-Camberos *et al.* (2001) reported that the EIL of *B. tabaci* (biotype B) in melon ranged from 0.02 to 3.92 adults per leaf, depending on the density and biological stage of the insect, control cost, crop cultivar and season. Also, onion can tolerate up to 10 or 30 nymphs of *Thrips tabaci* per plant before and after formation of the bulb, respectively, without impairing productivity (Gonçalves, 1998). An indication of the low practicality and acceptance of EILs and ETs is that no papers have been published on that matter from 2007 to 2017 in the four main Mexican agricultural journals, which is probably true for other Latin American and Caribbean countries.

An interesting alternative to formal ET are the so-called action thresholds, which are insect pest levels arising from a combination of scientific data taken from the literature and practical experience of researchers, extension agents and farmers; they are conceptually simpler for growers, as they consist of a single figure which is rather constant, but flexible enough as to be easily modified by experience, according to crop season and location. They have been used in Costa Rica and Nicaragua in IPM validation plots in farmers' fields, to manage key tomato and potato insect pests, and have proven successful in largely reducing insecticide applications and production costs (Hilje *et al.*, 2003).

Moreover, in Brazil, only large vegetable farmers use action thresholds, as it occurs in melon (4 larvae or 10 adults, on 20 leaves) for *Liriomyza sativae* and *L. trifolii*

in warmer regions (northeast), as well as *L. huidobrensis* in mild areas (central-south) (Guimarães *et al.*, 2009). Likewise, thresholds are utilized in tomato for both *T. absoluta* (45 adults/day/pheromone trap) (Benvenga *et al.*, 2007) and *B. tabaci* (4 nymphs on basal leaves, or 1 adult per catch, when shaking and beating apical leaves on a tray). The vast majority of small farmers apply control measures when estimated levels of 10% are reached, for defoliating larvae and leaf miners; an insect per plant, for sucking insects; and 5% of flowers or fruits attacked, for borers like *D. nitidalis*, *T. absoluta* and *Helicoverpa armigera*. In the case of pheromone traps used by small farmers, an average of a single insect per trap justifies the release of *Trichogramma pretiosum* against *T. absoluta* and *H. zea* in tomato and maize, respectively.

Finally, an important decision-making criterion is the use of critical periods, in recognition that crop plants can vary in their response to pests, according to their physiological state or phenological stage. Yet, although critical periods are conceptually and operationally simpler than any kind of threshold, there has been limited research towards their development and implementation. For instance, the impact of begomoviruses transmitted by *B. tabaci* on tomato yields is highest during the first five weeks after germination, so that by using seedbeds covered with fine-mesh netting for 22 days, complemented with living ground covers of perennial groundnuts (*Arachis pintoi*), 'cinquillo' (*Drymaria cordata*) or coriander (*Coriandrum sativum*) for another 35 days, it was possible to significantly reduce whitefly numbers, delay the onset of tomato yellow mottle virus (ToYMoV) and decrease disease severity, resulting in higher yields and profits (Hilje and Stansly, 2008); such results were widely validated subsequently in farmers' fields in Costa Rica, as well as in research centres of EMBRAPA in Brazil.

10.6.3 Monitoring, scouting and prediction

If decision-making criteria based on insect pest density are to be applied as a component of any IPM programme for vegetables, it is essential to count upon efficient and reliable methods for sampling and/or monitoring insect population abundance in the field. Ideally, sampling methods should be easy to use, require a minimum of effort and cost, and provide accurate estimates of insect abundance and/or plant injury level.

A pertinent example comes from CATIE's efforts to support resource-poor farmers in Nicaragua, through an IPM project that lasted from 1989 to 2010. Farmers were initially trained to use action thresholds to deal with fruitworms (*Heliothis* spp. and *Spodoptera* spp.) in tomato; thresholds were taken from US literature and CATIE's researchers had previously validated them elsewhere in Central America. Farmers even scouted their fields and recorded pest levels (eggs or egg masses, plus young larvae, and injured fruits in 30 plants per plot), to make decisions by themselves. They were willing to do this, as they realized that it was cost-effective, since sampling took only 2 h a week and savings from reducing insecticide applications were significant.

Unfortunately, in contrast to high-input annual export crops, like pineapple, melon and watermelon – owned mainly by transnational companies – for which there are standard protocols for monitoring insect pests on a continuous basis and which rely on trained personnel for such tasks, farmers involved in vegetable production do not normally use any sampling procedures. Also, as in the case of decision-making criteria, only one paper has been published on sampling from 2007 to 2017 in Mexican agricultural journals, and probably also in other Latin American and Caribbean countries. However, prior to that, efforts were made to develop sampling plans for *B. tabaci* in tomato (Nava *et al.*, 2008) and beans (Pereira *et al.*, 2004), for *Trialeurodes vaporariorum* in tomato (Bernal *et al.*, 2008), for *M. persicae* in potato (Narváez and Notz, 1995) and for *B. cockerelli* in husk tomato, *Physalis ixocarpa* (Crespo-Herrera *et al.*, 2012).

A less time-consuming tool for sampling insects are traps baited with pheromones, which allow monitoring of population density for adequately timed insecticide sprayings, the release of natural enemies, and even the application of pheromones themselves as a direct control method. In Mexico, there have been positive experiences with vegetable pests like *Keiferia lycopersicella*, *P. xylostella* and *P. operculella*, for which monitoring with pheromone traps, in combination with action thresholds based on moth catches, allowed opportune insecticide applications or the deployment of pheromones for disrupting mating (Nava *et al.*, 2006). Fortunately, there are local companies involved in pheromone production, as occurs with ChemTica International in Costa Rica, which commercializes pheromones of some 40 insect vegetable pests) in almost all Latin American and Caribbean countries.

Finally, prediction of insect population peaks can be attained by using phenological models that incorporate temperature thresholds and insect developmental rates, based on the calculation of degree-days (DD), so that control methods can be applied prior to reaching those peaks. The key parameters for using a degree-day prediction model for a particular insect stage are the lower development threshold (T_L), the upper developmental threshold (T_U) and the thermal constant (K), the latter expressed in DD. Recent examples of such parameters in the American tropics are 6.3°C, 35°C and 312.5 DD for *P. xylostella* (Marchioro and Foerster, 2011), as well as 8.0°C, 37.3°C and 416.7 DD for *T. absoluta* (Krechemer and Foerster, 2015).

Despite the fact that research for determining such parameters is quite laborious and requires expensive environmental chambers, a way to partially circumvent this constraint is to access databases elsewhere and to consult available compilations. This is the case of the University of California-Davis website (UCD, n.d.) and Ruiz *et al.* (2013a), where key parameters for about 50% of the vegetable insect pests listed in Table 10.2 can be found. This information, coupled with field counts, may be very helpful for developing prediction protocols for specific vegetable pests in the American tropics.

10.6.4 Management tactics

Even though, historically, conventional chemical control has been the first line of defence against vegetable insect pests, it will not be included here, as IPM emphasizes alternative tactics, thus relying on insecticides only as the last resource, in case other methods fail to deal with a particular pest situation. Such tactics will be analysed one at a time, in a sequence that encompasses several biological and agroecological levels, beginning with the intrinsic abilities of the crop to defend itself against pests, through existing natural mortality factors, and culminating with agroecosystem manipulation to adversely affect insect pests. In closing this subdivision, factors that either favour or hinder adoption and implementation of IPM programmes in the American tropics will be discussed.

Host-plant resistance

Throughout history, humans have strived for selecting vegetable cultivars aimed at increasing yields and ease of care during the cropping season and at harvest time, but also to render more tasty, nutritious and better-looking produce. However, such domestication processes have often entailed the suppression of natural defensive characteristics (leaf pubescence, waxes and colour, toxic or anti-nutritional secondary metabolites, etc.), which formerly precluded insect establishment on plants – a desired trait against disease-vectoring insects – or impaired reproduction and survival in the field.

At any rate, developing a commercial variety of any crop is a complex and time-consuming process; even for vegetables, which are short-lived crops, it normally takes 8–14 years to obtain a new cultivar. Such process has mainly focused on the search for varieties resistant to extreme climatic and edaphic conditions, as well as on characteristics such as precocity, firmness and shelf-life. In the case of resistance to plant-injuring organisms, examples for insects are scanty, in contrast to pathogens, due to a number of both biological and logistic factors that hamper the development of new resistant materials. Evidence of that comes not only from catalogues of seed companies, but also from the sustained efforts of both national research centres and universities in many Latin American and Caribbean countries, as well as from an international institution like The World Vegetable Center, formerly the Asian Vegetable Research and Development Center (AVRDC), based in Taiwan.

Nevertheless, amid a bitter debate, in recent years biotechnology has opened new avenues for dealing with insects, including vegetable pests, particularly through genetically modified (GM) crops, which involve the selection and transfer of genes that express specific defensive characteristics. In consequence, a key trait providing insect resistance is the production of *Bacillus thuringiensis* (Bt) insecticidal proteins in Bt-transgenic plants; they include both Cry (crystalline) and Vip (vegetative insecticidal proteins) toxins, aimed at controlling several lepidopteran and coleopteran species that attack vegetables like sweetcorn and cabbage. For instance, in Brazil, where an average of 30,000 ha of sweetcorn are planted every year, the seed market is overwhelmingly dominated by transgenic varieties, including Bt cultivars.

Technologies based upon genomics have certainly enabled rapid advances in the selection of new cultivars of vegetables, in comparison to conventional breeding programmes. However, overreliance on GM vegetables is not advisable, since the selection process keeps going in nature, as insects have biochemical mechanisms that may overcome plant defences. This co-evolutionary process is of particular importance in the tropics, because insect pests are active all year round, thus exerting a continuous pressure on crops, but also may express unprecedented ways, either behavioural or biochemical, to circumvent plant defences. Therefore, this sort of awareness should lead IPM practitioners to maintain constant efforts in searching for innovative mechanisms for plant resistance, as well as improving other management tactics.

Biological control

Even though, at the farm level, the short cycle of most vegetables creates favourable conditions for opportunistic herbivorous insects, so that biological control does not seem to be feasible, in some areas, irrigation allows farmers to continually grow vegetables throughout the year in the American tropics, often with overlapping cycles, so that natural enemies can get established. Therefore, from a landscape perspective, there is continuous availability of food resources not only for pests, but also for their natural enemies. This is commonly witnessed in vegetable systems involving tomato, cabbage and pepper in Brazil, with predators like ladybirds (*Cycloneda sanguinea, Eropis conexa, Hippodamia convergens, Harmonia axyridis, Scymnus* sp., *Diomus* sp. and *Coleomegilla* sp.), chrysopids (*Chrysoperla externa*), earwigs (*Doru luteipes*), bugs (*Orius insidiosus, Nabis* sp. and *Geocoris* sp.), as well as several species of carabids, staphylinids, robber flies, syrphids and spiders (Medeiros *et al.*, 2009). In addition, cultural practices like the ones mentioned in the following section can help entomophagous insects and even entomopathogens to perform well under field conditions.

From a practical viewpoint, biological control in Latin America and the Caribbean goes back some 100 years. For instance, in Mexico, formal biological control programmes started in the 1940s, for which mass-rearing laboratories were established, particularly for *Trichogramma* spp.; currently, there are about 60 laboratories, in which 34 species of natural enemies are reared. So far, there have been 49 successful cases of biological control, including 14 primary vegetable pests, all of which are well recorded in comprehensive books (e.g. Arredondo and Rodríguez, 2008, 2015) published by the Mexican Society of Biological Control, founded in 1989.

In Brazil, where there are huge areas planted with soybean, maize, cotton and sugarcane, large biological control programmes, involving mass-rearing and release of both native and exotic natural enemies, have been developed since the mid-1900s. In addition, export vegetables such as melon, tomato, potato and sweetcorn have received special attention by the development of biocontrol agents such as *Trichogramma* spp. and bioinsecticides – the latter discussed in another section – against fruit borers, aphids, leafworms and whiteflies. Parra *et al.* (2002) compiled successful cases of biological control by parasitoids and predators. Moreover, as low-input and organic farming have increased their coverage, so far comprising about 950,000 ha, it is expected that this production system will give rise to high and continuous demands for biological control agents.

However, to be incorporated into IPM programmes, it is essential to devote more efforts to the inventory, conservation, evaluation and utilization of natural enemies. Even though their diversity is very high in the tropics, their potential has been only partially exploited. In fact, most parasitoid groups are poorly known on taxonomic and ecological grounds, mainly due to logistical factors. For instance, a preliminary inventory of Hymenoptera in Costa Rica estimated 20,000 to 40,000 species, 70–80% of which are not even described (Hanson and Gauld, 1995). Moreover, Fernández-Triana *et al.* (2014) found 205 species – 186 of them new to science – of *Apanteles* (Braconidae) that parasitize lepidopteran larvae, in the Conservation Area of Guanacaste, Costa Rica; by the way, huge areas of melon, watermelon, rice and sugarcane are regularly planted not very far away from this preserved area.

Other evidence of parasitoid diversity comes from horticultural systems involving potato, crucifers or maize in Mexico, where 84, 90 and 263 parasitoid species of *P. operculella, P. xylostella* or *S. frugiperda*, respectively, were detected (Bahena-Juárez, 2008; Salazar-Solís and Salas-Araiza, 2008; Bahena and Cortez, 2015). Also, in central Brazil, *P. xylostella* is parasitized by *Diadegma leontiniae* (Ichneumonidae), *Apanteles piceotrichosus* (Braconidae), *Oomyzus sokolowskii* (Eulophidae), *Cotesia plutellae* (Ichneumonidae), *Conura pseudofulvovariegata* and *Conura* sp. (Chalcididae), and *Actia* sp. (Tachinidae), the first three being the dominant species.

Finally, there are several ecological factors intimately related to the population dynamics of a pest, such as natural enemies, weather and mass spread of populations, affecting mortality and recruitment of individuals. A correct understanding and management of cultural practices and selection of other control methods are crucial for biological control to function properly. This has been a main constraint to extensive adoption of biological control in vegetables IPM.

Cultural practices

Preventative in essence, these are probably the oldest means of pest control, though sometimes inadvertently. They are aimed at creating a less favourable environment for insect pest reproduction and survival by deliberately manipulating some current or new components of the agroecosystem (soil, external inputs, associated plants, and the crop *per se*). Ideally, these practices should contribute to build structurally robust agroecosystems, less vulnerable to insect direct attack and insect-borne diseases, or more resilient to other sorts of biotic and abiotic disturbances, but the truth is that quite often they have to be complemented with other tactics. Moreover, in systems with higher crop diversity it is believed that, in general, pest populations are better controlled by their natural enemies.

There is a wide range of cultural practices that some vegetable farmers in the American tropics have adopted to diversify vegetation at different scales within the farm landscape. They include intercropping, cover crops, trap crops, living barriers, crop rotations, use of green manure, establishment of corridors between fragments of native vegetation, and maintaining weeds in and around the plantations. Although vegetation diversification can provide benefits in terms of ecological services such as biological control, pollination, as well as the production and maintenance of organic matter and water, more scientific data are still needed to better substantiate its effects in reducing insect pest populations.

A good example comes from tomato in central Brazil, where by planting tomato intercropped with coriander during the dry season (with organic fertilization and sprinkler irrigation), plus sorghum barriers around plots, establishment of *B. tabaci* and *T. absoluta* was hindered, due to behavioural interference in plant–insect interactions, and their natural enemies' populations increased as well (Medeiros *et al.*, 2009; Togni *et al.*, 2009).

In addition, a large number of cultural practices have been successfully tested worldwide, and some of them have been incorporated as part of routine agronomic practices in a number of vegetable production systems in the American tropics, along with crop-free periods, crop residue disposal, planting dates, weed removal, solarization, use of seedbeds protected with fine-mesh netting, trap crops, plastic mulches, floating row covers and overhead irrigation. Nevertheless, some practices with different limitations, as occurs with crop-free periods, require implementation at a regional scale to be successful. Meanwhile, the use of living barriers, living ground covers, high planting densities and trap crops, and even floating covers and plastic mulches, entail substantial modifications in the conventional cropping practices.

Fortunately, cultural practices are fully compatible with organic agriculture. Along with widely planted crops, like coffee and bananas, organic vegetable production has increased over the years in Latin American and Caribbean countries, as a result of the premium value that organic products carry in the international markets, in comparison with the price paid for conventional produce.

Since the core of organic agriculture is the crop plant's health, resulting from its interaction with the soil as a living entity, cultural practices fit quite well in organic farming schemes, along with resistant plant cultivars, biological control and bioinsecticides. Of particular relevance is proper nutrition of crops, based on chemical analysis of soil and plant, to avoid imbalances in the availability of nutrients, which may cause plant stress and in turn favour pest species. Due to this coincidence in principles and practices, an increasing number of IPM practitioners are becoming involved in promoting low-input schemes, so as to make a smooth transition from conventional to organic vegetable production systems.

Bioinsecticides

In contrast to past attitudes from agrochemical companies towards IPM, which were rather sceptical and even confrontational, nowadays they have recognized IPM practitioners as partners more than as adversaries in promoting IPM programmes, as they understand that the economic 'sustainability' of their products depends upon their rational use. This belief explains their involvement in large-scale development of bioinsecticides *sensu lato* (toxicants, insect growth regulators, deterrents, repellents, etc.), made from other organisms (viruses, protozoa, bacteria, fungi, nematodes, marine worms and plants) or different types of substances (oils, detergents, waxes, etc.); such materials are more or less selective to non-target organisms, as well as more environmentally innocuous.

In the case of microbial products, there is a comprehensive compilation of achievements and experiences (Alves and Lopes, 2008), which needs to be updated, as this area continues growing at a rather steady pace.

For instance, toxins from the bacterium *Bacillus thuringiensis* (Bt), especially Cry and Vip, are the commercial bioinsecticides most widely marketed in the world, mainly for controlling lepidopteran larvae. The HD-1 strain of Bt *kurstaki* excels in use for control of vegetable pests such as *P. xylostella*, *Ascia monuste*, *H. zea*, *S. frugiperda*, *T. absoluta* and *Diaphania* spp. Besides this, other products from Bt *aizawai* and Bt *thuringiensis* are also registered for lepidopteran pests.

Virus of the Baculovirus and Granulovirus groups stand out as insecticides, due to their specificity and high efficiency, mainly against vegetable pests such as *P. xylostella*, *H. zea*, *P. operculella*, *Dione juno juno*, and *Spodoptera* spp. However, the difficulty of large-scale production, with consequent high manufacturing costs, has been a limiting factor for their use in minor crops such as vegetables. Only a Granulovirus has been commercialized for controlling *P. operculella* and *T. absoluta* in Colombia, Peru and Bolivia.

Entomopathogenic fungi differ from other entomopathogens in their ability for infection and horizontal transmission, thus penetrating the integument and exoskeleton openings without the need to be ingested. Fungal strains of *Beauveria bassiana*, *Metarhizium anisopliae* and *Lecanicillium* are registered and commercialized in several countries for controlling vegetable pests such as aphids, thrips, whiteflies and lepidopteran larvae.

In Latin American and Caribbean countries, bioinsecticides have been widely accepted not only by farmers, but there has been an increasing interest of local small and medium-size companies to get involved in bioinsecticide production, such as neem (*Azadirachta indica*) products in Nicaragua, as well as entomopathogens in several countries. Because of their acceptance by farmers, from 2000–2006 CATIE developed the project Promotion of Non Synthetic Phytosanitary Products, aimed not only at promoting the involvement of small and medium-size companies in manufacturing non-conventional products, but also at establishing particular regulations and protocols for registering them in Central America, which are in place today.

A noteworthy example comes from Brazil, where the Brazilian Association of Biological Control Companies acts as a network of 22 enterprises spread over six states, with a portfolio of 53 commercial biological products, including pheromones, biopesticides and even sterile males. They also maintain permanent links with research institutions like EMBRAPA, which keeps a bank of microorganisms (bacteria, fungi and viruses) with thousands of isolates, which is continuously and cautiously enriched through explorations of local biodiversity.

10.6.5 Adoption and implementation

Nowhere in the world are there extensive, permanent and fully successful IPM programmes for any crop, so that it is not expected to occur in vegetable production systems in the American tropics. In essence, this is so because IPM is multi-tactic by its own nature, in contrast to single tactics like a conventional insecticide, a bioinsecticide, a resistant variety, or even a parasitoid obtained from a mass-rearing facility, all of which can be bought and deployed immediately, and whose performance can be readily observed in a farm or a greenhouse.

In other words, for its implementation IPM requires both more agroecological (insect response to climate conditions and agronomic practices, crop phenology, critical periods, natural enemies, etc.) as well as economic knowledge (insect injury potential, action thresholds, input and labour costs, expected produce prices, etc.) for the farmer to make wiser decisions about which tactics to use and when, as well as their complementarity; in addition, the compatibility with tactics deployed to deal with other phytosanitary problems (diseases and weeds) in his cropping system has to be appraised.

In recognition of this inherent complexity, and in contrast to conventional extension approaches, different models and platforms of participatory research, training and learning have been tested and applied in Latin America and the Caribbean over the years. One of the most outstanding experiences was developed by CATIE in Nicaragua for some 20 years (1989–2010) with farmers involved in producing vegetables and coffee; they were required to belong to cooperatives and other types of organizations, in order to accomplish a multiplicative effect. Its final goal was to achieve farmer empowerment, meaning that farmers' capabilities and abilities for agroecological reasoning, planning and economic decision-making were continuously increased, regardless of the crops or pests they have to deal with. The novel conceptual approach arising from that experience (Staver, 2008) has been widely shared through a variety of publications aimed at different sectors, including farmers, extension agents, researchers and policy makers.

Nevertheless, despite this and other positive experiences in additional countries, farmer's involvement in the processes of generation, validation and IPM technology transfer has lagged behind expectations. This is so because such activities, which should be permanent, depend to a large degree on extension agents from public agricultural institutions. However, due to political decisions associated with particular economic development models of privatization, imposed by structural adjustment programmes from entities like the International Monetary Fund and The World Bank, extension services have considerably weakened in the last 25 years, to the point of collapsing in many countries. Under such circumstances, it is not easy for IPM alternatives to compete with conventional insecticide usage, as agrichemical companies invest huge amounts of money in promoting their products and also have their own personnel to provide continuous technical assistance to both large farmers and retailers in rural areas.

10.7 Concluding Remarks

Since its coining almost 60 years ago, the concept of IPM has evolved considerably, and today it is fully embodied in the paradigm of agricultural and environmental sustainability, as it was bluntly expressed in the 1992 Earth Summit, the United Nations Rio Conference on Environment and Development. Moreover, its application in real-world situations has been a source of continuous and relevant lessons of both practical and theoretical significance. Fortunately, such knowledge has been compiled in books, as is the case for Central America, a small but meaningful region where comprehensive IPM practically-oriented projects have taken place for more than 40 years (Hilje and Saunders, 2008).

The reasons explaining the lack of extensive and permanent IPM programmes for vegetables and other crops in the American tropics, were just mentioned. But, disappointing as it may be, it should be emphasized that in any crop system there is more IPM, on either conceptual or practical grounds, than is commonly believed. This is so because, in general, now farmers are more aware of agroecological processes that occur in their production systems, including natural biological control. Moreover, in addition to deliberately using cultural practices to prevent the risk of pest injury and protect their natural enemies, they are willing to pay for more expensive bioinsecticides (especially *B. thuringiensis*, fungal formulations, and neem extracts) and spray them only when really needed. At any rate, farmers are quite cautious, and reluctant to blindly adopt technology packages as a whole, thus preferring to select specific well-proven tactics and incorporate them in their vegetable production systems.

As expressed before, IPM is deeply rooted in universities, ministries and agricultural research institutes in almost all Latin American and Caribbean countries, and research efforts are ongoing despite usual and chronic personnel or financial shortcomings. In recent years, it has been gradually converging with other sustainable alternatives, such as both low external-input and organic agriculture systems, thus showing its plasticity to adapt and continue evolving. Nevertheless, the global threat of climate change, which could disrupt vegetable production systems, including insect reaction to climate itself as well as to the crop's response to weather conditions, imposes a very serious challenge, which is already forcing IPM practitioners to adapt and reshape their research agendas, in order to wisely adapt to unprecedented agroecological scenarios.

Acknowledgements

To Mario Saborío (Syngenta Seeds), Lilliana González and Carlos Rodríguez (ChemTica International), Charles Staver (Bioversity International), Amílcar Aguilar (CATIE – Nicaragua), Juan Tigrero (Ecuador), Luis Vázquez and Yamila Martínez (Cuba) and Francis Geraud (Venezuela), for providing useful information. To Felipe Abarca Fedullo, for redrawing the illustrations. To Paul Hanson, for kindly reviewing the final draft.

References

Álvarez, P. and Abud-Antún, A. (1995) Reporte de República Dominicana. *Ceiba* 36, 39–47.

Alves, S.B. and Lopes, R.R. (2008) *Controle microbiano de pragas na América Latina: avanços e desafíos.* FEALQ, Piracicaba, Brazil.

Araujo, E.L., Nogueira, C.H., Menezes Netto, A.C. and Bezerra, C.E.S. (2013) Biological aspects of the leafminer *Liriomyza sativae* (Diptera: Agromyzidae) on melon (*Cucumis melo* L.). *Ciência Rural* 43 (4), 579–582.

Argüello, H., Lastres, L. and Rueda, A. (eds) (2007) *Manual MIP en cucúrbitas.* Escuela Agrícola Panamericana, El Zamorano, Honduras.

Arredondo, H.C. and Rodríguez del Bosque, L.A. (eds) (2008) *Casos de control biológico en México.* Vol. I. Mundi Prensa México, Mexico.

Arredondo, H.C. and Rodríguez del Bosque, L.A. (eds) (2015) *Casos de control biológico en México*, Vol. II. Biblioteca Básica de Agricultura, Guadalajara, Mexico.

Ávila, C.J. and Parra, J.R.P. (2002) Desenvolvimento de *Diabrotica speciosa* (Germar) (Coleoptera: Chrysomelidae) em diferentes hospedeiros. *Ciência Rural* 32 (5), 739–743.

Bahena, F. and Cortez, E. (2015) Gusano cogollero del maíz, *Spodoptera frugiperda* (Lepidoptera: Noctuidae). In: Arredondo, H.C. and Rodríguez del Bosque, L.A. (eds) *Casos de control biológico en México*, Vol. II. Biblioteca Básica de Agricultura, Guadalajara, Mexico, pp. 181–250.

Bahena-Juárez, F. (2008) Palomilla de la papa, *Phthorimaea operculella* (Lepidoptera: Gelechiidae). In: Arredondo, H.C. and Rodríguez del Bosque, L.A. (eds) *Casos de control biológico en México*, Vol. I. Mundi Prensa México, Mexico, pp. 33–46.

Benvenga, S.R., Fernandes, O.A. and Gravena, S. (2007) Tomada de decisão de controle da traça-do-tomateiro através de armadilhas com feromônio sexual. *Horticultura Brasileira* 25 (2), 164–169.

Bernal, L., Pesca, L., Rodríguez, D., Cantor, F. and Cure, J.R. (2008) Plan de muestreo directo para *Trialeurodes vaporariorum* (Westwood) (Hemiptera: Aleyrodidae) en cultivos comerciales de tomate. *Agronomía Colombiana* 26, 266–276.

Boiça Junior, A.L., Tagliari, S.R.A., Pitta, R.M., de Jesus, F.G. and Braz, L.T. (2011) Influência de genótipos de couve (*Brassica oleracea* L. var. *acephala* DC.) na biologia de *Plutella xylostella* L. (1758) (Lepidoptera: Plutellidae). *Ciência e Agrotecnologia* 35 (4), 710–717.

Byrne, D.N. and Blackmer, J.L. (1996) Examination of short-range migration by *Bemisia tabaci*. In: Gerling, D. and Mayer, R.T. (eds) *Bemisia 1995: Taxonomy, Biology, Damage, Control and Management*. Intercept, Andover, UK, pp. 17–28.

Capinera, J.L. (2014) Corn earworm (*Helicoverpa zea*). Featured creatures. University of Florida/IFAS. Available at: http://entnemdept.ufl.edu/creatures/veg/corn_earworm.htm (accessed 5 March 2016).

CATIE (1990a) *Guía para el manejo integrado de plagas del cultivo de repollo*. Serie Técnica. Informe Técnico No. 150. CATIE, Turrialba, Costa Rica.

CATIE (1990b) *Guía para el manejo integrado de plagas del cultivo de tomate*. Serie Técnica. Informe Técnico No. 151. CATIE, Turrialba, Costa Rica.

CATIE (1993) *Guía para el manejo integrado de plagas del cultivo de chile dulce*. Serie Técnica. Informe Técnico No. 201. CATIE, Turrialba, Costa Rica.

Cave, R.D. (1996) Parasitoides y depredadores. In: Hilje, L. (ed.) *Metodologías para el estudio y manejo de moscas blancas y geminivirus*. CATIE, Turrialba, Costa Rica, pp. 69–76.

Coto, D., Saunders, J.L., Vargas, C.L. and King, A.B.S. (1995) *Plagas invertebradas de cultivos tropicales, con énfasis en América Central; un inventario*. CATIE, Turrialba, Costa Rica.

Crespo-Herrera, L., Vera-Graziano, J., Bravo-Mojica, H., López-Collado, J., Reyna-Robles *et al.* (2012) Distribución espacial de *Bactericera cockerelli* (Sulc) (Hemiptera: Triozidae) en tomate de cáscara (*Physalis ixocarpa* (Brot.)). *Agrociencia* 46, 289–298.

Daxl, R. (1989) Manejo integrado de plagas del algodonero. In: Andrews, K.L. and Quezada, J.R. (eds) *Manejo integrado de plagas insectiles en la agricultura. Estado actual y futuro*. Escuela Agrícola Panamericana, El Zamorano, Honduras, pp. 397–421.

Fernández-Triana, J.L., Whitfield, J.B., Rodriguez, J.J., Smith, M.A., Janzen, D.H. *et al.* (2014) Review of *Apanteles sensu stricto* (Hymenoptera, Braconidae, Microgastrinae) from Área de Conservación Guanacaste, northwestern Costa Rica, with keys to all described species from Mesoamerica. *ZooKeys* 383, 1–565.

Gonçalves, P.A.S. (1998) Determinação do nível de dano econômico de tripes em cebola. *Horticultura Brasileira* 16 (2), 128–131.

Greathead, A.H. (1986) Host plants. In: Cock, M.J.W. (ed.) *Bemisia Tabaci – A Literature Survey*. CAB International Institute of Biological Control, Egham, UK, pp. 17–26.

Guimarães, J.A., Michereff, M., Oliveira, V.R., Liz, R.S. and Araújo, E.L. (2009) *Biologia e manejo de mosca minadora no meloeiro*. Comunicado Técnico No. 77, EMBRAPA Hortaliças, Brasília.

Hanson, P.E. and Gauld, I.D. (eds) (1995) *The Hymenoptera of Costa Rica*. Oxford University Press, Oxford, UK.

Hart, R.D. (ed.). (1985) *Conceptos básicos sobre agroecosistemas*. CATIE, Turrialba, Costa Rica.

Herrera, J.M. (2010) Primera experiencia a nivel mundial del manejo integrado de plagas: el caso del algodonero en el Perú. *Revista Peruana de Entomología* 46 (1), 1–10.

Hilje, L. and Saunders, J.L. (eds) (2008) *Manejo integrado de plagas en Mesoamérica: aportes conceptuales*. Editorial Tecnológica de Costa Rica, Cartago, Costa Rica.

Hilje, L. and Stansly, P.A. (2008) Living ground covers for management of *Bemisia tabaci* (Gennadius) (Homoptera: Aleyrodidae) and tomato yellow mottle virus (ToYMoV) in Costa Rica. *Crop Protection* 27 (1), 10–16.

Hilje, L., Araya, C.M. and Valverde, B.E. (2003) Pest management in Mesoamerican agroecosystems. In: Vandermeer, J. (ed.) *Tropical Agroecosystems*. CRC Press, Boca Raton, Florida, pp. 59–93.

Holdridge, L.R. (1978) *Ecología basada en zonas de vida*. IICA, San José, Costa Rica.

Janzen, D.H. (ed.) (1983) *Costa Rican Natural History*. The University of Chicago Press, Chicago, Illinois.

Kogan, M. and Bajwa, W.I. (1999) Integrated pest management: a global reality? *Anais da Sociedade Entomológica do Brasil* 28 (1), 1–25.

Kranz, J., Schmutterer, H. and Koch, W. (eds) (1982) *Enfermedades, plagas y malezas de los cultivos tropicales.* Verlag Paul Parey, Berlin.

Krechemer, F. da S. and Foerster, L.A. (2015) *Tuta absoluta* (Lepidoptera: Gelechiidae): thermal requirements and effect of temperature on development, survival, reproduction and longevity. *European Journal of Entomology* 112, 658–663.

León, J. (1987) *Botánica de los cultivos tropicales.* IICA, San José, Costa Rica.

MacArthur, R.H. and Wilson, E.O. (1967) *The Theory of Island Biogeography.* Monographs in Population Biology No.1. Princeton University Press, Princeton, New Jersey.

Marchioro, C.A. and Foerster, L.A. (2011) Development and survival of the diamondback moth, *Plutella xylostella* (L.) (Lepidoptera: Yponomeutidae) as a function of temperature: effect on the number of generations in tropical and subtropical regions. *Neotropical Entomology* 40 (5), 533–541.

Medeiros, M.A., Sujii, E.R. and Morais, H.C. (2009) Effect of plant diversification on abundance of South American tomato pinworm and predators in two cropping systems. *Horticultura Brasileira* 27 (3), 300–306.

Michereff, M. and Nagata-Inoue, A.K. (2015) *Guia para o reconhecimento e manejo da mosca-branca, da geminivirose e da crinivirose na cultura do tomateiro.* Circular Técnica No. 142, EMBRAPA Hortaliças, Brasília.

Narváez, Z. and Notz, A. (1995) Disposición espacial del áfido verde del ajonjolí, *Myzus persicae* (Sulzer) (Homoptera: Aphididae), en un cultivo de papa, *Solanum tuberosum* L., en Saman Mocho, Edo. Carabobo, Venezuela. *Boletín de Entomología Venezolana* 10, 77–89.

Nava, U., Avilés, M.C., Fu, A.A. and Navarro, L. (2006) Feromonas para el monitoreo y control de plagas de cultivos hortícolas. In: Barrera, J.F. and Montoya, P. (eds) *Trampas y atrayentes en detección, monitoreo y control de plagas de importancia económica.* Sociedad Mexicana de Entomología y El Colegio de la Frontera Sur, Manzanillo, Colima, Mexico, pp. 59–70.

Nava, U., Avilés, M.C. and Fu, A.A. (2008) Disposición espacial, muestreo y umbrales de acción. In: Ortega, L.D. (ed.) *Moscas blancas. Temas selectos sobre su manejo.* Colegio de Postgraduados/ Mundi Prensa México, Texcoco, Mexico, pp. 43–56.

Nava-Camberos, U., Riley, D.G. and Harris, M.K. (2001) Density–yield relationships and economic injury levels of *Bemisia argentifolii* (Homoptera: Aleyrodidae) in cantaloupe. *Journal of Economic Entomology* 94, 180–189.

Núñez, E. (2005) *Espárrago peruano. Manejo integrado de plagas.* Servicio Nacional de Sanidad Agraria (SENASA), Lima.

Oliveira, M.R.V., Amancio, E., Laumann, R.A. and Gomes, L.O. (2003) Natural enemies of *Bemisia tabaci* (Gennadius) B biotype and *Trialeurodes vaporariorum* (Westwood) (Hemiptera: Aleyrodidae) in Brasília, Brazil. *Neotropical Entomology* 32 (1), 151–154.

Parra, J.R.P., Botelho, P.S.M., Corrêa-Ferreira, B.S. and Bento, J.M.S. (2002) *Controle biológico no Brasil: parasitóides e predadores.* Editora Manole, São Paulo, Brazil.

Pereira, M.F.A., Boiça Junior, A.L. and Barbosa, J.C. (2004) Amostragem seqüencial (presença-ausência) para *Bemisia tabaci* (Genn.) biótipo B (Hemiptera: Aleyrodidae) em feijoeiro (*Phaseolus vulgaris* L.). *Neotropical Entomology* 33 (4), 499–504.

Pianka, E.R. (1970) On *r* and *K* selection. *The American Naturalist* 104, 592–597.

Pratissoli, D., Polanczyk, R.A., Holtz, M.A., Cocheto, J.G., Tamanhoni, T. and Milanez, A.M. (2007) Desenvolvimento da broca-das-cucurbitáceas em diferentes tipos de substratos alimentares. *Horticultura Brasileira* 25 (4), 598–601.

Purseglove, J.W. (1974) *Tropical Crops: Dicotyledons.* Longman, London.

Purseglove, J.W. (1975) *Tropical Crops: Monocotyledons.* John Wiley & Sons, New York.

Ruiz, C., Bravo, E., Ramírez, G., Báez, A.D., Álvarez, M. *et al.* (2013a) *Plagas de importancia económica en México: aspectos de su biología y ecología.* Libro Técnico No. 2. Instituto Nacional de Investigaciones Forestales Agrícolas y Pecuarias (INIFAP), Campo Experimental Centro Altos de Jalisco, Jalisco, Mexico.

Ruiz, J.A., Medina, G., González, I.J., Flores, H.E., Ramírez, G. *et al.* (2013b) *Requerimientos agroecológicos de cultivos*, 2nd edn. Libro Técnico No. 3. Instituto Nacional de Investigaciones Forestales Agrícolas y Pecuarias (INIFAP), Campo Experimental Centro Altos de Jalisco, Jalisco, Mexico.

Salazar-Solís, E. and Salas-Araiza, M.D. (2008) Palomilla dorso de diamante, *Plutella xylostella* (Lepidoptera: Plutellidae). In: Arredondo, H.C. and Rodríguez del Bosque, L.A. (eds) *Casos de control biológico en México*, Vol. I. Mundi Prensa México, Mexico, pp. 155–166.

Santos, L.M. dos, Redaelli, L.R., Diefenbach, L.M.G. and Efrom, C.F.S. (2004) Fertilidade e longevidade de *Spodoptera frugiperda* (J.E. Smith) (Lepidoptera: Noctuidae) em genótipos de milho. *Ciência Rural* 34 (2), 345–350.

Saunders, J.L., Coto, D. and King, A.B.S. (1998) *Plagas invertebradas de cultivos anuales alimenticios en América Central*. CATIE, Turrialba, Costa Rica.

Schmutterer, H. (1990) *Plagas de las plantas cultivadas en el Caribe*. GTZ, Eschorn, Germany.

Staver, C. (2008) Aprendizaje de agricultores vinculado con procesos ecológicos, para un mejor manejo de plagas: retos para el CATIE y sus socios. In: Hilje, L. and Saunders, J.L. (eds) *Manejo integrado de plagas en Mesoamérica: aportes conceptuales*. Editorial Tecnológica de Costa Rica, Cartago, Costa Rica, pp. 493–517.

Stern, V.M., Smith, R.F., van den Bosch, R. and Hagen, K.S. (1959) The integrated control concept. *Hilgardia* 29 (2), 81–101.

Strong, D.R. (1974) Rapid asymptotic species accumulation in phytophagous insect communities. *Science* 185, 1064–1066.

Strong, D.R., McCoy, E.D. and Rey, J.R. (1977) Time and the number of herbivore species: the pests of sugarcane. *Ecology* 58 (1), 167–175.

Thomazini, A.P.B.W., Vendramim, J.D., Brunherotto, R. and Lopes, M.T.R. (2001) Efeito de genótipos de tomateiro sobre a biologia e oviposição de *Tuta absoluta* (Meyrick) (Lep.: Gelechiidae). *Neotropical Entomology* 30 (2), 283–288.

Togni, P.H.B., Frizzas, M.R., Medeiros, M.A., Nakasu, E.Y.T., Pires, C.S.S. and Sujii, E.R. (2009) Dinâmica populacional de *Bemisia tabaci* biótipo B em tomate monocultivo e consorciado com coentro sob cultivo orgânico e convencional. *Horticultura Brasileira* 27 (2), 183–188.

UCD (n.d.) University of California-Davis website. Available at: www.ipm.ucdavis.edu/WEATHER (accessed 17 July 2017).

Villar, A., Gómez, E., Morales, F. and Anderson, P. (1998) *Effect of Legal Measures to Control* Bemisia tabaci *and Geminiviruses in the Valley of Azua*. Report from the National Integrated Pest Management Program. Santo Domingo, Dominican Republic.

11 Integrated Pest Management and Good Agricultural Practice Recommendations in Greenhouse Crops

Abdelhaq Hanafi[1],* and Carmelo Rapisarda[2]

[1]*International Fund for Agricultural Development, Rome, Italy; [2]Dipartimento di Agricoltura, Alimentazione e Ambiente, Università degli Studi, Catania, Italy*

11.1 Introduction

In many tropical and subtropical regions, and parallel to the development of economic activities in these countries and of the international trade of their agricultural products, greenhouses are increasingly used for vegetable production as well as for cultivation of ornamental plants, due to their ability to allow a qualitative and quantitative control of production through a direct management of the growing environment. Especially in these areas of the world, where seasonal excursions of temperature are almost negligible and no need exists for artificially heating protected crops, greenhouses may be realized with relatively simple means, which allow obtaining higher yields with comparatively modest investments.

Of course, compared to similar situations in countries of the temperate areas of the world, greenhouse productions face a series of totally different challenges in tropical and subtropical regions (Shamshiri and Wan Ismail, 2013). First of all, high temperatures often represent a limiting factor for vegetable protected cultivation and, if heating the greenhouses is not necessary in these areas, cooling may become a priority (Mutwiwa *et al.*, 2006; Kumar *et al.*, 2009;

Ganguly and Ghosh, 2011). Therefore, site selection and correct orientation of the greenhouses become important choices, in order to manage positively the incidence of solar radiation and the flow direction of prevailing winds (El-Maghlany *et al.*, 2015). Ventilation capacity of the greenhouses is also a critical factor, aimed at reducing the humidity and temperature inside the structure (Mutwiwa *et al.*, 2008). Also, an infrared blocking plastic film on the greenhouse roof may become important sometimes and may act positively for replacing expensive cooling systems (Abdel-Ghany *et al.*, 2012). Maintenance of the greenhouse is also imperative, as regards both its structures (such as covers, nets, etc.) and all environmental elements surrounding the greenhouse (such as wild vegetation, drainage, etc.).

In tropical areas, greenhouses also fulfil the important role of protecting vegetable crops from heavy rains, winds and severe storms, which adversely affect plants in open field cultivation. To this aim, greenhouse construction has to withstand these weather conditions and the use of locally made materials and traditional skills and techniques is highly recommended, in order to minimize the costs of construction.

* Corresponding author e-mail: abdelhaq.hanafi@icloud.com

Moreover, through a control of environmental factors and the application of simple methods for achieving a mechanical exclusion, greenhouses also play an important protective function from many biotic agents (pests and diseases) attacking vegetable crops and reducing their production. Promotion of greenhouses in tropical regions is considered as a component of a broader approach for 'Integrated Production and Protection Management' (IPPM), with the final objective of reducing the spraying of pesticides and increasing quantity and quality of crop production all year round.

Pests and diseases management in greenhouse crops is probably the practice with the greatest impact, not only on the environment and consumers' health, but recently, also on public opinion in many countries. Due to the increasing development of protected cultivation (especially for vegetables), crops are becoming progressively susceptible to pests and diseases for multiple reasons, including monoculture cultivation and the use of selected cultivars with high yields, which may also stimulate pest/disease development. Therefore, various new control methods are being developed to face the increasingly aggressive pests and diseases, and new management practices are being incorporated into the production systems.

In the past, pest control in greenhouse crops has been based largely on the use of synthetic pesticides, but, towards the end of the 20th century, a major shift in crop protection practices began to take place also in these agro-ecosystems, mainly with biologically based technologies or cultural control used in integration with conventional chemical pesticides (Wearing, 1988; van Lenteren, 1989, 1990, 1995, 1998; Alebeek and van Lenteren, 1992; Wardlow and O'Neill, 1992; Gullino and Garibaldi, 1994; Elad and Shtienberg, 1997; Shtienberg and Elad, 1997). Nowadays, many of the environmentally friendly technologies account for an important part of the crop protection market in the greenhouse cultivations, particularly with regard to host-plant resistance and biological control but also novel biotechnological solutions (Castañe and

Hanafi, 2003; van Lenteren, 2003; Hanafi et al., 2004; Csizinszky et al., 2005; Bueno and van Lenteren, 2010), in spite of the feared reductions in the application possibilities of biological control due to recent ecological constraints (van Lenteren *et al.*, 2011).

The approach of using other management techniques in integration with rational chemical control is known as 'Integrated Pest Management' (IPM). Since 1997, the concept of IPM has been extended to 'Integrated Production and Protection Management' (IPPM), incorporating a range of practices: crop rotation, cultivation, fertilization, pesticide use, cultural control measures, biological control and other alternatives to conventional chemical control (Hanafi *et al.*, 1999; Hanafi and Papasalomontos, 1999; Hanafi, 2003; Hanafi and Schnitzler, 2004). IPPM is a sustainable, environmentally and economically justifiable system, in which damage caused by pests, diseases and weeds is prevented mainly through the use of natural factors and the application of good agricultural practices (GAP) which limit the population growth of these organisms and, where necessary, supplementing them with appropriate control measures.

11.2 IPM in Protected Cultivation

The occurrence, development and management of pests in greenhouse crops is undoubtedly impacted by the fact that the crops are grown in a closed or semi-closed environment. Greenhouses were initially designed to maintain an optimal environment for the crop, both economically and physiologically. The same conditions, unfortunately, provide a favourable environment for pests and diseases with optimal humidity and temperature, excluding the impact of rain and winds. A number of radically new systems for growing plants in soil and soilless systems have been used in recent years. As cultivation method changed, pest and disease organisms become exposed to a new ecosystem and this, in turn, required changes in the pest and diseases methods of control.

Rotation cannot normally be practised in greenhouses; due to economic factors related to the high level of expertise required for commercial success. As a consequence, the same crop or type of crops are grown year after year leading to pest and pathogens build up. Particular care must therefore be exercised, especially for pests, that persist in the soil or on the structure of the greenhouse.

At present, IPPM is seen as the standard for modern crop protection technology. Compared to other sectors, more of the new, non-chemical control techniques are already being used in greenhouses. A multifaceted approach to greenhouse crop protection, with integration of chemical, cultural, biological, mechanical and physical control of pests and diseases, will be more successful and make adaptive changes in pests and pathogens less likely.

11.3 Host-plant Resistance

Plant resistance to biotic factors can be based on different mechanisms: antixenosis (characteristics deterring or reducing colonization by herbivores) or antibiosis (characteristics leading to the killing or reduction of herbivores after landing and during feeding).

Breeding for improved quality and higher yields has been practised for centuries. In the early days of breeding, seeds were harvested selectively from plants with fewer symptoms of diseases or pests, resulting in some level of field resistance.

While the use of resistant cultivars is, in theory, the most convenient way of achieving pest control, there are no cultivars with resistance to all the pests of the crop. Host-plant resistance against insects and mites still requires further development, whereas resistance against diseases or nematodes is commercially available for many cultivars. In tropical regions, resistance is largely applied to control root nematodes in vegetable protected crops

(Luc *et al.*, 2005). Resistance as a management tool remains the most effective way of combating viral diseases and resistant varieties exist for a number of viruses (Dhaliwal and Sharma, 2016). However, complete and durable resistance is difficult to achieve in breeding programmes.

11.4 Biological Control of Insect and Mite Pests

During the past 50 years, more than 30 species of natural enemies have been introduced against more than 22 pest species of protected crops. The greenhouse area on which classical biological control is applied has increased from 400 ha in 1970 to over 70,000 ha nowadays. At present, biological control of key pests in greenhouses is applied in more than 25 countries out of a total of 35 countries with a greenhouse industry.

11.4.1 Biocontrol agents

Modern biological control depends upon the use of specific natural enemies of the target pest which are carefully selected and screened to control species which could pose a threat to other useful organisms.

Predators

Individual predators consume a number of prey during their lifetime and actively seek their food. Some species are polyphagous and consume a wide range of prey; others are oligophagous (narrow range) or monophagous (extreme specialists). Polyphagous arthropod predators do not concentrate their attention on target pests and tend to feed on the most abundant and easily captured prey. Monophagous and oligophagous predators are more likely to be suitable for biological control. Many different species of predators are used in greenhouse biological control programmes (Albajes *et al.*, 2002; Perdikis *et al.*, 2008), also in tropical and subtropical areas (Silveira *et al.*, 2004).

Parasitoids

These insects (many of which are monophagous) develop parasitically in a single host which is eventually killed. They include a remarkably diverse group of small wasps and flies. The adults are free-living and highly mobile, so that they are able to actively search for hosts in/on which to lay eggs. The larvae live in (endoparasitoid) or on (ectoparasitoid) a host until they are fully grown (egg, larvae and pupal stage); they generally kill their host at the moment of pupation. A number of parasitoid species are used for pest control in greenhouses.

Pathogens

Parasitic microorganisms, often killing their host outright, are used in biological control programmes in tropical greenhouses (Alves, 1998). Dead hosts liberate millions of individual microbes, which are dispersed by wind and vectors. With protozoans, the effect on pests is generally more long term. Pathogens can easily be mass-produced and release methods are similar to the application of chemical pesticides.

Bacteria. Almost all bacteria in microbial insecticides currently in production are species of the genus *Bacillus.* There are a number of other bacteria which are pathogenic to insects, but they are also potentially harmful to man or difficult to mass-produce. In most cases, bacteria affect their hosts after being ingested with food, often producing toxic metabolites that damage the gut wall. The best known and most successful species is *Bacillus thuringiensis* Berliner (*Bt*). The length of time from ingestion of a lethal dose to death varies from some hours to one or two days. However, even at sub-lethal doses, the insect stops feeding in a matter of hours. The toxin produced by *Bt* is extensively used in greenhouse biocontrol for controlling caterpillars, and has been the subject of development of several transgenic plants (essentially field crops) with resistance to some key lepidopteran and coleopteran pests.

Viruses. Viruses are not free-living and can only replicate in living host cells. There are at least seven families of viruses containing insect-pathogenic representatives. One family, the Baculoviridae, is unique, in that it is exclusive to invertebrates and they bear no resemblance, structural or biochemical, to invertebrate pathogens. Given the safety considerations and thanks to their known pesticidal potential, Baculoviruses are particularly suitable for use in biological control. Baculoviridae can be applied to control insect pests caused by the larvae of lepidoptera. The larvae will die 4–8 days after infection, and millions of virus baddies are set free by the putrefying cadaver. In greenhouses, an NPV (nuclear polyhedrosis virus) is used to control larvae of *Spodoptera exigua.* Many of these viruses are rapidly inactivated by UV radiation between 280 and 320 nm. However, this portion of the light spectrum is absorbed by glass and plastic sheet, which can make the application of viruses very successful in greenhouses.

Fungi. Fungi are the only insect pathogens capable of invading the insect by penetrating the cuticle – the most common means of infection. Therefore, those insects that feed by sucking, such as aphids and scales, are attacked by fungal pathogens. As a result, fungi are very dependent on environmental conditions, in particular high humidity, to achieve infection (Butt *et al.*, 2001). Most fungi successes have been obtained with Deuteromycetes, which cause epizootic on foliage-feeding insects in tropical environments (Maniania, 1991; Wraight *et al.*, 2000). In greenhouses, two fungal products are applied against whiteflies: *Verticillium lecanii* (Zimm.) Viégas and *Aschersonia aleyrodis* Webber; the more general fungus, *V. lecanii*, is also used for control of aphids and thrips.

Nematodes

The most promising nematodes belong to the family Steinernematidae – *Heterorhabditis* and *Steinernema* (*Neoplectana*). These nematodes are characterized by their association with bacteria of the genus *Xenorhabdus.* The infective juveniles of the nematode carry mutualistic bacteria in their intestines, and on entering the insect host (through natural openings), they release the bacterial cells that propagate and kill the insect within 48 hours. These nematodes are virulent, kill hosts quickly and are easily mass-produced *in vivo* and *in vitro*; furthermore, they have a very broad host range. Both groups are used in greenhouse biocontrol programmes.

11.4.2 Methods of release of biocontrol agents

Parasitoids, predators and pathogens can be used in different types of biological control programmes, described below.

Inoculative biological control

This is the first widely practised form of biological control, also called 'classical' biological control. Beneficial organisms are collected in an exploration area and introduced where there is a pest occurrence, releasing a limited number only. The aim is long-term suppression of pest populations. This method is typically used against introduced pests (invasive species), presumed to have arrived in a new area without their natural enemies. No examples exist for greenhouse crops, as permanent biocontrol is impossible in protected cultivation: the crop, its pests and natural enemies are removed from the structure at the end of each growing season.

Inundative biological control

Indigenous or exotic beneficial organisms are mass-reared in the laboratory or acquired from specialized biological control companies and periodically released in large numbers to obtain immediate control of pests for one or two generations, with no anticipation of the effects on subsequent generations. An example of this approach is the frequent application of high numbers of predatory mites (*Amblyseius* spp.) against thrips (*Frankliniella* sp.) in protected cultivation (Gerson and Weintraub, 2007).

Seasonal inoculative biological control

Native or exotic natural enemies are mass reared or acquired from the international market and periodically released in short-term crops (3–10 months) against pests, with control effects expected for several generations. A large number of natural enemies are released to obtain both an immediate control effect and also a buildup of the natural enemy population for control later during the same season. This method is essentially different from inundative control, because it aims to achieve a control effect over several generations and therefore resembles inoculative control. An illustration of this technique is the biological control of the recently introduced invasive species *Tuta absoluta* (Meyrick) to Europe (2006), north Africa (2008) and the Middle East (2009–2010), with the exotic parasitic wasp *Trichogramma acheae* Nagaraja & Nagarkatti (Cabello *et al.*, 2012; Vila and Cabello, 2014).

Conservation

Conservation is an indirect method where measures are taken to conserve natural enemies; it may result in a richer diversity of beneficial species, as well as in larger populations of each species, leading to better control of pests (Messelink *et al.*, 2014). For example, where parasitoids of leafminers and aphids occur naturally in the fields surrounding greenhouses, they may immigrate into protected structures and give adequate control. Proper management of the crops and the surroundings of greenhouses may therefore stimulate or restore natural control.

Shipment of natural enemies

Entomophagous insects and mites can be brought into the greenhouse in different stages of their development: eggs (e.g. *Chrysoperla*), larvae or nymphs (e.g. *Orius, Phytoseiulus*), pupae (e.g. *Trichogramma, Encarsia, Eretmocerus*), adults (e.g. *Aphidius, Diglyphus*), or all stages together (e.g. *Amblyseius*) (van Lenteren and Tommasini,

2003). The stage at which they are introduced depends mainly on transportation and manipulation in the greenhouse; usually, the stage at which they are least vulnerable to mechanical handling is chosen for transport and release, i.e. often the egg or pupa stage.

Release of natural enemies

There are a variety of methods for introduction in the greenhouse. Eggs and pupae may be distributed over the greenhouse on their normal substrate (leaves of the host plant, e.g. *Chrysoperla* and *Encarsia)* or glued on paper or cardboard cards (e.g. *Encarsia, Trichogramma, Eretmocerus*). At these stages, natural enemies can also be collected and put into containers, which are then brought into the greenhouse (e.g. *Nesidiocorris*).

Natural enemies at a mobile stage – larvae, nymphs or adults – can be put into the greenhouse in containers (e.g. many adult parasitoids and predators) or the grower can distribute natural enemies at these stages over the crop, for example, by sprinkling or distributing them onto the plant. The advice of the biological control companies, the distributors and the extension service are of paramount importance for the correct handling (especially after pick up from airport at arrival of shipment), timely delivery (time between airport and farm delivery) and release (practical handling of containers) of the beneficial insects. This is probably the weakest point in the chain of biological control, especially in countries adopting this technology only very recently and where sufficient advice in biological control applications are seldom available.

11.4.3 Practical cases of classical biological control of insects and mites

Biological control, which includes seasonal inoculative releases and techniques of augmentation and conservation of natural enemies, is used worldwide against major greenhouse pests and some diseases. The decision threshold and the rate of natural enemies required for the control of different pests are generally provided by companies that commercialize biological control agents. A huge experience is acquired in temperate areas, especially North America and Europe; but most techniques can also validly be applied in tropical and subtropical regions, and have been positively implemented in various integrated control programmes in Latin America, Africa or southeast Asia (Waterhouse, 1988; Altieri and Nicholls, 1999; Bueno, 1999, 2005a,b; Bueno and van Lenteren, 2002; Bueno *et al.*, 2003; van Lenteren and Bueno, 2003; Alene *et al.*, 2005).

Whiteflies

Biological control of *Trialeurodes vaporariorum* (Westwood) in greenhouse with seasonal inoculative releases of the parasitoid *Encarsia formosa* Gahan is widely used in the temperate greenhouse area and to a lesser extent in warmer regions. However, *E. formosa* is not very efficient under cool, cloudy conditions and inoculative release of the predator *Macrolophus caliginosus* Wagner is also made as a complementary measure. In warm areas, initial populations of *T. vaporariorum* are usually higher than in cold areas. Whitefly migration between crops also occurs. Thus, higher densities of natural enemies are required, but for shorter growing seasons *E. formosa* does not control *B.* spp. gr. *tabaci* sufficiently in winter greenhouse crops. At present, a combination of the predatory mite *Amblyseius swirski* Athias-Henriot, the predatory bug *Nesidiocorris tenuis* (Reuter) and the parasitoids *Eretmocerus eremicus* (Rose and Zolnerowich) and *E. mundus* Mercet are applied to control *T. vaporariorum* and *B.* spp. gr. *tabaci* in greenhouse crops in warm regions.

Leafminers

Inoculative releases of *Diglyphus isaea* (Walker) are commercially made for biological control of leafminers in greenhouse crops. In cold areas, it is applied together with *Dacnusa sibirica* Telenga. In warm areas, natural populations of leafminer parasitoids are abundant year-round and natural parasitism (up to 60%) controls leafminers for free. Therefore, further releases of *D. isaea* are to be made in these areas only when natural parasitism is low, especially after application of fine-mesh screens to the greenhouse.

Aphids

Suitable natural enemies are available to control all aphid species that attack greenhouse crops. These include the parasitoid *Aphidius colemani* Viereck, and predators like *Aphidoletes aphidimiyza* (Rondani) and *Chrysoperla carnea* Stephens. Natural enemies are usually released in temperate regions to control aphids in greenhouse crops; but in warmer regions, as in the tropics or subtropics, indigenous populations of their natural enemies are present almost all year round and aphids do not normally reach economic thresholds in greenhouses, especially if fine-mesh screens are applied and broad-spectrum insecticides are not used.

Mites

Biological control of spider mites with *Phytoseiulus persimilis* Athias-Henriot on tomato crops has been largely ineffective and is not widely employed. However, a new species of predatory mite, *Amblyseius swirski* Athias-Henriot, and a new strain of *P. persimilis* (T strain) have produced better results on greenhouse tomatoes.

Caterpillars

Chrysodeixis chalcites (Esper), *Autographa gamma* (L.) and *Spodoptera littoralis* (Boisduval) are kept well under control by *Bacillus thuringiensis* treatments. *Helicoverpa armigera* (Hübner) is also well controlled if the treatment is applied when eggs or young larvae are present. Inoculative releases of the parasitoid *Trichogramma evanescens* Westwood are also made for biological control of *C. chalcites* in some greenhouse crops.

11.4.4 Biological control through interference with the mating mechanisms

Greenhouse applications of these methodologies mainly regard the control of the Tomato leaf miner, *Tuta absoluta* (Meyrick). The attractive power of the sex pheromone of *T. absoluta* has been positively tested both in greenhouse crops and in open field (Filho *et al.*, 2000). Various tests performed by using this pheromone in mating disruption of the insect showed a satisfactory containment level (Navarro Lopis *et al.*, 2010; Martí Martí *et al.*, 2010; Cocco *et al.*, 2013), but only if applied in greenhouses adequately protected from accidental introductions of adults from the outside. Of course, the economic sustainability of this technique is to be better defined.

11.5 Physical Control

Control of pests and diseases by means of heat treatment or radiation is called physical control. Heat treatment is applied to control harmful organisms in soil and water. Production systems where there has been incidence of root and stem diseases should be rigorously sanitized. Moreover, all production houses benefit from passive solarization during the non-cropping period in summer: structures are sealed completely after wetting surfaces; temperatures should exceed 50°C to assist in the eradication or at least the reduction of pathogens and other pests in the production area.

To start crops with pest-/pathogen-free soil, chemical soil disinfection can be carried out if necessary, but cultural and physical methods, such as steam sterilization and solarization, should be first adopted when at all possible.

11.5.1 Sterilization of soil

Heat treatment of the soil is known as soil sterilization, but this is not strictly correct, as even the most effective treatment does not eliminate all living organisms and the soil is not sterile after treatment. Partial sterilization or pasteurization are more accurate terms and are achieved through steam sterilization or solarization.

11.5.2 Solarization

Solarization is easily combined with other control methods to reduce the necessity of chemical control. Together with reduced dosages of chemical fumigants it allows better management of soilborne pathogens, which are otherwise difficult to control. Solarization followed by use of biocontrol agents has good potential, facilitating the introduction of antagonists, especially in warm regions.

Solarization is a form of soil pasteurization whereby solar energy is trapped beneath plastic sheets spread on the soil surface. It is a cheap and effective means of pasteurizing, and controls soilborne pathogens, weeds, nematodes and other pests. However, it is only effective when at least 30 days are available during a period of high solar radiation and can therefore only be applied in warmer climates, during the hot season, when no protected crop situation is present. The soil has to be wetted (water-holding capacity) to ensure that the absorbed energy goes beneath the top layer of the soil. A polyethylene cover prevents transpiration and escape of the heat into the air. Only transparent polyethylene is suitable for solarization, as black plastic absorbs too much of the energy. The depth of heat penetration (and hence the efficacy) is improved by prolonging the period of solarization. The temperature increase is about 10°C in the topsoil and decreases with depth.

To control weeds effectively, a solarization period of 4–6 weeks (depending on radiation) is necessary. Annual weeds, especially Gramineae and parasitic weeds, seem to respond to this treatment; with perennial weed species, the results are not as good.

Soil solarization is a promising technique and may have an important future in many countries (especially with the phasing out of methyl bromide under the Montreal protocol). Initially used only in hot regions during the summer, solarization is spreading to cooler areas and cooler seasons thanks to technological advances. Soil solarization controls numerous pathogens, including *Colletotrichum coccodes* (Wallr.) S. Hughes, *Fusarium oxysporum f.* sp. *lycopersici* Snyder & Hansen, *Verticillium dahliae* (Cooke) Wint., *Pyrenochaeta lycopersici* (Cooke) Wint. and *R. solani* (Cooke) Wint. Solarization also decreases the population of *Meloidogyne* spp., as well as many insect pests of greenhouse crops, which pupate in the soil (*Liriomyza* spp., *Tuta absoluta*, *Frankliniella occidentalis*, etc.) in greenhouse crops. When used adequately, soil solarization decreases significantly the use of fumigants. Solarization of tomato stakes and hooks is a successful control method of some diseases such as Didymella stem canker, and could easily be achieved by storing this agricultural material in empty plastic greenhouses during the hot months of the year. In warmer areas, a significant kill of pathogens and insect and mite pests could be achieved simply by closing the greenhouse in the off-season (space solarization). Solarization of container media, such as peat, offers potential as it makes these products recyclable.

Unfortunately, only a limited number of tropical countries (especially in the warm regions) really take full advantage of this free available natural control technique (Kumar *et al.*, 1993; Adetunji, 1994; Wang and Sipes, 2009; Stevens *et al.*, 2012).

11.5.3 Steam treatment

Soil disinfestation using steam has been adopted by greenhouse growers for almost a century. Plant pathogens are eliminated by steaming and even seeds of weeds are annihilated. Furthermore, steaming the soil before planting stimulates the growth of the crops.

Heat treatment of the soil at 100°C is the most effective method of soil sterilization. Bacterial and fungal pathogens, nematodes and even soilborne viruses are killed at this temperature. A temperature of 70°C maintained for half an hour, on the other hand, is sufficient for the control of fungal pathogens, bacterial pathogens and nematodes. Methods for soil steam treatment are: sheet steaming, drain steam system and negative pressure steaming.

11.6 Mechanical Control

Mechanical control of insect pests include the use of exclusion insect nets and trapping of adults by means of various types of traps.

11.6.1 Using insect screens for mechanical exclusion of insect pests and vectors

Insects may enter the greenhouse from outside through the ventilation openings. Installation of screens on the ventilation openings will exclude or diminish the entry of pests (Teitel *et al.*, 1999). Screens with a mesh size of 0.15 mm exclude thrips. For whitefly and aphids a mesh size of 0.35 mm is enough; and for leafminers 0.8 mm. Screens do not suppress or eradicate pests, they merely exclude most of them; therefore, they must be installed prior to their appearance, and supplementary pest control measures, such as biocontrol or intelligent chemical control, are still required. Insect parasitoids and predators that are smaller than their prey can still immigrate through screens into the greenhouse, but larger ones are unfortunately excluded.

It should be noted, however, that the use of screens might impede ventilation, resulting in overheating and increased humidity (Kittas *et al.*, 2002; Soni *et al.*, 2005; Katsoulas *et al.*, 2006) which promote plant stress and susceptibility to pests and diseases, which may result in negative impact on plant growing and fruit quality (Harbi *et al.*, 2015). Increased humidity necessitates more frequent fungicide sprays than were required previously in an unscreened greenhouse. Moreover, screens reduce light transmission, so compromises in the management of light, temperature and humidity are necessary to avoid adverse effects on crops and their susceptibility to diseases. To minimize these harmful effects, it is possible to use forced ventilation, but this only helps to pull small insects through the screen. Thus, while screens can reduce immigrant populations of pests, they also reduce the immigration of beneficial arthropods. In neither case is exclusion total. The unfortunate fact is that the use of fine mesh screens, when coupled with dusty conditions prevalent in warm regions, seriously impedes greenhouse ventilation.

11.6.2 Types of screens

Various types of screens have been developed to protect crops from insects; the challenge for the grower is to match the proper type of screen to local conditions (climate and insect populations).

Woven screens

The conventional woven screens are made from plain woven plastic. In commercial screens, the slot is rectangular and width must be smaller than the whitefly's body size (i.e. about 0.2 mm), but it must allow maximum air and light transmission. Screens designed to exclude *Bemisia tabaci* still permit some to penetrate, and they fail to exclude *Frankliniella occidentalis*. However, they exclude most larger insects such as moths, beetles, leafminers, aphids and leafhoppers, and they retain bee pollinators inside the greenhouse.

Recently, screens and nets impregnated with insecticides have been tested to control important pests such as *Bemisia tabaci* or *Tuta absoluta* and gave good results (Martin *et al.*, 2014; Biondi *et al.*, 2015).

Unwoven screens

These are made of porous, unwoven polyester and polypropylene or of clear, microperforated, polyethylene fabric. All are very light materials which can be applied loosely and directly over transplants or seeded soil, without the need of mechanical support. They have been used primarily in the open field as floating covers in early spring, to enhance earliness and to protect against early virus infection.

UV-absorbing screens

Through a modification of the insect behaviour, this kind of screen may achieve a protection of crops from insect pests and, especially, from virus diseases vectored by insects (Antignus and Ben-Yakir, 2004; Rapisarda *et al.*, 2006). As a secondary effect, some disturbance may be also extended to pollinators and natural enemies.

Whitefly exclusion

Whitefly-proof (50 mesh) woven screens are by far the most efficient mechanical barrier. Whitefly species of the group *Bemisia tabaci* are small insects, about 0.2 mm wide, which transmit many viruses and have become the limiting factor in vegetable production in greenhouse crops. Their physical exclusion from greenhouses is crucial, and accordingly whitefly-proof screens have been developed (Berlinger *et al.*, 2002). While the rate of whitefly exclusion is generally proportional to the screen's mesh, the insect's ability to pass through any barrier could not be predicted solely from thoracic width and mesh size. There is an unexpectedly high rate of whitefly penetration, resulting from a great variability among the samples of the same screen resulting from uneven and slipping weave.

11.6.3 Double door system or SAS

It is of paramount importance that all greenhouses be equipped with a double door or safety access system (SAS). The air lock type entrance 'SAS' into the nursery or the production area of the greenhouse prevents the easy entry of insects into the main plant-growing areas. The SAS could also be supplied with traps and sticky yellow bands along both sides.

11.6.4 Mass trapping of insects

Colour sticky bands for insect mass trapping

Specific colours attract certain day-flying insects. For example, this is the case of yellow, which attracts winged aphids, leafminer flies or adult whiteflies, and is frequently used in sticky traps for these insects, while blue is attractive to thrips.

Yellow and blue cards or bands coated with adhesives are used to attract and capture small flying insects in greenhouses. For the purpose of mass trapping, sticky bands 10–40 cm wide are used for greenhouse crops (about 100 m/ha). The bands are installed at a height of about 1 m between greenhouse pools, 2 weeks before transplanting; they must be maintained throughout the crop cycle.

Mass trapping using yellow sticky bands is widely used in greenhouse crops. However, mass trapping is not recommended in greenhouses where natural enemies are released for biological control as many flying beneficial insects will be attracted and killed by yellow sticky bands.

Water traps for insect mass trapping

Water (pheromone-baited) water traps are used successfully for the mass trapping of the Tomato leaf miner *Tuta absoluta* (Meyrick), in tomato greenhouses in the Mediterranean region and the Middle East (Abbes *et al.*, 2012; Cocco *et al.*, 2012; Siscaro *et al.*, 2013). These traps, baited with the sex pheromone of the insect, are made by a vessel of about 40 cm in diameter and 20 cm in height, containing water mixed with a mild surfactant (oil and/or detergent) to prevent males that have fallen into the water from escaping; the pheromone dispenser is placed slightly above the water level. Since pheromones give to traps high effectiveness in catching male insects, numerous traps (over 10 per ha) placed throughout a greenhouse can sometimes remove sufficient insects to substantially depress reproduction, reduce the local population and limit the ensuing damage. Nevertheless, the greenhouse must be fully equipped with insect screens to avoid attracting any additional males of *T. absoluta* from the environment outside the greenhouse.

Light traps for insect mass trapping

Also for control of *T. absoluta*, different types of light traps are used, differing from each other in both the light source and the operation manner. They are generally based on fluorescent lamps or sources of ultraviolet light, the latter ones being generally combined with electrified grids or aspirators for the elimination of adults attracted to the trap. These traps are normally activated during the night and, taking advantage of the positive phototaxis of both sexes of the Tomato leaf miner, allow a massive elimination of adults. Experimental tests carried out to evaluate the effectiveness of these traps have provided varying results as a function of environmental and climatic factors (Cocco *et al.*, 2012). Compared to pheromone traps, that only capture males, the light traps also capture a high percentage of females, thus being generally more effective, especially at low population densities of the pest.

11.6.5 Mechanical control of diseases

Many viruses and airborne fungi and bacteria spread within a crop from one or several sources, including crop plants, but also weeds occurring within or around a crop. By eliminating the source, an epidemic may be avoided. However, elimination of plants in the area surrounding a greenhouse is not simple when commercial crops are grown adjacent to private gardens, abandoned or desolated crop fields, or when virus reservoirs occur in the natural environment.

Removal (roguing) of infected plants in a crop is effective, particularly in young crops where a small number of plants form foci of infection for secondary spread. Any disease development in the vicinity of the removed plant must be followed up. Roguing is particularly effective for viruses transmitted mechanically (Pepino mosaic virus, for example) or by vectors (TYLC virus group, for example).

Dead leaves and flowers on a crop plant should be removed before they become a massive saprophytic base for inoculum. Pruning should always be done with a sharp knife, leaving no snags. Disease can often be avoided simply by reducing damage to roots, stems and foliage during cultural operations.

Plants surviving from a previous crop (volunteers) form another potential reservoir of infection within a new crop. Prevention is the best solution, adopting adequate harvesting techniques and soil cultivating practices.

11.7 Monitoring

Monitoring involves systematically checking the greenhouse crop at regular intervals and at critical times to gather information not only about the crop, pests and their natural enemies but also diseases and their antagonists. Visual observation of symptoms, laboratory analyses of soil or plant parts, weather data, sticky colour traps and pheromone traps, can all be used to collect the maximum information necessary for an informed decision. The more often a crop is monitored, the more information a grower has about what is happening in his greenhouse.

11.7.1 Insect monitoring using colour sticky traps

In greenhouses both using and not using biological control, yellow and blue sticky cards of various dimensions are recommended to monitor the flying insect population. A minimum of eight sticky cards (8×20 cm) per ha are required and they should be distributed so as to cover the various climate zones in the greenhouse (especially in non-climate-controlled greenhouses).

While colour sticky bands are set at a fixed height and must be maintained throughout the crop cycle, sticky traps need to be monitored and changed every week and their height must be adjusted to the top of the plant canopy.

Sticky colour traps are undoubtedly excellent monitoring tools of small flying insect pests, but requires a minimum of expertise for the recognition of captured insect species.

11.7.2 Insect monitoring using attractant-baited traps

Attractant-baited traps are extensively adopted in pest management (Witzgall *et al.*, 2010; Prasad and Prabhakar, 2012; Caparros Megido *et al.*, 2013; Elimem *et al.*, 2014; Ettaib *et al.*, 2016) because they capture only one species or a narrow range of species, simplifying the identification and counting of target pests. Also, this trapping method is usually very sensitive and can capture pest insects occurring at very low densities, that could not be detected using other inspection methods. This specificity and sensitivity make attractant-baited traps efficient and labour-saving tools compared with colour traps, which are non-specific and require a specialist to carry out the monitoring.

Attractant-baited traps are used in integrated control programmes with the following aims:

- to detect the presence of an insect pest;
- to estimate the relative density of a pest population in a given greenhouse;
- to indicate the first emergence or peak flight activity of a pest species, so as to time an insecticide application, biological control release or to signal the need for additional scouting;
- to carry out mass trapping of male adults.

Most (but not all) of the attractant-baited traps which are used presently in greenhouse cultivations are pheromone traps, from where the active chemical attractant (contained in a rubber plastic lure) is released slowly over a period of up to several weeks. Traps containing these lures are made from paper, plastic or other materials. Most traps use an adhesive-coated surface (with the chemical attractant sometimes contained also in the coat) or a funnel-shaped entrance to capture the target insect. Pheromone application in greenhouses requires special skills; in fact, air currents in protected crops are different from those in the field and, as a result, insect attraction to the traps could be negatively affected by active ventilation.

11.8 Chemical Control

In Integrated Pest Management programmes, chemical control must be based on informed decisions and used only after considering the impact of all factors regulating the populations of pests, being sure that there are no other effective management tools available. In line with what has been mentioned above, and in order to maximize the success of chemical control, the following five major steps are suggested when using pesticides.

11.8.1 Step 1: proper identification and risk assessment of the pest and its life stage

Proper identification and risk assessment of the pest is mandatory as a first step, and a false identification may lead to possible misunderstanding of the pest's behaviour and its potential dangers. Correct identification allows the grower to understand the pest, seek out additional specific information about potential dangers to the crop and know how to best manage it.

Populations of pests must be monitored regularly following scientifically established methods appropriate to the region or locality. Existing and validated forecasting models for diseases should be implemented and adequate insect monitoring tools should be used as available.

11.8.2 Step 2: choosing the proper pesticide

Only after the grower has properly identified the pest can he select the best pesticide. Pesticides are sometimes effective against one pest but useless against other closely related ones. Also, one pesticide may be effective against a specific developmental stage while other pesticides may be effective against a different or perhaps all developmental stages. Correctly identifying the pest and understanding its biology and life cycle allows the grower to choose the best pesticide. Extension agents may be of extreme importance to be consulted by the growers in this step.

In various countries of the world, including many tropical ones, old conventional pesticides are progressively being abandoned, to leave space for new active ingredients having a greater sustainability in both their mode of action and their degradation pathways. Among the latter ones, various kind of biopesticides are increasingly used in vegetable production systems, also in tropical and subtropical greenhouse cultivations (Srinivasan, 2012).

11.8.3 Step 3: proper usage of pesticides

After selecting the pesticide, growers must decide on the amount to use. The information is contained on the label, but in general different options are given. To make the correct choice, as much knowledge as possible about the biology of the pest must be gathered. Factors such as pest size and population play an important role in insect management. For example, small worms may require the lowest recommended label rate, while large worms may require the highest label rate; however, continuous use of the higher rate can lead to insect resistance, not only to that particular pesticide but often to other pesticides as well.

The proper measuring devices are needed to make sure that the correct amounts of insecticide are used. Greenhouse growers frequently have smaller areas to spray than growers in the field and therefore need smaller amounts of pesticides. For example,

a field tomato farmer may use 1000 cc/ha of a material in 500 l of water; this is an easy quantity to measure. However, in a greenhouse, only 20 l of spray may be needed to do the greenhouse block; this means that very small amounts (even less than 1 teaspoon) must be measured, but a spoon is a non-graduated container and is a very poor, inaccurate and dangerous way to measure pesticides. Growers should therefore have a set of graduated cylinders marked in millilitres and a set of good-quality measuring cups available. Glass can be used, but plastic is often preferred to avoid breakage. Measuring devices, such as graduated cylinders, have pouring lips and graduated measuring markings that allow for accurate measurement and preparation of spray in quantities from 1 to 500 l or more. Appropriate measuring devices can guarantee accurate measurement, thus allowing for an effective control of the pest together with a safe range of pesticide residues on the crop, a more efficient use of chemicals and money, and a sharp reduction of phytotoxicity.

If excessive quantities of pesticide are used, the following problems can arise:

- the crop can have high residue levels, posing health hazards to consumers and potentially preventing the crop from entering those markets where these safety levels are defined by law;
- re-entry by workers into overdosed areas is potentially dangerous and can lead to illness, medical costs and liability to the grower;
- overdoses can speed up the pests' resistance process;
- production costs increase without the benefit of added profits;
- phytotoxicity is more likely to occur.

It is important not to exceed the label rates. If the maximum labelled rate does not achieve the desired results, other reasons for failure must be sought. It is unlikely that additional amounts of the same material will improve the situation. The old cliché 'If a little is good, a lot is better' can have disastrous consequences.

Safety of the overall handling of pesticides need special care and spillage of concentrated materials must be prevented. Pesticide concentrates are usually handled when the sprayer is loaded and dilute sprays are being prepared, and special handling precautions are necessary at this time. The applicator must be particularly careful when handling finished sprays, but even more so when dealing with concentrated material.

11.8.4 Step 4: proper timing

The chosen pesticide should be applied at the correct time and this is not an easy task. Determining the best time to apply chemical control is a dynamic and comprehensive process, and failure to treat at or near the correct time is one of the major reasons for unsuccessful pest management. Despite the difficulties involved, steps may be taken to help make a reasonable decision:

- regularly and thoroughly inspect the crop, noting the presence of pests and any increase in incidence;
- know the pest, its behaviour, and its ability to damage the crop;
- be aware of economic threshold when available;
- know the biology of the pest so that pesticide application can be aimed at the weakest, most vulnerable stage. Eggs are scarcely vulnerable to insecticides. Usually, young stages (larvae or nymphs) are more easily controlled and require less insecticide than older stages. Pupal stages are generally not affected by insecticides (large larvae nearing this stage are also difficult to control).

Once populations reach high numbers, even if 95% are controlled, the remaining 5 percent can still be a significant number according to their reproductive rate. Further considerations must be made with regard to timing of greenhouse insect control in particular:

- mites and whiteflies should be controlled as soon as they are observed and their populations should not be allowed

to grow; moreover, sprays for their control should be spaced at no longer than 1-week intervals until control is achieved;
- worms and caterpillars should be controlled when 14 days old;
- it is generally best to apply insecticides in the late afternoon or evening hours when temperatures start to fall: this also allows for maximum exposure time before the sprayed area is aired for employees; also, many insects are most active at night; the risk of phytotoxicity (burning) is greater when applications are made during the middle of the day; on the other hand, it has been reported that better mite control can be obtained by spraying early in the morning hours; as a rule, insecticide or miticide applications should be made while temperatures are low; a spray application time should be chosen when control, phytotoxicity, irrigation, temperature and worker re-entry considerations best fit the overall operation;
- insecticides should not be applied when plants are water-stressed, since water-deficient plants are more subject to phytotoxicity damage.

11.8.5 Step 5: proper application of pesticides

Proper application, like proper timing, is one of the most important steps in pest-control efforts. It does little good to complete the first four steps properly and then fail to deliver the material to the target area. There are numerous factors to be taken into account and various spray methods to consider for proper application of pesticides.

Spray equipment must be properly calibrated. A calibration error can quickly result in underdosing (not obtaining control) or overdosing (harmful and often illegal). Application of the proper amount of material is closely related to calibration.

Growers should purchase the specific equipment for the operation and the target pest, since each pest differs in habits and behaviour and one piece of equipment may not meet all needs, so that results may vary when using a high-volume sprayer, or a rotary atomizer, or a pulse jet applicator.

High-volume sprayers have been used for years in greenhouses and most pesticides are labelled for high-volume application, but they are usually work- and time-consuming and their efficiency is not always satisfactory, especially when using insecticides having a physical mode of action (i.e. oils or soaps) and against pests that live on the underside of leaves.

Low-volume methods for application of insecticides (as with aerosol generators, foggers, rotary atomizers, electrostatic applicators, mist blowers or pulse jet applicators) are widely tested in greenhouses and their use is increasing everywhere, in spite of their high technical requirement in terms of special formulations to be applied and specific accessory equipment to be used. Low-volume systems can be efficient in delivering pesticides, as they include contact, fumigant or residual control, but their greatest advantage is the time saved; however, some low-volume spray equipments are not legal for use in some countries on greenhouse crops.

For the best results, knowledge of the pest and its biology but also of the plant structure should be combined with the equipment's specific capabilities; for example, an equipment producing a driving, directed spray may be required to reach the underside of leaves in a plant having a thick canopy, but a rolling fog, or an atomizer, or an electrostatic applicator may be appropriate if the canopy is thin.

For correct application of pesticides, proper maintenance of spray equipment and its numerous parts is essential, in order to ensure all those physical characters (such as operating pressure or droplet size) influencing the effectiveness. Growers should regularly check nozzles and discs for wear and tear (greater when suspensions and wettable powder formulations are used) and replace them when they do not meet specifications. Moreover, most pesticides are highly corrosive and react with hoses, lines, nozzles, tanks and other components,

affecting the spray patterns and leading to the formation of foreign particles that clog the equipment.

In general, the following practical rules should be followed when applying a chemical treatment:

- spray should be used as soon as it is mixed; once mixed, the pesticide soon starts to change (usually faster the higher is the pH of the mixture) and its effective life can be just hours;
- water is the most commonly used dilutent (carrier) for pesticide sprays and it must be clean and free from contaminants; growers should filter water as many times as possible: between the source of water, the spray tank and where the water enters the tank; also between the tank and the final nozzle, so that the spray can flow and be delivered in the pattern according to equipment specifications;
- workers must mix only sufficient spray for the job; if not all used on the crop, disposal of the residual fraction must be done in line with the environmental requirements, such as for the proper clean-up and rinsing of equipments immediately after spraying;
- spray equipment must be properly stored after cleaning to keep it free of dust, dirt and other foreign materials (e.g. pieces of rubber line) that may enter the system, blocking it and causing poor spray patterns, particularly when pressure is applied;
- if a spray job has to be repeated because of inaccurate pest identification, or poor application or timing methods, potential profits can be reduced and the crop can be vulnerable to pests; of all the factors and measures necessary for pest control, none are more important than the overall proper delivery methods.

11.8.6 Storage of pesticides

Only fresh pesticides should be used. Growers should try to purchase the quantity required and not plan to store materials for longer than necessary, and in any case for not more than a single season.

Pesticides must be stored in a safe, dry location. The best storage temperatures are generally room temperature (20–30°C), and temperatures over 40°C are not recommended. Certain pesticides undergo undesirable changes if the storage temperature drops below freezing. Applicators must follow the label instructions for specific storage information and the local government laws with regard to storage sheds, locks and warning signs.

11.8.7 Safety issues

Pesticides can create serious problems when common sense and rules of safe use and handling are not used. The pesticide label is considered a legal document: 'the label is the law'. It is the duty and legal responsibility of the user to read and understand all the directions and information on the label, and to seek interpretation of any unclear part. Pesticide dealers, manufacturers and their representatives, as well as the extension services, can aid in interpreting pesticide labels. Lack of understanding of the label or misuse of a pesticide can have serious consequences. The label contains information on the safe use of its contents, protective clothing, worker contact, poisoning symptoms, disposal and other information. The user is encouraged to become familiar with all safety and other aspects of the label before use.

11.9 GAP Recommendations and Conclusion

According to the definition given by FAO (2003), 'Good Agricultural Practices are practices that address environmental, economic and social sustainability for on-farm processes, and result in safe and quality food and non-food agricultural products'.

In line with these GAP standards, maintaining a healthy greenhouse crop is essential for an economic and sustainable greenhouse vegetable and ornamental production. This requires first a consistent monitoring, which involves systematically checking the greenhouse crop at regular intervals and at critical times to gather information about the crop, pests and their natural enemies, diseases and their antagonists. Visual observation of symptoms, laboratory analyses of soils or plant parts, weather data, sticky colour traps and pheromone traps, are some of the available tools which should be used to collect the maximum information necessary for an informed decision, to make sure there is a clear distinction between non-parasitic afflictions (plant stress and other physiological disorders) and those caused by pests.

If a grower does not have the technical expertise to identify properly the problem, it is strongly recommended that he seeks special diagnostician advice. Many countries operate plant clinics and proper identification by experienced diagnosticians can go a long way where routine diagnosis of fairly familiar pests and diseases is concerned.

The more often a crop is monitored, the more information a grower has about what is happening in his greenhouse. For effective GAP programme, greenhouse workers have to be trained to recognize nutrient deficiencies, disease, nematode, mites and insect symptoms. In this regard, training of greenhouse workers in identifying visible symptoms is of paramount importance in the early detection of abnormalities. Unlike most crops, workers in greenhouse crops have the opportunity to visit every plant of the greenhouse on several occasions (pruning, training, harvesting, etc.). Their observations could be of great help in the early detection of spot infections/infestations by pathogens and pests. Personal protective gear, disinfectants, disposal bins, markers, etc. have to be made available to greenhouse workers so that they can play adequately their role in GAP. In large operations, it is suggested to have a site map of the greenhouses and a good record keeping system so that diseases and pests outbreaks as well as management actions that have been taken can be noted for the information of all greenhouse staff.

11.9.1 Simple questions to be answered before engaging in GAP

Good pest and disease management decisions can be made after answering the following questions appropriately.

- What pests and diseases are causing problems, and what is their incidence, numbers and stage of development in the specific greenhouse crop?
- What specific conditions might have a direct or indirect effect on the increase or decrease of pests and diseases in question?
- What is the status of natural enemies/ antagonists of pests and diseases, and are they playing an important role in the regulation of pests and diseases concerned?
- What is the stage of development, condition of the crop, and will it be economical to engage in a management programme?
- What management options are available, and will their implementation justify the economic cost of the IPM programme?
- If pesticides are the unique option, how will intelligent and effective chemical control following the steps outlined earlier in this chapter be implemented?

Again, if a grower does not have sufficient expertise, it is strongly recommended that he seeks advice from extension staff or plant clinics.

11.9.2 General guidelines for GAP in greenhouse crops

The careful integration of all GAP strategies and the implementation of Good Agricultural Practices, taking into consideration the various aspects of pests and diseases management discussed extensively in this chapter, should follow the following guidelines.

Pests and diseases management at nursery level

The nursery is the first source of healthy or contaminated planting material with many pests and diseases of greenhouse crops (insects, mites, nematodes, foliar pathogens, soil pathogens, etc.), this is why it is extremely important to:

- use certified high-quality seed;
- use resistant/tolerant varieties and/or rootstocks as available in the market;
- produce seedlings in conditions that ensure healthy and vigorous plantlets.

It is instrumental to ensure a healthy start of the greenhouse crop which makes implementation of GAP much easier, further down the road. If seedlings are contaminated with such diseases like viruses or soilborne pathogens at an early stage, very little can be done later, after transplanting, to save the crop, even with the best GAP programme. With this in mind, it is important to:

1. produce plantlets in a separate greenhouse devoted especially for this purpose and equipped with very fine mesh on all ventilation openings and a reliable Safety Access System (SAS) equipped with a foot pad (sponge impregnated with a disinfectant) as well as a disinfectant solution to be used systematically by nursery workers and visitors for disinfecting hands prior to access to the nursery;

2. use only clean or virgin substrate for the production of plantlets;

3. use adequate irrigation and fertilization management to avoid exposing plantlets to stress that could be conducive to diseases and pests;

4. avoid water saturation and provide the adequate environment for plantlets and unfavourable for pests/pathogens;

5. follow a stringent monitoring of pests and diseases as well as physiological disorders at all stages of seedling production using techniques described in this chapter;

6. implement strict sanitary measures at all stages of nursery activities:

    a. only specialized personnel can be authorized to enter the nursery space and prohibition of cigars, cigarettes and chewing tobacco to prevent viral diseases;

    b. no one should touch the plantlets unless it is necessary;

    c. make sure that all equipment used in nurseries (trays, tools, etc.) are disinfected to avoid any source of contaminants: there is a big risk of contaminants (pests but most importantly pathogens) in reused trays;

    d. spray growing nursery areas (walkways and benches) with chlorine solutions;

7. apply only registered pesticides as required by GAP protocols;

8. it is unsound hygiene to raise seedlings directly on the ground or alongside production crops because of the high risks of contamination of transplants from the soil and the crop. Such practice is quite common among small greenhouse farmers in some countries and jeopardizes the clean start of a crop which is instrumental to any management programme;

9. if proper infrastructure is not available or inadequate nursery expertise is lacking at farm level, it is strongly advised that the grower acquires good planting material from a certified nursery;

10. it is not enough to only produce or obtain healthy plantlets from a certified nursery but all precautions should be taken to make sure not to expose plantlets to any risks of contamination by pests and diseases during transfer of plantlets from nursery to the production greenhouse.

Pests and diseases management before planting

Apply strict sanitary measures: sanitation is by far the most effective and cheapest way of escaping disease epidemics and pest outbreaks. The cliché 'It is much cheaper to stay clean than to become clean' applies here. The following rules are of a basic importance:

- crop residues from the previous crop cycle should be destroyed immediately after the final harvest. The greenhouse should be thoroughly cleaned before planting a new or first-time crop. This means burning, burying or hauling away all leftover roots and other plant material from the previous crop;

- adequate solarization or soil disinfection only if required should be performed during the crop-free period especially in warm regions. If the above fails to bring soilborne pathogens and nematodes to a manageable level, the grower is then advised to move to soilless culture (when and where this is economically sustainable);

- elimination of all weeds from the greenhouse space before planting as they might harbour some pests and diseases and become the source of contamination of clean planting material;

- adequate washing of soil and plant debris from farm equipment when moving from one greenhouse to another to avoid any spread of pests and diseases to clean crops;

- before planting the new crop, growers should thoroughly clean or disinfect the greenhouse structure;

- apply fine mesh (thrips proof) insect nets to all aeration openings of the greenhouse and use a SAS at the entry of the production greenhouse to mechanically reduce the chances of insect pests and vectors from accessing the greenhouse space. The SAS should be equipped with a foot pad as well as a disinfectant solution to be used systematically by greenhouse workers and visitors for disinfecting hands prior to access to the greenhouse.

Pests and diseases management during crop cycle

Crop rotation is hardly used in any greenhouse production systems and in most greenhouses continuous cropping is practised, without or with very short fallow crop-free interval.

1. Crop scheduling:
   a. where there is a risk of disease being more destructive in cool soils (Fusarium crown and root rot and corky root rot, for example), transplanting should be delayed until the root zone has warmed up;
   b. where two or more crops are grown each year, overlapping of planting dates between the different greenhouses and uncontrolled movement of workers between these greenhouse crops means that pests and pathogens populations are spread from the old greenhouse crops to the young crops, unless special care is taken regarding worker movement between crops.
2. Use adequate cultural practices to maximize biological/natural prevention of pests and diseases, including: choice of the location of the greenhouse production; using adequate greenhouse structure with appropriate climate conditions, ensuring adequate soil management; using adequate quality water and irrigation management; ensuring adequate and balanced fertilization; enforcing the necessary sanitary measures before planting, during production and after the end of the crop cycle, etc.
3. Proper sanitary measures are to be observed in and around the production greenhouse before planting and during all the crop cycle.
4. Proper disposal of infected plants or infected plant debris (in bins or plastic bags).
5. Rigid hand-scrubbing rules for workers involved in pruning, pollinating, tying or harvesting activities.
6. Use the appropriate tools (visual observation, insect colour traps, pheromone traps, etc.) for a reliable and regular scouting of pests and diseases before and after planting of the crop in the greenhouse to take informed decisions (using ET as available) when it comes to pest and disease management, taking also into consideration the quantitative and qualitative assessment of the balance status between pests and diseases and their natural enemies.
7. Use all possible methods of pests and diseases management discussed earlier (biological control, physical control, mechanical control, cultural control, bio rational pesticides, etc.), taking into consideration the short- and long-term impact on greenhouse crop productivity and quality, as well as the impact on the environment and human health, with the aim to minimize the use of toxic pesticides.
8. Apply safe and intelligent chemical control as discussed in this chapter. Only registered pesticides as required by GAP protocols are advised. While other control tactics require fewer precautionary measures, chemical pesticides when necessary require the diligent application of the following precautionary measures:
   a. the pesticides to be used must be registered with the relevant national registration authority and be approved for use on specific greenhouse crop and in line with GAP protocol;
   b. the pesticides should be handled, stored, applied and disposed of in accordance with the instructions on the label and in line with the international conventions on pesticides;
   c. pesticides, passengers (human and animal) and food should never be transported in the same vehicle;
   d. the pesticide is to be used only when needed and only at the dose prescribed. Using the minimum recommended dose allows not only cost reduction, but also reduces risks of pesticide residues in produce and the contamination of the environment;
   e. the instructions for application of a particular pesticide should be read before its use. Information such as: restrictions for its use (beneficial insects, pollinators, etc.), application rate, approved doses, compatibility with other substances, mixing properties, and minimal intervals between application and harvesting (pre-harvest period) should all be carefully considered;
   f. special attention should be paid to spray equipment, pumps and nozzles used to apply pesticides. To minimize any potential risks of over- or underdosage, accidents and spills, all

equipment should be calibrated for accuracy and checked on a regular basis for any malfunction. Spray equipment should be washed after each treatment to prevent contamination of produce with compounds not authorized for the specific crop (GAP protocol);

g. all pesticides should be stored safely away from children, anyone who might misuse them, animals and water sources. Pesticides should be stored in clearly labelled containers, even though storage in the original containers is advised. In line with GAP protocol, pesticide containers should be kept in a safe storehouse that is well ventilated and can be closed off to prevent any unauthorized entry. The pesticide storehouse should be established away from populated areas, on well-drained land, and away from domestic water supplies. It should also be constructed with non-combustible material, and have a leakproof floor and an emergency exit. Small greenhouse farmers should keep the pesticides in a locking ventilated cupboard or cabinet made preferably from iron;

h. any pesticide spillage should be absorbed with sand or sawdust and then swept up and properly disposed of. Thereafter, the floor should be cleaned with detergent and water;

i. pesticide operators are advised to use appropriate masks and other protective gears as recommended on the label. Personnel cleaning and bathing after application are also essential;

j. pesticides should not be sprayed in strong wind conditions, especially in naturally ventilated greenhouses, to avoid pesticide drift;

k. strict adherence to Maximum Residue Limit (MRL) for each pesticide used and each specific crop. The MRL is the maximum level of residue which is legally allowed in or on greenhouse produce to provide reasonable assurance that no adverse effects will result to the consumer over a lifetime of dietary exposure;

l. pesticide disposal: empty pesticide containers should be washed multiple times and then taken to an appropriate place until disposed of correctly. Pesticide containers should never be disposed of in unused wells or near water sources;

m. personnel training: adequate training of personnel responsible for using and applying pesticides is critical. They should be trained in the appropriate use of the products and be aware of the dangers that could result from the improper use of pesticides. They should also be trained in the use of safety equipment and application devices. Greenhouse workers should be informed that adverse health effects caused by inappropriate use of pesticides are most often not noticeable immediately or in the short term (acute toxicity), but can develop over time (chronic toxicity) if exposure is not reduced. Growers and farm managers have the obligation to ensure that personnel involved in the manipulation of pesticides are provided with the necessary protective gear and appropriate spraying equipment that meet GAP;

n. avoiding crop injury: certain chemicals may cause phytotoxicity to certain crops under certain conditions. It is strongly recommended to always consult the label for such limitations. Before applying any pesticide, operators should take into account the stage of plant development, the pH of the water used for mixture, the soil type and conditions, the temperature, moisture and wind conditions. Operators should know that phytotoxicity often results from mixing incompatible materials;

o. personnel safety: label directions should be followed carefully, avoiding splashing, spilling, leaks, spray drift and contamination of clothing. Operators are advised to never eat, smoke, drink, or chew while using pesticides. Provision for emergency medical care should be made in advance as required by regulations;

p. in line with GAP standards and protocol, records of pesticide applications should be kept including the following information:

    i. date of application;

    ii. greenhouse location or number;

    iii. information on the pesticide (commercial name, active ingredient);

    iv. dosage and volume of spray applied per unit area;

    v. operator name for monitoring purpose;

    vi. common name of target pest, disease or weed;

    vii. pest/disease/weed level of infestation or risk as justification for the treatment;

    viii. first permitted harvest date.

Pests and disease management in organic greenhouse production

With the exception of chemical pesticides, all the other management techniques discussed in this chapter apply almost equally to conventional or GAP-certified as well as to organic-certified greenhouse crop production. While chemical pesticide use is strictly prohibited in organic greenhouse production systems, a wide range of bio rational pesticides are allowed. Therefore, many of the above recommendations concerning pesticide use also apply to the following compounds authorized in organic greenhouse production:

- minerals including sulfur, copper and diatomaceous earth;
- botanicals including neem, pyrethrum and other new plant extracts;
- oils and soaps including a number of commercially available soap-based products;
- pheromones used for monitoring and mass trapping or in sexual disruption techniques;
- microbials, including the fast-growing biopesticides products.

Appropriate adoption and monitoring of GAP helps improve the safety and quality of food and other agricultural products, as well as promote sustainable agriculture, contributing to meet national and international environment and social development objectives. It may help reduce the risk of non-compliance with national and international regulations, standards and guidelines (in particular of the Codex Alimentarius Commission, World Health Organisation (WHO) and the International Plant Protection Convention IPPC, regarding permitted pesticides, maximum levels of contaminants (including pesticides and mycotoxins) in food and non-food agricultural products, as well as other chemical, microbiological and physical contamination hazards.

Awareness-raising is needed of 'win-win' practices which lead to improvements in terms of yield and production efficiencies, as well as environment and health and safety of workers. One such approach is Integrated Production and Protection Management (IPPM).

References

Abbes, K., Arbi, A. and Chermiti, B. (2012) The tomato leaf miner *Tuta absoluta* (Meyrick) in Tunisia: current status and management strategies. *Bulletin OEPP/EPPO Bulletin* **42** (2), 226–233.

Abdel-Ghany, A.M., Al-Helal, I.M., Alzahrani, S.M., Alsadon, A.A., Ali, I.M. and Elleithy, R.M. (2012) Covering materials incorporating radiation-preventing techniques to meet greenhouse cooling challenges in arid regions: a review. *The Scientific World Journal*, Article ID 906360. DOI:10.1100/2012/906360

Adetunji, I.A. (1994) Response of onion to soil solarization and organic mulching in semi-arid tropics. *Scientia Horticulturae* **60** (1–2), 161–166.

Albajes, R., Gullino, M.L., van Lenteren, J.C. and Elad, Y. (eds) (2002) *Integrated Pest and Disease Management in Greenhouse Crops*. Kluwer Academic Publishers, New York.

Alebeek, F.A.N. and van Lenteren, J.C. (1992) *Integrated Pest Management for Protected Vegetable Cultivation in the Near East*. FAO Plant production and protection paper 114, Rome.

Alene, A.D., Neuenschwander, P., Manyong, V.M., Coulibaly, O. and Hanna, R. (2005) *The Impact of IITA-led Biological Control of Major Pests in Sub-Saharan African Agriculture: A Synthesis of Milestones and Empirical Results*. International Institute of Tropical Agriculture, Ibadan, Nigeria.

Altieri, M.A. and Nicholls, C. (1999) Classical biological control in Latin America. In: Bellows, T.S., Fisher, T.W., Caltagirone, L.E., Dahlsten, D.L., Gordh, G. and Huffaker C.B. (1999) *Handbook of Biological Control: Principles and Applications of Biological Control*. Elsevier/Academic Press, San Diego, California, pp. 975–991.

Alves, S.B. (1998) *Controle Microbiano de Insetos*. Ceres, Piracicaba, Brazil.

Antignus, Y. and Ben-Yakir, D. (2004) Ultraviolet-absorbing barriers, an efficient integrated pest management tool to protect greenhouses from insects and virus disease. In: Horowitz, R. and Ishaaya, A. (eds) *Insect Pest Management: Field and Protected Crops*. Springer, New York.

Berlinger, M.J., Taylor, R.A.J., Lebiush-Mordechi, S. and Shalhevet, S. (2002) Efficiency of insect exclusion screens for preventing whitefly transmission of tomato yellow leaf curl virus of tomatoes in Israel. *Bulletin of Entomological Research* 92 (5), 367–373.

Biondi, A., Zappalà, L., Desneux, N., Aparo, A., Siscaro, G., Rapisarda, C., Martin, T. and Tropea Garzia, G. (2015) Potential toxicity of α-cypermethrin-treated net on *Tuta absoluta* (Lepidoptera: Gelechiidae). *Journal of Economic Entomology* 108 (3), 1191–1197.

Bueno, V.H.P. (1999) Protected cultivation and research on biological control of pests in greenhouse in Brazil. *IOBC/WPRS Bulletin* 22, 21–24.

Bueno, V.H.P. (2005a) *Implementation of Biological Control in Greenhouses in Latin America: How Far Are We?* Second International Symposium on Biological Control of Arthropods, Davos, Switzerland, 12–16 September 2005. USDA Forest Service Publications, 531–537.

Bueno, V.H.P. (2005b) IPM and biological control of protected cropping in some developing greenhouse regions. *IOBC/WPRS Bulletin* 28, 23–26.

Bueno, V.H.P. and van Lenteren, J.C. (2002) The popularity of augmentative biological control in Latin America: history and state of affairs. *First International Symposium on Biological Control of Arthropods*, 14–18 January 2002. Honolulu, Hawaii, pp. 180–184.

Bueno, V.H.P. and van Lenteren, J.C. (2010) Biological control of pests in protected cultivation: implementation in Latin America and successes in Europe. In: *XXXVII Congreso Sociedad Colombiana de Entomologia*, 2–7 July 2010, Bogotá, pp. 261–269.

Bueno, V.H.P., van Lenteren, J.C., Silveira, L.C.P. and Moraes, S.M.M. (2003) An overview of biological control in greenhouse chrysanthemums in Brazil. *IOBC/WPRS Bulletin* 26, 1–5.

Butt, T.M., Jackson, C. and Magan, N. (eds) (2001) *Fungi as Biocontrol Agents*. CAB International, Wallingford, UK.

Cabello, T., Gallego, J.R., Fernandez, F.J., Gamez, M., Vila, E., del Pino, M. and Hernandez-Suarez, E. (2012) Biological control strategies for the South American tomato moth (Lep.: Gelechiidae) in greenhouse tomatoes. *Journal of Economic Entomology* 105, 2085–2096.

Caparros Megido, R., Haubruge, E. and Verheggen, F.J. (2013) Pheromone-based management strategies to control the tomato leafminer, *Tuta absoluta* (Lepidoptera: Gelechiidae): a review. *Base* 17(3), 475–482.

Castañe, C. and Hanafi, A. (eds) (2003) Integrated control in protected crops: Mediterranean climate. *IOBC/WPRS Bulletin* 26.

Cocco, A., Deliperi, S. and Delrio, G. (2012) Potential of mass trapping for *Tuta absoluta* management in greenhouse tomato crops using light and pheromone traps. *IOBC/WPRS Bulletin* 80, 319–324.

Cocco, A., Deliperi, S. and Delrio, G. (2013) Control of *Tuta absoluta* (Meyrick) (Lepidoptera: Gelechiidae) in greenhouse tomato crops using the mating disruption technique. *Journal of Applied Entomology* 137 (1–2), 16–28.

Csizinszky, A.A., Schuster, D.J., Jones, J.B. and van Lenteren, J.C. (2005) Crop protection. In: Heuvelink, E. (ed.) *Tomatoes*. Crop Production Science in Horticulture 13. CAB International, Wallingford, UK, pp. 199–233.

Dhaliwal, M.S. and Sharma, A. (2016) Breeding for resistance to virus diseases in vegetable crops. In: Peter, K.V. (ed.) *Innovations in Horticultural Sciences*. New India Publishing Agency, New Delhi, pp. 303–327.

Elad, Y. and Shtienberg. D. (1997) Integrated management of foliar diseases in greenhouse vegetables according to principles of a decision support system Greenman. *IOBC/WPRS Bulletin* 20 (4), 71–76.

Elimem, M., Teixeira da Silva, J.A. and Chermiti, B. (2014) Double-attraction method to control *Frankliniella occidentalis* (Pergande) in pepper crops in Tunisia. *Plant Protection Science* 50 (2), 90–96.

El-Maghlany, W.M., Teamah, M.A. and Tanaka H. (2015) Optimum design and orientation of the greenhouses for maximum capture of solar energy in North Tropical Region. *Energy Conversion and Management* 105, 1096–1104.

Ettaib, R., Belkadhi, M.S., Ben Belgacem, A., Aoun, F., Verheggen, F.J. and Caparros Megido, R. (2016) Effectiveness of pheromone traps against *Tuta absoluta*. *Journal of Entomology and Zoology Studies* 4 (6), 841–844.

FAO (2003) Development of a Framework for Good Agricultural Practices. Committee on Agriculture, Rome, 31 March–4 April 2003. FAO Corporate Document Repository, Rome. Available at: www.fao.org/docrep/meeting/006/y8704e.htm (accessed 5 July 2017).

Filho, M.M., Vilela, E.F., Attygalle, A.B., Meinwald, J., Svatoš, A. and Jham, G.N. (2000) Field trapping of tomato moth, *Tuta absoluta*, with pheromone traps. *Journal of Chemical Ecology* 26 (4), 875–881.

Ganguly, A. and Ghosh, S. (2011) A review of ventilation and cooling technologies in agricultural greenhouse application. *Iranica Journal of Energy & Environment* 2 (1), 32–46.

Gerson, U. and Weintraub, P.G. (2007) Mites for the control of pests in protected cultivation. *Pest Management Science* 63 (7), 658–676.

Gullino, M.L. and Garibaldi, A. (1994) Influence of soilless cultivation on soilborne diseases, *Acta Horticulturae* 361, 341–354.

Hanafi, A. (2003) Integrated production and protection: today and in the future in greenhouse crops in the Mediterranean region. *Acta Horticulturae* 614, 755–765.

Hanafi, A. and Papasalomontos, A. (1999) Integrated production and protection in protected cultivation in the Mediterranean region. *Biotechnology Advances* 17, 183–203.

Hanafi, A. and Schnitzler, W.H. (2004) Integrated production and protection in greenhouse tomato in Morocco. *Acta Horticulturae* 659, 323–330.

Hanafi, A., Achouri, M. and Baudoin, W.O. (1999) Integrated Production et Protection intégrées des cultures. *Proceedings du Symposium International sur la Production et la Protection intégrées des cultures Horticoles*, Agadir, Morocco, 6–7 Mai 1997.

Hanafi, A., Gressel, J., Head, G., Marasas, W., Obilana, B. *et al.* (2004) Major heretofore intractable biotic constraints to African food security that may be amenable to novel biotechnological solutions. *Crop Protection* 23, 661–689.

Harbi, A., Abbes, K., Dridi-Almohandes, B. and Chermiti, B. (2015) Efficacy of insect-proof nets used in Tunisian tomato greenhouses against *Tuta absoluta* (Meyrick) (Lepidoptera: Gelechiidae) and potential impact on plant growth and fruit quality. *Journal of Entomological and Acarological Research* 47 (3), 109–116.

Katsoulas, N., Bartzanas, T., Boulard, T., Mermier, M. and Kittas, C. (2006) Effect of vent openings and insect screens on greenhouse ventilation. *Biosystems Engineering* 93 (4), 427–436.

Kittas, C., Boulard, T., Bartzanas, T., Katsoulas, N. and Mermier, M. (2002) Influence of an insect screen on greenhouse ventilation. *Transactions of the ASAE* 45 (4), 1083–1090.

Kumar, B., Yaduraju, N.T., Ahuja, K.N. and Prasad, D. (1993) Effect of soil solarization on weeds and nematodes under tropical Indian conditions. *Weed Research* 33 (5), 423–429.

Kumar, K.S., Tiwari, K.N. and Madan, K.J. (2009) Design and technology for greenhouse cooling in tropical and subtropical regions: a review. *Energy and Buildings* 41 (12), 1269–1275.

Luc, M., Sikora, R.A. and Bridge, J. (eds) (2005) *Plant Parasitic Nematodes in Tropical and Subtropical Agriculture*, 2nd edn. CAB International, Wallingford, UK.

Maniania, N.K. (1991) Potential of some fungal pathogens for the control of pests in the tropics. *International Journal of Tropical Insect Science* 12 (1–3), 63–70.

Martí Martí, S., Muñoz, C. and Casagrande, E. (2010) El uso de feromonas para el control de *Tuta absoluta*: primeras experiencias en campo. *Phytoma España* 217, 35–40.

Martin, T., Kamal Wijaya, A., Gogo, E.O., Saidi, M., Deletre, E., Bonafos, R., Simon, S. and Ngouajio, M. (2014) Repellent effect of an alphacypermethrin treated net against the whitefly *Bemisia tabaci* Gennadius (Hemiptera: Aleyrodidae). *Journal of Economic Entomology* 107, 684–690.

Messelink, G.J., Bennison, J., Alomar, O., Ingegno, B.L., Tavella, L., Shipp, L., Palevsky, E. and Wäckers, F.L. (2014) Approaches to conserving natural enemy populations in greenhouse crops: current methods and future prospects. *BioControl* 59 (4), 377–393.

Mutwiwa, U.N., Tantau, H.J. and Salokhe, V.M. (2006) Cooling of greenhouses in the humid tropics – problems and solutions. *Poljoprivredna Tehnika* 31, 73–84.

Mutwiwa, U.N., von Elsner, B., Tantau, H.J. and Max, J.F.J. (2008) Cooling naturally ventilated greenhouses in the tropics by near-infra red reflection. *Acta Horticulturae* 801, 259–266.

Navarro Lopis, V., Alfaro, C., Vacas, S. and Primo, J. (2010) Application de la confusion sexual al control de la polilla del tomate *Tuta absoluta* Povolny (Lepidoptera: Gelechiidae). *Phytoma Espană* 217, 35–40.

Perdikis, D., Kapaxidi, E. and Papadoulis, G. (2008) Biological control of insect and mite pests in greenhouse solanaceous crops. *The European Journal of Plant Science and Biotechnology* 2, 125–144.

Prasad, Y. and Prabhakar, M. (2012) Pest monitoring and forecasting. In: Shankar, U. and Abrol, D.P. (eds) *Integrated Pest Management: Principles and Practice*. CAB International, Wallingford, UK, pp. 41–57.

Rapisarda, C., Tropea Garzia, G., Cascone, G., Mazzarella, R., Colombo, A. and Serges, T. (2006) UV-absorbing plastic films for the control of *Bemisia tabaci* (Gennadius) and Tomato Yellow Leaf Curl Disease (TYLCD) on protected cultivations in Sicily (South Italy). *Acta Horticulturae* 719, 597–604.

Shamshiri, R. and Wan Ismail, W.I. (2013) A review of greenhouse climate control and automation systems in tropical regions. *Journal of Agricultural Science and Applications* 2 (3), 176–183.

Shtienberg, D. and Elad, Y. (1997) Incorporation of weather forecasting in integrated, biological, chemical management of *Botrytis cinerea*. *Phytopathology* 87, 332–340.

Silveira, L.C.P., Bueno, V.H.P. and van Lenteren, J. (2004) *Orius insidiosus* as biological control agent of Thrips in greenhouse chrysanthemums in the tropics. *Bulletin of Insectology* 57, 103–109.

Siscaro, G., Biondi, A., Haddi, K., Rapisarda, C., Tropea Garzia, G. and Zappalà, L. (2013) Orientamenti di lotta integrata per il contenimento di *Tuta absoluta* (Meyrick) in Italia. *Atti Accademia Nazionale Italiana di Entomologia* 60, 111–124.

Soni, P., Salokhe, V.M. and Tantau, H.J. (2005) Effect of screen mesh size on vertical temperature distribution in naturally ventilated tropical greenhouses. *Biosystems Engineering* 92 (4), 469–482.

Srinivasan, R. (2012) Integrating biopesticides in pest management strategies for tropical vegetable production. *Journal of Biopesticides* 5 (Supplementary), 36–45.

Stevens, C., Khan, L., Victor, A., Rodriguez-Kabana, R., Rhoden, E.G., Bartlett, J.R., Fyffe, A.E. and Willian, K.E.B.J. (2012) Soil solarization in a subtropical and humid climate. In: *Soil Solarization: Theory and Practice*, American Phytopathological Society, St. Paul, Minnesota, pp. 193–197.

Teitel, M., Barak, M., Berlinger. M.J. and Lebiush-Mordechai, S. (1999) Insect proof screens: their effect on roof ventilation and insect penetration. *Acta Horticulturae* 507, 29–37.

van Lenteren, J.C. (1989) World situation of integrated pest management in greenhouses. *Proceedings of the Symposium on Insect Control Strategies and the Environment*, Amsterdam, pp. 32–50.

van Lenteren, J.C. (1990) Integrated pest and disease management in protected crops: the inescapable future. *IOBC/WPRS Bulletin* 13 (5), 91–99.

van Lenteren, J.C. (1995) Integrated pest management in protected crops. In: Dent, D. (ed.) *Integrated Pest Management*. Chapman & Hall, London, pp. 311–343.

van Lenteren, J.C. (1998) Sustainable and safe crop protection: a reality? *Mededelingen van de Faculteit landbouwwetenschappen, Rijksuniversiteit Gent* 63/2b, 409–414.

van Lenteren, J.C. (ed.) (2003) *Quality Control and Production of Biological Control Agents – Theory and Testing Procedures*. CAB International, Wallingford, UK.

van Lenteren, J.C. and Bueno, V.H.P. (2003) Augmentative biological control of arthropods in Latin America. *BioControl* 48, 123–139.

van Lenteren, J.C. and Tommasini, M.G. (2003) Mass production, storage, shipment and release of natural enemies. In: van Lenteren, J.C. (ed.) *Quality Control and Production of Biological Control Agents – Theory and Testing Procedures*. CAB International, Wallingford, UK, pp. 181–190.

van Lenteren, J.C., Cock, M.J.W., Brodeur, J., Barratt, B.I.P., Bigler, F. *et al.* (2011) Will the convention on biological diversity put an end to biological control? *Revista Brasileira de Entomologia* 55 (1), 1–5.

Vila, E. and Cabello, T. (2014) Biosystems engineering applied to greenhouse pest control. In: Torres, I. and Guevara, R. (eds) *Biosystems Engineering: Biofactories for Food Production in the XXI Century.* Springer, Berlin, pp. 99–128.

Wang, K.-H. and Sipes, B.S. (2009) Solarization and cover cropping as alternatives to soil fumigants for nematode management in Hawaii's pineapple fields. *Soil and Crop Management.* College of Tropical Agriculture and Human Resources, Honolulu, Hawaii, pp. 1–4.

Wardlow, L.R. and O'Neill, T.M. (1992) Management strategies for controlling pests and diseases in glasshouse crops. *Pest Management Science* 36, 341–347.

Waterhouse, D.F. (1988) *Biological Control of Insect Pests: Southeast Asian Prospects.* Monograph no. 51. Australian Centre for International Agricultural Research (ACIAR), Canberra.

Wearing, C.H. (1988) Evaluating the IPM implementation process. *Annual Review of Entomology* 33, 17–38.

Witzgall, P., Kirsch, P. and Cork, A. (2010) Sex pheromones and their impact on pest management. *Journal of Chemical Ecology* 36 (1), 80–100.

Wraight, S.P., Carruthers, R.I., Jaronski, S.T., Bradley, C.A., Garza, C.J. and Galaini-Wraight, S. (2000) Evaluation of the entomopathogenic fungi *Beauveria bassiana* and *Paecilomyces fumosoroseus* for microbial control of the silverleaf whitefly, *Bemisia argentifolii*. *Biological Control* 17, 203–217.

12 Integrated Pest Management in Banana and Plantain

Daniel L. Coyne[1,*], Thomas Dubois[2] and Mieke S. Daneel[3]

[1]International Institute of Tropical Agriculture, Nairobi, Kenya; [2]The World Vegetable Center, Eastern and Southern Africa, Arusha, Tanzania; [3]ARC-Institute for Tropical and Subtropical Crops, Nelspruit, South Africa

12.1 Background

Bananas and plantains (*Musa* spp.) are of major importance as a food crop across tropical and subtropical regions throughout the world. After grapes, citrus and apples, bananas are also the fourth most important fruit crop. Bananas and plantains consequently serve various purposes from subsistence foods to dessert bananas for export markets. The vast proportion of produce is consumed locally with just 10% used for commercial purposes, while a third of production is generated in sub-Saharan Africa, where it provides 25% of the food energy requirements for over 100 million people (FAOSTAT, 2014). The varieties that are used for commercial or subsistence production are predominantly triploid cultivars that are evolutionarily derived from crosses within and between diverse accessions of two diploid ancestor species, *Musa acuminata* Colla (A genome) and *M. balbisiana* Colla (B genome). The existing cultivars are usually classified into three genome groups, i.e. predominantly AAA for dessert and east African highland cooking bananas, AAB for plantains, and ABB for cooking bananas (Tenkouano and Swennen, 2004).

Major sources of losses, and barriers to increased production, stem from pests and diseases. As banana is grown widely across tropical and subtropical regions, it attracts a wide range of associated pests, which may vary according to geography and banana type. Changes in cropping practices or the introduction of new or unfamiliar cultivars may alter the equilibrium, stimulating new pest or disease problems. Some may affect the cosmetic appearance of fruit, which are of limited importance to subsistence cooking bananas, but can disrupt export shipments of dessert bananas. Within the concept of IPM, and particularly for smallholder systems, ill-linked management options, some of which are often unsuitable, should be avoided; rather, a focus on key basic issues such as pest identification, cultural management options, and deployment of basic training for farmers and extension workers should prevail. Numerous pests and diseases affect banana, but for the purposes of this chapter, the six key constraints that ravage banana and plantain production in most production areas will be the focus of discussion, describing symptoms, management strategies and pest management options.

---

* Corresponding author e-mail: d.coyne@cgiar.org

12.2 Aspects of IPM in Banana and Plantain

12.2.1 Technology transfer

Once present in the field, many of the economically important pests and diseases of banana and plantain are difficult to control. Prevention should therefore be considered a critical aspect of any IPM strategy. In preventing the introduction and transfer of pests and diseases to new fields and areas, a key component is knowledge and awareness by farmers and agricultural staff. Awareness of the possible threats, how to identify them and how to manage them with available management options is an important step in the whole IPM process. This is as much a concern in commercial as well as smallholder systems, although the knowledge gap is much greater in smallholder farming systems. Bridging this information gap and providing sensible and appropriate information transfer should therefore be a major consideration within IPM packages. Diagnostic services are not always accessible for many farmers and especially in rural subsistence settings. Among a raft of innovative initiatives, Global Plant Clinics in particular deserve merit, aimed at improving access to effective plant health services through regular advisory services made available in local communities (Bentley *et al.*, 2009). Such innovative mechanisms of knowledge and technology transfer are vital to safeguarding our crops and our ability to deal with outbreaks early.

Healthy planting material

As with most crops, but especially for vegetatively propagated crops, the health and quality of planting (seed) material is paramount. All of the major pests and diseases detailed below can be transferred through infected planting material. For some, this has been and continues to be the principal mode of transmission. Ensuring the availability, access to and use of healthy, uninfected planting material therefore provides the foundation to any banana IPM strategy.

Through the use of micropropagation, tissue culture plantlets are produced, which are certified pest- and disease-free, and used routinely in commercial banana production. For smallholder farmers this practice has yet to gain a strong foothold, for various reasons (Dubois *et al.*, 2006, 2013). One obstacle is the lack of producers to supply tissue culture plants, partly due to limited demand. For demand to increase, farmers need to understand the benefits of using more costly tissue culture plants. In any approach to supply and use tissue culture planting material, it is important to consider the education and training of farmers in handling and maintenance of these delicate plants. This should additionally include training for and establishment of propagation nurseries (Dubois and Coyne, 2011). However, limited virus indexing, suboptimal weaning procedures, accidental cultivar mixing during production, inappropriate farmer handling, and lack of sustainable and market-driven distribution channels combine to delay the uptake of tissue culture technology among smallholder farmers (Dubois *et al.*, 2006, 2013). Smallholder farmers across the globe use suckers, side shoots detached from the mother plant, as their primary source of planting material. Any pest or disease affecting the mother plant will consequently be transferred, along with the infected sucker, to the new field. Farmers will undoubtedly continue to use sucker planting material, even once tissue culture technology becomes widely adopted. It is therefore necessary to develop and disseminate practices for the selection of healthy material, using simple techniques to routinely sanitize suckers and to enable farmers to understand the benefits of establishing healthy plantations using healthy material. Tenkouano *et al.* (2006) provide a comprehensive account of the principles and techniques for healthy seed material.

Although tissue culture technology is commonplace within commercial systems, research to further improve this continues. By biologically enhancing plants with endophytic microorganisms at an early stage, plant host protection can be improved against certain pests and diseases (Backman and Sikora, 2008; Sikora *et al.* 2008; Dubois *et al.*, 2010; Viaene *et al.* 2013). A number of mechanisms are responsible, including enzymatic activation of host-plant defence mechanisms following inoculation of certain endophyte species or strains. With increasing restrictions on chemical pesticide use, the bio-enhancement of planting material presents a promising and environmentally responsible option for pest and disease management. A range of microorganisms have been identified, such as non-pathogenic *Fusarium oxysporum* Schlect., *Trichoderma* spp. and *Bacillus* spp., with specific strains or populations often proving more suitable than others. Arbuscular mycorrhizal fungi have also proved effective in providing pest and disease protection (Dubois and Coyne, 2011). In addition, traditional biopesticides have been assessed and shown to provide plant protection as 'non-typical' endophytes. *Beauveria bassiana* (Balsamo) Vuillemin, an entomopathogenic fungus, active against a range of insect pests, was successfully introduced into bananas, where it provided protection against the cryptic banana weevil without penalty to the host (Akello *et al.*, 2007; Akello *et al.*, 2008). One of the advantages of bio-enhancement with endophytes is that farmers need not be involved in the treatment but receive colonized plants. The search for more effective isolates, which also persist over banana generations, continues.

Good agronomic practices

Intrinsic to IPM are good agricultural practices, which need to be followed and should be considered as components of the IPM strategy itself. Cleaning farm tools and equipment should be standard; mulching and suitable fertilizer regimes should be administered, as should sanitary practices for removal of pest and disease inoculum. Basic principles of crop and resource management are intrinsic for crop health. In Hawaii's IPM scheme, for instance, incorporation of crop residue and fallowing fields for 6–8 months is common (Hawaiian

Banana Industry Association, 2010). In Uganda, male flower buds are routinely removed using clean tools as a strategy to limit bacterial Xanthomonas wilt (BXW) infection (Blomme *et al.*, 2005).

Breeding

It is clear that genetic host-plant resistance to a number of the key bio-constraints would be a highly effective and economic mode of management. Resistant, agronomically acceptable cultivars would provide a strong foundation, upon which additional IPM components could be built, in both export and subsistence situations. This is especially important for reducing the use of and reliance upon chemical pesticides. However, breeding bananas is not straightforward, but marred by high levels of sterility and complicated genetics, and further confounded by difficulties with embryo rescue, to name a few. Commercial dessert bananas, for example, are characterized by few landraces with an extremely narrow genetic base (Ortiz *et al.* 1995). Various banana breeding programmes, strategically located around the world, have made significant progress in developing cultivars with resistance against one or more of the main bio-constraints. The process requires long-term commitment in order for success to be realized, the rewards of which have started to show dividends. In addition to traditional breeding, significant efforts have focused on the genetic modification of banana for host resistance. Ironically, some of the obstacles that complicate traditional breeding, such as high levels of sterility, make banana a useful candidate for transformation, as the perceived environmental risks, such as contamination through cross-fertilization, are lessened. Progress to date in genetically transforming bananas has been positive, with field containment trials already underway using local cultivars modified for BXW and nematode resistance in Uganda, while creating cultivars with multiple resistance is a future goal (Tripathi, 2009; Tripathi *et al.*, 2015). A key advantage of this procedure is the ability to use locally adapted and preferred cultivars. Given the difficulties and long-term requirements for breeding bananas, this strategy offers significant promise, not only to the banana industry but also to the millions of people who depend on banana as a staple food and source of income.

Quarantine, monitoring and surveillance

Several pests and diseases remain regional in nature and have not yet reached pandemic, global proportions. Some of these may be of minor significance where they currently occur, but could prove disastrous under new conditions, partly due to fewer natural enemies. Others, such as Fusarium wilt Tropical Race 4, are a major threat where currently present and feared for its potential devastating global impact if spread. Preventing the entry of a serious pest or disease is crucial and financially preferable to managing it post-entry. Quarantine measures are therefore a key component in the continuous struggle against biotic constraints. However, surveillance, monitoring and detection measures are equally required to support quarantine policies and procedures.

The International Plant Protection Convention (IPPC) is an international agreement on plant health with the aim to protect cultivated and wild plants by preventing the introduction and spread of pests and diseases. The IPPC uses Pest Risk Analysis (PRA) as the basis by which countries prepare the requirements to trade commodities and protect plant resources. The PRAs underscore the importance of surveillance and diagnostics, which need to be effective at the governmental as well as the subgovernmental level. A global strategy is dependent upon quality information, such as current distributions and key dispersal pathways, but also on accurate diagnosis and early detection. However, a false announcement, through imperfect diagnostics, may be disastrous to trade (and political relations), while delayed detection may compromise containment and result in escalating costs to counter the outbreak.

For risk-based security measures to be effective, supportive policies also need to

be in place, while careful attention to deployment of surveillance activities are necessary to ensure optimal cost effectiveness (Kroschel *et al.*, 2013). Detection levels and the intensity of assessment required for this need to be carefully determined. For example, detection of a pest may be dependent upon or affected by the intensity of assessment. This, in turn, may be affected by technical capacity and finances. Should a high-risk pest be detected, however, having in place a contingency plan, which outlines the course of action needed to be taken, will prove highly beneficial. Strategies, based on IPPC guidelines, should also consider related occurrences, taking into account successful as well as less successful aspects of the programme. A useful recent example, for instance, is the outbreak of banana BXW in east and central Africa and the programme employed to contain it and raise awareness with farmers. Although this was highly damaging to the banana industry and farmers, it is a model example of combating a serious disease outbreak through regional harmonization (Karamura *et al.*, 2005).

Implementation of policies to restrict the movement of plant material from infected areas are critical to reducing the spread of a major disease or pest. In South Africa, the spread of *Radopholus similis* Cobb Thorne was contained by implementing a strict quarantine system on the movement of plant material from areas where *R. similis* was present (Willers *et al.*, 2002). Australia successfully implemented similar restrictions to prevent the spread of banana bunchy top virus (BBTV) (Daly, 2006), with similar tactics being employed to eradicate banana freckle disease (*Phyllosticta cavendishii* Wong & Crous) in the Northern Territory.

12.3 Plant Parasitic Nematodes

Nematodes are insidious pests, microscopic unsegmented worms, that attack the roots and rhizomes causing necrosis and root death. Nematode damage reduces root efficiency causing stunting, delayed flowering and poor yields; uptake of nutrients and water is disrupted leading to symptoms of leaf yellowing. With severe infestations, root systems can become completely undermined resulting in the toppling of the whole plant, especially when bunches are approaching maturity or during strong winds (Jones, 2009). Snapping of plant stems can occur due to reduced turgidity during periods of low water availability, which is exacerbated under nematode infection (Coyne *et al.*, 2013). Toppling and snapping lead to extreme losses since the bunch has no value.

Although a wide range of species occur on banana, a few key species are the most damaging: *R. similis* or burrowing nematode, *Pratylenchus coffeae* Sher and Allen and *P. goodeyi* Sher and Allen or root lesion nematodes, *Helicotylenchus multicinctus* (Cobb) Golden or spiral nematodes and *Meloidogyne* spp. or root knot nematodes. Other species found in more localized areas are *Rotylenchulus reniformis* Linford and Oliviera, *Hoplolaimus pararobustus* Schuurmans, Stekhoven and Teunissen and *Paratylenchus minor* Sharma, Sharma & Khan 1986. As with most plants, bananas are invariably affected by a combination of species (Dubois and Coyne, 2011). A major difficulty, however, concerns partitioning losses between the various species when present in combination (Gowen *et al.*, 2005). Traditionally, *R. similis* has been viewed as the most damaging species on banana (Sarah, 2000; Gowen *et al.*, 2005). This is likely based on the nuisance it posed in lowland tropical commercial dessert banana plantations, which resulted in extensive use of carbamate and organophosphate based pesticides for its management (Cianco and Mukerji, 2009). However, *R. similis* does not occur in the higher, more temperate climates at altitudes >1400 m above sea level (asl) in the east African Highlands (Price, 2006) or at high latitudes such as the Canary Islands and Taiwan (Jones, 2009). In west Africa, *R. similis* was previously the key nematode pest species (Spejier and Fogain, 1999) but indications now indicate that *P. coffeae* is becoming more widespread and in places replacing *R. similis* (Coyne, 2009). Uncertainty over *Pratylenchus* species identification has highlighted the need for caution on determining species distribution and prevalence, and consequently for the use of host resistance. For instance, an aggressive *P. coffeae* population in Ghana has subsequently been identified as *P. speijeri*, a morphologically similar, but different species (De Luca *et al.*, 2012). *Pratylenchus goodeyi*, indigenous to Africa, is prevalent at altitudes >1400 m asl in the east Africa highlands and Mt Cameroon, where *R. similis* does not persist (Spejier and Fogain, 1999). However, its pest status is unclear and although may be present in high densities (Peregrine and Bridge, 1992; Speijer and Bosch, 1996), cannot always be correlated with yield losses (Gaidoshova *et al.*, 2009; Dubois and Coyne, 2011). *Pratylenchus goodeyi* has been found on banana in numerous locations, from the Canary Islands, northeast Africa and sub-Saharan Africa to Australia, where it has been reported as damaging (Gowen *et al.*, 2005).

Helicotylenchus multicinctus is found in all banana-producing areas, often in combination with *R. similis* as well as other species, such as *Meloidogyne* spp. (Gowen *et al.*, 2005). Since *H. multicinctus* is mostly present in combination with other species, it is difficult to estimate its effect, and has generally been overlooked in favour of the damage caused by *R. similis*. However, Ssango *et al.* (2004) were able to separate the damage effects of nematode species, demonstrating *H. multicinctus* as a pest in its own right. Accumulating evidence is also indicating that *H. multicinctus* is responsible for large proportions of the damage caused to banana production even if other species are present (Coyne *et al.*, 2013).

Meloidogyne spp. has been found in all banana-producing areas although is always reported in combination with other nematode species. For similar reasons as for *H. multicinctus*, their importance is likely underestimated (Gowen *et al.*, 2005). In some instances, however, they dominate the nematode populations and contribute significantly to production losses.

12.3.1 IPM for nematode management

In essence, the use of clean, healthy planting material is critical to effective nematode management, as this is the primary cause of infections in new plantations. Hot water treatment is a technique that successfully disinfects sucker material and is mostly used in commercial settings. Suckers must be pared or cleaned of infected root and rhizome material. Thereafter, suckers are submerged in hot water (53°C) for 20 min (Colbran, 1967). This technique has since been adapted for ease of implementation and appeal to smallholder farmers, using a dipping period of 30 s in boiling water (Coyne *et al.*, 2010; Hauser and Coyne, 2010). Treatment of pared suckers is an easy and effective technique for sanitizing plant material, and can be used in both commercial and smallholder situations.

The use of sterile tissue cultured plants is an ideal source of clean material. In commercial systems such certified pest-, disease- and virus-free plants are routinely used. However, for smallholder farmers, the use of tissue culture plants is often not an option and may not be available, or may be of suboptimal quality (Dubois and Coyne, 2006; Dubois *et al.*, 2006). Alternatively, macropropagation of plantlets can readily provide healthy planting material, providing correct procedures are respected, such as sterile potting media and disinfecting the corm prior to incubation (Tenkouano *et al.*, 2006). The technique involves the removal of apical dominance and incubation of a banana corm in a plastic covered frame. These produce small plantlets that are then potted for weaning and transfer to farmers.

Post-nematode control consists of applying nematicides to the banana plants. Previously, most nematicides were labelled 'Class 1' pesticides, i.e. extremely hazardous (class 1a) or highly hazardous (class 1b) (WHO, 2006). However, with the removal of these from the market, less hazardous and more environmentally sensitive products have been sought and developed (Zum Felde *et al.*, 2009), such as furfuraldehydes and biologically based solutions, such as endophytes, mycorrhizae and biopesticides

(Meyer and Roberts, 2002; Sikora *et al.*, 2008; Viaene *et al.*, 2013). Addition of manures and compost is also recommended, since this enhances microbial activity in the root zone and boosts plant and root growth. In general, agronomic practices that improve and ameliorate the soil microbial community and soil health are encouraged, and have been shown to reduce nematode damage and losses (Tenkouano *et al.*, 2006).

Nematicides can keep nematode densities below threshold levels, but following replanting, additional methods should be considered, as nematodes are never completely eradicated. Soil sanitation can be obtained by injecting the herbicide glyphosate into banana plants before uprooting. Removal of plant residues, complete fallow and crop rotation also help to reduce nematode densities in the field before replanting (Risède *et al.*, 2009).

For smallholder farmers, post-plant applications are often not an option, underlining the need for healthy planting material. The use of locally grown nematode-resistant cultivars together with healthy planting material is highly desirable (Coyne, 2009). Varying levels of resistance occurs among traditional banana cultivars against nematodes, although the genetic resource base is limited. Breeding for resistance is underway, although complications are encountered due to the range of nematode species to contend with and difficulties with banana sterility (Tenkouano and Swennen, 2004; Lorenzen *et al.*, 2010). Resistance against *R. similis*, for example, will not necessarily confer resistance against *P. coffeae* and *vice versa*. The genetic modification of traditional cultivars for nematode resistance, however, is showing promise (Roderick *et al.*, 2012; Tripathi *et al.*, 2015).

To establish the potential damage of nematodes to a banana plantation, it is important to monitor the population levels and take avoiding action once these pose a threat. Under commercial systems this is often routine, with samples sent to analytical laboratories. For smallholder farming systems, this is not possible. Therefore, depending on the type of nematodes, incidence and severity of root lesions and damage can be used to provide an indication of the level of the nematode threat. If farmers can identify root lesions and damage and appreciate the cause and consequences, they will be more prepared to take action.

12.4 Banana Weevil

The banana weevil (*Cosmopolites sordidus* Germar) feeds solely on bananas. Adults are mostly found between the leaf sheaths and at the base of the banana plant in the soil or in crop residues. The life cycle, distribution and the damage they cause is extensively covered in the review by Gold *et al.* (2001). Dissemination is mostly through infested plant material and similarly for nematode management, the use of healthy planting material is critical to avoid distribution of weevils to new plantations.

Weevil damage results from larvae burrowing through the corm (rhizome), which they prefer, or through the true stem and the pseudostem. The adult beetle may live for up to two years but will most likely die younger, and produces an average 50–70 eggs per year. Larval development rates are temperature-dependent and duration from egg to adult is between 5 and 7 weeks in the tropics, extending as mean temperatures fall. Egg development will not occur below 12°C, limiting distribution to lower altitudes. After hatching, the larvae tunnel into the stem and corm until they pupate close to the surface, from which the adult weevil emerges. Consequently, it is important to chop up fallen pseudostems to hasten their drying and prevent weevils multiplying on it (Robinson and De Villiers, 2007). Weevils are nocturnal and highly susceptible to desiccation (Gold *et al.*, 2001).

Damage to the rhizome and pseudostem is particularly damaging to young banana plants. Older, more established plants are more able to withstand damage, but with extensive tunnelling are prone to stem-snapping and total bunch loss. Disruption to the vascular system, interference with root initiation and root death reduces nutrient and water uptake, reducing plant vigour, delaying flowering, and increasing susceptibility to other pests and diseases. Yield losses are partly due to snapping of the plant and reduced bunch weights. Yield losses of up to 40% have been recorded (Gold *et al.*, 2001).

The banana weevil is cosmopolitan: it occurs in both commercial and smallholder systems, and is present in both tropical and subtropical regions. The banana weevil is considered the most important insect pest of bananas (Jones, 2009). In commercial dessert banana systems, the problem seems less severe than in smallholder systems, possibly because beer, cooking and roasting bananas are more susceptible than dessert bananas, such as cv. Cavendish (Gold and Messiaen, 2000).

Banana weevils are attracted to volatiles released from cut surfaces, and so the freshly cut rhizomes of suckers appear especially susceptible to infections (Gold *et al.*, 2001). Following harvest of a bunch, the pseudostem is usually cut and set aside as a mulch, releasing volatiles which attract the adult banana weevil. Providing the psuedostem is cut up sufficiently to dry out before a life cycle can be completed, this poses no problem, and indeed acts as a trap limiting multiplication through incomplete life cycles. Furthermore, it enables the manual collection of adult weevils from beneath the cut pieces, reducing populations in the field.

12.4.1 IPM for banana weevil

Banana weevils are primarily dispersed through contaminated planting material, accentuating the importance of using healthy planting material, although adult weevils can disperse and infest new plantations when left untreated. Equally important is field sanitation by removing infected plant debris and cutting open pseudostems to hasten drying. In the subtropics, banana weevils tend to hibernate when temperatures fall below 12°C. This facilitates collection of adult weevils manually during this period, which will reduce populations. Pseudostem traps, made using cut pieces, further facilitate collection of weevils and enable producers to monitor populations. In Australia, traps are used for this purpose with five weevils per trap indicating severe damage and the need to employ additional control measures (Mau and Lessing, 2007).

Chemical control methods include chemical sprays around the psuedostem and stem injections but with variable success, and on baiting with pheromones. Including the use of entomopathogenic fungi, especially *B. bassiana*, has also provided some but often variable success; timing of the application appears to be a key determining factor (Robinson and De Villiers, 2007). The endophytic use of *B. bassiana*, however, offers promise (see above). No known effective parasitoids have yet been identified, despite several surveys. In smallholder farming systems, cultural management remains the dominant means of reducing weevil populations (Gold *et al.*, 2001).

12.5 Banana Bunchy Top Virus (BBTV)

BBTV (*Babuvirus*, Nanoviridae) causes banana bunchy top disease (BBTD) and is one of the most economically important diseases of bananas across production areas of Asia, Africa and the Pacific (Hu *et al.*, 1996; Dale and Harding, 1998). Bunched, emerging leaves are symptomatic of infected plants. The new leaves struggle to emerge, are narrower than normal, are wavy and have yellow chlorotic leaf margins. Severely infected plants will not produce a bunch or will have distorted and twisted hands and fingers. Initial symptoms include dark green streaks in the veins of the lower part of the leaf, midrib and leaf stem, while dark green hornlike extensions of the leaf lamina veins can be seen in the narrow, light green zone between midrib and lamina (Magee, 1927). The virus spreads to suckers through the rhizome, resulting in the complete mat becoming infected (Dale and Harding,

1998). As soon as the disease is present, it is difficult to prevent it spreading further. Movement of infected plant material ensures the spread of the virus to new fields and areas.

The virus is transmitted by the banana aphid, *Pentalonia nigronervosa* Coquerel (Robson *et al.*, 2006). Following feeding on an infected plant, it passes the disease to the next plant it feeds on. To break the disease cycle, it is necessary to manage both the aphids and remove infected plants. Banana aphids are also known to feed on *Heliconia* spp. a close relative of banana and flowering ginger (*Zingiber* spp.), although neither seem to harbour the virus.

12.5.1 IPM for BBTD

Once established, the virus is extremely difficult to eradicate, let alone manage. It is therefore of paramount importance to ensure that any movement of plant material involves only uninfected material. In Hawaii, a plant quarantine system prohibits importation of banana planting material into the state without a permit, while movement between several islands is also prohibited (Ferreira *et al.*, 1997). In Australia, strict quarantine procedures and a zero-tolerance policy have helped prevent the establishment of the aphid and BBTV (Thomas *et al.*, 1994; Robson *et al.*, 2006).

Infected plant material should be eradicated immediately. In commercial systems, insecticides and herbicides can be used to treat both the aphids and infected plants, while smallholder systems need to rely on removal of the entire infected mats. Infected material must be disposed of by chopping and drying, decomposing, burning or burial in landfills. It is important to eradicate the aphids to prevent spread of the virus to uninfected plants. It is recommended that alternate aphid hosts, when present in the vicinity of banana plants, are treated with insecticides to control aphids feeding on them (Ferreira *et al.*, 1997). In commercial systems, pesticides are applied regularly while ongoing research ensures the use of newly released chemicals (Robson *et al.*, 2007). In subsistence farming systems, alternatives, such as the application of detergents or soapy water are recommended to manage aphid populations (Ferreira *et al.*, 1997). To prevent rapid re-infestation of virus-tested planting material, control should be executed meticulously across whole production areas (Thomas *et al.*, 1994). Effective management of BBTD is dependent on early detection so that the source can be destroyed (Hooks *et al.*, 2008). Monitoring for aphids is similarly important.

Although no sources of resistance are known in the banana genome, some cultivars are less susceptible than others due to aphid preferences or host morphological factors, with dessert banana cv. Cavendish highly susceptible. When establishing new plantations, planting material should be obtained from BBTV-free areas, while tissue culture plants should be derived from mother plants that have been indexed for BBTV, preferably grown in insect-proof screenhouses (Ferreira *et al.*, 1997).

12.6 Fusarium Wilt of Banana (Panama Disease)

Banana Fusarium wilt (Panama disease), caused by the soilborne fungus *Fusarium oxysporum* f. sp. *cubense* (Foc), is the most destructive disease of banana globally and a particular threat to dessert and beer banana production. The disease was first noticed in Australia, but became endemic in Panama in 1890 and eventually devastated the banana production in Central America and the Caribbean Islands based on dessert cv. Gros Michel (AAA) in the 1950s and 1960s (Pegg *et al.*, 1996). The disease is widespread in banana-growing areas of Asia, Africa, Australia, the South Pacific and Latin America (Robinson and De Villiers, 2007).

Four races have been classified, of which 1, 2 and 4 are considered banana pathogens; race 3 is a problem only on *Heliconia* spp. Race 1 was responsible for the epidemics in dessert banana cv. Gros Michel but also attacks cv. Lady's Finger (AAB), Pome, Maqueno, Silk (AAB) and the hybrid IC2. Race 2 will affect cooking banana, such as cv. Bluggoe (ABB) and some tetraploids, while race 4 affects mostly cv. Cavendish (AAA) but can also attack cultivars affected by race 1 and 2 (Ploetz, 2000). Race 4 has a tropical and subtropical strain of which the tropical is the more virulent. Originally it was only found in the subtropics, but has since been detected in tropical southeast Asia (Ploetz and Pegg, 1999), also northern Australia and recently in isolated locations in Mozambique (AC4TR4, 2014). Race 4 affects not only cv. Cavendish but also threatens plantains (Ploetz, 2000).

Fusarium wilt of banana is a vascular wilt disease invading the vascular tissue through the roots causing discoloration and wilting. The fungus enters the plant through the root tip while rhizome surfaces are regarded as a minor infection site (Beckman, 1990). The fungus then progresses to the rhizome and is most prominent where the stele joins the cortex. The vessels in the roots and rhizomes turn reddish-brown to maroon as the fungus spreads through the tissue. As infection continues, the lines of discoloration are visible along the length of the pseudostem. In severe cases, it may enter the leaf petioles and peduncle, but has yet to be observed in the fruit (Ploetz, 2000; Daly, 2006). External symptoms include wilting and a light yellowing of the lower leaves, being most prominent around the leaf margins. Eventually the leaves turn bright yellow, skirted with dead leaf margins. Progressively more leaves become yellow and die. This often ends with a wreath of dead leaves circling the pseudostem (Daly, 2006).

The disease is transmitted through infected plant material. Suckers are traditionally used for planting and infected material often does not present visual symptoms, creating difficulties when selecting for healthy material. The pathogen can also spread through soil and running water, and on farm tools, footwear and machinery (Daly, 2006).

12.6.1 IPM for Fusarium wilt

Few effective and/or reliable management options exist for this lethal disease. Effective management requires the integration of available options (Blomme *et al.*, 2011), although exclusion and the use of host resistance are considered the most effective (Jones, 2000). A number of cultivars with resistance against one or other race have been developed in the various breeding programmes, especially in Honduras, Brazil and India, with cv. Goldfinger (AAAB) resistant to race 1 and 4 a notable output from Honduras (Moore *et al.*, 1999; Jones, 2000; Kumar *et al.*, 2009). Several Cavendish clones resistant to various races have also been released by the Taiwan Banana Research Institute (Hwang and Ko, 2004). Chemical control has had limited success as the fungus recolonizes the treated areas again. Methyl bromide did reduce the disease incidence significantly in South Africa but was effective for three years only, after which the disease re-occurred (Herbert and Marx, 1990). Injecting the pseudostem with carbendazim and potassium phosphonate also provided some limited success. Treatment of infected soils with calcium compounds and phosphate salts were found to inhibit chlamydospore germination, although care is required as ammonia nitrogen was found to increase the disease (Blomme *et al.*, 2011). Disease-suppressive soils have been identified, where high production levels were maintained with the pathogen present. However, our understanding of the mechanisms for this remain limited, with difficulties experienced when trying to transfer to disease-conducive soils (Ploetz, 2000). Interest in the use of biological control agents continues to rise with some reasonable successes. Applications of endophytic non-pathogenic *F. oxysporum* strains and *Trichoderma* spp., in combination with silicon and organic waste, and rhizospheric and endophytic *Pseudomonas* and *Streptomyces* spp. bacteria hold great promise (Sikora *et al.*, 2008; Kidane and Lang, 2010).

Restricting the movement of infected plant material and use of healthy propagation material and clean farming tools are essential to reduce the spread of the disease. Abandoning infected areas, where possible, will also help in reducing the rate of spread. Planting of less susceptible, or non-host crops, such as macadamia (*Macademia* spp.) and avocado (*Persea americana* Mill.), is practised in the infected areas in South Africa as re-infestation makes banana farming unprofitable (Kidane and Lang, 2010). However, reinfection will easily occur through irrigation water and infected soil (Ploetz, 2000).

12.7 Banana Xanthomonas Wilt

Banana Xanthomonas Wilt (BXW), or banana bacterial wilt, is caused by *Xanthomonas campestris* pv. *musacearum*. The disease originated in Ethiopia on Enset (*Ensete ventriculosum* (Welw.) Chessman), a close relative of *Musa* (Yirgou and Bradbury, 1968). It has since resulted in a major outbreak in Uganda (Tushemereirwe *et al.*, 2003), which spread across the region (Tripathi *et al.*, 2009), where it affects all commonly grown cultivars, although some appear more susceptible than others, such as cv. Pisang Awak (ABB) (Eden-Green, 2004; Tripathi *et al.*, 2009). Transmission is caused by insects, bats and birds that visit infected flowers/bunches from which they contaminate healthy plants (Eden-Green, 2004). Insects are attracted by the fresh scar after the bract drops, on both male and female flowers. Cultivar, growth stage and method of transmission may have an effect on symptoms (Brandt *et al.*, 1997). Cultivars with persistent bracts and flowers, where cushions on which insects could land are not exposed, seem less susceptible to contracting the disease through insect transmission. When insects visit an infected male bud, they become contaminated with oozing bacteria from the wounds and thus transmit the pathogen to the next plant they visit with similar wounds. Insects associated with transmission of BXW are stingless bees (*Plebeina* spp.), honeybees (*Apis mellifera* L.), fruit flies (Drosophilidae), grass flies (Chloropidae) (Tinzaara *et al.*, 2006) and wasps (Fiaboe *et al.*, 2008).

Some cultivars do not produce exudates on the cushions when they shed the male bract, enabling them to evade insect transmitted infection by BXW through the inflorescence, as they remain dry and unattractive to insects. Other cultivars have persistent male flowers and so also evade insect-mediated transmission, but may not possess cell-mediated resistance. These include the east African highland cooking bananas cv. Nakitembe and cv. Mbwazirume. Germplasm with similar inflorescence traits is understood to be present in Indonesia (Tripathi *et al.*, 2009). It is unlikely that infections occur through the female inflorescence since plants are not infected when the male flower is removed as part of routine BXW management practice (Blomme *et al.*, 2005). Above 1700 m asl insect transmission is lower, probably due to reduced insect activity at lower temperatures (Ndungo *et al.*, 2006).

Under more commercial systems greater attention to tool disinfection is observed, to reduce transmission through contaminated tools and soil. However, this is a transmission pathway in less commercial plantations, and at the higher altitudes contaminated tools are a major cause of disease spread (Eden-Green, 2004). Harvesting of leaves for sale and domestic use further contributes to the spread of the disease. Additionally, bacteria also enter the plant through injured roots caused by nematodes, banana weevils or other soilborne pests (Mwangi *et al.*, 2007).

The first symptoms after insect transmission are shrivelling of the male bud bracts and decaying rachis, followed by fruit discoloration and rotting (Eden-Green, 2004). Leaves turn yellow, wilt and eventually become brown and die, along with the plant (Tushemereirwe *et al.*, 2003). Fruit ripens unevenly and prematurely. The pulp of the rotting fruit has rusty brown stains. By splitting open the pseudostem, a yellow-orange discoloration of the vascular bundles and dark brown tissue scarring can be observed, together with a yellow bacterial ooze. Symptoms develop rapidly under favourable field (3–4 weeks) and greenhouse (2–3 weeks) conditions (Tripathi *et al.*, 2009).

Movement of infected plant material is responsible for the spread of the disease over short or long distances.

12.7.1 IPM for Banana Xanthomonas Wilt

Once established, both viral and bacterial diseases are very difficult to eradicate, due to the lack of efficient chemical or curative control options. For BXW, no resistant cultivars have yet been identified. Genetic modification of traditional cultivars for BXW resistance has made excellent progress, however, with several transformed lines showing good resistance and looking promising under field evaluation (Tripathi *et al.*, 2010, 2014). However, a combination of management measures are recommended for durable management, including exclusion, eradication, host resistance and protection. De-budding is an important tool in prevention of the spread of BXW. The bud should be removed following formation of the last hand. This will prevent flower infection and more evenly developed fruit (Blomme *et al.*, 2005).

Tools must be disinfected before working with new planting material, between plantations and between activities. The tools can be sterilized by heating until too hot to touch or by soaking in a sodium hypochlorite solution (one cup of household bleach in 5 cups of water).

Further to initial detection of infections in plants, it is essential to remove and destroy infected plants without delay (Karamura *et al.*, 2005). Roguing and burying, removal of plant material to the outskirts of the plantation to rot and/or using herbicides such as 2,4 D or glyphosate will ensure death of the infected plant material. For this to be effective, people need to be sensitized to the situation and be trained in the identification and early detection of the disease and how to implement suitable control measures to prevent its spread.

12.8 Leaf Spot Diseases

There are numerous fungal leaf spot diseases reported from banana, caused by species of *Mycosphaerella*. Just two cause considerable damage: *Mycosphaerella musicola* Leach (Sigatoka or yellow Sigatoka) and *Mycosphaerella fijiensis* Morelet (black leaf streak (BLS) or black Sigatoka), that results in significant leaf necrosis and yield losses between 33% and 76% (Mobambo *et al.*, 1993). Their names are derived from the Sigatoka valley in Fiji, where severe outbreaks on cv. Cavendish plantations occurred and where *M. fijiensis* was first detected (Leach, 1964). The two diseases are similar, but BLS develops more rapidly, is more virulent, affects a wider range of cultivars (such as plantain) and is more difficult to control than yellow Sigatoka, which is often replaced by the former (Leach, 1964; Jones, 2009). The diseases have effectively reached all banana-growing areas. Their spread is thought to have resulted initially from the uncontrolled movement of infected planting material and leaves followed by the subsequent dispersal of wind-blown spores (ascospores).

Symptoms of the two diseases are similar, and difficult to distinguish with the naked eye. Faint reddish-brown specks appear on the lower surface of third or fourth unrolled leaves. These specks elongate and widen to create reddish-brown streaks, which develop parallel to the leaf veins. Streaks remain more visible on the lower surface, but as they continue to develop they may aggregate and overlap. Streaks become quite visible on the upper surface and darken or blacken, sometimes with a purple tinge on the upper surface, but remain brown on the underside. If streaks are numerous, the whole leaf may blacken, but if less dense, broad streaks may have water-soaked borders or a yellow halo. Eventually the whole leaf blackens; streak centres fade to a clear grey and in time the whole leaf dries up. It usually takes 3–4 weeks after symptoms appear for a leaf to die. Symptom development is dependent on a number of factors, including intensity of infection and climate, host resistance (delays development), and plant age: younger plants are affected more quickly.

12.8.1 IPM for leaf spot leaf streaks

In commercial plantations, a strong emphasis has traditionally focused on the use of chemical fungicides, with non-chemical measures insufficient to provide satisfactory management. Reliance and overuse of chemical pesticides has, however, led to serious concerns over health and environmental issues, providing an impetus to more integrated strategies with cultural practices, and the search for more biologically based options. Knowledge of the disease epidemiology and factors, such as sensitivity to fungicides help in the selection of appropriate fungicides, which should be used in a rotational manner to prevent resistance build-up. Alternating the use of protectant fungicides, such as chlorothalonil and mancozeb with systemic fungicides, provides the basis for fungicide regimes. A sterol biosynthesis inhibitor, tridemorph, several different sterol demethylation inhibitors, especially propiconazole, and the methoxyacrylate, azoxystrobin, are commonly used systemics (Ploetz, 2013). The use of chemicals, however, needs to be integrated, involving good sanitation practices, suitable drainage systems and good agronomic practices. Reducing inoculum levels within the farm is a key factor. By simply removing affected leaves or leaf areas from the plant disease development can be suppressed. Destruction of the leaf outside the plantation, or placing affected leaves top-side down further suppresses spore dispersal. Reducing humidity within the plantation, through the use of an efficient irrigation, canopy aeration and drainage system is also important. Where irrigation is necessary, overhead sprinkling should be avoided in high-risk areas. Interplanting with other non-susceptible crops, and planting in partial shade will result in less severe disease development. In smallholder systems, where pesticides are rarely used for BLS, good sanitation practices together with good agronomy will help. Until an acceptable BLS-resistant cultivar is developed, management of BLS is dependent on reducing the spore density by cultural means, followed by suitable fungicide applications in commercial settings. Unfortunately, resistance to BLS among pre-existing banana genotypes is poor, with Cavendish cultivars particularly susceptible. Resistant cultivars that could be used in subsistence situations are available, but often less productive or desirable. The situation is changing however, with resistance to BLS a key focus for banana-breeding programmes (IITA, 2017). A number of tetraploid hybrids with resistance to BLS have been developed through the hybridization of resistant diploids with triploids (Jones, 2000). Using hybrids with improved resistance to BLS in varietal mixtures with farmers' cultivars, reduced the inoculum load of BLS, enabling the susceptible landraces to perform better (IITA, 2008). Suitable biological control products are yet to make the market.

References

AC4TR4 (2014) *Stellenbosch Declaration on addressing the threat of* Fusarium oxysporum *f. sp.* cubense *tropical race 4 (Foc TR4) to banana production in Africa.* An output from The First Workshop of the African Consortium for Fusarium oxysporum f. sp. cubense Tropical Race 4 (AC4TR4), 10 June. AC4TR4, Stellenbosch, South Africa.

Akello, J., Dubois, T., Gold, C.S., Nakavuma, J. and Paparu, P. (2007) *Beauveria bassiana* Balsamo (Vuillemin) as a potential endophyte in tissue culture banana (*Musa* spp.). *Journal of Invertebrate Pathology* 96, 34–42.

Akello, J., Dubois, T., Coyne, D. and Kyamanywa, S. (2008) Effect of endophytic *Beauveria bassiana* on populations of the banana weevil, *Cosmopolites sordidus*, and their damage in tissue-cultured banana plants. *Entomologia Experimentalis et Applicata* 129, 157–165.

Backman, P.A. and Sikora, R.A. (2008) Endophytes: an emerging tool for biological control. *Biological Control* 46, 1–3.

Beckman, C.H. (1990) Host response to the pathogen. In: Ploetz, R.C. (ed.) *Fusarium Wilt of Banana.* APS Press, St. Paul, Minnesota, pp. 93–105.

Bentley, J.W., Boa, E., Danielsen, S., Franco, P., Antezana, O. *et al.* (2009) Plant health clinics in Bolivia 2000–2009: operations and preliminary results. *Food Security* 1, 371–386.

Blomme, G., Mpiira, S., Ssemakadde, R. and Musaka, H. (2005) Controlling banana *Xanthomonas* wilt through debudding. *InfoMusa* 14, 46.

Blomme, G., Eden-Green, S., Mustaffa, M., Nwauzoma, B. and Thangavelu, R. (2011) Major diseases and their management. In: Pillay, M. and Tenkouano, A. (eds) *Banana Breeding: Constraints and Progress.* CRC Press, Boca Raton, Florida, pp. 85–120.

Brandt, S.A., Spring, A., Hiebsch, C., McCabe, J.T., Tabogie, E. *et al.* (1997) *The Tree Against Hunger: Ensete-based Agricultural Systems in Ethiopia.* American Association for the Advancement of Science, Washington, DC.

Cianco, A. and Mukerji, K.G. (2009) *Integrated Management of Fruit Crops and Forest Nematodes.* Springer, Heidelberg, Germany.

Colbran, R.C. (1967) Hot water tank for treatment of banana planting material. Advisory leaflet. Division of Plant Industry no. 294. Queensland Department of Primary Industries, Brisbane, Australia.

Coyne, D.L. (2009) Pre-empting plant-parasitic nematode losses on banana in Africa: Which species do we target? *Acta Horticulturae* 828, 227–235.

Coyne, D.L., Wasukira, A., Dusabe, J., Rotifa, I. and Dubois, T. (2010) Boiling water treatment: a simple, rapid and effective technique for producing healthy banana and plantain (*Musa* spp.) planting material. *Crop Protection* 29, 1478–1482.

Coyne, D.L., Omowumi, A., Rotifa, I. and Afolami, S.O. (2013) Pathogenicity and damage potential of five plant-parasitic nematode species on plantain (Musa spp., AAB genome) cv. Agbagba. *Nematology* 15, 589–599.

Dale, J.L. and Harding, R.M. (1998) Banana bunchy top disease: current and future stratified for control. In: Hadidi, A., Khetarpal, R.K. and Koganezawa, H. (eds) *Plant Virus Disease Control*. APS Press, St. Paul, Minnesota, pp. 659–669.

Daly, A. (2006) Fusarium *Wilt of Bananas (Panama Disease)* (Fusarium oxysporum *f. sp.* cubense). Agnote No. 151. Department of Primary Industry, Fisheries and Mines, Northern Territory Government, Australia.

De Luca, F., Troccoli, A., Duncan, L.W., Subbotin, S.A., Waeyenberge, L., Coyne, D.L., Brentu, F.C. and Inserra, R.N. (2012) *Pratylenchus speijeri* n. sp., a new root-lesion nematode pest of plantain in West Africa. *Nematology* 14, 987–1004.

Dubois, T. and Coyne, D.L. (2006) Endophytes: natural biodiversity bolstered to combat banana pests. *Geneflow* 2006, 52.

Dubois, T. and Coyne, D.L. (2011) Integrated pest management of banana. In: Pillay, M. and Tenkouano, A. (eds) *Banana Breeding: Constraints and Progress*. CRC Press, Boca Raton, Florida, pp. 121–144.

Dubois, T., Coyne, D.L., Kahangi, E., Turoop, L. and Nsubuga, E.N. (2006) Endophyte-enhanced banana tissue culture: technology transfer through public–private partnerships in Kenya and Uganda. *African Technology Development Forum Journal* 3, 18–23.

Dubois, T., Coyne, D.L. and Zum-Felde, A. (2010) *SP-IPM Technical Innovation Brief 9*. Enhanced protection for tissue cultured banana plants. IITA, Ibadan, Nigeria.

Dubois, T., Dusabe, Y., Lule, L., Van Asten, P., Coyne, D.L. *et al.* (2013) Tissue culture banana (*Musa* spp.) for smallholder farmers: lessons learnt from East Africa. *Acta Horticulturae* 986, 51–60.

Eden-Green, S. (2004) Focus on bacterial wilt: How can the advance of banana Xanthomonas wilt be halted? *InfoMusa* 13, 148–155.

FAOSTAT (2014) Crops. Available at: www.fao.org/faostat/en/#data/QC (accessed 26 July 2017).

Ferreira, S.A., Trujillo, E.E. and Ogata, D.Y. (1997) Banana Bunchy Top Virus. In: *Plant Disease PD-12*. College of Tropical Agriculture and Human Resources, Honolulu, Hawaii, pp. 1–4.

Fiaboe, K.K.M., Beed, F., Mwangi, M., Katembo, M. and Ndungo, V. (2008) Survey of insects visiting banana male buds in eastern Democratic Republic of Congo and their contamination with the bacterium causing wilt. In: *Banana 2008: Banana and Plantain in Africa: Harnessing International Partnerships to Increase Research Impact*. IITA, Ibadan, Nigeria, pp. 97–98.

Gaidoshova, S.V., Van Asten, P., De Waele, D. and Delvaux, B. (2009) Relationship between soil properties, crop management, plant growth and vigour, nematode occurrence and root damage in East African highland banana-cropping systems: a case study in Rwanda. *Nematology* 11, 883–894.

Gold, C.S. and Messiaen, S. (2000) The banana weevil *Cosmopolites sordidus*. *Musa* pest fact sheet no. 4. INIBAP, Montpellier, France.

Gold, C.S., Pena, J.E and Karamura, E.B. (2001) Biology and integrated pest management for the banana weevil *Cosmopolites sordidus* (Germar) (Coleoptera: Curculionidae). *Integrated Pest Management Review* 6, 79–155.

Gowen, S.C., Quénéhervé, P. and Fogain, R. (2005) Nematode parasites of bananas and plantains. In: Luc, M., Sikora R.A. and Bridge, J. (eds) *Plant Parasitic Nematodes in Subtropical and Tropical Agriculture*, 2nd edn. CAB International, Wallingford, UK, pp. 611–643.

Hauser, S. and Coyne, D. (2010) *SP-IPM Technical Innovation Brief 10*. Boiling water treatment of banana and plantain: a hot bath cleans all. IITA, Ibadan, Nigeria.

Hawaiian Banana Industry Association (2010) *Banana IPM Protocol Supporting Documentation*. Available at: www.extento.hawaii.edu/IPM/Certification/banana/banprotocol1.pdf (accessed 26 July 2017).

Herbert, J.A. and Marx, D. (1990) Short-term control of Panama disease in South Africa. *Phytophylactica* 22, 339–340.

Hooks, C.R.R., Wright, M.G., Kabasawa, D.S., Manandhar, R. and Almeida, R.P.P. (2008) Effect of banana bunchy top virus infection on morphology and growth characteristics of banana. *Annals of Applied Biology* 153, 1–9.

Hu, J.S., Wang, M., Sether, D., Xie, W. and Leonhardt, K.W. (1996) Use of polymerase chain reaction (PCR) to study transmission of banana bunchy top virus by the banana aphid (*Pentalonia nigronervosa*). *Annals of Applied Biology* 128, 55–64.

Hwang, S.C. and Ko, W.H. (2004) Cavendish banana cultivars resistant to Fusarium wilt acquired through somaclonal variation in Taiwan. *Plant Disease* 88, 580–588.

IITA (2008) *Annual Report 2007.* International Institute of Tropical Agriculture. IITA, Ibadan, Nigeria.

IITA (2017) Work package 2: pest and disease control. Available at: http://bananabreeding.iita.org/index.php/2017/01/24/work-package-2-pest-and-disease-control (accessed 17 July 2017).

Jones, D.R. (2000) *Diseases of Banana, Abaca and Enset.* CAB International, Wallingford, UK.

Jones, D.R. (2009) Diseases and pest constraints to banana production. *Acta Horticulturae* 828, 21–36.

Karamura, E., Osiru, M. and Blomme, G. (2005) Containing banana *Xanthomonas* wilt. *InfoMusa* 14, 45–46.

Kidane, E.G. and Lang, M.D. (2010) Integrated control of Fusarium wilt of banana (*Musa* spp.). *Acta Horticulturae* 879, 315–322.

Kroschel, J., Beed, F., Garrett, K., Coyne, D., van Etten, J. *et al.* (2013) *Management of Critical Pests and Diseases through Enhanced Risk Assessment and Surveillance and Understanding Climate Impacts through Enhanced Modelling.* CCAFS and CRP-RTB Workshop Report. CGIAR Research Program on Climate Change, Agriculture and Food Security (CCAFS) and on Roots, Tubers and Banana. Copenhagen and Lima. Available at: www.ccafs.cgiar.org (accessed 26 July 2017).

Kumar, N., Damodaran, T. and Krishnamoorthy, V. (2009) Breeding banana for combined resistance to Fusarium wilt and nematodes in Tamil Nadu, India. *Acta Horticulturae* 828, 323–327.

Leach, R. (1964) A new form of banana leaf spot in Fiji, black leaf streak. *World Crops* 16, 60–64.

Lorenzen, J., Tenkouano, A., Bandyopadhyay, R., Vroh, B.I., Coyne, D.L. and Tripathi L. (2010) Overview of banana and plantain (*Musa* spp.) improvement in Africa: past and future. *Acta Horticulturae* 879, 595–604.

Magee, C.J. (1927) *The Control of Banana Bunchy Top.* South Pacific Community Technical Paper no. 150. South Pacific Commission, Nouméa, New Caledonia.

Mau, R.F.L. and Kessing, J.L.M. (2007) Cosmopolites sordidus (Germar). Available at: www.extento.hawaii.edu/kbase/crop/type/cosmopol.htm (accessed 26 July 2017).

Meyer, S.L.F. and Roberts, D.P. (2002) Combinations of biocontrol agents for management of plant-parasitic nematodes and soilborne plant-pathogenic fungi. *Journal of Nematology* 34, 1–8.

Mobambo, K.N., Gauhl, F., Vuylesteke, D., Ortiz, R., Pasberg-Gauhl, C. and Swennen, R. (1993) Yield loss in plantain from black Sigatoka leaf spot and field performance of resistant hybrids. *Field Crops Research* 35, 35–42.

Moore, N.Y., Pegg, K.G., Bentley, S. and Smith, L.J. (1999) Fusarium wilt of banana: global problems and perspectives. In: Molina, A.B., Masdek, N.H.N. and Liew, K.W. (eds) *Banana Fusarium Wilt Management: Towards Sustainable Cultivation.* INIBAP, Kuala Lumpur, pp. 11–30.

Mwangi, M., Bandyopadhyay, R., Ragama, P. and Tushemereirwe, R.K. (2007) Assessment of banana planting practices and cultivar tolerance in relation to management of soil-borne *Xanthomonas campestris* pv. *musacearum. Crop Protection* 26, 1203–1208.

Ndungo, V., Eden-Green, S., Blomme, G., Crozier, J. and Smith, J. (2006) Presence of banana *Xanthomonas* wilt (*Xanthomonas campestris* pv. *musacearum*) in the Democratic Republic of Congo (DRC). *Plant Pathology* 55, 294.

Ortiz, R., Ferris, R.S.B. and Vuylesteke, D.R. (1995) Banana and plantain breeding. In: Gowen, S. (ed.) *Bananas and Plantains.* Chapman and Hall, London, pp. 110–146.

Pegg, K.G., Moore, N.Y. and Bentley, S. (1996) Fusarium wilt of banana in Australia: a review. *Australian Journal of Agricultural Research* 47, 637–650.

Peregrine, W.H.T. and Bridge, J. (1992) The lesion nematode *Pratylenchus goodeyi,* an important pest of ensete in Ethiopia. *Tropical Pest Management* 38, 325–326.

Ploetz, R.C. (2000) Panama disease: a classic and destructive disease of banana. *Plant Health Progress.* Available at: www.plantmanagementnetwork.org/pub/php/management/bananapanama (accessed 26 July 2017).

Ploetz, R.C. (2013) Black Sigatoka of banana: the most important disease of a most important fruit. The American Phytopathological Society. Available at: www.apsnet.org/publications/apsnetfeatures/Pages/BlackSigatoka.aspx (accessed 26 July 2017).

Ploetz, R.C. and Pegg, K.G. (1999) Fusarium Wilt. In: Jones, D.R. (ed.) *Diseases of Bananas, Abaca and Enset.* CAB International, Wallingford, UK, pp. 143–159.

Price, N.S. (2006) The banana burrowing nematode, *Radopholus similis* (Cobb) Thorne in the Lake Victoria region of East Africa: its introduction, spread and impact. *Nematology* 8, 801–817.

Risède, J.M., Chabrier, C., Dorel, M., Rhino, B., Lakhia, K., Jenny, C. and Quénéhervé, P. (2009) Recent and up-coming strategies to counter plant-parasitic nematodes in banana systems of the French West Indies. *Acta Horticulturae* 828, 117–127.

Robinson, J.C. and De Villiers, E.A. (2007) *The Cultivation of Bananas*. ARC-Institute for Tropical and Subtropical Crops, Nelspruit, South Africa.

Robson, J.D., Wright, M.G. and Almeida, R.P.P. (2006) Within-plant distribution and binomial sampling plan of *Pentalonia nigronervosa* (Hemiptera, Aphididae) on banana. *Journal of Economic Entomology* 99, 2185–2190.

Robson, J.D., Wright, M.G. and Almeida, R.P.P. (2007) Biology of *Pentalonia nigronervosa* (Hemiptera, Aphididae) on banana using different rearing methods. *Environmental Entomology* 36, 46–52.

Roderick, H., Tripathi, L., Babirye, A., Wang, D., Tripathi, J., Urwin, P.E. and Atkinson, H.J. (2012) Generation of transgenic plantain (*Musa* spp.) with resistance to plant pathogenic nematodes. *Molecular Plant Pathology* 13, 842–851.

Sarah, J.L. (2000) Burrowing nematode. In: Jones, D.R. (ed.) *Diseases of Banana, Abaca and Enset*. CAB International, Wallingford, UK, pp. 295–303.

Sikora, R.A., Pocasangre, L., Zum Felde, A., Niere, B., Vu, T.T. and Dababat, A.A. (2008) Mutualistic endophytic fungi and *in-planta* suppressiveness to plant parasitic nematodes. *Biological Control* 46, 15–23.

Spejier, P.R. and Bosch, C.H. (1996) Susceptibility of *Musa* cultivars to nematodes in Kagera region, Tanzania. *Fruits* 51, 217–222.

Spejier, P.R. and Fogain, R. (1999) *Musa* and *Ensete* nematode pest status in selected African countries. In: Frison, E., Gold, C., Karamura, E. and Sikora, R.A. (eds) *Mobilizing IPM for Sustainable Banana Production in Africa*. INIBAP, Montpellier, France, pp. 99–108.

Ssango, F., Spejier, P.R., Coyne, D.L. and De Waele, D. (2004) Path analysis: a novel approach to determine the contribution of nematode damage to East African Highland banana (*Musa* spp., AAA) yield loss under two crop management practices in Uganda. *Field Crops Research* 90, 243–253.

Tenkouano, A. and Swennen, R.L. (2004) Progress in breeding and delivering improved plantain and banana to African farmers. *Chronica Horticulturae* 44, 9–15.

Tenkouano, A., Hauser, S., Coyne, D. and Coulibaly, O. (2006) Clean planting materials and management practices for sustained production of banana and plantain in Africa. *Chronica Horticulturae* 46, 14–18.

Thomas, J.E., Iskra-Carvana, M.L. and Jones, D.R. (1994) Banana bunchy top disease. *Musa* Disease Fact Sheet No. 4. INIBAP, Montpellier, France.

Tinzaara, W., Gold, C.S., Ssekiwoko, F., Bandyopadhyay, R., Abera, A. and Eden-Green, S.J. (2006) Role of insects in the transmission of banana bacterial wilt. *African Crop Science Journal* 14, 105–110.

Tripathi, L. (2009) *Biotechnology and Nematodes*. IITA R4D Review, Edition 2. IITA, Ibadan, Nigeria.

Tripathi, L., Abele, S., Aritua, V., Tushemereirwe, W.K. and Bandyopadhyay, R. (2009) Xanthomonas wilt: a threat to banana production in East and Central Africa. *Plant Disease* 93, 440–451.

Tripathi, L., Mwaka, H., Tripathi, J.N. and Tushemereirwe, W.K. (2010) Expression of sweet pepper *Hrap* gene in banana enhances resistance to *Xanthomonas campestris* pv. *musacearum*. *Molecular Plant Pathology* 11, 721–731.

Tripathi, L., Tripathi, J.N., Kiggundu, A., Korie, S., Shotkoski, F. and Tushemereirwe, W.K. (2014) Field trial of *Xanthomonas* wilt disease-resistant bananas in East Africa. *Nature Biotechnology* 32, 868–870.

Tripathi, L., Babirye, A., Roderick, H., Tripathi, J., Changa, C. *et al.* (2015) Field resistance of transgenic plantain to nematodes has potential for future African food security. *Scientific Reports* 5, 8127. DOI:10.1038/srep08127

Tushemereirwe, W.K., Kangire, A., Smith, J., Ssekiwoko, F., Nakyanzi, M., Kataama, D., Musiitwa, C. and Karyeija, R. (2003) An outbreak of bacterial wilt on banana in Uganda. *InfoMusa* 12, 6–8.

Viaene, N., Coyne, D.L. and Davies, K. (2013) Biological and cultural control. In: Moens, M. and Perry, R. (eds) *Plant Nematology*, 2nd edn. CAB International, Wallingford, UK, pp. 346–369.

WHO (2006) *Recommended Classification of Pesticides by Hazard, and Guidelines to Classification*. World Health Organization, Geneva, Switzerland.

Willers, P., Daneel, M.S. and De Jager, K. (2002) Nematodes: banana. In: Van den Berg, M.A., de Villiers, E.A. and. Joubert, P.H. (eds) *Pest and Beneficial Arthropods of Tropical and Non-Citrus Subtropical Crops in South Africa*. Ad Dynamics, Nelspruit, South Africa, pp. 34–43.

Yirugo, D. and Bradbury, J.F. (1968) Bacterial wilt of Enset (*Ensete ventricosum*) incited by *Xanthomonas musacearum* sp. nov. *Phytopathology* 58, 111–112.

Zum Felde, A., Mendoza, A., Cabrera, J.A., Kurtz, A., Schouten, A., Pocasangre, L. and Sikora, R.A. (2009) The burrowing nematode of banana: strategies for controlling the uncontrollable. *Acta Horticulturae* 828, 101–108.

13   Integrated Pest Management in Citrus

Giuseppe E. Massimino Cocuzza* and Carmelo Rapisarda

Dipartimento di Agricoltura, Alimentazione e Ambiente, Università degli Studi, Catania, Italy

13.1 Introduction

Citrus is a typical cultivation of the tropical and subtropical areas of the world. In 2014, citrus was grown in 9,080,780 ha, with a total production of 139,796,997 t (FAOSTAT, 2016), 75% of which is concentrated in ten countries (Table 13.1). The main citrus productions are sweet oranges, mandarins (with tangerines and clementines), grapefruits and lemons/limes, with prevalence and relative importance depending on the geographic area. For instance, in China the production of tangerines, mandarins and clementines is strongly prevalent, whereas the production of sweet oranges is the most widespread in Brazil, USA, Mediterranean countries and South Africa. In Brazil, most orange production is for the production of juice, whereas in Spain and Italy it is for fresh consumption. Mexico and India are the main producer countries of lemons and limes.

Such a diversity of citrus species causes differences in techniques of production, methods of cultivation, management of pests and diseases that often are a direct consequence of climatic variable effects on tree phenology (Davies, 1997). In the tropics, due to the limited temperature fluctuations, flowering and fruiting is almost continuous throughout the year, whereas in the subtropics the seasons are well defined and the flowering is normally limited to once a year (Smith and Peña, 2002). A citrus orchard is a rather stable agroecosystem (Vacante and Bonsignore, 2012), in which diverse organisms are strictly linked to each other, and pest management, which may alter the stability of ecosystems, should be performed in full respect of ecological balances. As in other crops, an irrational use of chemicals may lead to long-term negative effects, both in terms of a pest's explosion or increment of environmental pollution.

Table 13.1. Citrus production and harvested areas in the main producing countries. (From FAOSTAT, 2016.)

Country	Production (in tons)	Harvested area (ha)
China	34,926,500	2,378,650
Brazil	19,073,914	778,003
India	11,146,630	1,018,991
United States	8,537,514	314,724
Mexico	7,823,498	564,063
Spain	7,055,427	300,838
Egypt	4,404,979	185,217
Nigeria	3,783,826	791,524
Turkey	3,783,517	130,497
South Africa	2,715,212	77,615
World	**139,796,997**	**9,080,780**

* Corresponding author e-mail: cocuzza@unict.it

Similarly to other crops, also for citrus IPM has been adopted by an increasing number of producers in several countries. The percentage of citrus orchards in which this strategy is applied has reached 70–80% in USA, Spain and Italy, and 30% in Turkey (Franco *et al.*, 2006). In other main citrus-producing countries, conventional chemical control of pests remains the most common method. Here, IPM is slowly but progressively expanding for reasons including the severe pesticide residue regulations that force exporter countries to adopt alternative systems of pest management, the rising of pesticide costs, the increasing problem of insecticide resistance, the growth of secondary pests and the increasing environmental awareness in the growers. Frequently, the adoption of IPM is supported by guidelines and financial helps by governmental organizations that frequently play a fundamental role in spreading out this strategy.

The citrus agroecosystem hosts numerous and various organisms, whose variability depends on climatic conditions, the cultural and agronomic practices adopted, the citrus species and variety, the stability of the environment and the pest management strategies applied. Most of the pests occurring on citrus develop their life cycle causing no relevant damage, whereas in some cases their injuries to crops are the consequence of environmental modifications that alter the stability of the ecosystem. Among them, in addition to the misuse of pesticides (Hardin *et al.*, 1995; Dutcher, 2007), the introduction of alien species (Franco *et al.*, 2006) and the alteration of natural habitat for planting new crops can be mentioned.

An approach to citrus IPM should be developed following the guidelines defined almost in each country and adapting their indications to the habitat of interest. At present, a number of data are available (i.e. extension service official guidelines, meteorological previsions, pictures for pests and natural enemies identification), which are easy to consult and greatly help in management decisions. A deep knowledge of environmental conditions in which the orchard is located is extremely important in applying the more suitable agricultural practices (appropriate cultivar, tillage, weed management and pruning, etc.).

13.2 Nematodes

Nematodes are worm-shaped organisms of microscopic dimension and invisible to the naked eye. They are naturally present in all soils, living freely as saprophytes, as predators of bacteria, fungi and other nematodes or as phytophagous associated with plants. Most of them occur at multiple trophic levels and are fundamental in the soil food chains (Ferris and Matute, 2003), being considered important in maintaining and enriching soil fertility (Ferris and Bongers, 2006). Nematodes prefer to live in sandy soils, rich of moisture and organic material. Plant parasitic nematodes may infest all parts of the plant, but the majority attack the roots by piercing tissue cells by means of the stylet present in their mouthparts. They may feed from outside (ectoparasite) or inside (endoparasite) the plant tissues. About 200 species of nematode can develop on citrus (Ferguson and Grafton-Cardwell, 2014). However, most of them are polyphagous and their presence on citrus is occasional and not harmful (Verdejo-Lucas and McKenry, 2004).

On citrus, the presence of harmful nematodes on roots is visually evidenced by necrosis, lesions, injuries, and abnormal development of secondary roots. In case of strong attack, the root system reduces its capacity to satisfy the plant's physiological needs (water, minerals, etc.). Symptoms of heavy infestation on adult trees are manifested by a general state of suffering of the tree, lack of vigour, leaves smaller and sometimes chlorotic, stunted growth and limited fruit production; these symptoms are non-specific and attributable to other problems (deficiency of mineral nutrients, water stress, disease caused by pathogens, etc.). Sufferance symptoms are more evident during periods of environmental stress (Duncan *et al.*, 2016).

The nematode species which are more frequently reported as citrus pests are the Citrus nematode, *Tylenchulus semipenetrans* Cobb, and the Burrowing nematode, *Radopholus similis* (Cobb), whereas other species only occasionally and in limited geographic areas may rise to the role of pests (Duncan, 1999, 2005). The economic importance of these parasites is greatly variable and difficult to quantify (Chen *et al.*, 2004). This variability is determined by the age of the tree, the soil texture, the susceptibility of the rootstocks, the abundance of the nematode populations, the presence of opportunistic pathogens and the agricultural practices adopted. Normally, mature and vigorous trees tolerate heavy attacks without showing sufferance symptoms. On the other hand, young trees, when infested by nematodes, may develop significantly more slowly as a consequence of reduction in primary root biomass and total leaf surface (Kallel and B'Chir, 2006; Ferguson and Grafton-Cardwell, 2014). As stated by Duncan *et al.* (2016), in IPM for nematodes it is essential to determine whether and how many nematodes are present in the grove and whether their management is cost-effective. In mature citrus orchards, control of nematodes may start by ascertaining the real size of their populations and correlate it to the damage caused to fruit production (Verdejo-Lucas and McKenry, 2004). For example, for *T. semipenetrans*, vigorous trees can sustain heavy infestations without exhibiting symptoms of sufferance. The abundance of nematodes may be simply determined by sampling soil and root tissues (on 30–45 cm depth) of suspected plants or at 5–10 random points across the field if the orchard is going to be planted (Ferguson and Grafton-Cardwell, 2014). In the case of *R. similis*, the survey must focus on fibrous roots. The best period for sampling on citrus is during the active growing season. A specialized laboratory should carry out the analysis of the samples collected. However, very useful information on practical plant nematology is provided by Coyne *et al.* (2007). Damage thresholds indicated for various citrus growing areas of the world are extremely different, ranging from the low values suggested in South Africa (100 females/g root) and California (400–700 females/g root, depending on the period of the year) to the highest densities recommended in countries of the Mediterranean basin (4000 females/g root). These great differences suggest that the damage threshold should be assessed by correlating manifested symptoms with nematode densities ascertained through the analysis.

In lands in which nematodes are historically a problem, only resistant rootstocks and certified nematode-free plants should be used (Ferguson and Grafton-Cardwell, 2014). Some rootstock hybrids show a good tolerance to nematodes as well as to *Phytophthora* spp. and Citrus Tristeza virus (Jude and Gmitter, 2013). This simple practice has been successful in controlling nematodes (Lee *et al.*, 1999; Verdejo-Lucas *et al.*, 2003), demonstrating that young trees could be preserved during the susceptible period of rapid growth. Controlling the introduction of infested soils or avoiding the movement of large volumes of water with irrigation is equally useful to stop or slow the diffusion of nematodes (Inserra *et al.*, 2005). In adult citrus orchards, the eradication of nematodes is practically impossible. An indirect method to control nematode infestation is to apply all agronomic practices (watering, fertilization, etc.) to maintain trees vigorous and capable to tolerate the nematode populations. Moreover, agronomic practices should be aimed to encourage the development of antagonist organisms of nematodes (bacteria, fungi and predacious nematodes) which are naturally present in the soil.

Chemical control of nematodes is difficult and often the results are unsatisfactory. Moreover, with repeated use of nematocides the risk of groundwater contamination is high (Ferguson and Grafton-Cardwell, 2014). Fumigation is an effective method to apply pre-plantating to control nematodes and other undesirable soil parasites. However, the convenience of this practice should be assessed both in its economic and agronomic aspects. For instance, it has been reported that soil fumigation could lead to stout growth of the plant as a consequence of the suppression of mycorrhiza (Menge *et al.*, 1978). The use of the few nematocides still available commercially should be carefully planned so as to maximize effectiveness. Pesticides must be distributed in the soil around the tree and incorporated mechanically. Local irrigation by a drip system allows nematocides a uniform distribution into the first 30–40 cm of soil. With heavy infestations, repeated treatments are required at 1–3-year intervals (Van Gundy *et al.*, 1982; Duncan *et al.*, 2016).

13.3 Mites

In total, more than 100 mite species (nearly all of them polyphagous, with only a very few exceptions) have been recorded on citrus, but the majority of them have a scarce impact on plants and only a dozen species can be considered as pests (Gerson, 2003; Vacante, 2010a,b). Damage they cause to plants is strictly related to their trophic activity on leaves, stems, twigs, buds and fruits.

Among the rust mites (Acari, Eriophyidae) living on citrus, the citrus rust mite, *Phyllocoptruta oleivora* (Ashmead), and the pink citrus rust mite, *Aculops pelekassi* (Keifer), can cause severe damage in many areas of the world (Vacante, 2016). The two species are very similar. They attack leaves, twigs and fruits, destroying epidermal cells by injecting saliva and sucking out their contents. As a consequence, part of the surface of infested fruits become bronzed and rough, without altering the quality of content. In the same family, the citrus bud mite, *Aceria sheldoni* (Ewing), is worthy of mention for the frequent attacks on lemon (especially in humid, coastal areas), on which it causes typical deformations to flowers and fruits. Within IPM programmes, control is applied by using selective chemicals alone or with white oils after exceeding the threshold of 2–3% of infested fruits (Vacante, 2016).

The broad mite or citrus silver mite, *Polyphagotarsonemus latus* (Banks) (Acari, Tarsonenidae), is a polyphagous species that damages leaves and fruits. At first, terminal leaves, on which the toxic saliva is injected, appear twisted and distorted. Infested fruits show discoloration and bronzing areas on pericarp, especially on the shaded side. In severe cases, attack on premature fruits causes their drop. Population densities of this mite increase especially in areas of the orchard where humidity approaches saturation point.

More than 20 species of Tenuipalpidae are known to live on citrus, yet they are usually not considered as pests. The only exceptions are a few species of the genus *Brevipalpus*, which are primary pests of citrus in the American continent, where they transmit the leprosis viral disease (Childers *et al.*, 2003). The reddish black flat mite, *B. phoenicis* (Geijskes), is among the most important of this mite group. It infests the lower surface of mature leaves, buds, fruits and branches. Feeding may cause yellowish spots on leaves that turn to brown, or may damage fruits with a variety of symptoms (greyish scabby patches, cracked epidermis, lesions of oleiferous glands, etc.). Other flat mites, such as *B. californicus* (Banks), *B. lewisi* McGregor and *B. obovatus* Donnadieu, have similar biology but less pest importance than *B. phoenicis*.

Tetranychidae are the most represented spider mites on citrus, with up to 60 species recorded. The citrus red mite, *Panonychus citri* (McGregor), is an important mite pest on citrus, with almost worldwide spread. It may affect all cultivated citrus, infesting green bark, leaves and fruits. On leaves, it causes a pale stippling on the upper surface, that may enlarge with severe infestation, causing dry or necrotic areas. On fruits, stippling on green fruits tends to disappear with maturation, but when large mite populations attack nearly mature fruits, the stipples become permanent (Vacante, 2016). The cosmopolitan two-spotted spider mite, *Tetranychus urticae* Koch, is a polyphagous species, which infests all cultivated citrus species. In temperate areas, this mite tends to remain at low densities during winter, mostly concealed in protected sites of the trees, and its activity increases in late spring, peaking during summer. Development occurs all year round in tropical and subtropical environments. On citrus leaves, feeding activity reduces photosynthesis and negatively affects transpiration and metabolism of the plant. Infested leaves progressively curl upwards, producing a yellowish chlorotic blistering at the upper side, with a light rusty-brown concavity on the leaf underside. On fruits, the mite causes reddish-brown rusty areas on the epicarp, which may turn to necrosis. Typical attacks on lemon, around the style extremity, create the so-called 'iron nose' of the fruit. Damage is particularly serious under a hot and dry climate, since the mite activity increases water stress effects on plants and could result in leaf and fruit drop. The oriental red mite, *Eutetranychus orientalis* (Klein), and the Texas citrus mite, *E. banksii* (McGregor), attack all cultivated citrus (especially in hot and dry regions) causing similar effects to those by the citrus red mite. On the contrary, the six-spotted spider mite, *E. sexmaculatus* (Riley), is adversely influenced by the dry hot weather and mainly develops in humid coastal regions, where it causes damage to citrus similar to *T. urticae.*

IPM of mites on citrus is primarily based on cultural practices (balanced fertilization, irrigation, new varieties plant cultural diversity and controlled alternative host weeds), aiming at avoiding pest outbreaks and promote the activity of beneficials (Vacante, 2016). Phytoseiids predaceous mites, such as species of the genera *Amblyseius, Euseius, Galendromus, Typhlodromus, Typhlodromalus* and *Typhlodromina*, are among the most important natural enemies living in citrus orchards (Childers and Denmark, 2011; Maoz *et al.*, 2016). However, these predators are particularly effective at low citrus-mite densities rather at moderate to high populations (Childers, 1994). Again, several predaceous insects (coccinellids, staphilinids, thrips, lacewings, dustywings) are voracious and effective to contrast mite infestations (Abad-Moyano *et al.*, 2009). Pathogen fungi (mainly of the genera *Entomophthora* or *Hirsutella*) are frequent mortality factors on citrus mites, whose incidence is particularly relevant during rainy summer periods or foggy autumn weather (Vacante, 2016).

Although various factors are probably involved (fertilizers, irrigation, local climatic condition, etc.), it has been observed that outbreaks of pest mites often follow the use of broad-spectrum pesticides (Vacante, 2016). Therefore, a general reduction of chemical control or the use of selective pesticides should be promoted. Treatments with oil sprays, while controlling other pests (especially scale insects), are usually effective against most citrus mites. Eriophyds and tenuipalpids are susceptible to sulfur (both sprays and dusts). In any case, the correct use of chemicals should be preceded by a precise monitoring of the mite population, based on innovative sampling methods and updated threshold values (Peña *et al.*, 2002; Song *et al.*, 2003; Martinez Ferrer *et al.*, 2006; Childers *et al.*, 2007; Hall *et al.*, 2007; Vacante, 2010b).

13.4 Thrips

Thrips (order Thysanoptera) are small, thin, fringe-winged insects, just visible to the naked eye. With their asymmetrical mouthparts, compacted within a short cone-shaped rostrum, they feed by piercing the plant tissues (leaves, buds, flowers and fruits) and sucking up the cellular fluids. Longo (1985) reports 40 thrips species as phytophagous or zoophagous on citrus worldwide. Of these, *Scirtothrips citri* Moulton, *S. aurantii* Favre and *S. dorsalis* Hood may be considered serious pests, whereas *Pezothrips kellyanus* (Bagnall), *Caliothrips fasciatus* Pergande, *Heliothrips haemorroidalis* (Bouché) and *Frankliniella bispinosa* (Morgan) are reported as pests sporadically. *Frankliniella occidentalis* (Pergande) and *Thrips tabaci* Lindeman, although frequently recovered in the flowers, does not cause major damage.

On citrus, fruits are susceptible to be attacked by thrips from bloom until the fruits reach a 3.5 cm diameter. All thrips stages may attack the fruits, but the second instar larvae are the most damaging, because they tend to take refuge under the sepals from where they carry out their feeding punctures. Damage is greater on citrus varieties that retain the sepals for longer (Webster *et al.*, 2006). Initial symptoms are visible on small fruits after the fall of the sepals, as blackening or dents in the area under the stalk. With the growth of the fruit, damage evolves as scabby, greyish or silvery scars on the rind. These latter ones, for *Scirtothrips* species and *P. kellyanus*, assume a typical shape of ring around the stem or as patch in the contact points between fruits. Scars caused by *H. haemoroidalis* occur rather irregularly from the stem to the central part of the rind (Reuther, 1989), on which spots of black excrement are also clearly visible. Damage caused by thrips on fruits is wholly cosmetic, reducing the quality and market value. Scars should be accurately evaluated before attributing responsibility to thrips. In fact, the rubbing of little fruits (about 7–8 mm in diameter) with leaves or twigs due to the wind causes similar scars; nevertheless, they are of transversal or diagonal shape and irregularly distributed across the rind. Grafton-Cardwell *et al.* (2003) furnished a practical guide to distinguish fruit scarring.

The consistence of the thrips population can greatly change year to year as a consequence of environmental conditions (temperature, humidity, etc.), abundance of flowering (in years of low flowering, thrips tends to concentrate in the few flowers available and damage can be unsustainable), alternative host plants and natural enemies (Navarro-Campos *et al.*, 2013). These latter are numerous and diverse (mites, spiders and insects), but their contribution is controversially evaluated in the control of thrips populations, mainly in years when the pest density is extremely high (Morse and Grafton-Cardwell, 2012). In Spain, a negative correlation between abundance of predatory mites population and fruit damages caused by thrips was recorded (Navarro-Campos *et al.*, 2012). Monitoring should start at blooming until the fruits reach 2 cm in diameter (about 3 weeks). Twice a week, 100 flowers (or fruits)/ha may be chosen and checked with a 5–10× magnifying lens for the presence of immature forms of thrips. The economic threshold varies with regions, citrus species, cultivar and market request. Roughly, the threshold ranges from 5% for Navel Orange (the most susceptible cultivar), lemon and grapefruit to more than 20% for Valencia Orange. In South Africa, yellow traps are used to time chemical treatments against *S. aurantii* (Grout and Richards, 1990), whereas in Cyprus and Turkey the blue or white traps were the most attractive for *P. kellyanus* (Vassiliou, 2010; Elekcioğlu, 2013). However, the main problem is that traps catch also thrips species which are not harmful for citrus and this gives a wrong perception of the real situation in the orchard. When chemical control is necessary, it is recommended to use non-toxic insecticides for beneficial and non-target insects (i.e. bees), with short residual activity (3–7 days). One chemical treatment is normally enough, with an additional one only if strictly necessary and after a careful monitoring (Planes *et al.*, 2015). A study conducted from 1981 to 2003 evidenced that during 1981–1992 the mean economic scarring caused by *S. citri* was 30.2%, whereas from 1993 to 2003 it was only 4.4% (Morse and Grafton-Cardwell, 2012). The authors explain these conflicting data with the decrement between the two periods in chemical treatments adopted to control thrips, both in quantity (numbers of treatments) and quality (less use of broad-spectrum insecticides).

13.5 Psyllids

Commonly known as 'jumping plant-lice', these are sap-sucking hemipterous insects whose small adults possess two pairs of membranous wings and a hind pair of saltatorial legs. Only two psyllid species are known to live on citrus, but they are very active vectors of huanglongbing (HLB) or citrus greening disease (known in three forms, namely *Candidatus* Liberibacter asiaticus, *C. L.* africanus and *C. L.* americanus), the most destructive disease of citrus (Aubert, 1987; da Graca, 1991). This is why they must be regarded carefully and adequate quarantine measures must be applied in order to reduce risks of their spread in non-infested citrus regions.

The Asian citrus psyllid (ACP), *Diaphorina citri* Kuwayama, has mottled brown adults, with forewings almost rounded at the apex and with a brown band along nearly the whole upper margin. Nymphs are initially light yellow and become progressively darker during development. The species is presently distributed in Asia, Africa (Mauritius and Reunion islands), the American continent and some areas of Oceania (CABI, 2016). Adults of the African citrus psyllid (AfCP), *Trioza erytreae* (Del Guercio), have almost similar dimensions of ACP, are orange-yellow in colour, turning to brown with age, and differ from the previous species for the apically pointed and almost uniformly transparent forewings; newly hatched nymphs are yellowish but progressively turn to greenish up to light brown and show a marginal fringe of waxy filaments. The species is of African origin and has been recorded recently for the Iberian peninsula (CABI, 2016; Massimino Cocuzza *et al.*, 2016).

Direct damage to plants is caused by sucking sap and producing honeydew on which sooty mould develops; moreover, the saliva of *T. erytreae* injected into the plant tissues causes the formation of small pit-galls on the leaves. The importance of *D. citri* and *T. erytreae* as pests is related to their efficacy in transmitting HLB. On plants affected by the bacterial disease, leaves turn chlorotic, resembling zinc deficiency, and fruits show an irregular size and colour. The disease leads trees to decline rapidly within a few years.

In citrus areas where the psyllids are present, chemical control is the most commonly used method to suppress psyllid population and to slow down the spread of HLB (van den Berg *et al.*, 1987; Le Roux *et al.*, 2012; Qureshi *et al.*, 2014). However, the environmental cost of treatments should be carefully considered, as well as their side-effects on beneficial insects; also, the rotation of different chemicals to avoid the development of insecticide resistant populations should be evaluated (Qureshi *et al.*, 2014). For both species, a number of predators have been reported, belonging to Hemiptera (Reduviidae, Pentatomidae, Anthocoridae, Miridae), Neuroptera, Coleoptera (Coccinellidae) and Syrphidae (van den Berg *et al.*, 1987; Rakhshani and Saeedifar, 2013; Kondo *et al.*, 2015). However, these are not very effective in avoiding the spread of the disease. In Florida, for biological control of ACP, the eulophid wasps *Tamarixia radiata* (Waterston), the encyrtid wasp *Diaphorencyrtus aligarhensis* (Shafee, Alam and Agarwal) and the convergent ladybirds *Hippodamia convergens* Guérin-Méneville have been introduced (Qureshi *et al.*, 2009; Qureshi and Stansly, 2011; Rohrig *et al.*, 2012). Discrete possibilities of natural control are provided by the eulophid *Tamarixia dryi* (Waterston) on *T. erytreae* in South Africa (van den Berg and Greenland, 2000) and the encyrtid *Psyllaephagus pulvinatus* Waterston (Tamesse *et al.*, 2002). In citrus regions free from these pests, application of severe quarantine measures is necessary. Eradication with insecticides is very difficult and probably not successful; however, it can keep the population as low as possible (Massimino Cocuzza *et al.*, 2016).

13.6 Whiteflies

Whiteflies are tiny insects, with body and wings covered by fine wax. Young stages are flattened, almost subelliptical and frequently covered with wax. After hatching, first instar nymphs initially look for a site to settle, then become sedentary for the rest of their nymphal life. Basically thermophilic, whiteflies are especially diffused in the tropics and subtropics and only a few species live in temperate areas. About 20 species may infest citrus worldwide and some of them are considered serious pests. Damages by whiteflies are caused by suction of phloem sap and production of abundant honeydew on which sooty mould develops. Sometimes, heavy infestation can cause the decay of plants, whereas sooty mould on fruits makes them unmarketable.

The main species that occasionally can reach the pest status on citrus are *Aleurothrixus floccosus* (Maskell) (woolly whitefly), *Dialeurodes citri* (Ashmead) (citrus whitefly), *Singhiella citrifolii* (Morgan) (cloudy-winged whitefly), *Parabemisia myricae* (Kuwana) (Japanese bayberry whitefly) and *Paraleirodes minei* Iaccarino (nesting whitefly) (Mound, 2007; Martin *et al.*, 2000; Longo and Rapisarda, 2014; CABI, 2016). The adults of these species have body and wings covered with white wax. In the genus *Aleurocanthus*, the orange spiny whitefly, *A. spiniferus* (Quaintance), and the citrus blackfly, *A. woglumi* Ashby, both have adults with orange-reddish body and dark blue to grey wings, whereas nymphs are black and covered with many stout dorsal spines. For its wide geographical distribution and broad host plants range, the spiralling whitefly, *Aleurodicus dispersus* Russell, can be also mentioned among potentially important citrus whiteflies (Martin *et al.*, 2000).

On citrus, several natural enemies attack whiteflies and provide a good biological control when undisturbed by insecticide treatments. The harmfulness of the citrus whiteflies is evident mainly in the first years of their introduction into a new area. Afterwards, the control carried out by many natural enemies (both introduced or indigenous) usually brings whitefly populations back to a non-damaging level. The most important and effective biocontrol agents of citrus whiteflies are the parasitic wasps of the hymenopterous families Aphelinidae (especially *Cales* spp., *Encarsia* spp. and *Eretmocerus* spp.) and Platygasteridae (*Amitus* spp.) (Smith and Peña, 2002). In addition, a weak but fair control is provided by generic predators, mainly represented by common lacewings, ladybirds and syrphid flies. Successful biological control of citrus aleurodids programmes are well documented for *D. citri*, by means of *Encarsia lahorensis* (Howard) (Ru & Sailer, 1979; Argov *et al.*, 2000), *A. floccosus* with *Cales noacki* (Howard) and *Amitus spiniferus* (Brethes) (Miklasiewicz and Walker, 1990; Debach and Rose, 1991), *P. myricae* with *Eretmocerus debachi* (Rose and Debach,

1992), *A. woglumi* with *Encarsia perplexa* Huang & Polaszek and *Amitus hesperidum* Silvestri (Lopez *et al.*, 2009; Nguyen *et al.*, 2016), and *A. spiniferus* with *Encarsia smithi* (Silvestri) (van den Berg and Greenland, 1997). Chemical applications are usually not effective, since they only ensure a temporary suppression of whitefly populations and, moreover, depress their natural enemies. When chemical treatment must be applied, the use of simple spray oils is preferred where possible, at a concentration ensuring 1% actual oil in the spray mixture. Particular attention should be given, in such a case, to carefully treat the underside of the leaves, since usually the greatest number of nymphs is found there. Insecticidal soaps are another choice to reduce whitefly populations before introduction of natural enemies.

13.7 Aphids

Aphids or 'plant lice', are small sap-sucking insects having various morphological forms, a relatively complicated life cycle, comprising a parthenogenetic reproduction with which they rapidly colonize host plants (Williams and Dixon, 2007). Damage they cause on trees may be direct (loss of sap, leaves curling and deformation, production of honeydew on which fungi develop) and/or indirect (virus transmission) (Quisenberry and Ni, 2007; Katis *et al.*, 2007).

On citrus, 18 aphid species are recorded worldwide (Blackman and Eastop, 2000). Out of these, only *Aphis spiraecola* Patch, *Aphis gossypii* Glover, *Toxoptera citricidus* (Kirkaldy) and occasionally *T. aurantii* (Boyer de Fonscolombe) are to be considered as pests (Barbagallo *et al.*, 1997), while the remaining species are of secondary importance. The cotton or melon aphid, *A. gossypii*, has a variable coloration, with apterous form that varies from ochreous brown to dark blue-green, whereas in winged forms, head and thorax are blackish and abdomen as in the apterous form. *A. gossypii* is a cosmopolitan and polyphagous species, reported on about 700 host plants (Holman, 2009). Orange, tangerine and clementine are the most infested citrus species, while lemon is less attacked (Barbagallo *et al.*, 2017). Young shoots and leaves infested by colonies of *A. gossypii* appear without serious distortions, and direct damage consists mainly of a possible developmental delay of young plants. In orange crops, the species is an efficient vector of Citrus Tristeza Virus (CTV) on citrus trees propagated on sour orange. *A. spiraecola* is slightly larger than the previous species, green or yellowish-green in colour with cornicles and cauda blackish. In winged form, head and thorax are blackish and abdomen as in the apterous form. Shoots infested by this aphid appear strongly curled and the leaves deformed, as a consequence of the toxic saliva injected. *A. spiraecola* is reported on more than 450 botanical species (Holman, 2009). On young plants or grafted scions, high infestations may considerably retard growth, whereas it is a moderate vector of CTV. The black citrus aphid, *T. citricidus*, is the most efficient vector of CTV (6–25 times more efficient than *A. gossypii*) (Yokomi *et al.*, 1994). Apterous *T. citricidus* are shining reddish-brown or blackish in colour. The species is reported on about 70 host plants (Holman, 2009). The direct damage it causes is usually considered of secondary importance. The brown citrus aphid, *T. aurantii*, is distinguishable from *T. citricidus* by morphological features on antennae, cauda and wings (Blackman and Eastop, 2000), polyphagous (about 260 host plants) (Holman, 2009) and the damage on citrus consists of slight distortion of leaves and impaired growth of shoots, while its ability in CTV transmission is scarce.

A first step in aphids IPM is an accurate and weekly direct field inspection to evaluate the consistence of colonies during the flushing periods of plants. Normally, vigorous mature trees may tolerate infestations on about 40% of new flushes. The infestation period is rather limited in time (2–3 weeks), because the hardening of shoots and leaves induces the differentiation of the winged form that abandon the trees for other suitable hosts. In this way, the trees may easily recover their normal growth without negative consequences. Some cultural practices (in particular, pruning and nitrogen fertilization), stimulating the growth of new flushes, indirectly encourage the increase of aphid population and make tender shoots available for a longer time (Barbagallo *et al.*, 2017). Consequently, these cultural practices should be carried out in a balanced way. Numerous natural enemies (lacewings, ladybirds, syrphid flies and hymenopterous parasitoids) have a detrimental effect on colonies and greatly contribute to reduce infestations. Control of ants that protect and disseminate aphids in new shoots, increasing their infestations in the orchard, is a collateral useful action. In most cases, chemical treatments are not necessary in mature trees; on the contrary, they may be harmful for beneficial insects and economically not justifiable. Differently, for young trees or in new orchards, after an evaluation of 10% of plants per ha, the chemical threshold is reached with 25% of infested shoots for *T. aurantii* and *A. gossypii* or 5–10% for *A. spiraecola* (Barbagallo *et al.*, 2017). The aphicide to apply should be of short persistence and selective, in order to minimize negative side-effects on beneficial and non-target arthropods (Barbagallo *et al.*, 2017).

Crop protection in areas interested by CTV is solely preventative. Nursery rootstock must be virus-free (or inoculated with mild strains of CTV in case of cross-control method) and certified. In areas where CTV is present, the old orchards propagated on sour orange (highly susceptible to CTV) should be quickly replaced with tolerant certified rootstocks (Citrange Troyer or Carrizo, *Poncirus trifoliata*). The control of CTV by chemical treatments on aphid vectors may be useful in protecting nursery stocks, while it is completely ineffective in the field.

13.8 Scale Insects

Scale insects are the most important pests attacking citrus worldwide, with over 330 species reported on these plants (García Morales *et al.*, 2016), with at least a dozen species occurring at damaging level wherever citrus are cultivated. Though mainly based on the feeding activity, through suction of sap and (apart from the armoured scales) production of honeydew, their specific interactions with plants and ecological characters may vary greatly from one group to another, as well as their practical importance and control strategies, which consequently need to be treated separately.

13.8.1 Margarodid scales

The cottony-cushion scale, *Icerya purchasi* Maskell, is the most important and widespread margarodid scale known to feed on citrus. Native to Australia, this species reached California during the late 19th century and from there it rapidly spread throughout all tropical and subtropical regions (CABI, 2016).

Adult females of *I. purchasi* have an orange-red body, covered by white waxy secretion; posteriorly, a large, white ovisac is clearly visible with 16 longitudinal grooves containing 800–1000 eggs. Males are very rare, since the females are usually hermaphroditic and males derive from unfertilized eggs. Newly hatched nymphs are orange-light red with a caudal anal tube where the drops of honeydew are conveyed.

Apart from citrus, *I. purchasi* attacks plants belonging to a wide range of genera, such as *Acacia, Genista, Laurus, Pittosporum, Robinia, Rosa* and many others (García Morales *et al.*, 2016). Newly hatched nymphs leave the ovisac and settle initially on small twigs and leaves, moving to larger twigs in the following stages. When reaching the adult stage, they tend to settle on lignified parts such as branches or trunks. In addition to the suction of sap and production of honeydew, *I. purchasi* injects a toxic saliva into the plant tissues while feeding, causing defoliation, fruit drop, bark lesions, deformations and a general loss of vigour to the plants. The cottony-cushion scale is

actively preyed upon by the vedalia beetle *Rodolia cardinalis* Mulsant, and to a lesser extent by the endoparasitic fly *Cryptochaetum iceryae* (Williston) (Quezada and DeBach, 1973; Caltagirone and Doutt, 1989). Yet, when kept under biological control by its natural enemies, it rarely reaches high infestation level. Due to the extreme diffusion and efficacy of the vedalia beetle, chemical control is discouraged because of its negative effects on the predator (Grafton-Cardwell and Gu, 2003); white mineral oil, eventually activated with organophosphates, may be used only in case of high infestation of the scale insect and during overwintering of the vedalia beetle.

13.8.2 Mealybugs

Mealybugs (Pseudococcidae) are dimorphic, with females often covered by mealy wax and showing typical waxy filaments arranged around their soft and broadly oval body. Males are winged, elongated with reduced mouth parts.

Among the several species recorded on *Citrus* worldwide (García Morales *et al.*, 2016), the citrus mealybug, *Planococcus citri* (Risso), is considered worldwide a primary pest, whereas other species are of secondary or local importance, such as the long-tailed mealybug, *Pseudococcus longispinus* Targioni, the cryptic mealybug, *Pseudococcus cryptus* Hempel (=*P. citriculus* Green), the citrophilus or scarlet mealybug, *Pseudococcus calceolariae* (Maskell), the pink hibiscus mealybug, *Maconellicoccus hirsutus* (Green), the obscure mealybug *Pseudococcus viburni* (Signoret), and the spherical mealybug, *Nipaecoccus viridis* (Newstead) (Franco *et al.*, 2004). Mealybugs attack plants by sucking sap, producing large amounts of honeydew that cause the growth of sooty mould. The fruit infested during the phase of swelling, exhibits yellowing and soiling by sooty mould, whereas the leaves appear yellowed chlorotic. A very important role is played by ants in disseminating these insects to optimize honeydew productions.

Females of the citrus mealybug, *P. citri*, may be recognized by the presence of 18 pairs of lateral waxy filaments having increasing length from the head to the abdomen. On citrus, nymphs usually overwinter on the trunks, under the bark, and move to the canopy as soon as average temperature exceeds 15°C. In early summer, little colonies of the mealybug may be noticed near to the calyx of little fruits or inside fruit clusters, especially in some susceptible varieties, and its population may reach high densities within only a few weeks (Franco *et al.*, 2001). When not fertilized, the females begin to emit the sex pheromone. After hatching, the nymphs wander (from a few hours to three days) searching for a place to settle and start to feed. In this phase, mortality caused by abiotic factors (rain, wind, temperature changes) is high.

The long-tailed mealybug, *P. longispinus*, derives its common name from the caudal waxy filaments in the adult females, which are usually longer than the other ones surrounding the body. It is cosmopolitan and extremely polyphagous. Damage on citrus is similar to that caused by *P. citri*, but it is generally considered as a minor pest (Franco *et al.*, 2004).

P. cryptus is reported as a damaging mealybug in Israel and Turkey (Blumberg *et al.*, 1999; Holat *et al.*, 2014), particularly on grapefruit and lemon. This species, morphologically similar to *P. citri* (with which it often shares the same feeding sites), preferably settles on leaves, fruits, sprouts, trunks and even roots, causing trees to decline when occurring at very high levels (Blumberg *et al.*, 1999). *P. calceolariae* is often found in mixed populations with *P. citri*, having similar biological and ecological characters. *Maconellicoccus hirsutus* is a polyphagous species, considered of secondary importance on citrus (Chong *et al.*, 2015).

Outbreaks of mealybugs are correlated to various factors, with pesticides playing a key role but with other ones being importantly involved, such as darkness and humidity inside a dense canopy, unbalanced nitrogenous fertilization or different susceptibility among varieties (Franco *et al.*, 2004). To control citrus mealybugs, IPM strategies must be applied, mainly based on cultural methods and biological control through natural enemies. Chemicals should be used only in special situations. A wide range of predators and parasitoids are reported in the literature to attack citrus mealybugs and their efficacy may be determinant to control infestations (Franco *et al.*, 2004). They occur in many insect groups, such as dustywings (Neuroptera Coniopterygidae), ladybirds (Coleoptera Coccinellidae), flies (Diptera Camaemydae) and wasps (Hymenoptera Encyrtidae). For instance, predators such as the exotic mealybug ladybird, *Cryptolaemus montrouzieri* Mulsant, artificially introduced into many citrus-growing areas infested by *P. citri*, give a satisfactory predatory activity also on other citrus mealybug species (Hodek and Honěk, 2009). Adults of *C. montrouzieri* should be released as close as possible to the infestation sites of the mealybugs, to permit a quick contact with the prey. Better results can be obtained when the action of predators is completed through augmentative release of parasitoids, which, unlike predators, have a great specificity. Important results for *P. citri* control is given by the encyrtid wasp *Leptomastix dactylopii* Howard, a parasitoid easily available from most biological control suppliers (Franco *et al.*, 2004). It is also worth mentioning the satisfactory activity shown in Israel by the encyrtid *Clausenia purpurea* Ishii, introduced from Japan to control *P. cryptus* (Blumberg *et al.*, 1999), as well as the activity of *Tetracnemoidea peregrina* (Compere) and *Anagyrus fusciventris* (Girault) on *P. calceolariae* (Laudonia and Viggiani, 1986). Facilitating aeration within plants through regular pruning is another method that contributes to the reduction of mealybug populations.

Chemical control may be justified only in case of failure of the abovementioned methods. The availability of the sex pheromone of *P. citri* and *P. cryptus* allows the monitoring of the population dynamic of mealybugs and ensures direct chemical treatment against the most vulnerable nymphal stages. The best time to intervene is summer, when the percentage of infested fruits (counted collecting 10 fruits on 10% of the citrus trees) is over 10% of total examined fruits. In this case, light mineral oil (1–2%) with or without an organophosphate, can be sprayed, taking into consideration the possibility of applying a side control of ants and their nests (Franco *et al.*, 2004).

13.8.3 Soft scales

Soft scales (Coccidae) develop mainly on perennial plants of which they infest almost any live organ, including leaves, twigs, trunk and roots. Host specificity is not the rule in this insect group, as they are widely polyphagous, with about 90 species occurring on citrus (García Morales *et al.*, 2016). Soft scales cause damage to plants by sucking sap and through copious honeydew production, on which sooty mould develops and reduces photosynthesis and respiration, and affects the quality and marketability of fruits (Gill, 1997). Moreover, the abundant honeydew produced attracts ants, which enhance infestations disturbing the numerous natural enemies or by spreading the scales among plants (Buckley and Gullam, 1991).

The most common soft scales of citrus are the brown soft scale, *Coccus hesperidum* L., the green coffee scale, *Coccus viridis* (Green), the grey citrus scale, *Coccus pseudomagnoliarum* (Kuwana), the black scale, *Saissetia oleae* (Olivier), the chinese wax scale, *Ceroplastes sinensis* Del Guercio, the fig wax scale, *C. rusci* (L.), the Florida wax scale, *C. floridensis* Signoret, the pyriform scale, *Protopulvinaria pyriformis* (Cockerell) and the cottony camellia scale, *Pulvinaria floccifera* (Westwood). They are all cosmopolitan species and can sometimes achieve a role of important pests of citrus especially in some tropical and subtropical areas (Camacho and Chong, 2015). Among citrus-feeding *Ceroplastes*, the pink wax scale, *C. rubens* Maskell and the white wax scale, *C. destructor* Newstead, are worthy of mention, in spite of their minor distribution. In most areas, these species are potential major pests

of citrus, but ecological conditions and especially non-biotic factors play an important role in regulating their populations and influencing their demography.

Normally, high damage levels by these insects are reached when the ecological balance is altered, with the consequent reduction of efficacy of the numerous natural biotic mortality factors. The first step to control soft scales is the continuous monitoring of their populations by periodic visual scouting of plants or by using sticky traps. Frequently, populations of soft scales gradually increase as a result of minor crop care (pruning, unbalanced fertilization, etc.). Pruning is an effective method in creating unfavourable conditions for pests and reducing infestation. All over the world, many natural enemies (both native and artificially introduced in the frame of various programmes) can provide efficient control of soft scales. A more significant action is shown by predators, as for the chrysopids, coccinellids (*Chilocorus* spp., *Exochomus* spp., *Rhyzobius lophanthae* Blaisdell), larvae of noctuid moths, or parasitoid chalcidoid wasps (especially *Coccophagus* spp., *Diversinervus* spp., *Metaphycus* spp., *Microterys* spp., *Scutellista* spp., *Tetrastychus* spp.) (Lampson and Morse, 1992; Tena *et al.*, 2007; Chamacho and Chong, 2015). Therefore, until the population density of soft scales on citrus is lower than the intervention threshold, the activity of entomophagous insects should be preserved. The control of ants favours the activity of beneficial insects and consequently provides a reduction of soft scales populations (Vanek and Potter, 2010). During certain seasons of the year, entomopathogenic fungi may play an important role as natural mortality factors on citrus, as in the case of *Verticillium lecanii* (Zimmermann) Viégas. However, natural enemies do not always satisfactorily control soft scales (Grafton-Cardwell and O'Connell, 2006) and the use of chemicals may be sometimes necessary. As a general rule, for soft scales the economic threshold of intervention is defined as 5–10% or more of green twigs infested (with at least 1 female per 10 cm of twig). In this case, and with a scarce occurrence of natural enemies, an oil spray applied in the season when young stages or crawlers are prevalent is usually enough for undertaking a satisfactory control.

13.8.4 Armoured scales

About 112 species of armoured scales (Diaspididae) are associated with citrus (García Morales *et al.*, 2016). These insects can severely damage plants through the injection of toxic saliva, which seriously interacts with plant physiology. Armoured scales usually are firmly attached to the infested branches, leaves and fruits, on which they can develop dense colonies. On fruits, the feeding sites of the scales do not ripen and form lighter spots which affect their marketability. Heavy infestation causes foliar yellowing and, in some case, the death of the plant (Grafton-Cardwell *et al.*, 2015). Most species of the armoured scales are polyphagous and widely spread in almost all the world citrus areas.

The California red scale, *Aonidiella aurantii* (Maskell), is the most important and harmful armoured scale on citrus. Its mature females have an almost round-shaped armour, clear brown, almost translucent. Reproduction is ovoviviparous and the crawlers, after their emergence from under the female, move around until they find a suitable place to settle and secrete a white circular cover ('white cap' stage). From the second instar, development starts to differ with sex and males form an elongated and usually clearer cover. The adult males are thin-winged insects, yellowish-orange in colour. The yellow scale, *Aonidiella citrina* (Coquillet), is very similar to *A. aurantii*, with which it can be confused. In California, *A. citrina* appears to prefer citrus growing in the arid, warmer valleys and foothills of the interior, whereas *A. aurantii* infests citrus groves in the coastal regions (Gill, 1997); *A. citrina* is considered as a secondary citrus pest (Longo *et al.*, 1994).

The chaff scale, *Parlatoria pergandii* Comstock, and the black parlatoria scale,

Parlatoria ziziphi (Lucas), are worth being mentioned for their diffusion and for their occasional damages in some localized areas, mostly on orange, tangerine and clementine (Miller and Davidson, 2005). In Israel, it has been observed that *P. pergandii* is limited by low relative humidity and high temperatures (Gerson, 1977). An almost cosmopolitan distribution is shown also by the citrus mussel scale, *Lepidosaphes beckii* (Newman), and the mussel-shell scale, *Lepidosaphes gloverii* (Packard) (García Morales *et al.*, 2016), characterized by the elongated and brownish female armour. These polyphagous pests may infest all species and varieties of citrus. In some countries (e.g. South Africa, Spain and Greece), *L. beckii* is considered a major pest (Miller and Davidson, 2005; Stathas *et al.*, 2015; García Morales *et al.*, 2016). Attacks to citrus by the most polyphagous scale insect, the oleander scale, *Aspidiotus nerii* Bouché, are mainly limited to lemon, on which it can be recognized by the round-shaped, pale brown armour of the females, with the yellow nymphal exuviae located centrally or slightly laterally.

The Florida or circular red scale, *Chrysomphalus aonidum* (L.), and the Spanish red scale, *Chrysomphalus dictyiospermi* Morgan, are widespread species considered as primary pests in some citrus areas and minor in others (Rose, 1990; Danzig and Pellizzari, 1998); females' armour in these species is nearly circular, reddish-brown to shiny black in colour for *C. aonidum* and greyish or reddish-brown for *C. dictyiospermi*. The female armour of the citrus snow scale, *Unaspis citri* (Comstock), is brownish to black with a grey border, oystershell-shaped and bearing a central longitudinal ridge. Its common name derives from the white armour of males and the visual effect given by its aggregations on tree trunks or main branches. *U. citri* is an oliphagous species with a remarkable preference for *Citrus* spp., on which it is regarded as a pest in nearly all tropical and subtropical regions (CABI, 2016). A very similar species, the arrowhead or Japanese citrus scale, *Unaspis yanonensis* (Kuwana), is an important citrus pest in Asia and is now spreading to Mediterranean Europe and the Middle East (Campolo *et al.*, 2013). The pest feeds almost exclusively on *Citrus* spp., on which it may cause serious damage. The rufus scale, *Selenaspidus articulatus* (Morgan), is a serious pest of citrus with tropicolitan distribution (Rose, 1990). The armour of the adult female is semicircular, greyish-white and with exuviae in the centre (García Morales *et al.*, 2016).

Natural control of armoured scales by non-biotic (rain, wind, temperature, mainly during the dispersion phase of crawlers) and biotic factors (beneficial insects, entomopathogens) plays a fundamental role and often is enough to maintain their population under the threshold of economic damage. Among parasitoids (Hymenoptera Aphelinidae and Encyrtidae) and predators (Coleoptera Coccinellidae, Neuroptera Crysopidae, Diptera Cecidomyidae and predatory mites of the families Hemisarcoptidae and Phytoseiidae) are worth mentioning as the main natural enemies of armoured scales. Usually, numerous parasitoids are associated with each armoured scale species and their efficacy varies according to the differences of climatic and orchard conditions more or less favourable to their activity. Various beneficial insects are considered to be more effective than others and employed in biological control programmes especially by means of augmentative releases. For example, for *Aphytis melinus* DeBach (Hymenoptera Aphelinidae), which is used worldwide to control *A. aurantii*, the technique consists of an augmentative release of about 180,000 parasitoids/ha per year at several dates from early spring to late summer, at bi-weekly intervals (Moreno and Luck, 1992; Zappalà *et al.*, 2008) and at different points 40 metres distant (Zappalà *et al.*, 2012). Just to mention a few additional cases, effectiveness has been reported also for *C. aonidum* with *Aphytis holoxanthus* De Bach (Steinberg *et al.*, 1987), for *U. citri* with *Aphytis lingnanensis* Compere and the coccinellid *Chilocorus circumdatus* Gyllenhal (Rogers, 2012), and for *L. beckii* with *Aphytis lepidosaphes* Compere (Dean, 1975). Cultural control may be important

too, starting with a choice of an appropriate site of cultivation; regular pruning (to expose crawlers to the action of non-biotic factors), quick removal of heavily infested branches, balanced fertilization and watering are also recommended to reduce infestation risks by these insects (Miller and Davidson, 2005). Heavy infestations (up to 1 female/cm of twig or 4 specimens/fruit) may require chemical control. In this case, mineral oils are recommended for their selectivity and low toxicity to natural enemies, with the possibility of adding organophosphate compounds or insect growth regulator (IGR) if essential. For some species, such as in the case of the California red scale, integrated control strategies must be supported by a monitoring system using pheromone traps, whose capture enables identification of the flight of male scales and the infested areas within the orchard, and can determine the right time for applying both chemical and biological control. Two to four pheromone traps per 4-ha block are used, adding two traps for each additional 4 ha. A critical point is represented by the control of ants which can severely disrupt armoured scale parasitoids (Dao *et al.*, 2014).

13.9 Lepidoptera

Worldwide, numerous species of Lepidoptera may use citrus to develop their larval stages. Damage is due to the feeding activity on tender vegetation or flowers or, in some cases, on fruit peel; yet, on the whole, these pests are not very harmful to mature trees (more than four years old), which are able to tolerate their attacks without negative consequences to growth or fruit production. On the contrary, in nursery and young orchards attacks by larvae to tender vegetation may compromise the normal growth of plants and, in the case of massive infestation, may lead to their death.

On young vegetation, the citrus leafminer (CLM), *Phyllocnistis citrella* Stainton, is considered the most harmful species to all citrus. Native from southeast

Asia, it is distributed worldwide (CABI, 2016). Larvae (yellowish in colour) develop by mining under the cuticle of tender leaves and, less frequently, stems or young fruits. Mines produced by this species are easily recognized by eye, for their progressive width and tortuous shape. As a result, the leaves are deformed, rolled and their photosynthetic activity is reduced. An indirect damage has been recorded in Florida, where the injuries caused by larval mining activity favour citrus canker (*Xanthomonas axonopodis* pv. *citri*) attack on trees (Graham *et al.*, 2004). Adults of *P. citrella* are minute moths, with fringed, silvery wings, on which a dark spot is evident on the terminal edge. The species is controlled by several generalist predators (spiders, lacewings, ants and bugs) and hymenoptera parasitoids (more than 80 species have been recorded worldwide by Shauff *et al.*, 1998). Field observation in mature orchards has demonstrated that natural enemies have a significant role in regulating and stabilizing the CLM population at a role of secondary pest (Heppner and Fasulo, 2016). In orchards, cultural techniques (reduced irrigation and fertilization) advise avoiding an excessive production of new vegetation during the susceptible periods (i.e. summer–autumn), are effective practices to control the pest. Differently, on young and re-grafted trees, attacks by *P. citrella* should be controlled accurately. A good practice is to cover the plants during the more susceptible periods with non-woven fabric, so as to interpose physical barriers between the vegetation and the pest. As alternative or integration, chemical treatments may be considered with cytotropic or systemic insecticides (Heppner and Fasulo, 2016).

The citrus flower moth, *Prays citri* (Millière) is a damaging pest to flowers, buds and young fruits of lemon and lime. In the Mediterranean basin, with favourable conditions (i.e. flowering in late summer), this moth may cause the falling of a high percentage of flowers. Normally, the high flower production of lemon and lime allows them to tolerate up to 50% infestation without significant economic losses (Mineo *et al.*, 1980).

All other butterflies and moths recorded on citrus are normally considered as minor pests or without relevant economic importance, due to the activity of natural enemies or to correct cultural practices which keep their populations under control.

13.10 Coleoptera

All over the world, numerous Coleoptera species may adapt to live on citrus. Normally, in mature citrus orchards, Coleoptera are considered of secondary importance and no control measure is required. In most cases, a sudden population increase is due to the occurrence of particular environmental factors (e.g. the change of cultural practices), and the problem usually stops with the restoration of the ordinary conditions. The exceptions are represented by *Diaprepes abbreviatus* (L.) and *Anoplophora chinensis* (Förster) (= *malasiaca* (Thomson)).

The root weevil, *D. abbreviatus* is a polyphagous species (Simpson *et al.*, 1996), native to the Caribbean area and presently spreading into Florida and other localized areas of Texas and California (Grafton-Cardwell *et al.*, 2004; Stuart *et al.*, 2006). Major damage on trees is caused by the feeding activity of its larvae on roots of all dimensions; prolonged attacks progressively reduce the capacity of the tree to absorb water and nutrients and the plants appear stunted. In young trees, attacks by the weevil may lead to death (Rogers and Dewdney, 2014). Secondary damage may derive from *Phytophthora* spp., whose infection of roots is promoted by the wounds caused by the beetle. Adults feed on leaves, producing typical notches, but the damage is negligible. Experiences in Florida and California indicate that once *D. abbreviatus* is established in a new area its eradication is particularly difficult and pest control is highly expensive (Jetter and Godfrey, 2009). The presence of adults in orchards may be detected by the typical semicircular notches produced with feeding activity on young leaves, together with the abundant excrement scattered on their surface. Subsequently, adults can be easily detected by shaking the vegetation and observing the leaves early in the morning or in the late afternoon. An accurate investigation on the roots may allow recovering the large, white larvae. Healthy maintenance of the trees, through regular irrigation and fertilization, permits to the plants to better tolerate the pest attacks. Another measure is to avoid movement of ground and vegetal material to prevent further spread of the pest (McCoy, 1999; McCoy and Duncan, 2000). This weevil has several antagonists, mostly occasional predators (stink bug, spiders and ants) and larval parasitoids but their activity is rather weak. The hymenopterous egg parasitoids *Aprostocetus vaquitarum* Wolcott and *Quadrastichus haitiensis* (Gahan), native of the Caribbean and introduced in Florida, appear the most effective (Castillo *et al.*, 2006). Similarly valid seems to be the control by *Steinernema riobravis* (Cabanillas *et al.*, 1994), an entomopathogenic nematode of larval instars of the weevil. An efficacious chemical control may be obtained using mineral oil or/and growth regulators against the eggs laid on the leaves. Oil is also useful to prevent female oviposition. Soil insecticides plus a fungicide to control *Phytophthora* spp. can be applied at the base of the trunks. Promising is the research on transgenic citrus rootstock expressing *Bacillus thuringiensis* toxins to control *D. abbreviatus* (Weathersbee *et al.*, 2006; Ben-Mahmoud *et al.*, 2016).

In its native areas (China and other fareast Asian countries), the citrus longhorned beetle, *A. chinensis* (Förster) is considered a serious pest of citrus. Recently, the beetle has been recovered in the Mediterranean basin (van der Gaad *et al.*, 2010). The pest attacks young or mature trees and the damage is due to the feeding activity of its larvae at the base of the trunks or exposed roots, on which they produce large and deep tunnels in the wood, that compromise the physiological activity of the plant (Chambers, 2002). Plants attacked are more susceptible to fungal or bacterial diseases and breakage of plant parts caused by strong wind. Adults feed on leaves and are considered not harmful. In case of infestation on

citrus orchards, plants heavily infested should be quickly removed, whereas, with initial attack, they may be sanitized by cutting and burning the infested parts to eliminate the larvae. A protective barrier with a fine wire mesh placed all around the trunk is useful to prevent female oviposition (Adachi, 1990). In Japan, the pathogenic fungus *Beauveria brognartii* strongly affected the populations of the longhorn beetle (Kobayashi *et al.*, 1999). In some eradication programmes, systemic insecticides (imidacloprid) injected at the base in both infested and uninfested trees or directly into the holes have been successfully used (Komazaki *et al.*, 1989; USDA, 2002).

13.11 Diptera

Probably, Tephritidae belonging to the genera *Ceratitis, Anastrepha, Bactrocera* and *Dacus*, commonly known as 'fruit flies', are the most harmful pests for citrus (White and Elson-Harris, 1992). Their importance derives from direct damage to fruits, the high costs for control and/or eradication and, finally, the severe quarantine restrictions that limit exports to countries where these flies are absent or apply expensive disinfestation programmes (Da Lima *et al.*, 2007; Goergen *et al.*, 2011). Most fruit flies are polyphagous with similar biological behaviour (Liquido *et al.*, 1991). They develop all year round, are long-lived and their females may lay a high number of eggs in the fruit rind during their lifetime. Adults belonging to the abovementioned genera can be easily discriminated from each other by the colour patterns of their body and wings as well as for some typical morphological features. Larvae are of cylindrical maggot-shape appearance, legless and whitish; they usually lack good discriminatory morphological characters so that species identification through larvae is very difficult. Lastly, the introduction of DNA sequences technique has made possible the rapid identification of species (Haymer *et al.*, 1994; Barr *et al.*, 2012). Adults of the

Mediterranean fruit fly or medfly, *Ceratitis capitata* (Wiedmann), on citrus varieties with thin peel, laid their eggs directly in the pulp, where the larvae develop feeding on it and causing the decay of the fruit. Instead, on varieties with thick peel, the oil emitted by glands devitalizes the eggs laid. However, punctures are equally noxious because they cause at first a change in colour of the oviposition area, a shallow depression, and subsequent acceleration in the ripening process of peel and a premature drop of the fruits. The species is widely spread almost everywhere thanks to its great adaptation capacity. As for other fruit flies, the control of *C. capitata* is very difficult. Various hymenopterous parasitoids are reported worldwide but their efficacy to control the pest is rather low. More effective in limiting the medfly population growth are some cultural practices, such as avoiding crop consociations and destroying the infested fruits collected on tree or on ground. Presently, the control of medfly is based on monitoring the numeric consistence of its populations and their dynamics throughout the year by using chromatic traps that release attractive substances (synthetized chemical molecules like siglure, medlure and trimedlure) and chemical insecticides when populations reach the economic thresholds. Another method to control the pest consists of spraying localized spots with a powerful attractant added to spinosad. The adult flies are attracted, feed on it and subsequently die. This method has the advantage of a very limited quantity of insecticides released in the environment and the absence of negative effects for the biocenosis.

Most of *Bactrocera* spp. are native from southeast Asia, Australia or South Pacific and only some species have spread in other areas of the world. The flies of this genus have similar morphology and biology, with an elevated fecundity (1200–1500 eggs/ female) and a rapid juvenile development (about 16 days). All species are highly polyphagous (about 150 fruit and vegetable species) and the injuries produced on attacked fruits are similar to those caused by *C. capitata*. The most serious pests for citrus are the oriental fruit fly *B. dorsalis* (Hendel), the chinese citrus fly *B. minax* (Enderlein), the Queensland fruit fly *B. invadens* Drew, Tsurata & White, and *B. tryoni* (Froggatt). Once *Bactrocera* spp. are introduced and established in a new area their eradication is very difficult and expensive, as well as their control. The application of severe quarantine measures is important to prevent their introduction in new areas. For their control, indications given above for *C. capitata* are valid.

Anastrepha spp. comprises about 200 species endemic to the American continent (Aluja, 1994). *A. ludens* (Loew) is reported as the most damaging for citrus, except lemon and lime. Basic life cycle is very similar among all species of this genus, with eggs laid in groups below the rind of the fruit, facilitated by the typical long ovipositor (Browning *et al.*, 1995). Their control is very difficult. For years, it was based on catching the adults with McPhail traps and applying broad-spectrum insecticides with hydrolyzed proteins. However, these control methods are not very efficacious. As for other Tephritid, a recent method based on attractants with spinosad has been introduced to control these pests. The collection and destruction of infested fruits is difficult and expensive. With regards to natural enemies, several parasitoids have been introduced in Central and South America and some of these, such as *Fopius arisanus* (Sonan) and *Diachasmimorpha longicaudata* (Ashmead) appear to be promising (Ovruski *et al.*, 2000; Cancino *et al.*, 2009).

13.12 Rodents

Voles, or meadow mice, are a group of important pests in many citrus-growing regions due to their serious attacks to bark around the root crown and the collar area of the trunks, which lead plants to rapidly weaken and even die. Occasionally, *Microtus* spp. may reach high population densities in citrus agrosystems in consequence of changes in cultural practices (application of minimum- or non-tillage techniques or the substitution of traditional surface irrigation with drip rrigation) which can indirectly favour demographic increase of these rodents (Ranchelli *et al.*, 2016). Dense wild vegetation around the base of citrus plants provides the voles with protection along the runways that connect the many shallow burrows of their tunnels, and provide food. Attacks on the bark or roots take place during periods in which the wild vegetation is sparse and these animals search for alternative foods (Ranchelli *et al.*, 2016). In citrus orchards, the reversal of unfavourable environmental conditions (shallow tillage and wild weed control) is sufficient to lower drastically the population of the voles. In highly infested citrus orchards, control of these rodents requires the use of poisoned baits. Wheat kernels poisoned with Chlorophacinone (as 0.25% concentrated oil, aiming at obtain a final product with 0.005% of the anticoagulant) are indicated (Santini, 1997). In order to reduce risks of possible ingestion by other vertebrates, such poisoned baits must be applied directly within the burrows produced by the voles on the ground surface. Treatment of infested areas is to be recommended in autumn, when the bait has the highest attractiveness for the voles. It can be repeated weekly for a few weeks, until the presence of new burrows (which are a valid activity indicator of these rodents) will be noticeably reduced (Rapisarda, 2007).

References

Abad-Moyano, R., Pina, T., Dembilio, O., Ferragut, F. and Urbaneja, A. (2009) Survey of natural enemies of spider mites (Acari: Tetranychidae) in citrus orchards in eastern Spain. *Experimental and Applied Acarology* 47, 49–61.

Adachi, I. (1990) Control methods for *Anoplophora malasiaca* (Thomson) (Coleoptera: Cerambycidae) in citrus groves. II. Application of wire netting for preventing oviposition in the mature grove. *Applied Entomology and Zooolgy* 25, 79–83.

Aluja, M. (1994) Bionomics and management of *Anastrepha. Annual Review of Entomology* 39, 155–178.

Argov, Y., Rössler, Y., Voet, H. and Rosen, D. (2000) Introducing *Encarsia lahorensis* against *Dialeurodes citri* in Israel: a case of successful biological control. *BioControl* 45, 1–10.

Aubert, B. (1987) *Trioza erytreae* Del Guercio and *Diaphorina citri* Kuwayama (Homoptera: Psylloidea), the two vectors of citrus greening disease: biological aspects and possible control strategy. *Fruits* 42, 149–162.

Barbagallo, S., Cravedi, P., Pasqualini, E. and Patti, I. (1997) *Aphids of the Principal Fruit-Bearing Crops.* Bayer/L'Informatore Agrario, Milan, Italy.

Barbagallo, S., Massimino Cocuzza, G., Cravedi, P. and Komazaki, S. (2017) IPM case studies: tropical and subtropical fruit trees. In: van Emden, H.F. and Harrington, R. (eds) *Aphids as Crop Pests.* 2nd edition. CAB International, Wallingford, UK, pp. 663–676.

Barr, N.B., Islam, M.S., De Meyer, M. and McPherson, B.A. (2012) Molecular identification of *Ceratitis capitata* (Diptera: Tephritidae) using DNA sequences of the COI barcode region. *Annals of Entomological Society of America* 105, 339–350.

Ben Mahmoud, S., Ramos, J.E., Shatters Jr, R.G., Hall, D.G., Lapointe, S.L., Niedz, R.P., Rougé, P., Cave, R.D. and Borovsky, D. (2016) Expression of *Bacillus thuringiensis* cytolytic toxin (Cyt2Ca1) in citrus roots to control *Diaprepes abbreviatus* larvae. *Pesticide Biochemistry and Physiology.* DOI:10.1016/j.pestbp.2016.07.006

Blackman, R.L. and Eastop, V.F. (2000) *Aphids on the World's Crops: An Identification and Information Guide,* 2nd edn. Wiley, Chichester, UK.

Blumberg, D., Ben-Dov, Y. and Mendel, Z. (1999) The citriculus mealybug, *Pseudococcus cryptus* Hempel in Israel: history and present situation. *Entomologica* 33, 233–242.

Browning, H., McGovern, R., Jackson, L., Calvert, D. and Wardoski, W. (1995) *Florida Citrus Diagnostic Guide.* Florida Science Source, Lake Alfred, Florida.

Buckley, R. and Gullan, P. (1991) More aggressive ant species (Hymenoptera: Formicidae) provide better protection for soft scales and mealybugs (Homoptera: Coccidae, Pseudococcidae). *Biotropica* 23, 282–286.

Cabanillas, H.E., Poinar, G.O. and Raulston, J.R. (1994) *Steinernema riobravis* n. sp. (Rhabditida: Steinernematidae) from Texas. *Fundamental and Applied Nematology* 17, 123–131.

CABI (2016) *Invasive Species Compendium.* CAB International, Wallingford, UK. Available at: www.cabi.org/isc

Caltagirone, L.E. and Doutt, R.L. (1989) The history of the vedalia beetle importation to California and its impact on the development of biological control. *Annual Review of Entomology* 34, 1–16.

Camacho, E. and Chong, J.-H. (2015) General biology and current management approaches of soft scale pests (Hemiptera: Coccidae). *Journal of Integrated Pest Management* 7, 17. DOI:10.1093/jipm/pmv016

Campolo, O., Malacrinò, A., Maione, V., Laudani, F., Chiera, E. and Palmeri, V. (2013) Population dynamics and spread of *Unaspis yanonensis* in Calabria, Italy. *Phytoparasitica* 41, 151–157.

Cancino, J., Ruiz, L., Montoya, P. and Harris, E. (2009) Biological attributes of three introduced parasitoids as natural enemies of fruit flies, genus *Anastrepha* (Diptera: Tephritidae). *Journal of Applied Entomology* 133, 181–188.

Castillo, J., Jacas, J.A., Peña, J.A., Ulmer, B.J. and Hall, D.G. (2006) Effect of temperature on life history of *Quadrastichus haitiensis* (Hymenoptera: Eulophidae), an endoparasitoid of *Diaprepes abbreviatus* (Coleoptera: Curculionidae). *Biological Control* 36, 189–196.

Chambers, B. (2002) *Citrus Longhorned Beetle Program, King County, Washington: Environmental Assessment, April 2002.* Available at: https://www.aphis.usda.gov/plant_health/ea/downloads/clb.pdf (accessed 26 October 2017).

Chen, Z.X., Chen, S.Y. and Dickson, D.W. (2004) *Nematology: Advances and Perspectives.* CAB International, Wallingford, UK.

Childers, C.C. (1994) Biological control of phytophagous mites on Florida citrus utilizing predatory arthropods. In: Rosen, D., Bennett, V. and Capinera, J.L. (eds) *Pest Management in the Subtropics: Biological Control, a Florida Perspective.* Intercept, Andover, UK.

Childers, C.C. and Denmark, H.A. (2011) Phytoseiidae (Acari: Mesostigmata) within citrus orchards in Florida: species distribution, relative and seasonal abundance within trees, associated vines and ground cover plant. *Experimentalis Applied Acarology* 54, 311–371.

Childers, C.C., French, J.V. and Rodrigues, J.C.V. (2003) *Brevipalpus californicus, B. obovatus, B. phoenicis* and *B. lewisi* (Acari: Tenuipalpidae): a review of their biology, feeding injury and economic importance. *Experimental Applied Acarology* 30, 5–28.

Childers, C.C., McCoy, C.W., Nigg, H.N., Stansly, P.A. and Rogers, M.E. (2007) *Florida Citrus Pest Management Guide: Rust Mites, Spider Mites and Other Phytophagous Mites.* ENY-603, UF, University of Florida, IFAS Extension. Available at: http://edis.ifas.ufl.edu/cg002 (accessed 27 July 2017).

Chong, J.-H., Aristizábal, L.F. and Arthurs, S.P. (2015) Biology and management of *Maconellicoccus hirsutus* (Hemiptera: Pseudococcidae) on ornamental plants. *Journal Pest Management* 6. DOI:10:1093/jipm/pmv004

Coyne, D.L., Nicol, J.M. and Claudius-Cole, B. (2007) *Practical Plant Nematology: A Field and Laboratory Guide.* SP-IPM Secretariat, IITA, Cotonou, Benin.

da Graça, J.V. (1991) Citrus greening disease. *Annual Review of Phytopathology* 29, 109–136.

da Lima, C., Jessp, A.G., Cruickshank, L. and Mansfield, E.R. (2007) Cold disinfestation of citrus (Citrus spp.) for Mediterranean fruit fly (*Ceratitis capitata*) and Queensland fruit fly (*Bactrocera tryoni*) (Diptera: Tephritidae). *New Zealand Journal of Crop and Horticultural Science* 35, 39–50.

Danzig, E.M. and Pellizzari, G. (1998) Diaspididae. In: Kozár, F. (ed.) *Catalogue of Palaearctic Coccoidea.* Hungarian Academy of Sciences. Akaprint Nyomdaipari Kft, Budapest, pp. 172–370.

Dao, H.T., Meats, A., Beattie, G.A.L. and Spooner-Hart, R. (2014) Ant-coccids mutualism in citrus canopies and its effect on natural enemies of red scale, *Aonidiella aurantii* (Maskell) (Hemiptera: Diaspididae). *Bulletin of Entomological Research* 104, 137–142.

Davies, F.S. (1997) An overview of climatic effects on citrus flowering and fruit quality in various parts of the world. In: *Citrus Flowering and Fruiting Short Course,* 1. CREC/IFAS, Lake Alfred, Florida.

Dean, H.A. (1975) Complete biological control of *Lepidosaphes beckii* on Texas citrus with *Aphytis Lepidosaphes. Environmental Entomology* 4, 110–114.

Debach, P. and Rose, D. (1991) *Biological Control by Natural Enemies.* Cambridge University Press, Cambridge, UK.

Duncan, L.W. (1999) Nematode diseases of citrus. In: Timmer, L.W. and Duncan, L.W. (eds) *Citrus Health Management*. APS Press, St. Paul, Minnesota, pp. 136–148.

Duncan, L.W. (2005) Nematode parasites of citrus. In: Luc, M., Sikora, R.A. and Bridge, J. (eds) *Plant Parasitic Nematodes in Subtropical and Tropical Agriculture*. CAB International, Wallingford, UK, pp. 437–466.

Duncan, L.W., Noling, J.W. and Inserra, R.N. (2016) *Florida Citrus Management Guide*, Ch. 14: Nematodes. EENY-606. Institute of Food and Agricultural Sciences, University of Florida, Gainesville, Florida. Available at: http://edis.ifas.ufl.edu (accessed 30 September 2016).

Dutcher, J.D. (2007) A review of resurgence and replacement causing pest outbreaks in IPM. In: Ciancio A., Mukerji K.G. (eds) *General Concepts in Integrated Pest and Disease Management. Integrated Management of Plants Pests and Diseases*, vol. 1. Springer, Dordrecht, The Netherlands.

Elekcioğlu, N.Z. (2013) Color preference, distribution and damage of thrips associated with lemon and orange in Adana, Turkey. *Pakistan Journal of Zoology* 45, 1705–1714.

FAOSTAT (2016) *Statistics Division*. Available at: http://www.fao.org/faostat/en/#home (accessed 26 October 2017).

Ferguson, L. and Grafton-Cardwell, E.E. (2014) *Citrus Production Manual*. University of California Agriculture and Natural Resources Publications.

Ferris, H. and Bongers, T. (2006) Nematode indicators of organic enrichment. *Journal of Nematology* 38, 3–12.

Ferris, H. and Matute, M. (2003) Structural and functional succession in the nematode fauna of a soil food web. *Applied Soil Ecology* 23, 93–110.

Franco, J.C., Russo, A., Suma, P., Silva, E.B., Dunkelblum, E. and Mendel, Z. (2001) Monitoring strategies for the citrus mealybug in citrus orchards. *Bollettino Zoologia Agraria, Bachicoltura, Ser. II* 33, 297–303.

Franco, J.C., Suma, P., Borges da Silva, E., Blumberg, D. and Mendel, Z. (2004) Management strategies of mealybug pests of citrus in Mediterranean countries. *Phytoparasitica* 32, 507–522.

Franco, J.C., Garcia-Marì, F., Ramos, A.P. and Besri, M. (2006) Survey on the situation of citrus pest management in Mediterranean countries. *IOBC/WPRS Bulletin* 29, 335–346.

García-Morales, M., Denno, B.D., Miller, D.R., Ben-Dov, Y. and Hardy, N.B. (2016) ScaleNet: a literature-based model of scale insect biology and systematic (database). Available at: http://scalenet.info (accessed 5 October 2016).

Gerson, U. (1977) The scale-insect *Parlatoria pergandii* Comstock and its natural enemies in Israel. *Boletin del Servicio de Defensa contra Plagas y Ispecion Fitopatologica* 3, 21–53.

Gerson, U. (2003) Acarine pests of citrus: overview and non chemical control. *Systematic & Applied Acarology* 8, 3–12.

Gill, R.J. (1997) *The Scale Insects of California. Part 3: Armored Scales*. Technical Series in Agriculture Biosystematics and Plant Pathology no. 3. California Department of Food and Agriculture, Sacramento, California.

Goergen, G., Vayssières, J.F., Gnanvossos, D. and Tindo, M. (2011) *Bactrocera invadens* (Diptera: Tephritidae), a new invasive fruit fly pest for the Afrotropical region: host plant range and distribution in West and Central African region. *Environmental Entomology* 40, 844–854.

Grafton-Cardwell, E.E. and Gu, P. (2003) Conserving vedalia beetle, *Rodolia cardinalis* (Mulsant) (Coleoptera: Coccinellidae), in citrus: a continuing challenge as new insecticides gain registration. *Journal of Economic Entomology* 96, 1388–1398.

Grafton-Cardwell, E.E. and O'Connell, N.V. (2006) The status of biological control in San Joaquin Valley Citrus. In: Hoddle, M. and Johnson, M. (eds) *The Fifth California Conference on Biological Control*. Riverside, California, pp. 8–15.

Grafton-Cardwell, E.E., O'Connell, N.V., Kallsen, C.E. and Morse, J.G. (2003) Photographic guide to citrus fruit scarring. *Agricultural Natural Resources* 8090, 1–8.

Grafton-Cardwell, E.E., Godfrey, K.E., Peña, J.E., McCoy, C.W. and Luck, R.F. (2004) *Diaprepes Root Weevil*. Publication 8131. Available at: http://anrcatalog.ucanr.edu/Details.aspx?itemNo=8131 (accessed 11 July 2017).

Grafton-Cardwell, E.E., Morse, J.G., O'Connell, N.V., Philips, P.A., Kallsen, C.E. and Haviland, D.R. (2015) *UC Pest Management Guidelines: Citrus*. UC ANR Publication 3441. Available at: www.ipm.ucdavis.edu/PMG/r107301111.html (accessed 10 October 2016).

Graham, J.H., Gottwald, T.R., Cubero, J. and Achor, D.S. (2004) *Xanthomonas axonopodis* pv. *citri* factors affecting successful eradication of citrus canker. *Molecular Plant Pathology* 5, 1–15.

Grout, T.G. and Richards, G.I. (1990) Monitoring citrus thrips, *Scirtothrips aurantii*, with yellow card traps and the effect of latitude on treatment thresholds. *Journal of Applied Entomology* 109, 385–389.

Hall, D.G., Childers, C.C. and Eger, J.E. (2007) Binomial sampling to estimate citrus rust mite (Acari: Eriophyidae) densities on orange fruit. *Journal Economic Entomology* 100, 233–240.

Hardin, M.R., Benrey, B., Coll, M., Lamp, W.O., Roderick, G.K. and Barbosa, P. (1995) Arthropod pest resurgences: an overview of potential mechanisms. *Crop Protection* 14, 3–18.

Haymer, D.S., Tanaka, T. and Teramae, C. (1994) DNA probes can be used to discriminate between tephritid species at all stages of the life cycle. *Journal of Economic Entomology* 87, 741–746.

Heppner, J.B. and Fasulo, T.R. (2016) Citrus Leafminer, *Phyllocnistis citrella* Stainton (Insecta: Lepidoptera: Phyllocnistinae). ISAF Extension, University of Florida Publ. EENY38. Available at: https://edis.ifas.ufl.edu/pdffiles/IN/IN16500.pdf (accessed 28 July 2017).

Hodek, I. and Honěk, A. (2009) Scale insects, mealybugs, whiteflies and psyllids (Hemiptera, Sternorrhyncha) as prey of ladybirds. *Biological Control* 5, 232–243.

Holat, D., Kaydan, M.B. and Muştu, M. (2014) Investigations on some biological characters of *Pseudococcus cryptus* Hempel (Hemiptera: Pseudococcidae on four *Citrus* species. *Acta Zoologica Bulgarica* 6, 35–40.

Holman, J. (2009) Host plant catalogue: Palaearctic region. Springer, Dordrecht, The Netherlands.

Inserra, R., Stanley, J.D., O'Bannon, J.H. and Esser, R.P. (2005) Nematode quarantine and certification programmes implemented in Florida. *Nematologia Mediterranea* 33, 113–123.

Jetter, K.M. and Godfrey, K. (2009) Diaprepes root weevil, a new California pest, will raise costs for pest control and trigger quarantines. *California Agriculture* 63, 121–126.

Jude, W.G. and Gmitter Jr, F.G. (2013) Breeding disease-resistant citrus for Florida: adjusting to the canker/HLB world – part 2: rootstocks. *Citrus Industry* 94 (3), 10–16.

Kallel, S. and B'Chir, M.M. (2006) Effect of citrus nematode, *Tylenchulus semipenetrans*, on morphogenesis of Mailtaise douce orange grafted on sour orange (*Citrus reticulata* Blanco; *Citrus aurantium* L.). *Nematologia Mediterranea* 34, 115–128.

Katis, N.I., Tsitsipis, J.A., Stevens, M. and Powell, G. (2007) Transmission of plant viruses. In: van Emden, H.F. and Harrington, R. (eds) *Aphids as Crop Pests*. CAB International, Wallingford, UK, pp. 353–390.

Kobayashi, S., Onomoto, N., Negoro, M., Okada, H., Nakai, M. *et al.* (1999) Control of white-spotted longicorn beetle, *Anoplophora malasiaca* (Thomson) on citrus by *Beauveria brongniartii*. *Proceedings of the Kansai Plant Protection Society* 41, 65–66.

Komazaki, S., Sakagami, Y., Jolly, G.M. and Seber. G.A.F. (1989) Capture-recapture study on the adult population of the white spotted longicorn beetle, *Anoplophora malasiaca* (Thomson) (Coleoptera: Cerambycidae), in a citrus orchard. *Applied Entomology and Zoology* 24, 78–84.

Kondo, T., González, F.G., Tauber, C.S., Guzmán, Y.C., Mondragon, A.F. and Forero, D. (2015) A checklist of natural enemies of Diaphorina citri Kuwayama (Hemiptera: Liviidae) in the department of Valle del Cauca, Colombia and the world. *Insecta Mundi* 966, 1–14.

Lampson, L.J. and Morse, J.G. (1992) A survey of black scale, *Saissetia oleae* (Hom.: Coccidae) parasitoids (Hym.: Chalcidoidea) in southern California. *Entomophaga* 37, 373–390.

Laudonia, S. and Viggiani, G. (1986) Natural enemies of the citrophilus mealybug (*Pseudococcus calceolaria* Mask) in Campania. *Bollettino Laboratorio Entomologia Agraria 'Filippo Silvestri'* 43, 167–171.

Le Roux, H.F., van Vuuren, S.P., Pretotius, M.C. and Buitendag, C.H. (2012) Short-term systems (African HLB). *Acta Horticulturae* 965, 876–878.

Lee, R.F., Lehman, P.S. and Navarro, L. (1999) Nursery practices and certification programs for budwood and rootstocks. In: Timmer, L.W. and Duncan, L.W. (eds) *Citrus Health Management*. APS Press, St. Paul, Minnesota, pp. 35–46.

Liquido, N.J., Shinoda, L.A. and Cunningham, R.T. (1991) Host plants of Mediterranean fruit fly (Diptera: Tephritidae): an annotated world review. *Miscellaneous Publications of the Entomological Society of America* 77.

Longo, S. (1985) Thrips on citrus groves. In: Cavalloro, V. and di Martino, E. (eds) *Integrated Pest Control in Citrus-groves: Proceedings of the Expert's Meeting – Acireale (Italy), 26–29 March 1985*. Balkema Publishers for the Commission of the European Commission, Rotterdam, The Netherlands, pp. 121–125.

Longo, S. and Rapisarda, C. (2014) Spreading of the citrus nesting whitefly, *Paraleyrodes minei* Iaccarino, in Italian citrus groves. *Bulletin OEPP/EPPO Bulletin* 44 (3), 529–533. DOI:10.111/epp.12146

Longo, S., Mazzeo, G., Russo, A. and Siscaro, G. (1994) *Aonidiella citrina*, nuovo fitofago degli agrumi in Italia. *Informatore fitopatologico* 34, 19–25.

Lopez, F., Kairo, M.T.K., Pollard, G.V., Pierre, C., Commodore, N. and Dominique, D. (2009) Post-release survey to assess impact and potential host range expansion by *Amitus hesperidum* and *Encarsia perplexa*, two parasitoids introduced for the biological control of the citrus blackfly, *Aleurocanthus woglumi* in Dominica. *BioControl* 54, 497–503.

Maoz, Y., Gal, S., Argov, Y., Domeratzky, S., Coll, M. and Palevsky, E. (2016) Intraguild interactions among specialized pollen feeders and generalist phytoseiids and their effect on citrus rust mite suppression. *Pest Management Science* 72, 940–949.

Martin, J.H., Mifsud, D. and Rapisarda, C. (2000) The whiteflies (Hemiptera: Aleyrodidae) of Europe and the Mediterranean Basin. *Bulletin of Entomological Research* 90, 407–448.

Martinez Ferrer, M.T., Jacas Miret, J.A., Ripolles Moles, J.L. and Aucejo Romero, S. (2006) Sampling plans for *Tetranychus urticae* (Acari: Tetranychidae) for IPM decisions on clementines in Spain. *Bulletin IOBC/WPRS* 29, 303.

Massimino Cocuzza, G.E., Urbaneja, A., Hernández-Suárez, E., Siverio, F., Di Silvestro, S., Tena, A. and Rapisarda, C. (2016) A review on *Trioza erytreae* (African citrus psyllid), now in mainland Europe, and its potential risk as vector of huanglongbing (HLB) in citrus. *Journal of Pest Science* 90 (1), 1–17. DOI:10.1007/s10340-016-0804-1

McCoy, C.W. (1999) Arthropod pests of citrus roots. In: Timmer, L.W. and Duncan, L.W. (eds) *Citrus Health Management*. APS Press, St. Paul, Minnesota, pp. 149–156.

McCoy, C.W. and Duncan, L.W. (2000) IPM: an emerging strategy for *Diaprepes* in Florida citrus. In: *Diaprepes Short Course*. Lake Alfred, Florida Agricultural Experimental Station, pp. 90–140. Available at: http://irrec.ifas.ufl.edu/flcitrus/short_course_and_workshop/diaprepes/IPM.shtml (accessed 27 July 2017).

Menge, J.A., Johnson, E.L.V. and Platt, R.G. (1978) Mycorrhizal dependency of several citrus cultivars under three nutrient regimes. *New Phytologist* 81, 553–559.

Miklasiewicz, T.J. and Walker, G.P. (1990) Population dynamics and biological control of the woolly whitefly (Homoptera: Aleyrodidae) on citrus. *Environmental Entomology* 19, 1485–1490.

Miller, D.R. and Davidson, J.A. (2005) Armored scale insect pests of trees and shrubs (Hemiptera: Diaspididae). Comstock Publishing Associates, Cornell University Press, Ithaca, New York.

Mineo, G., Mirabello, E., Del Busto, T. and Viggiani, G. (1980) Catture di adulti di *Prays citri* Mill. (Lep. Plutellidae) con trappole a feromoni e andamento delle infestazioni in limoneti della Sicilia occidentale. *Bollettino Laboratorio Entomologia Agraria 'Filippo Silvestri'* 37, 177–197.

Moreno, D.S. and Luck, R.F. (1992) Augmentative releases of *Aphytis melinus* (Hymenoptera: Aphelinidae) to suppress California red scale (Homoptera: Diaspididae) in Southern California lemon orchards. *Journal of Economic Entomology* 85, 1112–1119.

Morse, J. and Grafton-Cardwell, B. (2012) Management of citrus thrips to reduce the evolution of resistance. *Citrograph* 2, 22–30.

Mound, L.A. (2007) Thysanoptera (thrips) of the world – a checklist. Available at: www.ento.csiro.au/thysanoptera/worldthrips.html

Navarro-Campos, C., Pekas, A., Moraza, M.L., Aguilar, A. and Garcia-Marí, F. (2012) Soil-dwelling predatory mites in citrus: their potential as natural enemies of thrips with special reference to *Pezothrips kellyanus* (Thysanoptera: Thripidae). *Biological Control* 63, 201–209.

Navarro-Campos, C., Pekas, A., Aguilar, A. and Garcia-Marí, F. (2013) Factor influencing citrus fruit scarring caused by *Pezothrips kellyanus*. *Journal of Pest Management* 86, 459–467.

Nguyen, R., Hamon, A.B. and Fasulo, T.R. (2016) Citrus blackfly, *Aleurocanthus woglumi* Ashby (Homoptera: Aleyrodidae). EENY-0042. Institute of Food and Agricultural Sciences, University of Florida, Gainesville, Florida. Available at: http://edis.ifas.ufl.edu (accessed 4 October 2016).

Ovruski, S., Aluja, M., Sivinski, J. and Wharton, R. (2000) Hymenopteran parasitoids on fruit-infesting Tephritidae (Diptera) in Latin America and the Southern United States: diversity, distribution, taxonomic status and their use in fruit fly biological control. *Integrated Pest Management Review* 5, 81–107.

Peña, J.E., Sharp, J.L. and Wysoki, M. (2002) *Fruit Pests and Pollinators: Biology, Economic Importance, Natural Enemies and Control*. CAB International, Wallingford, UK.

Planes, L., Catalán, J., Jaques, J.A., Urbaneja, A. and Tena, A. (2015) *Pezothrips kellyanus* (Thysanoptera: Thripidae) nymphs on orange fruit: importance of the second generation for its management. *Florida Entomologist* 98, 848–855.

Quezada, J.R. and Debach, P. (1973) Bioecological and population studies of the cottony-cushion scale, *Icerya purchasi* Mask., and its natural enemies, *Rodolia cardinalis* Mul. and *Cryptochaetum iceryae* Will., in Southern California. *Hilgardia* 41, 631–688.

Quisenberry, S.S. and Ni, X. (2007) Feeding injury. In: van Emden, H.F. and Harrington, R. (eds) *Aphids as Crop Pests*. CAB International, Wallingford, UK, pp. 331–352.

Qureshi, J.A. and Stansly, P.A. (2011) Three homopteran pests of citrus as preys for the convergent lady beetle: suitability and preference. *Environmental Entomology* 40, 1503–1510.

Qureshi, J.A., Rogers, M.E., Hall, D.G. and Stansly, P.A. (2009) Incidence of invasive *Diaphorina citri* (Hemiptera: Psyllidae) and its introduced parasitoid *Tamarixia radiata* (Hymenoptera: Eulophidae) in Florida citrus. *Journal of Economic Entomology* 102, 247–256.

Qureshi, J.A., Kostyk, B.C. and Stansly, P.A. (2014) Insecticidal suppression of Asian Citrus Psyllid *Diaphorina citri* (Hemiptera: Liviidae) vector of huanglongbing pathogens. *PLOS ONE* 9, e112331.

Rakhshani, E. and Saeedifar, A. (2013) Seasonal fluctuations, spatial distribution and natural enemies of Asian citrus psyllid *Diaphorina citri* Kuwayama (Hemiptera: Psyllidae) in Iran. *Entomological Science* 16, 17–25.

Ranchelli, E., Barfknecht, R., Capizzi, D., Riga, F., Mazza, V., Dell'Agnello, F. and Zaccaroni, M. (2016) From biology to management of Savi's pine vole (*Microtus savii*). *Pest Management Science* 72, 857–863.

Rapisarda, C. (2007) Roditori tellurici negli agrumeti: le arvicole. *Informatore Fitopatologico* 57, 28–30.

Reuther, W. (1989) Biology of citrus insects, mites and mollusks. *The Citrus Industry* V, Division of Agriculture and Natural Resources, University of California, Berkeley, California.

Rogers, M. (2012) Citrus pest spotlight: citrus snow scale. *Citrus Industry* 5, 12.

Rogers, M.E. and Dewdney, M.M. (2014) *Florida Citrus Pest Management Guide*. University of Florida Cooperative Extension Service, Institute of Food and Agricultural Services. SP-43. Available at: http://edis.ifas.ufl.edu/features/handbooks/CPMG.html (accessed 27 July 2017).

Rohrig, E., Hall, D.G., Qureshi, J.A. and Stansly, P.A. (2012) Field release in Florida of *Diaphorencyrtus aligarhensis* (Hymenoptera: Encyrtidae) an endoparasitoid of *Diaphorina citri* (Homoptera: Psyllidae) from mainland China. *Florida Entomologist* 95, 479–481.

Rose, M. (1990) Diaspidid pest problems and control in crops. Citrus. In: Rosen, D. (ed.) *World Crop Pests: Armoured Scale Insects, Their Biology, Natural Enemies and Control*. World Crop Pests, vol. 4B. Elsevier, Amsterdam, pp. 535–541.

Rose, M. and Debach, P. (1992) Biological control of *Parabemisia myricae* (Kuwana) (Homoptera: Aleyrodidae) in California. *Israel Journal of Entomology* XXV–XXVI, 73–95.

Santini, L. (1997) The problem of *Microtus* (*Pitymys*) voles in Italian orchards. *Bulletin OILB/SROP* 20, 21–24.

Shauff, M.E., Lasalle, J. and Wijesekara, G.A. (1998) The genera of Chalcidoidea parasitoids (Hymenoptera: Chalcidoidea) of citrus leafminer, *Phyllocnistis citrella* Stainton (Lepidoptera: Gracillaridae). *Journal of Natural History* 32, 1001–1056.

Simpson, S.E., Nigg, H.H., Coile, N.L. and Adair, R.A. (1996) *Diaprepes abbreviatus* (Coleoptera: Curculionidae): host plant associations. *Environmental Entomology* 25, 333–349.

Smith, D. and Peña, J.F. (2002) Tropical fruit pests. In: Peña, J.E., Sharp, J.L. and Wysoki, M. (eds) *Tropical Fruit Pests and Pollinators*. CAB International, Wallingford, UK.

Song, J.H., Kim, S.N. and Riu, K.Z. (2003) Spatial dispersion and sampling of adults of citrus red mite, *Panonychus citri* (McGregor) (Acari: Tetranychidae) in citrus orchards in autumn season. *Korean Journal of Applied Entomology* 42, 29–34.

Stathas, G.J., Skouras, P.J. and Kontodimas, D.C. (2015) Data on ecology of the purple scale *Lepidosaphes beckii* (Newman) on citrus in Greece. *Bulletin OEPP/EPPO Bulletin* 46, 128–132.

Steinberg, S., Podoler, H. and Rosen, D. (1987) Competition between two parasites of the Florida red scale in Israel. *Ecological Entomology* 12, 299–310.

Stuart, R.J., McCoy, C.W., Castle, W.S., Graham, J.H. and Rogers, M.E. (2006) *Diaprepes, Phytophthora*, and hurricanes: rootstock selection and pesticide use affect growth and survival of 'hamlin' orange trees in a Central Florida citrus grove. *Proceedings of Florida State Horticultural Society* 119, 128–135.

Tamesse, J.L., Messi, J., Silatsa Soufo, E., Kambou, J., Bosco Tiago, A., Ondoua Ndongoa, A. and Dzokou, V.J. (2002) Complexe des parasitoïdes de *Trioza erytreae* (Del Guercio) (Homoptera: Triozidae), psylle des agrumes au Cameroun. *Fruits* 57, 19–28.

Tena, A., Soto, A. and Garcia-Marí, F. (2007) Parasitoid complex of black scale *Saissetia oleae* on citrus and olives: parasitoid species composition and seasonal trend. *Biocontrol* 53, 473–487.

USDA (2002) The Asian Longhorned Beetle: an invasive tree pest. *Animal and Plant Health Inspection Service Program Aid* no. 2182. Available at: www.aphis.usda.gov/publications/plant_health/2016/book-alb.pdf (accessed 27 July 2017).

Vacante, V. (2010a) *Citrus Mites*. CABI International, Wallingford, UK.

Vacante, V. (2010b) Review of the phytophagous mites collected on citrus in the world. *Acarologia* 50, 221–241.

Vacante, V. (2016) *Mites of Economic Plants*. CAB International, Wallingford, UK.

Vacante, V. and Bonsignore, C.P. (2012) Implementation of IPM in citriculture. In: Vacante, V. and Gerson, U. (eds) *Integrated Control of Citrus Pests in the Mediterranean Region*. Bentham Science, Sharja, United Arab Emirates, pp. 28–55.

van den Berg, M.A. and Greenland, J. (1997) Classical biological control of *Aleurocanthus spiniferus* (Hem.: Aleyrodidae) on citrus in Southern Africa. *Entomophaga* 42, 459–465.

van den Berg, M.A. and Greenland, J. (2000) *Tamarixia dryi*, parasitoid of the citrus psylla, *Trioza erytreae*: a review. *African Plant Protection* 6, 25–28.

van den Berg, M.A., Deacon, V.E., Fourie, C.J. and Anderson, S.H. (1987) Predators of the citrus psylla, *Trioza erytreae* (Hemiptera: Triozidae), in the Lowveld and Rustenburg areas of Transvaal. *Phytophylactica* 19, 285–289.

van der Gaad, D.J., Sinatra, G., Roversi, P.F., Loomans, A., Herald, F. and Jukadin, A. (2010) Evaluation of eradication measures against *Anoplophora chinensis* in early stage infestations in Europe. *EPPO Bulletin* 40, 176–187.

van Gundy, S., Garabedian, S. and Nigh, E.L. (1982) Alternatives to DBCP for citrus nematode control. *Proceedings of the International Society of Citriculture*, vol. 1, 9–12 November 1981, Tokyo, pp. 387–390.

Vanek, S.J. and Potter, D.A. (2010) Ant-exclusion to promote biological control of soft scales (Hemiptera: Coccidae) on woody landscape plants. *Environmental Entomology* 39, 1829–1837.

Vassiliou, V.A. (2010) Ecology and behaviour of *Pezothrips kellyanus* (Thysanoptera: Thripidae) on citrus. *Journal of Economic Entomology* 103, 47–53.

Verdejo-Lucas S. and McKenry, M. (2004) Management of the citrus nematode, *Tylenchulus semipenetrans*. *Journal of Nematology* 36, 424–432.

Verdejo-Lucas, S., Galeano, M., Sorribas, F.J., Forner, F.B. and Alcaide, A. (2003) Effect on resistance to *Tylenchulus semipenetrans* of hybrid citrus rootstocks subjected to continuous exposure to high population densities of the nematode. *European Journal of Plant Pathology* 109, 427–433.

Weathersbee III, A.A., Lapointe, S.L. and Shatters, R.G. (2006) Activity of *Bacillus thuringiensis* isolates against *Diaprepes abbreviatus* (Coleoptera: Curculionidae). *Florida Entomologist* 89, 441–448.

Webster, K.W., Cooper, P. and Mound, L.A. (2006) Studies on Kelly's citrus thrips *Pezothrips kellyanus* (Bagnall) (Thysanoptera: Thripidae): sex attractants, host associations and country of origin. *Australian Journal of Entomology* 45: 67–74.

White, I.M. and Elson-Harris, M.M. (1992) *Fruit Flies of Economic Significance: Their Identification and Bionomics*. CAB International, Wallingford, UK.

Williams, I.S. and Dixon, A.F.G. (2007) Life cycle and polymorphism. In: van Emden, H.F. and Harrington, R. (eds) *Aphids as Crop Pests*. CAB International, Wallingford, UK, pp. 69–85.

Yokomi, R.K., Lastra, R., Stoetzel, M.B., Damsteegt, V.C., Lee, R.F. *et al.* (1994) Establishment of the brown citrus aphid (Homoptera: Aphididae) in Central America and the Caribbean Basin and transmission of citrus tristeza virus. *Journal of Economic Entomology* 87, 1078–1085.

Zappalà, L., Campolo, O., Saraceno, F., Grande, S.B., Raciti, E., Siscaro, G. and Palmeri, V. (2008) Augmentative releases of *Aphytis melinus* (Hymenoptera: Aphelinidae) to control *Aonidiella aurantii* (Hemiptera: Diaspididae) in Sicilian citrus groves. *IOBC/WPRS Bulletin* 38, 49–54.

Zappalà, L., Campolo, O., Grande, S.B., Saraceno, F., Biondi, A., Siscaro, G. and Palmeri, V. (2012) Dispersal of *Aphytis melinus* (Hymenoptera: Aphelinidae) after augmentative release in citrus orchards. *European Journal of Entomology* 109, 561–568.

14 Integrated Pest Management in Oil Palm Plantations in Malaysia

Norman Kamarudin*, Siti Ramlah A. Ali, Ramle Moslim, Zulkefli Masijan and Mohd Basri Wahid

Malaysia Palm Oil Board, Biological Research Division, Selangor, Malaysia

14.1 Introduction

Palm oil is the main commodity for Malaysia, highly contributing to the National GDP and providing huge economic gains. The land bank for oil palm plantation in Malaysia in 2015 was more than 5.23 million ha. Being a monocrop, oil palm is intermittently plagued with the attack of insect pests; some of them are primarily pests of other crops but have adapted to attack oil palm. Some of the common insect pests will be treated in this chapter, i.e. bagworms (Lepidoptera: Psychidae), rhinoceros beetle (Coleoptera: Scarabaeidae), nettle caterpillars (Lepidoptera: Limacodidae), termites (Isoptera: Rhinotermitidae) and bunch moth (Lepidoptera: Pyralidae). Each of these insect pests attacks different parts of the oil palm: fronds and leaflets (infested by bagworms and nettle caterpillars), spear/shoot (attacked by rhinoceros beetle), spear, trunk and bole region (by termites) and the fruit bunch (by bunch moth).

14.2 Bagworms

Bagworms (Lepidoptera: Psychidae) are common and major defoliating pests of oil palm (Wood, 1968; Basri *et al.*, 1988; Norman and Basri, 2007). The total infested area with bagworms was more than 37,000 ha throughout Malaysia towards the last decade of the last century (Basri *et al.*, 1988). In the new millennium, the area of infestation had increased by about one-third, to more than 49,000 ha, as reported between 2000 and 2005 (Norman and Basri, 2007). In Malaysia, bagworms associated with oil palm belong to two genera and eight species (Norman *et al.*, 1994), with three common species of bagworms, namely *Metisa plana* Walker, *Pteroma pendula* Joannis and *Mahasena corbetti* Tams. *M. plana* is considered more prevalent in Peninsular/west Malaysia, while *M. corbetti* is more prominent in Sabah and Sarawak/east Malaysia (Basri *et al.*, 1988). The dominance of *M. plana* and *P. pendula* in Peninsular Malaysia and *M. corbetti* in Sabah was similarly reported in a later survey conducted between 2000 and 2005 (Norman and Basri, 2007). However, based on the latter survey, *M. plana* had gained more importance as a pest in Sabah, compared to more than 20 years ago.

The larval stage of the bagworms develops within a bag constructed of leaf fragments (Norman *et al.*, 1994). The wingless adult females remain in their pupal bags

* Corresponding author e-mail: norman@mpob.gov.my

and await for the males, free-flying moths, for mating. The adult moth seeks the female via pheromonal attraction.

14.2.1 Economic impact

Crop loss from a moderate bagworm attack can be quite serious. Foliar damage of 50% could cause a 43% yield loss (Wood *et al.*, 1972) over two subsequent years. A 10–13% damage to leaflets could cause similar yield loss (Basri and Kevan, 1995).

14.2.2 Integrated pest management

Census and monitoring

Higher damage tends to occur in mature palms, more than eight years old. This phenomenon could be associated with the rapid dispersal of bagworms via the overlapping fronds of adjacent palms (Basri *et al.*, 1988). Frequent inspection is required to determine the population densities and life stages of bagworms populations. It is crucial that control measures are conducted as soon as the eggs hatch. This is because younger larvae are more susceptible and easier to control than the older ones. A follow-up census should be conducted every six months after a control operation, to monitor for any population resurgence (Chung and Sim, 1991).

The correct timing of biopesticide or pesticide application is important for effective treatment. Thus, once an outbreak is encountered, it is important to ascertain or predict the increase in population, so that timing for the control measures can be planned.

Beneficial plants

The oil palm environment can be more conducive for the populations of the bagworm's natural enemies. The establishment of nectar producing, beneficial plants serves to prolong the life span of natural enemies and this helps sustainable control of bagworms.

There are four species of nectar producing plants which have been identified as beneficial: *Cassia cobanensis, Antigonon leptopus, Euphorbia heterophylla* and *Turnera subulata* (Ho *et al.*, 2003). Beneficial plants like *C. cobanensis* provide honey as food for the parasitoids. For rapid establishment of natural enemies, beneficial plants can be grown in a mix of *C. cobanensis, A. leptopus* and *T. subulata*, with *E. heterophylla* planted in between the three slower-growing plants (Ho *et al.*, 2003).

Natural enemies – parasitoids and predators

The parasitoids associated with the bagworms are mainly small, wasp-like insects which oviposit within the host (Norman *et al.*, 1996). A previous study has shown that parasitoids are able to reduce the population of *M. plana* (Basri *et al.*, 1995). As for predators, the common predators include a reduviid bug, which preys on *M. plana* larvae, and a clerid beetle, of which both larval and adult stages feed on the bagworms (Norman *et al.*, 1998).

Pheromone trapping

Pheromone trapping can be used to reduce population of the male adults in Integrated Pest Management (IPM) programmes. In several oil palm smallholdings, there was a consistent reduction in the percentage of pupal bags with hatched eggs, after a large number of male moths were captured by sticky traps baited with live females as bait (Norman and Othman, 2006). Mating activities were hampered, which reduced the number of eggs and live larvae in the following generations. Oil palm yield (bunch weight) was recorded to be significantly higher in trapping plots compared to control (Norman *et al.*, 2010).

Chemical control

Bagworms can be controlled by spraying selective chemicals (trichlorfon, chlorantrinilaprole, cypermethrin) or by trunk injection (monocrotophos, methamidophos or acephate) (Lai and Tey, 2009; Chua *et al.*,

2012). Most importantly, chemicals should be applied at the beginning of a generation, as younger larvae are more susceptible than older ones. Therefore, the correct timing is of utmost importance to attain effective control.

For localized areas of bagworm attack, systemic chemicals can be applied by trunk injection of monocrotophos or methamidophos (Chung, 1998). Application of chemicals needs proper supervision and mandatory use of proper clothing and personal protective equipments. For infestations over larger areas (more than 400 ha), aerial spraying with biopesticides is advocated. An authorization from the Malaysian Pesticide Board is required for any aerial spraying operation.

Microbial control (Bacillus thuringiensis)

Unlike chemicals, *B. thuringiensis* is a target-specific microbial pesticide with a limited life span within the environment. This bacterial pathogen is virtually non-toxic to humans, mammals, animals, birds, fish and beneficial insects such as parasitoids, predators and oil palm pollinating weevils, *Elaeidobius kamerunicus* (Mohd Najib, *et al.*, 2007, 2009, 2011, 2012). This biopesticide can be aerially sprayed using aircraft for control of bagworms over a large hectarage (Siti Ramlah *et al.*, 2013). Unlike chemical insecticides, toxicity to non-target organisms is unlikely.

14.3 Rhinoceros Beetle

The rhinoceros beetle, *Oryctes rhinoceros* (L.) (Coleoptera: Scarabaeidae), has been recorded in Malaysia since 1889. With the rapid replanting of oil palm, after the oil palms reach the economic life span of 25 years, it has emerged as a pest of young oil palms in replanting areas. The Malaysian oil palm industry has adopted the zero-burning policy for replanting since the 1990s. This is to avoid the occurrence of haze within the southeast Asian region. The zero-burning replanting method involves shredding of oil palm trunks, stacking them along the inter rows, for the debris to rot

in-situ. In a survey (Norman and Basri, 1997), it was observed that the zero-burning replanting attracts this pest to breed and establish its population within the decomposing oil palm trunks (Norman and Basri, 1995). Another method of replanting, which is called underplanting, involves planting of young palms underneath the old palms due for replanting. However, this practice is even more attractive to the beetle when the poisoned palms are left to rot without proper felling. Standing rotten trunks attracts beetles to breed, thereby causing greater damage to the young palms below.

14.3.1 Mode of spread

Oryctes rhinoceros spreads mainly through the adult beetle stage, as its larvae are quite sessile and live confined within their breeding habitat (oil palm trunks/decomposing materials). Female beetles are attracted by the fresh smell of chipped trunks during the zero-burning replanting. Studies have shown that beetles are able to migrate and breed within the oil palm trunk chips at the onset of replanting (Norman and Basri, 2004).

14.3.2 Ecology

In the oil palm environment, rhinoceros beetles are found breeding in rotting materials, such as oil palm trunk heaps, standing, rotting palms and empty oil palm fruit bunches (Norman and Basri, 1995). The temperatures suitable for larval development are 27–29°C, with relative humidities of 85–95% (Bedford, 1980). As such, *O. rhinoceros* prefers to breed in semi-decayed trunk chips, where moisture content of the trunk plays a significant role in determining successful development of the beetles (Norman *et al.*, 2001).

14.3.3 Economic impact

Rhinoceros beetle primarily attacks immature oil palms, which inevitably will reduce the early yields. Considerably serious attack in mature oil palms has been observed in areas adjacent to a breeding site with a high beetle population. Young palms affected by the beetle have a delayed immaturity period (Liau and Ahmad, 1991). About 25% yield loss over the first two years of production is expected after a serious attack by rhinoceros beetles (Liau and Ahmad, 1991). This is possibly caused by the reduction in the canopy size of more than 15% for moderately serious to high damage levels (Samsudin *et al.*, 1993).

Adult beetles feed on the spear region by boring through petiole bases into the central cabbage. The attacks commonly result in fronds with wedge-shaped gaps or the typical characteristic of fan-shaped fronds. With repeated attacks, the palms can become susceptible to secondary infestation by the red stripe weevil, *Rhyncophorus bilineatus* (Montrouzier) (Coleoptera: Curculionidae) (Sivapragasam and Tey, 1994).

14.3.4 Integrated pest management

Census and monitoring

A census can be conducted on young palms attacked by the rhinoceros beetle. A method of assessing the damage severity levels is by counting the number of damaged or 'little leaves' symptoms which indicate the earlier attack by the beetle at the spear region (Samsudin *et al.*, 1993). Damage is considered to be slight with 1–4 little leaves attacked, but serious with 9–12. Control measures are only applied when 5% of the total area is severely damaged (Ho, *personal communication*). Pheromone traps can be used to monitor beetle population and to decide whether and when control measures have to be started. Captures of more than 3 beetles per trap per night are considered high and require control measures.

Cultural and mechanical methods

Leguminous cover crops (i.e. combination of *Centrosema pubescens*, *Calopogonium caeruleum* and *Pueraria javanica*) should be sown at 5–6 months before planting of the new palms. This combination has been shown effective for a quick and complete coverage of the oil palm residues. The ground covers hamper the flight of the beetle in search of breeding sites (Wood, 1968). Damage is lesser in plots with natural covers compared to those with bare ground. The cover crops also hasten the decomposition of the trunk chips. However, if successive replanting occurs within the same plantation, there is a high probability of a population build-up of this pest. Therefore, it is recommended that a large replanting programme be staggered over a two-year period, in order to avoid a constant and recurring attack by *O. rhinoceros*.

O. rhinoceros prefers breeding in semi-decayed trunk chips. The moisture of the trunk is significant in determining successful development for the beetles. High moisture content (>77%) is essential for the growth and complete development of the larva into pupal stage and adults (Norman *et al.*, 2001).

The beetle population can be reduced by creating unsuitable breeding sites. The trunk chips, normally heaped in between rows during replanting, should be spread out in a single layer to facilitate drying, thus preventing the beetle from breeding on them. In an agronomic trial where trunk chips were spread evenly along the planting rows as mulch (Khalid *et al.*, 1999), the population of *O. rhinoceros* breeding in the trunk chips was found insignificant.

However, if *O. rhinoceros* population is very high, especially in the heaped trunk chips where there is a high accumulation of moisture, pulverization of the trunk heaps may be required. The cost for pulverization is initially high at replanting but reduces as the trunk decomposes over time. The pulverized trunk chips must also be spread thinly so as to facilitate rapid decomposition and reduce the possibility of *O. rhinoceros* breeding.

Pheromone trapping

The Oryctes aggregation pheromone, ethyl-4-methyloctanoate, was confirmed 10 times more effective than ethyl chrysanthemumate in attracting the adults (Hallet *et al.*, 1995). Pheromone traps are useful for monitoring and mass trapping the adults at 1 trap for every 2 ha (Chung, 1997). A medium-sized *Oryctes* population (fewer than 5 individuals per square metre) can be reduced in immature oil palms via trapping. Female beetles were mainly caught at the fringes of a replanting block, suggesting their search for breeding sites (Norman and Basri, 2004). Beetles have been observed to immigrate at the onset of replanting. Therefore, trapping can be initiated earlier at the borders, to trap the incoming beetles, before replanting is conducted. The number of female beetles trapped correlated to the number of second instar larvae in the decomposing heaps at about 2 months of trapping (Norman and Basri, 2004), suggesting that trapping can be used as a monitoring tool for estimating population density in the trunk chips, provided the substrate and environmental conditions are similar.

In a 2005 survey (Norman and Basri, 2007), more than half (57%) of the plantations (N = 199) that responded had used pheromone traps for monitoring and controlling the pest. The average density gathered from this survey was 1 trap in 12 ha, with a majority of the plantations (48%) placing 1 trap for up to 10 ha.

Chemical control

Many chemicals have shown potential for the control of this pest in nursery and immature plantings (Chung *et al.*, 1995; Ho, 1996). For the control of *O. rhinoceros*, lambda-cyhalothrin, cypermethrin, fenvalerate, monocrotophos and chlorpyrifos were all effective in both nursery and field trials (Ho, 1996). These chemicals had significantly reduced *O. rhinoceros* damage after 11 weeks. Although cypermethrin and lambda-cyhalothrin can control *O. rhinoceros*, the use of synthetic pyrethroids are not recommended, since they are highly toxic to the oil palm pollinators. To protect the environment, chemicals are applied only when severe damage occurs (i.e. palms having many damaged fronds and few leaf symptoms). The recommended threshold for chemical control is 5% of the total area having severe damage (Ho, *personal communication*).

Microbial control (Metarhizium spp. and Oryctes rhinoceros virus)

Among all natural enemies (predators, parasitoids and pathogens), the entomopathogenic pathogens, such as fungi, have shown successful and long-term control of pests in many introduction programmes. The fungus *Metarhizium anisopliae* was classified into two varieties (var. *anisopliae* and var. *major*) based on the spore dimensions (Tulloch, 1979). The spore length for var. *anisopliae* is 5.0–8.0 µm while var. *major* is 9.0–15.0 µm and highly pathogenic. In the mid-to-late 1970s, many attempts were made to control *O. rhinoceros* with *M. anisopliae* on coconut palms in the Pacific (Bedford, 1980). Infection of *O. rhinoceros* by *M. anisopliae* in oil palm plantations was reported by Sivapragasam and Tey (1994). However, intensive research on the use of *M. anisopliae* as a potential biocontrol agent of rhinoceros beetle was only carried out in the late 1990s (Ramle *et al.*, 2006).

An attempt to use the adult beetles as a vector to spread the spores into natural populations was conducted by Ho (1996). Adult beetles were collected by pheromone traps, dusted with spores and later released at the centre of a 100-ha field of one-year-old palms. The percentage of infected beetles increased from 0.3% at the commencement of treatment to 22.0% at 93 days after treatment. A specially designed inoculation trap for auto dissemination of *M. anisopliae* spores was developed (Ramle *et al.*, 2011b). The inoculation trap used the aggregation pheromone to attract the adult beetles. The flying beetles would collide with the trap and then fall into a specially designed disc containing a spore solution of the fungus *M. anisopliae*. The design of the chamber and the disc allow the infected beetles to escape from the trap, carrying and disseminating the spores to their natural habitat. The fungus then spreads to other healthy beetles which come into contact with the contaminated materials.

Field results showed that both rates caused high mortality to the escaped beetles from the trap within a period of 15–30 days after trapping (DAT). Complete mortality of the beetles was recorded at 45 DAT. The density of viable spores collected from the soil in the trapping region increased, suggesting that *M. anisopliae* was established in the breeding sites of the beetle. However, a reduction in the population of *O. rhinoceros* was not observed. This was probably due to the low trap density, short study period and the migrating behaviour of the *O. rhinoceros* beetles.

In another trial, a pit measuring 2.0 m (W)×3.0 m (L)×0.2 m (D) was filled with rotting empty fruit bunches. To attract beetles into the pit, a pheromone trap connected to a PVC pipe was placed at the centre of the pit. A spore suspension was made and applied onto the pit by spraying Metarhizium and repeated 4 times a year. Attracted beetles would either stay in the pit, dead or infected, or would fly out with the spores and disseminate them to other breeding sites.

Oryctes rhinoceros virus is another microbial pathogen of the rhinoceros beetle. The virus originated from Malaysia and infects both larvae and beetles (Huger, 1966; Zelazny, 1972, 1973). The virus is rod-shaped and non-occluded; it has been originally described as *Rhabdionvirus oryctes*, but was later ascribed to a new genus under the Baculoviridae Subgroup C and is commonly cited as *Baculovirus oryctes*. The *O. rhinoceros* virus has recently been proposed to be assigned in a new established genus Nudivirus, together with the *Heliothis zae* virus 1 (HzV-1) (Wang *et al.*, 2007).

A pilot release study was then conducted by releasing the type B virus into young oil palm areas by a simple capture, inoculate and re-release method established by Zelazny *et al.* (1978). Parameters measured include the infection of virus on beetles collected from the pheromone trap and the level of damage on palms in the field. The type B virus was released in fields where the virus type A was already existing. The impact of the released type B virus in reducing the beetle population and palm damage was evaluated using molecular approaches

(Ramle *et al.*, 2011a). The virus infection had increased and peaked at 3–4 months after release (MAR) and maintained 90% infection to the end of the experiment. A high virus infection of the beetles had reduced the palm damages to below 5%. The type B virus had spread in the released area at 3 MAR, later spreading to the control plot, located 3 km away from the release point.

The type B virus was finally introduced into four estates with palms 3–4 years old, with high numbers of matured and infected beetles (Ramle *et al.*, 2011a). The higher proportion of male beetles in the population indicated a slower virus transmission, as the population of male beetles gradually reduced between 4 and 6 MAR; much slower compared to other studies in western Samoa, Maldives and Philippines (Zelazny *et al.*, 1992). This also suggests that the beetles may have adapted to virus infection. The low density of susceptible young beetles in the release sites could have minimized the virus transmission. Even if the virus had spread, the DNA concentration of the released virus might have been too low to be detected by the restriction endonuclease enzyme.

14.4 Termites

Initially, termite infestation in oil palm plantations was only reported on mineral soils (Wood, 1968). When more areas were opened up for oil palm, termites emerged as a major insect pest of oil palm on peat (Tayeb, 2005). There are about 22 species of the genus *Coptotermes* Wasmann recorded from the Indo-Malayan region, of which five were found in Malaysia, namely *Coptotermes curvignathus* Holmgren, *C. gestroi* (Wasmann), *C. kalshoveni* Kemner, *C. sepangensis* Krishna and *C. travians* (Haviland) (Tho, 1992).

14.4.1 Mode of spread

The presence of termites in oil palm plantations can be detected with the occurrence of a mud trail on the palm trunk and petiole. The common symptom on leaves is a yellowing of the lower fronds (Khoo *et al.*, 1991; Faszly *et al.*, 2003), whose recovery at this stage is already too late because most of the meristematic systems are already affected. Some infestations are so extensive that the palm spear will eventually be covered with a layer of soil. The accumulation of soil at the shoots will eventually cause death of the palm within a month, when the mudwork becomes dry. In young palms, termites infest the roots, stem bulbs and leaves. Termite damage accompanied by bacteria will lead to palm death within two weeks. Therefore, if treatment is delayed, the palms will die due to the rotting of the shoots.

The infestation usually starts from the remains of frond bases and the termites tunnel their way into the trunk. In some cases, the termites build their colony at the centre of the trunk, sometimes at rotting points caused by *Ganoderma* disease. It appears that termites can bore into the trunk and cause serious damage.

14.4.2 Economic impact

Termites infest and kill vulnerable palms in the field. The indirect losses include cost of control, supplying new palms and lesser standing palms at maturity. Termite infestation may reach 8–9% per ha, with 3% of palms killed at the early stage.

14.4.3 Integrated pest management

Census and monitoring

Termite identification is very important and provides vital information in the management of these pests. High number of termite species with a lack of taxonomic understanding may lead to incorrect and incomplete identification, thus making it difficult to understand the biology and importance of each species (Kirton, 2005). Some of the key identification of the *Coptotermes*

species refers to the width and length of the head of the soldier caste, in addition to their mandible shapes. The mandible of *C. sepangensis* is nearly straight, with apices slightly bent inwards, while *C. kalshoveni* has a sabre-shaped mandible, strongly curved inwards from the middle.

Monitoring and detection of the termite population can be conducted by using rubber wood stakes. The stakes are driven into the soil and examined after two weeks, at which time any damage will be visible. After 1 month, each stake can be collected to determine the termite species infesting the stakes. The differences in the weight of each stake could reflect the termite consumption of the wood stakes.

Cultural methods

Water-level management for high FFB yields on peat is between 50 and 70 cm from the soil surface (Hasnol *et al.*, 2010). However, at this depth, termites can still forage for their food on the remaining timber buried in the soil (Zulkefli *et al.*, 2011). Treatment by drenching chemicals at the palm base may not reach the termite population, which is located much deeper in the peat soil, especially during the dry season (Zulkefli, 2007). Thus, raising the water level in peat is considered a simple method which could force the termite colony up to the soil surface for easier control. The water level can be raised for a short period without affecting the palm growth. Relatively, more termites move up to the soil surface when the water levels are high (15 cm from soil surface), especially during the rainy period, as compared to the drier months when the water levels are low. The appropriate timing for increasing the water level is during the hot season, in order to avoid chemical washout after drenching. Seven days is sufficient to force the pest termite to move up to the surface for effective application of chemicals.

Chemical control

Most of the control methods are adopted from pre-construction and treatment of infested building. With the banning of organochlorine pesticides, organophosphates, carbamates and synthetic pyrethroids became the only alternatives for termite treatment (Langewald *et al.*, 2003). Currently, termite control in oil palm plantations is applied by using repellents or fast-acting termiticides. Spraying at the oil palm's shoots, frond bases and trunks by drenching around the base at the early stages of infestation would reduce termite infestation. Breaking the mud trail by scraping before spraying improves the termiticide penetration. However, these methods sometimes trigger the termite alarm system sealing off the treated area until there is no contact with the termiticides, protecting other termites in other parts of the plant. Infestation would then start again after several months.

Based on an MPOB survey, it was found that drenching and spraying with chlorpyrifos and fipronil seem to be the most effective methods, as currently implemented by most of the oil palm plantations.

14.5 Nettle Caterpillars

Nettle caterpillars (Lepidoptera: Limacodidae) are slow-moving caterpillars with stinging spines. The colour of the caterpillars is species-specific. Some species have different colorations for the immature and mature larvae. There are many types of nettle caterpillars found in Malaysia, but some of the most common species are *Darna trima* Moore, *D. diducta* (Snellen), *Setora nitens* Walker and *Setothosea asigna* van Eecke. The less common species are *Thosea vetusta* Walker, *T. bisura* Moore, *Susica pallida* Walker and *Birthamula chara* Swinhoe (Norman and Basri, 1992).

14.5.1 Economic impact

The damage caused by nettle caterpillars is certainly more devastating than bagworms. Their voracious feeding habit can eventually skeletonize the whole frond. In serious cases of *D. trima* outbreaks in Indonesia, up to 2000 caterpillars can be found per frond (de Chenon *et al.*, 2004). In Malaysia, outbreaks of nettle caterpillars are more frequent in young plantings in Sabah (Ang, *et al.*, 1997) compared to the Peninsular. *D. diducta* is more than a pest in young palms up to 5–6 years of age (de Chenon *et al.*, 2004).

In Indonesia, *S. nitens* damage is often associated with *Pestalotiopsis* occurence on the lower fronds. A single caterpillar can consume up to 100 cm² of foliage. Before pupating at the base of the fronds, the caterpillars also feed on the epidermis of the fruit, inducing additional damages (de Chenon *et al.*, 2004). Outbreaks of nettle caterpillars are more frequent on young oil palms aged between 2 and 8 years.

Severe nettle caterpillar outbreaks occurred mainly during the 1970s and 1980s. Between 1981 and 1990, there were a total of 49 outbreaks of nettle caterpillars in Malaysia, averaging about 5 outbreaks per year (Norman and Basri, 1992). Currently, there are very few cases of nettle caterpillar outbreaks and most of them are easily controlled. Between the years 2000 to 2005, 11% of the estates throughout Malaysia suffered nettle caterpillar attacks (more than 11,000 ha) (Norman and Basri, 2007).

14.5.2 Integrated pest management

Census and monitoring

Census and monitoring must be carried out at least twice a month, assessing on 1 palm/ha on the middle fronds (fronds 15–25), depending on the species (Ang *et al.*, 1997; de Chenon *et al.*, 2004). In Indonesia, the critical level is at 20–30 caterpillars per frond on young palms and 40–60 on older ones (de Chenon *et al.*, 2004). For example, *Darna diducta* is monitored on frond 17 or 25, according to the age of the palms. In Malaysia, the threshold is 10 larvae/frond. Due to the rapid increase in population of this species, early monitoring and immediate action must be applied when the pest is detected.

Natural enemies

Nettle caterpillars are also controlled by numerous insects that act as natural enemies. These include parasitoids, mainly from the Order Hymenoptera (braconid and ichneumonid wasps) (Norman *et al.*, 1998), and predators from the Order Hemiptera (pentatomid and reduviid bugs).

The presence of beneficial plants could increase the population of natural enemies of the insects, thereby reducing the attack of nettle caterpillars. These flowering beneficial plants provide nectar to the parasitoids and also provide shelter to the predators.

Chemical control

Nettle caterpillars can be controlled with chemicals such as monocrotophos, methamidophos, cypermethrin and acephate. Systemic insecticides like methamidophos and monocrotophos must be used for trunk injection and should never be sprayed. Monocrotophos applied via trunk injection is effective for controlling *Setora* and *Setothosea* (Hutauruk and Sipayung, 1978). Other trunk-injected chemicals like acephate and metamidophos gave varied responses. The use of triclorfon for nettle caterpillars may show adverse effects on its predators, *Cantheconidea* spp., which often predate on the nettle caterpillars. The predators could be affected by the residual effects of the chemicals inside the pests. A study on bagworms showed that it can last up to 20 days (Zulkefli, 1996).

Microbial control (fungal pathogens and Bacillus thuringiensis)

Cordyceps sp. is a fungal pathogen for nettle caterpillars. This fungal infection is common on pupae of the pest. The infected pupae produce fruiting bodies on their cocoons. Spraying of crushed infected pupae around the bases of oil palm during the wet weather kills the pupae effectively (Papierok *et al.*, 1993).

Bacillus thuringiensis is effective against *S. nitens*, *D. trima* and *S. asigna* (Wood *et al.*, 1977). A laboratory bioassay of *B. thuringiensis* against *D. trima* from Tawau (Sabah) showed that two commercial products of *B. thuringiensis* (*aizawai* and *kurstaki* ES) were effective against this pest, and resulted in 90% mortality within seven days.

Field trials on the use of commercial formulations of *B. thuringiensis* against nettle caterpillars have been reported by Basri *et al.* (1994), Chung (1998) and Ho (1988).

14.6 Bunch Moths

Bunch moths are known as occasional pests of oil palms and also known as the 'inflorescence moths' or 'fruit moths' (Basri and Norman, 2000). They belong to the genus *Tirathaba* Walker, which contains about 30 species, though some of them could be synonyms (Moore, 2001). The common species found in oil palm is *Tirathaba rufivena* (Walker), which has been referred earlier to as *Tirathaba mundella* Walker (Khoo *et al.*, 1991). This pest has been reported attacking *Nipah fruticans*, *Plectocomia* spp., *Pritchardia pacifica* and *Roystonea regia* (Moore, 2001). The incidences of bunch moth infestations were reported on young oil palms planted on peat in Sarawak. Young palms planted on peat are conducive for the breeding of bunch moth because more often fruit bunches are left unharvested and allowed to rot.

14.6.1 Mode of spread

The bunch moth is fast-spreading, with a short life cycle of about 1 month. It is attracted by poor sanitation, especially with the presence of rotten fruit bunches on the palms (Lim, 2012).

14.6.2 Symptoms and damage

The bunch moth, *T. rufivena*, is now becoming one of the important pests of oil palms, particularly planted on peat. The oil palm female inflorescences and fruit bunches are attacked by the caterpillars or larvae of

T. rufivena at various stages of development (Lim, 2012). The larvae feed on both male and female inflorescences (Moore, 2001; Turner and Gillbanks, 2003). The common damage is pitting and scoring of maturing fruits (Khoo *et al.*, 1991). Infested fruit bunches with lots of frass significantly reduce the quality of the fruit bunches. Damaged fruits fall prematurely or develop into fruits without the kernel. The average bunch weight is greatly reduced and there is increased free fatty acid (FFA) content, which reduces the quality of the fruit bunch (Lim, 2012).

14.6.3 Economic impact

Under serious attack, the fruit bunches will not develop fully and may abort prematurely. Detection of damage caused by the bunch moth is normally obtained during routine grading of harvested fruit bunches at the FFB platforms. In cases where the *T. rufivena* infestation on the harvested bunches is >5%, detailed census on the blocks is mandatory. Failure to implement proper management of *T. rufivena* infestation may lead to yield losses of >50% (Lim, 2012).

14.6.4 Integrated pest management

Census and monitoring

A visual inspection on the infested bunches and the percentage of infested palms will not represent the actual bunch moth population in the infested area. Therefore, infested bunches must be chopped and opened in order to count the actual number of live larvae and pupae within the bunch. Census of bunch moth population from the chopped bunches will indicate the actual number and stages of bunch moth at that time. Hence, the timing of treatment can be determined more precisely, to avoid spraying the adults and pupae which are no longer susceptible, hence avoiding a much higher dose of chemicals for control.

A survey on infested female inflorescences and bunches in Sarawak has indicated a high number of bunch moth larvae during post-anthesis and anthesizing female inflorescences of young palms. During pre-anthesis with open sheath, the mean larvae recorded were between 5 and 23 live larvae per inflorescence. In some samples, live larvae were absent. The infested bunches recorded 5 live larvae with the highest at 18 larvae per bunch. In an earlier study at Teluk Intan, Perak, the number can be up to 30 live larvae per bunch (Basri *et al.*, 1991).

Cultural control

Integrated pest management of bunch moth begins with early detection and regular census on newly harvested bunches. Cultural practice by removing female inflorescences from young palms, known as disbudding or ablation can reduce the pest population (Turner and Gillbanks, 2003). However, the cultural practice should be consistent, otherwise within three years the pest can build up in population because of the availability of ample food supply for the larvae.

Pheromones

Currently, no work has been conducted to control bunch moth using pheromones, despite having identified the male pheromonal structure many years ago (Sasaerila *et al.*, 2003). It would certainly be beneficial if this pheromone could be commercialized for control of this pest.

Chemical control

Treatment with insecticides should prevent more damage before bunches reach 3½ months old (Basri *et al.*, 1991). The immature bunches need less volume of chemicals and water during spraying compared to mature bunches and matured palms. Spray penetration into immature bunches is much easier than mature bunches. A wetting agent should be included into the mixture to improve penetration inside the infested bunches. Although bunch moth control with cypermethrin is fast-acting and cheaper than *B. thuringiensis*, visual observations showed that cypermethrin affected the population of oil palm pollinating weevils (*Elaeidobius kamerunicus* Faust) and earwigs (*Chelisoches morio* (Fabricius)), the latter one being a natural predator of the *Tirathaba* caterpillars (Lim, 2012). It is therefore important not to use cypermethrin in order to avoid adverse effects to the pollinating weevil, *E. kamerunicus.*

Microbial control (Bacillus thuringiensis)

The common control for bunch moth is by using *Bacillus thuringiensis (Bt)*, cyfluthrin and diflubenzuron (Basri *et al.*, 1991). The reductions in damage were observed 7–14 days after treatment (DAT) with *B. thuringiensis* and 49 DAT with cyfluthrin and diflubenzuron (Basri *et al.*, 1991).

Timely application of the biopesticide, *Bt*, coupled with good sanitation is an effective integrated approach to prevent outbreak of the bunch moth. A trial was conducted on 7-year-old oil palms with serious *T. rufivena* infestation (>50% palms and >50% bunches infested) (Lim, 2012). Spraying of cypermethrin (a.i. 5% at 1 ml/l litre water) was compared with *B. thuringiensis* variety kurstaki strain HD-7 (16,000 IU/mg) at 1 g product per litre of water, sprayed on infested bunches using a knapsack sprayer. Six continuous, 2-weekly rounds of *Bt* spraying over 3 months were able to bring down infestation to less than 15% on the bunches. In treatment plots, where cypermethrin was applied by spraying, infested bunches remained high at more than 60%. In the untreated control plot, infestation went up to more than 95% (Lim, 2012).

Integrated pest management combining good sanitation and timely spraying with *Bt*, based on census, can significantly reduce crop losses and minimize expenditures for *Tirathaba* control (Lim, 2012).

14.7 Discussion

Bagworms and nettle caterpillar are serious leaf-feeding pests of oil palm and should be properly managed. More often, pest outbreaks occur due to the unregulated and widespread use of broad spectrum chemicals, which poses detrimental effects to beneficial insects (i.e. parasitoids and predators). These natural enemies regulate the population of the leaf-feeding pests and halt them from manifesting an outbreak. Usually, broad spectrum, contact chemicals only give short-term control, and, with a reduced population of natural enemies, a serious outbreak of pests is expected.

Oryctes rhinoceros commonly attacks immature oil palms, especially during replanting. Careful and judicious placement of the oil palm trunk debris during replanting plays a crucial role in reducing its status as pest. Avoiding large stacked heaps of the oil palm trunk debris, proper EFB mulching and destruction of standing rotten old trunks in the underplanting method will largely reduce risks of the pest to breed. The use of biocontrol agents, such as *Metarhizium* fungi and '*Oryctes* virus', supports the effective management of this pest.

Unlike bagworms, outbreaks of rhinoceros beetles tend to occur within the availability of abundant breeding sites within oil palm plantations. Major replanting activities, which include zero-burning and sometimes underplanting, coupled with the application of multi-layered empty fruit bunches as mulch and rotting *Ganoderma*-diseased palms would lead to conducive breeding activities for the pest. The integrated control approach has relied heavily on the combined applications of a number of practices: the chipping of oil palm trunks during replanting and covering them with leguminous cover crops, the use of aggregation pheromone and the application of chemical insecticides. With this approach, the rhinoceros beetle numbers have been reduced in many instances. Currently, a higher emphasis is placed on the use of biological organism components, including *Metarhizium* spp. and baculovirus.

Termite occurrences in oil palm are associated more with young palms planted on peat, of which the termites thrive because of the availability of decomposing timber below the soil surface. The moisture and abundance of food materials creates an ideal habitat for the termites to propagate. As the palm matures, the quantity of decomposing timber declines and consequently the termite population would also diminish. Integrated control should be undertaken during this period to prevent severe crop loss and possible death of the oil palms. The current approach is still dependent on chemical pesticides. However, new approaches, particularly using microbial agents, should be pursued. The challenge is that microbial control should balance its repellent nature in order to largely spread within the termite colony.

The approach to bunch moth control should be integrated, starting from early detection and regular census, combined with timely application of insecticides to ensure effective control. Good agricultural practice and sanitation can reduce bunch moth infestation. However, the use of broad spectrum, contact pesticides should be avoided as the application will affect the population of oil palm pollinators.

14.8 Conclusions

The long-term protection of oil palm against attack by insect pests requires IPM approaches and unified strategies of utilizing the biological control components for each pest. By amalgamating biological control components, the oil palm production can be made more sustainable in the long term. With reduced risk of releasing toxic chemicals in the environment, the biodiversity of flora and fauna within the oil palm ecosystem is preserved, thus restoring the natural balance between the oil palm host, its pests and natural enemies.

References

Ang, B.N., Chua, T.H., Chew, P.S., Mohd, M.M. and Saserilla, Y. (1997) Distribution of *Darna trima* Moore and *D. bradleyi* Holloway larvae (Lepidoptera: Limacodidae) in oil palm canopy, in a single species and double species infestations. *The Planter* 73 (852), 107–118.

Basri, M.W. and Kevan, P.G. (1995) Life history and feeding behaviour of the oil palm bagworm, *Metisa plana* Walker (Lepidoptera: Psychidae). *Elaeis* 7(1), 18–35.

Basri, M.W. and Norman, K. (2000) Insect pests, pollinator and barn owl. In: Yusof, B., Jalani, S. and Chan, K.W. (eds) *Advances in Oil Palm Research*, vol. 1. Malaysian Palm Oil Board, Kuala Lumpur, pp. 466–541.

Basri, M.W., Hassan, A.H. and Zulkefli, M. (1988) Bagworms (Lepidoptera: Psychidae) of oil palm in Malaysia. *PORIM Occasional Paper*, No. 23. PORIM, Kuala Lumpur.

Basri, M.W., Mukesh, S. and Norman, K. (1991) Field evaluation of insecticides and a cultural practice against the bunch moth, *Tirathaba rufivena* (Lepidoptera: Pyralidae) in a mature oil palm plantation. *Elaeis* 3(2), 355–362.

Basri, M.W., Siti Ramlah, A.A. and Norman, K. (1994) Status report on the use of *Bacillus thuringiensis* in the control of some oil palm pests. *Elaeis* 6 (2), 82–101.

Basri, M.W., Norman, K. and Hamdan, A.B. (1995) Natural enemies of the bagworm *Metisa plana* (Lepidoptera: Psychidae) and their impact on host population regulation. *Crop Protection* 14 (8), 637–645.

Bedford, G.O. (1980) Biology, ecology, and control of palm rhinoceros beetles. *Annual Review of Entomology* 25, 309–339.

Chua, C.K., Eng, O.K., Razakz, A.R., Arshad, A.M. and Marcon, P.G. (2012) Susceptibility of Bagworm Metisa plana (Lepidoptera: Psychidae) to Chlorantraniliprole. *Pertanka Journal of Tropical Agricultural Science* 35 (1), 149–163

Chung, G.F. (1997) The bioefficacy of the aggregation pheromone in mass trapping of *Oryctes rhinoceros* (L.) in Malaysia. *The Planter* 73 (852), 119–127.

Chung, G.F. (1998) Strategies and methods for the management of leaf-eating caterpillars of oil palm. *The Planter* 74 (871), 531–558.

Chung, G.F. and Sim, S.C. (1991) Bagworm census and study: a case study. In: Yusof, B., Jalani, B.S., Chang, K.C., Cheah, S.C., Henson, I.E. *et al.* (eds) *Proceedings of 1991 PORIM International Palm Oil Conference – Agriculture (Module I).* Palm Oil Research Institute of Malaysia, Kuala Lumpur, pp. 433–442.

Chung, G.F., Basri, M.W. and Ariffin, D. (1995) Recent development in plant protection of the Malaysian oil palm industry – 1990 to 1995. In: Jalani, B.S., Ariffin, D., Rajanaidu, N., Mohd Tayeb, D. and Basri, M.W. (eds) *Proceedings of the 1995 PORIM National Oil Palm Conference – Technologies in Plantations – The Way Forward.* Palm Oil Research Institute of Malaysia, Kuala Lumpur, pp. 105–124.

de Chenon, D.R., Sudharto, P.S. and Poeloengan, Z.H. (2004) Nettle caterpillars and grasshoppers of oil palm in Indonesia. In: *Proceedings of the International Conference on Pests and Diseases of Importance to the Oil Palm Industry.* Malaysian Palm Oil Board, Kuala Lumpur, pp. 75–95.

Faszly, R., Idris, A.B., Sajap, A.S., Norman, K. and Basri, M.W. (2003) List of termite (Insecta: Isoptera) from an ex-felled peat soil oil palm plantation near Endau-Rompin Forest Reserve. *SERANGGA* 8 (1–2), 107–111.

Hallet, R.H., Perez, A.L., Gries, G., Greis, R., Pierce Jr, H.D. *et al.* (1995) Aggregation pheromone of the coconut rhinoceros beetle *Oryctes rhinoceros* L. (Coleoptera: Scarabaeidae). *Journal of Chemical Ecology* 21 (10), 1549–1570.

Hasnol, O., Tarmizi, A.M., Haniff, M.H., Farawahida, M.D. and Hasimah, M. (2010) Best management practice for oil palm planting on peat: optimum ground water table. *MPOB TT No. 472.* Malaysian Palm Oil Board. Kuala Lumpur.

Ho, C.H. (1988) Mechanised mistblowers for treatment of oil palm leaf pests. In: *Proceedings of National Oil Palm Seminar.* Palm Oil Research Institute of Malaysia, Kuala Lumpur.

Ho, C.T. (1996) The integrated management of *Oryctes rhinoceros* (L.) populations in the zero burning environment. In: Ariffin, D., Mohd Basri, W., Rajanaidu, N., Mohd Tayeb, D., Paranjothy, K., Cheah, S.C., Chang, K.C. and Ravigadevi, S. (eds) *Proceedings of the 1996 PORIM International Palm Oil Congress-Agriculture Conference.* PORIM, Kuala Lumpur, pp. 336–368.

Ho, C.T., Khoo, K.C., Yusof, I. and Dzolkifli, O. (2003) Comparative studies on the use of beneficial plants for natural suppression of bagworm infestation in oil palm. In: *Proceedings of the 2003 PIPOC: Agriculture.* Malaysian Palm Oil Board, Kuala Lumpur, pp. 372–424.

Huger, A.M. (1966) A virus disease of the Indian rhinoceros beetle *Oryctes rhinoceros* (Linnaeus), caused by a new type of insect virus, *Rhabdionvirus oryctes* gen. n., sp. n.. *Journal of Invertebrate Pathology* 8, 38–51.

Hutauruk, C.J. and Sipayung, A. (1978) Development of trunk injection of systemic insecticides against *Setora nitens* and *Thosea asigna* on oil palm in North Sumatra. In: Amin, L.L., Kadir, S.A., Lim, G.S., Singh, K.G., Tan, A.M. and Varghese, G. (eds) *Proceedings of Plant Protection Conference 1978.* Malaysian Plant Protection Society, Kuala Lumpur, pp. 265–278.

Khalid, H., Zin, Z.Z. and Anderson, J.M. (1999) Quantification of oil palm biomass and nutrient value in a mature plantation. I. Above-ground biomass. *Journal of Oil Palm Research* 11 (1), 23–32.

Khoo, K.C., Peter, A.C.O. and Ho, C.T. (1991) *Crop Pests and Management in Malaysia.* Tropical Press Sdn. Bhd., Kuala Lumpur.

Kirton, L.G. (2005) The importance of accurate termite taxonomy in the border perspective of termite management. In: Lee, C.L. and Robinson, W.H. (eds) *Proceeding of the Fifth International Conference on Urban Pests.* P&Y Design Network, Kuala Lumpur.

Lai, C.H. and Tey, C.C. (2009) Evaluation of systemic insecticides for trunk injection against bagworm. In: *Proceedings of Agriculture, Biotechnology and Sustainability Conference Vol. II. PIPOC 2009.* Malaysian Palm Oil Board, Kuala Lumpur, pp. 450–462.

Langewald, J., Mitchell, J.D., Maniania, N.K. and Kooyman, C. (2003) Microbial control of termite in Africa. In: Neuenschwander, P., Borgemeister, C. and Langewald, L. (eds) *Biological Control in IPM System in Africa.* CAB International, Wallingford, UK, pp. 227–242.

Liau, S.S. and Ahmad, A. (1991) The control of *Oryctes rhinoceros* by clean clearing and its effect on early yield in palm to palm replants. In: Yusof, B., Jalani, B.S., Chang, K.C., Cheah, S.C., Henson, I.E. *et al.* (eds) *Proceedings of the 1991 PORIM International Palm Oil Development Conference Module II – Agriculture.* PORIM, Kuala Lumpur, pp. 396–403.

Lim, K.H. (2012) Integrated pest management of *Tirathaba* bunch moth on oil palm planted on peat. *The Planter* 88 (1031), 97–104.

Mohd Najib, A., Siti Ramlah, A.A., Mohamed Mazmira, M.M. and Mohd Basri, W. (2007) Effect of *Bacillus thuringiensis*, Terakil-1™ on *Elaeidobius kamerunicus* and beneficial insect as compared to Cypermethrin. In: *Proceedings of Palm Oil International Congress (PIPOC): Palm Oil Empowering Change*, vol. 2. Malaysian Palm Oil Board, Kuala Lumpur, pp. 720–728.

Mohd Najib, A., Siti Ramlah, A.A., Mohamed Mazmira, M.M. and Mohd Basri, W. (2009) Effect of *Bacillus thuringiensis*, Terakil-1 and Teracon-1 against oil palm pollinator, *Elaeidobius kamerunicus* and beneficial insects associated with *Cassia cobanensis. Journal of Oil Palm Research* 21, 667–674.

Mohd Najib, A., Siti Ramlah, A.A., Mohamed Mazmira, M.M. and Mohd Basri, W. (2011) Ecotoxicity of *Bacillus thuringiensis*, Terakil-1R and Teracon-1R against freshwater fish, *Tilapia nilotica. Journal of Oil Palm Research* 23, 1036–1039.

Mohd Najib, A., Siti Ramlah, A.A., Mohamed Mazmira, M.M. and Mohd Basri, W. (2012) Effect of *Bacillus thuringiensis*, Lepcon-1, Bafog-1 (S) and Ecobac-1 (EC) against oil palm pollinator, *Elaeidobius kamerunicus* and beneficial insects associated with *Cassia cobanensis. Journal of Oil Palm Research* 24, 1442–1447.

Moore, D. (2001) Insect of palm flowers and fruits. In: Howard, F.W., Moore, D., Giblin-Davis, R.M. and Abad, R.G. (eds) *Insects on Palm*. CAB International, Wallingford, UK.

Norman, K. and Basri, M.W. (1992) *A Survey of Current Status and Control of Nettle Caterpillars (Lepidoptera: Limacodidae) in Malaysia (1981–1990)*. PORIM Occasional Paper No. 27.

Norman, K. and Basri, M.W. (1995) *Control Methods for Rhinoceros Beetle,* Oryctes rhinoceros *(L.) (Coleoptera: Scarabaeidae)*. PORIM Occasional Paper No. 35.

Norman, K. and Basri, M.W. (1997) Status of rhinoceros beetle, *Oryctes rhinoceros* (Coleoptera: Scarabaeidae) as a pest of young oil palm in Malaysia. *The Planter* 73 (850), 5–21.

Norman, K. and Basri, M.W. (2004) Immigration and activity of *Oryctes rhinoceros* in an oil palm replanting. *Journal of Oil Palm Research* 16 (2), 64–77.

Norman, K. and Basri, M.W. (2007) Status of common oil palm insect pests in relation to technology adoption. *The Planter* 83 (975), 371–385.

Norman, K. and Othman, A. (2006) Potentials of using the pheromone trap for monitoring and controlling the bagworm, *Metisa plana* Wlk (Lepidoptera: Psychidae) in a smallholder plantation. *Journal of Asia Pacific Entomology* 9 (3), 281–285.

Norman, K., Robinson, G.S. and Basri, M.W. (1994) Common bagworm pests (Lepidoptera: Psychidae) of oil palm in Malaysia with notes on related South-east Asian species. *Malayan Nature Journal* 48, 93–123.

Norman, K., Walker, A.K., Basri, M.W., LaSalle, J. and Polaszek, A. (1996) Hymenopterous parasitoids of the bagworm, *Metisa plana* and *Mahasena corbetti* on oil palm in Peninsular Malaysia. *Bulletin of Entomological Research* 86, 423–439.

Norman, K., Basri, M.W. and Zulkefli, M. (1998) *Handbook of Common Parasitoids and Predators Associated with Bagworms and Nettle Caterpillars in Oil Palm Plantations*. PORIM, Kuala Lumpur.

Norman, K., Zaidi, M.I., Maimon, A. and Basri, M.W. (2001) Factors affecting development of *Oryctes rhinoceros* in some substrates commonly found in the oil palm environment. *Journal of Oil Palm Research* 13 (1), 64–74.

Norman, K., Siti Nurulhidayah, A. and Basri, M.W. (2010) Pheromone mass trapping bagworm moths *Metisa plana* (Lepidoptera: Psychidae) for its control in mature oil palms in Perak, Malaysia. *Journal of Asia-Pacific Entomology* 13, 101–106.

Papierok, B., de Chenon, D.R., Freulard, J.M. and Suwandi, W.P. (1993) New perspectives in the use of Cordyceps fungus in the biological control of nettle caterpillars in oil palm plantations. In: Jalani, B.S., Ariffin, D., Rajanaidu, N., Mohd Tayeb, D. and Basri, M.W. (eds) *Proceedings of the 1993 PORIM National Oil Palm Conference Update and Vision*. Palm Oil Research Institute of Malaysia, Kuala Lumpur, pp. 706–710.

Ramle, M., Basri, W., Norman, K., Ramlah, S.A.A. and Noor Hisyam, H. (2006) Research into the commercialization of *Metarhizium anisopliae* (Hypomycetes) for biocontrol of oil palm rhinoceros beetle, *Oryctes rhinoceros* (Scarabaeidae), in oil palm. *Journal of Oil Palm Research* (Special Issue, April 2006), 37–49.

Ramle, M., Norman, K., Idris, A.B., Basri, W., Jackson, T.A., Tey, C.C. and Mohd Ahdly, A. (2011a) Molecular approaches in the assessment of *Oryctes rhinoceros* virus for the control of rhinoceros beetle in oil palm plantations. *Journal of Oil Palm Research* 23, 1096–1109.

Ramle, M., Norman, K. and Basri, W. (2011b) Trap for auto dissemination of *Metarhizium anisopliae* for control of rhinoceros beetle, *Oryctes rhinoceros*. *Journal of Oil Palm Research* 23, 1011–1017.

Samsudin, A., Chew, P.S. and Mohd, M.M. (1993) *Oryctes rhinoceros*: breeding and damage on oil palms in an oil palm to oil palm replanting situation. *The Planter* 69 (813), 583–591.

Sasaerila, Y., Gries, R., Gries, G., Khaskin, G., King, S. and Takacs, S. (2003) Sex pheromone components of male *Tirathaba mundella* (Lepidoptera: Pyralidae). *Chemoecology* 13, 89–93.

Siti Ramlah, A.A., Mohd Najib, A., Mohd Mazmira, M.M., Nor Shalina, A.T., Mohd Fahmi, K. and Norman, K. (2013) Microbial control for pest and disease and its challenges. In: *Proceedings of Palm Oil International Congress (PIPOC) 2013*. Malaysian Palm Oil Board, Kuala Lumpur. A27, pp. 1–13.

Sivapragasam, A. and Tey, C.C. (1994) Susceptibility of *Oryctes rhinoceros* (L.) larvae to three isolates of *Metarhizium anisopliae* (Metsch.) Sorokin. *MAPPS Newsletter* 18 (2), 13–14.

Tayeb, M.D. (2005) *Technologies for Planting Oil Palm on Peat*. Malaysian Palm Oil Board, Kuala Lumpur.

Tho, Y.P. (1992) Termites of peninsular Malaysia. In: Kirton, L.G. (ed.) *Malayan Forest Record* No. 36. Forest Research Institute of Malaysia, Kepong, Kuala Lumpur.

Tulloch, M. (1979) The genus of Metarhizium. *Transactions of the British Mycological Society* 66, 407–411.

Turner, P.D. and Gillbanks, R.A. (2003) *Oil Palm Cultivation and Management*, 2nd edn. The Incorporated Society of Planters, Kuala Lumpur.

Wang, Y., Van Oers, M.M., Crawford, A.M., Vlak, J.M. and Jehle, J.A. (2007) Genomic analysis of *Oryctes rhinoceros* virus reveals genetic relatedness to *Heliothis zea* virus. *Archives of Virology* 152 (3), 519–531.

Wood, B.J. (1968) *Pests of Oil Palms in Malaysia and Their Control*. Incorporated Society of Planters, Kuala Lumpur.

Wood, B.J., Corley, R.H.V. and Goh, K.H. (1972) Studies on the effect of pest damage on oil palm yield. In: Wastie, R.L. and Earp, D.A. (eds) *Advances in Oil Palm Cultivation*. Incorporated Society of Planters, Kuala Lumpur, pp. 360–379.

Wood, B.J., Hutauruk, C.H. and Liau, S.S. (1977) Studies on the chemical and integrated control of nettle caterpillars (Lepidoptera: Limacodidae). In: Earp, D.A. and Newall, W. (eds) *International Development in Oil Palm*. Incorporated Society of Planters, Kuala Lumpur, pp. 591–616.

Zelazny, B. (1972) Studies on *Rhabdionvirus oryctes*. I. Effect on larvae of *Oryctes rhinoceros* and inactivation of the virus. *Journal of Invertebrate Pathology* 20 (3), 235–241.

Zelazny, B. (1973) Studies on *Rhabdionvirus oryctes*. II. Effect on adults of *Oryctes rhinoceros*. *Journal of Invertebrate Pathology* 22 (1), 122–126.

Zelazny, B. (1978) Methods of inoculating and diagnosing the Baculovirus disease of *Oryctes rhinoceros*. *FAO Plant Protection Bulletin* 26 (4), 163–168.

Zelazny, B., Lolong, A. and Pattang, B. (1992) *Oryctes rhinoceros* (Coleoptera: Scarabaeidae) populations suppressed by a baculovirus. *Journal of Invertebrate Pathology* 59, 61–68.

Zulkefli, M. (1996) The effect of selected insecticides on the oil palm bagworm, *Metisa plana* Walker (Lepidoptera: Psychidae) and predatory bug *Cantheconidea* sp. (Hemiptera: Pentatomidae). MSc thesis, Universiti Pertanian Malaysia.

Zulkefli, M. (2007) Termite population in oil palm plantation and the effect of different water table on *Coptotermes curvignathus* in peat soil. MSc thesis, Universiti Malaysia Sabah.

Zulkefli, M., Norman, K., Basri, M.W., Hasnol, O., Afandi, A.A. and Jai, H. (2011) *Water Table Management for the Control of Termite in Peat*. MPOB TOT No. 485. Malaysian Palm Oil Board, Kuala Lumpur.

15 Integrated Pest Management in Tea, Cocoa and Coffee

Devid Guastella[1,*], Giuseppe E. Massimino Cocuzza[2] and
Carmelo Rapisarda[2]

*[1]Agrisudafrica (Pty) Ltd, Franklin, South Africa; [2]Dipartimento di Agricoltura,
Alimentazione e Ambiente, Università degli Studi, Catania, Italy*

15.1 IPM in Tea

After water, tea (*Camellia sinensis* (L.) Kuntze) is the most consumed drink in the world (Awasom, 2011; FAO, 2015). It is cultivated in about 35,534,721 ha, with a total production of 5,412,223 t (FAOSTAT, 2016). Despite the popularity of the beverage, tea cultivation is restricted in a relatively limited number of countries, as the plant requires very specific climatic conditions, such as temperatures ranging from 10 to 30°C, annual rainfall up to 1000 mm and an area of cultivation preferably up to 1000 m above sea level (FAO, 2015). Currently, tea is grown in 49 countries located on all continents, although the vast production takes place in Asia and Africa. China and India are the countries with the highest cultivated area (66% of the total) and production (42.7% of the total) in the world (FAOSTAT, 2016). The top 10 producing countries of tea are shown in Table 15.1.

In major production areas, tea is grown as a perennial monoculture, with crops persisting on the same soil for several decades (Banerjee, 1983). With regard to the adversities, more than a thousand arthropod species can live on tea (Chen and Chen, 1989), but only a small number (about 3%) can be considered pests, with a variable degree of harmfulness depending on climatic and

Table 15.1. Indicators for production and harvested areas of tea, cocoa and coffee in the world. (From: FAOSTAT, 2015.)

Crop	Area (ha)	Production (tonnes)	Total number of producing countries	Top 10 producing countries (descending order)
Tea	3,799,831	5,561,339	49	China, India, Sri Lanka, Kenya, Vietnam, Turkey, Indonesia, Iran, Myanmar, Argentina
Cocoa	10,434,201	4,450,263	59	Cote D'Ivoire, Ghana, Indonesia, Nigeria, Brazil, Cameroon, Ecuador, Colombia, Dominican Rep., Papua New Guinea
Coffee	10,142,836	8,923,006	78	Brazil, Vietnam, Colombia, Indonesia, Ethiopia, India, Honduras, Peru, Uganda, Mexico

* Corresponding author e-mail: devid.guastella@agrisudafrica.co.za

growing conditions (Lehmann-Danzinger, 2000; Ye *et al.*, 2014). For decades, pest control was carried out on tea crops with the almost exclusive use of chemicals (Hazarika *et al.*, 2009). Nowadays, as for other agricultural products intended for human consumption, for tea there is a growing interest in productions using a limited use of pesticides. Several prospects have been undertaken to promote organic or IPM tea production, in order to protect the environment as well as the growers and consumers. This has especially derived from some studies that have demonstrated how a variable amount of pesticides is transferred from the tea leaves to the infusion (Jaggi *et al.*, 2001; Gupta *et al.*, 2008). Consequently, the presence of pesticide residues represents an important factor of depreciation or rejection of product, especially among those consumers that require a pesticide residue-free tea, and producing a healthy tea is currently the major challenge for growers (Choudhary, 1999; Lehmann-Danzinger, 2000; Rabindra, 2012).

Pest species affecting tea crops and the magnitude of their damage depend on various factors, such as altitude, climatic conditions and agronomic practices. Moreover, resurgence or outbreaks of pests are frequently direct consequences of the abuse or incorrect use of pesticides, which favour undesirable arthropods and depress populations of the beneficials (Sivapalan, 1999; Lehmann-Danzinger, 2000; Hazarika *et al.*, 2001; Ye *et al.*, 2014). Thus, creating favourable conditions for a stable agroecosystem by applying adequate cultural techniques should be the first and basic step for a low pesticide residue production of tea (Rabindra, 2012). Interesting research has been carried out, especially in India, on natural plant extracts with antifeedant, repellent and/or insecticidal effects, which may represent a valid partial alternative to pesticides (Barthakur, 2011; Mamun and Ahmed, 2011). Moreover, the longstanding nature of tea plantations favours the formation of ecological niches that facilitate improving the variety of natural enemies and their efficiency in controlling the pests (Banerjee, 1983). This is far from an easy and complete achievement, albeit yield losses caused by pests can be quite reduced if a correct control strategy is applied (Rattan, 1992; Muraleedharan and Chen, 1997; Sivapalan, 1999). Finally, a deep knowledge of the agroecosystem, the effective incidence of pests on production, the activity of beneficials on pests and the environmental influence on their ecology are essential steps to improve quality and quantity of tea production.

15.1.1 Mites

Altogether, mites can be considered the most serious pests of tea (Muraleedharan, 1992; Hazarika *et al.*, 2009). By sucking cell contents from leaves, the main damage caused by mites on tea is the loss of photosynthetic capacity of leaves, which later show a rusty appearance or a reddish-brown colour, until they eventually dry. Mites increase their populations during dry periods; but frequently their outbreaks are a direct consequence of incorrect cultivation practices or the wrong use of chemicals (pesticide and fertilizers) (Roy *et al.*, 2014). The indiscriminate use of broad-spectrum pesticides (e.g. pyrethroids and organophosphates) can favour mite outbreaks by reducing their natural enemies (Bartlett, 1968; Sarma, 1979; Penman and Chapman, 1988; Szczepaniec *et al.*, 2011) or, in some cases, increasing the reproductive rate of pest mite populations (James and Price, 2002). Mites have numerous natural enemies, which keep their populations under economic threshold level if undisturbed; this is why the use of selective pesticides is a control measure to apply only when necessary (Lehnmann-Danzinger, 2000). The latter decision should be made only after an appropriate monitoring of numerical consistency of the mite population, by checking both the damage on leaves and the actual presence of mites by shaking the leaves over a white sheet of paper (Hazarika *et al.*, 2009; Roy *et al.*, 2014). Management of mite pests is possible by correct cultural practices, including pruning old leaves, irrigation during dry periods, moderate shade of the plantation, balanced supply of fertilizers (overall nitrogen and phosphate) and weed management (Hazarika *et al.*, 2009; Roy *et al.*, 2014).

Worldwide, the most harmful mite species attacking tea is the red spider mite, *Oligonychus coffeae* Nietmer (Acari, Tetranychidae), and the scarlet red mite, *Brevipalpus phoenicis* (Geijskes) (Acari, Tenuipalpidae). They are polyphagous species that occasionally can cause crop losses of 17–46% (Muraleedharan *et al.*, 2005; Hazarika *et al.*, 2009; Sudoi *et al.*, 2011). Other species have a primary role in some tea-growing areas, like the kanzawa spider mite, *Tetranychus kanzawai* Kishida (Acari, Tetranychidae), the yellow tea mite, *Polyphagotarsonemus latus* (Banks) (Acari, Tarsonemidae), the pink tea rust mite, *Acaphylla theae* (Watt), and the carinate tea mite, *Calcarus carinatus* (Green) (Acari, Eryophidae) (Rattan, 1992; Takafuji *et al.*, 2000). The most important natural enemies of mites are predatory mites (gen. *Phytoseiulus, Amblyseiulus, Neoseiulus, Typhlodromus, Anystatis*), antochorids (gen. *Orius*), lady beetles (gen. *Scymnus, Stethorus*, etc.) and lacewings (gen. *Chrysopa*) (Lehnmann-Danzinger, 2000; Hazarika *et al.*, 2001, 2009; Ye *et al.*, 2014).

15.1.2 Termites

Several species of termites can occasionally damage tea plantations. These insects attack both dead or living plants, using natural cavities of roots or of the collar at below ground level for building a nest inside the heartwood (Nadda *et al.*, 2013). The attack causes an initial weakening of the plants that can lead to a loss of branches or roots. In severe cases, termites can cause the death of the plant, especially at nursery stage. Plants do not show symptoms of the attack until the colony is already well established. However, the presence of termites can be detected during the pruning period, by localizing the termites' cavities, or observing sudden wilt of branches during dry periods, or their tendency to break off. Control measures include the early detection of attacked branches, their immediate cutting off and disinfection of the canopy after pruning. When plants are irretrievably compromised, they must be destroyed (including the whole root system). In areas in which termites are a recurrent problem, the use of tolerant tea varieties is suggested (TRI, 2003).

15.1.3 Feeding insects

Thrips

Normally, thrips frequently occur in tea ecosystems, but sometimes some species can become pests as a consequence of altered environmental conditions. The attack by these insects causes discolorations, dark spots, scars on leaves and deformations on shoots when they pierce unopened buds. For management of thrips, it is recommended to monitor their presence and abundance by using white/blue sticky traps, to protect their natural enemies (mostly zoophagous thrips, antochorids and chrysopids), to pluck frequently to remove the pests and to limit the use of broad-spectrum pesticides (Zeiss and den Braber, 2001).

Bugs

Several species of the genus *Helopeltis* (Hemiptera, Miridae) are known in all tea-growing areas as important pests. In particular, *H. schoutedeni* Reuter, *H. theivora* Waterhouse and *H. antonii* Signoret are tea mosquito bugs reported as the most damaging species in Africa and Asia (Muraleedharan, 1992; Rattan, 1992; Sundararaju and Sundara Babu, 1999; Zeiss and den Breber, 2001). Both nymphs and adults sting on shoot and tender leaves and, as a consequence, they cause typical dark-brown circular necrotic spots that make leaves unusable. Sometimes, the leaves curl and wither, whereas several punctures to the shoots lead to the death of the stem (Hazarika *et al.*, 2009). In India and

Bangladesh, losses of 15–25% have been recorded by *H. theivora* (Ahmed, 1998). Control of *Helopeltis* spp. is very difficult, considering the abundance of their populations and the lack of efficacious natural enemies (Hazarika *et al.*, 2009). Therefore, mirid bugs are mostly controlled by pesticides, whose use is frequently unnecessary, ineffective and uneconomical (Ahmed *et al.*, 1998, 2011), especially when it is carried out without an appropriate monitoring of the insect populations. Mirid population reduction could be obtained by intensifying plucking (every 10–15 days), by removing damaged buds to stimulate the growth of new ones and protecting the activity of natural enemies (Nadda *et al.*, 2013). Several studies have shown the antifeedant, repellent and insecticidal effectiveness of some natural extracts from plants against *H. theivora* (Deka *et al.*, 1998; Gogoi *et al.*, 2003; Roy *et al.*, 2009; Sarmah and Kumar Bhola, 2015). Selective insecticides should be used when the presence of mirids exceeds the economic damage threshold of 8 adults counted inside a frame of $0.25\,m^2$ held up against the canopy on one side of randomly selected plants (Lehnmann-Danzinger, 2000), or of 5% of leaves on which adults or nymphs of the mirids are present (Muraleedharan, 1992). Recently, positive results in control of mirids have been obtained by using potassium chloride and potassium sulphate in combination with selective insecticides (Rahman *et al.*, 2014).

Leafhoppers

Until a few years ago, leafhoppers or cicadellids were considered as 'minor or occasional' pests and only recently have they achieved a major role in several tea-growing areas (Mu *et al.*, 2102; Saha *et al.*, 2012). The most important species in the Indian subcontinent is *Empoasca flavescens* Fabr. (Saha *et al.*, 2012), whereas in east Asia *E. onukii* Matsuda have a primary role (Qin *et al.*, 2014; Shi *et al.*, 2015). Their increasing importance as pests is probably a consequence of improper use of insecticides, that caused the development of resistant populations or weakened populations of natural enemies (Gurusubramanian *et al.*, 2008; Saha *et al.*, 2012). Damage by cicadellids is due to the feeding activity of all life stages on leaf tissues, that initially appear as small yellow spots which lead to curling and drying (commonly known as 'rim blight' or 'hopperburn'). An abundant oviposition into shoots may lead to their growth arrest. Control of cicadellids is particularly difficult, partly because of the limited effectiveness of their natural enemies to keep them under the economic damage threshold (Hazarika *et al.*, 2001; Sudhakaran and Muraleedharan, 2006). Good results can be achieved through the combined use of adequate cultivation practices, based on regular pruning (once every year) and plucking (every 15 days), control of weeds and balanced fertilization, without excessive use of nitrogen (Hazarika *et al.*, 2009; Nadda *et al.*, 2013). For *Empoasca* spp. control, use yellow sticky cards positioned at the top of the tea canopy, to monitor the real population and their fluctuation all year round (Shi *et al.*, 2015).

Aphids

Although various species of aphids (Hemiptera, Aphididae) are recorded on tea (Holman, 2009), only *Toxoptera aurantii* Bois de Fonscolombe may be occasionally harmful (Hazarika *et al.*, 2001; Hamasaki *et al.*, 2008). Heavy infestation by this aphid on tea is a rare event, often the consequence of the wrong cultural practices or particular climatic conditions. Normally, the species is well controlled by numerous hymenopterous parasitoids (*Lysiphlebus* spp., *Aphidius* spp., etc.) and various kinds of predators (syrphids, coccinellids and lacewings) (Hazarika *et al.*, 2001; Ye *et al.*, 2014). Tender shoots attacked by *T. aurantii* curl downward; but the main damage is caused by the sooty mould that grows on honeydew produced by aphids, reducing the quality of the leaves. Large bushes tolerate moderate infestations and no specific control is required. More attention is needed for young plants, on which heavy infestations could retard development.

Whiteflies

Among whiteflies (Hemiptera, Aleirody-dae), *Aleurocanthus spiniferus* Quaintance and *Dialeurodes citri* Ashmead are reported as pests of tea in China (Chen *et al.*, 1997), as is *Aleurodicus dispersus* Russell in Hawaii (Hamasaki *et al.*, 2008). They are considered minor pests as they hardly ever develop populations capable of causing serious damage to tea crops.

Scale insects

Although more than 40 species of mealy-bugs and scale insects (Hemiptera, Coccoi-dea) have been recorded on tea (Das, 1978), only a few of them can occasionally reach an economic importance (Narasimhan, 1987). The pseudococcid *Nipaecoccus viridis* (Newstead) is considered an important pest in some areas of tea regions of India (Sharma and Kashyap, 2002). When uncontrolled, this species attacks shoots, stems and branches, on which it may cause drying, although the main damage is due to the honeydew production and the consequent development of sooty mould, that compromises the quality of tea leaves (Nadda *et al.*, 2013). Many natural enemies are effective on coccids (Hazarika *et al.*, 2001), especially if their activity is safeguarded through proper cultivation practices and with the careful use of pesticide (Sharma and Kashyap, 2002). Ants are important in spreading mealybugs and protect them from beneficials in tea-growing areas; their control is suggested to reduce infestations.

15.1.4 Lepidoptera

Torticids, belonging prevalently to the genera *Adoxophyer* and *Homona*, are considered the main pests of tea in all growing areas (Muraleedharan, 2002). Because of the similarity in wing patterns, coloration and morphology of male genitalia, an effective discrimination among *Adoxophyer* spp. is possible only with molecular analysis (Lee *et al.*, 2005). In Japan, the most important species are *A. homnai* and *H. magnanima*, whereas in India and Sri Lanka, *H. coffearia* has a primary role (Kawai, 1997; Hazarika *et al.*, 2009). Damage is due to the activity of larvae, that shortly after emergence disperse on the vegetation to feed on shoots and tender leaves, which are tangled with silky threads and rendered useless. As demonstrated in several studies, torticids are brought under control overall by parasitoids (Cranham, 1961; Takagi, 1978; Mao and Kunimi, 1991; Takahashi *et al.*, 2008) as well as by viruses which cause epizootics (Hong, 1998; Nakai, 2009; Ye *et al.*, 2014). Less effective seem to be *Bacillus thuringiensis* Berliner (Ye *et al.*, 2014) or entomopathogenic fungi (Mao and Kunimi, 1991). In any case, the populations of tortricids should be kept under control though a visual monitoring and sampling, especially during the dry season. Sex pheromone can be useful for mating disruption, or sticky traps used to follow the population dynamics and help to prevent possible outbreaks. Pheromones should be used with caution, since in the literature, cases are reported of resistance in *A. honmai* and *A. magnanima*, when pheromones were used for mating disruption (Mochizuki *et al.*, 2002). *Andraca bipunctata* Walker (Lepidoptera, Bombycidae), known as 'bunch caterpillar', can reach a primary pest role because of the feeding activity of larvae on leaves of tea. Severe infestation causes retardation of growth of young plants and loss of product (Banerjee, 1983; Ghorai *et al.*, 2010). Various entomopathogenic fungi have been reported as effective against the caterpillar (Panigrahi, 1997; Ghorai and Bera, 1999). In Tanzania, the carpenter moth, *Teragra quadrangula* Gaere (Lepidoptera, Cossidae), caused losses of up to 50% in young tea cultivations. The larvae burrow feeding holes that, in the case of a severe attack, can lead the young plants (<5 years old) to die back. Infestation on mature plants takes place on branches and it is tolerated if they are in good vegetative state (Rattan, 2005). In mature crops, control of this insect is possible by the precocious detection of attacked plants and removal of infested branches, whereas in young crops the use of selective chemicals is necessary.

15.1.5 Coleoptera

Several scolytids or bark beetles (Coleoptera, Scolytidae) can attack tea. These small beetles burrow feeding and oviposition tunnels in the trunk or branches of the plants. The damage is worsened by *Monacrosporium ambrosium* Gadd and Loos, a fungus carried by these insects inside the plant and whose growing occludes the lymphatic channels and blocks the circulation, leading the branches or the entire plant to wither. In some cases, bark beetles can cause serious damage, such as those recorded in Sri Lanka and caused by *Xyleborus fornicates* Eichhoff (Walgama and Pallemulla, 2005). The control is achieved through a careful monitoring of the plants and keeping the same in good vegetative state with regular pruning and balanced fertilization. In particular, this latter promotes the growth of new tissues and the production of saponins and sterol analogues, that inhibit the development of larvae (Wickremasinghe and Thirugnanasuntheram, 1980).

The Coleoptera *Holotrichia* spp. (Scarabeidae), *Astycus* spp. (Curculionidae) and *Basilepta melanopus* Lefevre (Chrisomelidae) are known as defoliators and on rare occasions may become harmful for tea production (Panda, 2011). Against *B. melanopus* effective control was achieved by the nematode *Steinernema carpocapsae* (Weiser) and the fungus *Beauveria bassiana* (Balsamo) Vuillemin (Liao *et al.*, 2007; Zeng *et al.*, 2012).

15.2 IPM in Cocoa

Cocoa, *Theobroma cacao* L., represents one of the main export commodities of many developing countries, together with tea and coffee (FAO, 2015). Cocoa is mainly cultivated by smallholder farmers in the tropics and around 90% of world cocoa production comes from small farms (Baah and Garforth, 2008; ICCO, 2013). West Africa is the largest producer and exporter of cocoa in the world (FAO, 2015; ICCO, 2012), where Côte d'Ivoire is the fulcrum and pivot of the entire world cocoa market. Among the first positions, there are also Indonesia and Malaysia (FAO, 2015). In South America, Brazil was the second largest exporter of cocoa in the world; however, after the 1980s, due to disease outbreaks and a larger internal consumption, exports dropped dramatically (Bowers *et al.*, 2001). Cocoa does not represent only an important agricultural output for the macro-economy, but also a pillar for almost 600 million small farms, for which this crop is the main source of income (Gotsch, 1997; Neilson, 2007; FAO, 2014). Given the lack of resources in terms of information, technology and capitals, cocoa farmers are greatly susceptible to constrictions such as pests, which represent one of the most important impediments to agricultural productivity (FAO, 2014; Curry *et al.*, 2015).

It is known that more than 1000 different herbivorous insects feed on cocoa; however, only a small percentage of these represent a serious threat for cocoa production (Wood and Lass, 2008). Among them, mirids are the most harmful and widespread insect pests of cocoa, followed by the cocoa pod borer, which is considered the main key pest in Asia. Mealybugs are well known to be vectors of diseases on cocoa. Other pests which can represent a threat to cocoa plantations are mainly caterpillars, moths, termites and thrips.

15.2.1 Mirids

Mirids, also known as Capsids, are among the major threats to cocoa plantations. They are ascribed to the family Miridae (in the order Hemiptera), which includes more than 10,000 species, mainly polyphagous (Gillot, 2005). Several species are known to attack cocoa, causing conspicuous losses up to 75% (Entwistle, 1972; Mahob *et al.*, 2011). Mirids are sap-sucking insects that feed on young pods and stems. They inject toxic saliva into the plants, causing tissue necrosis, desiccation of leaves, dieback, malformation and, in extreme cases, the annihilation of the entire plant structure. Often, injuries induced by their feeding can evolve into cankers, facilitating the colonization of harmful fungi and bacteria (Crowdy, 1947; Akpesse, 2014; Mahob *et al.*, 2015; Adu-Acheampong *et al.*, 2015). In west and central Africa, cocoa production is mainly threatened by two mirid species, *Sahlbergella singularis* Hagl. and *Distantiella theobroma* (Dist.). Alone, these two species cause up to 30% of cocoa losses each year (Sosan and Akingbohungbe, 2009). Species ascribed to the genus *Helopeltis* are the major pests of cocoa in Asia (Azhar, 1989). *Helopeltis theobromae* Mill. and *H. clavifer* (Walker) are the most common species ascribed to this genus (Azhar, 1989; Srikumar and Bhat, 2012). Other relevant genera, but restricted to limited areas, are *Pseudodoniella*, in Papua New Guinea, and *Platyngomiriodes*, in Borneo. Instead, species of the genus *Monalonion* represent important constraints in South and Central America (World Cocoa Foundation, 2003). Several approaches aimed at reducing the impact of mirids have been envisioned.

Chemical control

Generally, chemical control has been considered the most effective control method of cocoa mirids (Sonwa *et al.*, 2008; Mahob *et al.*, 2011). However, chemical control has not always been successful, especially due to the frequent development of pesticide resistance (Dittrich *et al.*, 1990; Bisseleua *et al.*, 2011). In addition, smallholder farmers in developing countries have few resources in terms of credit and technical information for the application and appropriate use of chemical pesticides (Abate *et al.*, 2000; Guastella *et al.*, 2015). Often, cocoa farmers apply insecticides that have been already banned, and generally do not follow a spraying programme (Babin *et al.*, 2010). This is why research efforts have been focused on developing appropriate, sustainable and integrated pest management solutions.

Cultural techniques

Cultural techniques, such as upkeep and sucker removal on farms and the maintenance of the tree canopy, have been widely applied, even in combination with chemicals, with the aim of reducing the pest damage to cocoa plantations (World Cocoa Foundation, 2003; Adu-Acheampong *et al.*, 2015; Azeez, 2015). Recently, an interesting technique which is becoming widely recommended in cocoa areas is shade management, considered as an effective pest management strategy against cocoa mirids (Rice and Greenberg, 2000; Bisseleua, 2013). This technique is inspired directly by the biology of miridis. In fact, mirids form aggregations called pockets, which often occur in sunny areas of plantations (Entwistle, 1972). Several studies have confirmed the efficiency of this method (Mahob *et al.*, 2011, 2015). Unfortunately, smallholder farmers do not include always this technique in their pest management programmes.

Biological control

The efficiency of biological control remains still questionable. The parasitoid *Euphorus sahlbergella* Wilk. (Braconidae) was recorded to control young stages of *S. singularis* from 6 to 20% (Collingwood, 1971) or to 40% in other areas (Azeez, 2015). The deployment of ants, such as *Dolichoderus thoracicus* (Smith) (Formicidae) is worthy of mention to control several species of mirids and other pests (van Mele *et al.*, 2001; Saripah, 2014). However, as for many other ant species, this species is associated with the cocoa mealybug *Cataenococcus hispidus* Morrison, from which it obtains nutrition thanks to the production of honeydew. Therefore, in areas where Cocoa Swollen Shoot Virus (CSSV) is present, this method could represent a threat to cocoa plantations since the mealybug is able to transmit the virus (World Cocoa Foundation, 2003; Saripah, 2014). A more complete list of biological control agents is presented in Table 15.2.

Plant resistance

Another way to implement IPM strategies on cocoa is to develop new varieties that are resistant to pests (host plant resistance). The use of hybrid is an option that has been

Table 15.2. Main control strategies of pests attacking tea, cocoa and coffee.

Crop	Pests		Biological control	Cultural approach	Chemical control
TEA	Mites	*Olygonicus coffeae* Nietner; *Brevipalpus phoenicis* (Geijskes); *Tetranychus kanzawai* Kishida; *Polyphagotarsonemus latus* (Banks); *Acaphylla theae* (Watt); *Calacarus carinatus* (Green)	*Phytoseiulus* spp., *Amblyseiulus* spp., *Neoseiulus* spp. and *Typhlodromus* spp. (Mites); *Orius* spp. (Anthocorids), *Scymnus* spp. and *Stethorus* spp. (Lady beetles); *Chrysopa* spp. (Lacewings); *Verticillium lecani* Zare & Gams (entomopathogenic Fungi)	Monitoring population; shade and weed management to remove alternative hosts; pruning and plucking; avoid excess of nitrogen	Sulphur formulations; Neonicotinoids; Spinosyns; Avermectins; Pyrazoles; Oxadizines; Pyrethroids; Pyrroles
	Termites	Several species		Early detection; pruning; eradication	
	Thrips	Several species	Zoophagus thrips; Antochorids; Chrysopids		Avermectins; Neonicotinoids; Spinosyns; Oxadizines; Pyrethroids
	Bugs	*Helopeltis* spp.		Intensify plucking; removal of damaged buds	Neonicotinoids; Organophosphates; Pyrethroids;
	Leafhoppers	*Empoasca* spp.		Pruning; plucking; weed control; avoid excess of nitrogen	Neonicotinoids; Spinosyns; Avermectins; Organophoshates; Pyrethroids
	Aphids	*Toxoptera aurantii* (Boyer de Fonscolombe)	*Lysiphlebus* spp. and *Aphidius* spp. (Hymenopteran); *Beauveria bassiana* (Bals.-Criv.) Vuill., *Paecilomyces* spp. and *Metarhizium* spp. (entomopathogenic Fungi)		Organophosphates and N-Methyl Carbamates; Neonicotinoids; Pyrethroids; Insect Growth Regulators (IGRs) such as (S)-Kinoprene or Methoprene
	Whiteflies	*Aleurocanthus spiniferus* (Quaintance), *Dialeurodes citri* (Ashmead) and *Aleurodicus dispersus* (Russell)	Several natural enemies (Hymenopteran): *Amitus hesperidum* Silvestri, *Encarsia guadelupe* Viggiani, *E. haitiensis* Dozier, *E. lahorensis* (Howard), *E. smithi* (Silvestri)		Avermectins; Neonicotinoids; soaps, neem oil; Pyrethroids; Ryanoid

	Scale insects	*Nipaecoccus viridis* (Newstead)	*Anagyrus agraensis* Saraswat (Hymenopteran)	Monitor field population	Mineral oils, Neonicotinoids; Pyrethroids
	Lepidoptera	*Adoxophyer* spp.; *Homona* spp.; *Andraca bipunctata* Walker	Virus causing epizootics; *Bacillus thuringiensis* Berliner (Bacteria); sex pheromones for mating disruption; *Zoophthora radicans* (Brefeld) Batko (entomopathogenic Fungi)		Neem formulations; Neonicotinoids; Spinosyns; Avermectins; Pyrazoles; Oxadizines; Pyrethroids
	Coleoptera	*Xyleborus fornicatus* Eichhoff; *Holotrichia* spp.; *Astycus* spp.; *Basilepta melanopus* Lefevre	*Steinernema carpocapsae* (Weiser) [Nematode]; *Beauveria bassiana* (Bals.-Criv.) Vuill. (entomopathogenic Fungi)	Regular pruning; balanced fertilization	Neonicotinoids; Spinosyns; Avermectins; Pyrazoles; Oxadizines; Pyrethroids
COCOA	Termites	Kalotermitidae; Rhinotermidae	Several species of Ants	Deep ploughing; flooding; burning nests	Pyrethroids
	Thrips	*Selenothrips rubrocinctus* (Giard)	Mites, Lacewings and Bugs		Avermectins; Neonicotinoids; Spinosyns; Oxadizines; Pyrethroids
	Mirids	*Sahlbergella* spp.; *Distantiella theobroma* (Dist.); *Helopeltis* spp.; *Pseudoniella* spp.; *Platyngomiriodes* spp.; *Monalonion* spp.	*Euphorus sahlbergella* Wilkinson (Hym. Braconidae); *Dolichoderus thoracicus* (Smith) (Ants); *Beauveria bassiana* (Bals.-Criv.) Vuill. (entomopathogenic Fungi)	Upkeep and sucker removal; shade management; host plant resistance	Carbamates; Organophosphates; Phenylpyrazoles; Pymetrozine Neonicotinoids
	Lepidoptera	*Conopomorpha cramerella* Snellen; *Eulophonotus myrmelon* Felder; *Zeuzera coffeae* Nietner	*Dolichoderus* spp. and *Oecophylla smaragdina* (Fabr.) (Ants); *Beauveria bassiana* (Bals.-Criv.) Vuill. (entomopathogenic Fungi); *Glyptomorpha* spp. and *Bracon zeuzerae* (Rohwer) (Hymenopteran)	Early harvesting; pruning attacked stems	Methoxyfenozide; Pyrethroids; Carbamates

continued

Table 15.2. *continued*

Crop	Pests		Biological control	Cultural approach	Chemical control
COFFEE	Mites	*Oligonychus* spp.; *Polyphagotarsonemus latus* (Banks); *Brevipalpus phoenicis* (Geijskes)	*Phytoseiulus* spp., *Amblyseiulus* spp., *Neoseiulus* spp. and *Typhlodromus* spp. (Mites); *Orius* spp. (Anthocorids), *Scymnus* spp. and *Stethorus* spp. (Lady beetles); *Chrysopa* spp. (Lacewings)		Sulphur formulations; Potassium salts; Buprofezin
	Hemiptera	*Antestiopsis* spp. (Pentatomid bugs); *Planococcus* spp. and *Ferrisia virgata* Cockerell (Mealybugs); *Coccus viridis* (Green) and *Saissetia oleae* (Olivier) (Scale insects)		Pruning; plant spacing	Azadirachtin; Borax; Buprofezin; Neonicotinoids; Mineral oil; Pyrethroids
	Lepidoptera	*Leucoptera coffeella* (Guérin-Méneville)	*Bacillus thuringiensis* Berliner		Avermectins; Diamides; Carbamate; Pyrethroids; Neonicotinoids; Organophosphates
	Coleoptera	*Hypothenemus hampei* (Ferrari); *Xylosandrus* spp.; *Xylotrechus quadripes* Chevrolat	*Prorops nasuta* Waterston; *Phymastichus coffea* LaSalle; *Cephalonomia stephanoderis* Betrem; *Heterospilus caffeicola* Schmiedeknecht (Hymenopteran); *Beauveria bassiana* (Bals.-Criv.) Vuill. (entomopathogenic Fungi)	Complete harvest; destruction of residuals; balanced fertilization; low nitrogen input	Azadirachtin; Vegetable oils; Mineral oils; Organophosphates; Pyrethroids

taken into consideration as new approach to control mirids on cocoa. However, none of these new varieties has been fully satisfactory (Adu-Acheampong *et al.*, 2007, 2015; Azeez, 2015).

15.2.2 Cocoa pod borer (CPB)

Conopomorpha cramerella Snellen (Lepidoptera: Gracillariidae) is one of the most injurious pests on cocoa, which can reduce production by 90% (Rosmana, 2010). The species attacks directly the cocoa pod by feeding on the placental tissues and the pulp around the beans. Thereafter, the beans are often unusable. Several chemical, biological and mechanical methods are used to control this pest. Chemical control involves the use of pyrethroids or carbamates, which are used to control also other pests. Often, farmers prefer to utilize chemicals due their rapid action (Azhar, 1992; Saripah, 2014). An important step on cultural control was made by Mumford (1986). He noticed a strong relationship between harvesting practices and the level of population of cocoa pod borer (CPB). In fact, the majority of CPB larvae emerge from the pods after they become ripe, and if the pods are picked at the earliest stage, the majority of larvae will still be inside the pods. If they are quickly destroyed, the larval mortality will be very high and a considerable degree of control can be achieved. As for mirids, biological control encourages the utilization of ants, such as *Dolichoderus* sp. and *Oecophylla smaragdina* (Fabricus), also known as weaver ants. This last ant species is known to be very aggressive and inflict painful bites (Gathalkar and Barsagade, 2016); for this reason it is not considered to be the best choice by farmers. However, several studies showed the great potential of ants to control CPB; in some cases, similar to insecticides (Shahabuddin and Pasaru, 2013; Saripah, 2014). Among pathogens, the most effective is *Beauveria bassiana* (Balsamo) Vuillemin, which can control efficiently CPB (World Cocoa Foundation, 2003; Posada *et al.*, 2010), especially when

CPB larvae are exposed to the fungus. A successful example of mechanical control of CPB has been applied in the Philippines, where a few cocoa plantations have been planted at high densities as hedges, accessible with small tractors between the rows, allowing an easy and complete harvest. By this way, infestations of pod borer were almost unimportant (World Cocoa Foundation, 2003). Synthetic pheromones were also evaluated in the mid-1980s, when field experiments gave good results and new hopes (Shahabuddin and Pasaru, 2013). Unluckily, in the late 1980s, a new pod borer race was found in west Malaysia, which is not sensitive to the pheromones (Shahabuddin and Pasaru, 2013). For this reason, semio-chemicals are no longer used to control CPB.

15.2.3 Stem borers

Several insects belonging to the orders Lepidoptera and Coleoptera are able to bore and feed into cocoa stems, and sometimes can inflict serious damages. In Africa, *Eulophonotus myrmeleon* Felder (Lepidoptera) is common and can also feed on other important economic plants such as pecan (*Carya illinoinensis* (Wangenh.) K. Koch) and coffee (*Coffea* spp.). Severe outbreaks by this species occurred recently in west Africa. *Zeuzera coffeae* Nietner (Lepidoptera), a polyphagous insect which has several hosts, is mainly spread in Asia and Papua New Guinea. Coleoptera belonging to the genus *Pantorhytes* are present in New Guinea, Solomon Islands, and only one species is found in Australia.

Cultural control of these species includes mainly pruning of attacked stems; it does reduce stem borer populations but is labour-intensive. Biological control involves the use of parasitoids. In Java, *Z. coffeae* is usually parasitized by *Bracon zeuzerae* (Rohwer) (Hymenoptera). In Malaysia, *Eulophonotus myrmeleon* larvae are parasitized by a *Glyptomorpha* sp. (Hymenoptera). On *Pantorhytes*, no parasites or predators have shown promising control. Regarding the chemical control, unfortunately, the only effective insecticides are extremely toxic and too expensive for smallholder farmers (Cocoasafe, 2016).

15.2.4 Termites

Many species of termites are found in cocoa plantations worldwide; however, only species belonging to three families represent a serious threat. The family Kalotermitidae includes dry- and damp-wood termites, which can live in cavities inside the wood with no connection with the soil. The Termitidae are wood-eaters and live underground. Rhinotermitidae are living mainly underground; they mainly attack dead and decaying wood but occasionally can attack living tissues. Symptoms of their attacks on young plants are more evident around the collar area and the plants can show severe wilt. This usually happens during the dry season. Cultural control involves deep ploughing, with the aim to open the underground nests and expose termites to drying-out. Flooding or burning nests can also kills the colonies. Ants can be used as biological control agents against termites. Chemical control includes the use of insecticides belonging to the pyrethroids group, mainly applied as barriers in the soil around roots (World Cocoa Foundation, 2003; Cocoasafe, 2016).

15.2.5 Thrips

Selenothrips rubrocinctus (Giard), known also as red-banded thrips, was discovered in Guadeloupe (West Indies) and nowadays is found in all cocoa areas of Africa, Asia, Australia and South America (Denmark and Wolfenbarger, 2010). This pest is able to cause substantial injuries on cocoa. The larvae and adults attack leaves and fruits by piercing the epidermis with their mouthparts and causing leaf distortion and drop as well as injuries to the fruits. Several biological control agents are known on the red-banded thrips, such as mites, lacewings and

bugs (Chin and Brown, 2008). If needed, chemical control includes a series of contact insecticides such as azadiractin, insecticidal soaps, narrow-range oil, neem oil and pyrethrins. To be effective, contact sprays must be applied as cover spray on buds, shoot tips and hidden parts. Spinosad seems to be as effective against thrips as the products above, but obviously more costly. Use of organophosphate insecticides, such as malathion, or carbamates, is not advised, since these chemicals are highly toxic and can even cause a dangerous increase of spider mite populations.

Chemical control

Smallholder farmers usually utilize pesticides because it is easier, less time-consuming and (when results are positive) provides a visual response in the short term. However, inaccurate dosage of chemicals, poor coverage due ineffective applications, wrong timing, and inappropriate methods of spraying or wrong calibration of the equipment very often result in failures (Asogwa and Dongo, 2009), which usually imply more sprays or overdosage. The full understanding of the equipment used is a crucial step for the farmer. For example, knapsacks (manual or motorized) are widely used by smallholder farmers, since they are efficient tools that can be used for different operations. However, they are not so easy to manipulate, since they require knowledge on the several interchangeable nozzles which will give the desired spray pattern and which should be adopted based on the type of operation, pest target, chemical and weather condition. Calibration of the equipment is an important step, but it is rarely part of the farmer's routine. Overdosing pesticides means incurring expensive applications, phytotoxicity and problems of pest resistance. Another important obstacle to rational pesticides use is their availability, especially in those rural areas which are distant from major cities. In these areas, it is always difficult to buy the most appropriate product; or, in the worst case, only expired or banned chemicals may be available.

Biological control

On cocoa, several natural enemies have been identified on the main pests, but at the moment biocontrol is mainly focused on ants and on fungi (Amin *et al.*, 2014). In addition, many parasitoids and predators have been described (Azhar, 1989), but information on them is still scarce. Despite all this, biocontrol agents are a pillar of IPM strategies. Compared with other control methods, their advantage is that, once established, their efficiency remains over time and there is almost no risk of resistance development (Bale *et al.*, 2008; Naranjo *et al.*, 2015). However, there are several reasons why smallholder farmers are sceptical about biological control agents. The first one is that biocontrol is much slower to control pest populations if compared to pesticides, although in the long term, this strategy can pay. Biological control is not appreciated by smallholder farmers also due to costs, limited conservation time and methods of distribution. The efficiency of commercial chains of biological control agents is still rather poor in countries where cocoa is mainly cultivated. And again, the application of these agents requires relevant technical knowledge, since the application timing (season, day temperature, etc.), the establishing time and the integration of biocontrol with chemical spray are fundamental for successful pest control.

15.3 IPM in Coffee

Coffea arabica L. and *C. canephora* Pierre (=*robusta*) (Rubiales, Rubiaceae) presently represent about 62% and 38% of commercial coffee cultivated in the world, respectively (FAS/USDA, 2015). Coffee is a typical tropical and subtropical plant, grown on about 10,142,836 ha distributed in 70 countries, with a total production of about 8,923,000 tons (FAOSTAT, 2016). *C. arabica* is endemic to the mountain regions of Ethiopia, from which it spread as a number of varieties in all tropical and subtropical regions of the world. The species grows from 1200 to 2200 m above sea level, with mean temperatures that range from 15 to 24°C and are fairly constant, with a long period (about 7 months) of rain (Pohlan and Janssens, 2012). *C. robusta* is native to central Africa and widespread in all the main areas of coffee cultivation between 15°N and 12°S. It is adapted to places with an altitude between 300 and 800 m above sea level, mean temperatures ranging from 22 to 30°C and high relative humidity (Pohlan and Janssen, 2012). A traditional method of cultivation (still followed or re-adopted in some areas of Mexico, Colombia, Ethiopia and other countries) contemplates the growth of the Arabica coffee plants under the shade of tall trees, originally thought to protect the plants from direct exposure to sunlight. Currently, most of the crops are in full field using 'sun-tolerant' hybrids, which allow intensive cultivation of coffee (Rice, 1996). Without discussing the advantages and disadvantages of the two methods, they undoubtedly lead to different environmental conditions and distinct control strategies to be adopted for insect pests. Some studies have shown that in 'shade-grown coffee' high biodiversity results in a reduction of damage caused by some pests. On the other hand, the greater productivity of the intensive cultivation, in addition to the environmental costs and the loss of biodiversity, requires high and expensive production inputs (Perfecto *et al.*, 1996; Moguel and Toledo, 1999; Takahashi and Todo, 2014; Jha *et al.*, 2014). Recently, thanks to the 'Shade coffee certification programme', that aims to preserve forest environments and allow farmers to get a higher economic return from the sale of coffee, there has been an increase of this traditional cultivation method in various parts of the world (Ruben and Fort, 2012; Takahashi and Todo, 2014). Regardless of the method of cultivation, the success in pest control is reached when it is applied through management of insects rather than a struggle against them, and through a joint consideration of all components of the agroecosystem, in order to achieve a balance that keeps pests below the threshold of harmfulness. On the other hand, sustainable production through the adoption of certified schemes is one of the main objectives of the coffee industry and big retail (Jaramillo *et al.*, 2006; Elder *et al.*, 2014). Cultural strategies are very effective but are expensive and require specialized personnel, and these reasons frequently hamper its spread.

Worldwide, there are more than 850 insects reported in the coffee agroecosystem, reflecting the ecological complexity of the environment (Le Pelley, 1973). Of course, only a small percentage of them reach a 'harmful' level and seriously affect the coffee crops.

15.3.1 Mites

On coffee, the main species reported as pests are the tea red spider mite, *Oligonychus ilicis* (McGregor), the coffee red spider mite, *O. coffeae* (Nietner) (Acari, Tetranychdae) and the broad mite *Polyphagotarsonemus latus* (Banks) (Acari, Tarsonenidae). They are polyphagous species, normally not considered as primary pests and distributed in all main geographical areas of cultivation of coffee (Franco *et al.*, 2009; Vacante, 2016). Spider mites pierce the cells of foliar mesophyll-destroying chloroplasts. As consequence, the leaves became yellowish-brown in colour and the photosynthetic activity can be seriously compromised (Fahl *et al.*, 2007). Another species, the false spider mite, *Brevipalpus phoenicis* (Geijskes) (Acari, Tenuipalpidae), is the feared vector of the 'coffee ringspot virus' that worsens the quality of coffee (Reis and Chagas, 2001). Mite populations are usually maintained below the economic threshold of damage by a number of predatory mites and other natural enemies (Franco *et al.*, 2010). The increase of pest mite populations is often the consequence of environmental imbalance (e.g. anomalous weather conditions, improved nutritional conditions of plants, improper use of chemicals) (Reis and Teodoro, 2000). The return to normal conditions could be sufficient to reduce mite populations to the levels of non-harmfulness. The damage threshold adopted in Brazil is 30–40 mites per leaf in the dry season, and when it is reached, one acaricide treatment is suggested on infested plants (Barrera, 2008). However, Fahl *et al.* (2007) observed that attacks of *O. ilicis* on coffee plants during the dry season (when the vegetative growth is reduced) do not affect crop yield. Alternatively, some products approved for organic/coffee production can be used to control pest mites and protect predatory Phytoseiid mites (Tuelher *et al.*, 2014).

15.3.2 Coleoptera

The coffee berry borer (CBB), *Hypothenemus hampei* (Ferrari) (Coleoptera, Scolytidae), is worldwide considered the most harmful insect pest of coffee, considering both its direct damages (coffee weight and quality loss) and the high costs involved in its control (Damon, 2000). The data from several worldwide producing countries indicate that the percentage of attacked coffee berries may exceed 80% (Vega, 2004). Native to central Africa (Baker, 1984; Murphy and Moore, 1990; Damon, 2000), *H. hampei* is reported currently in all coffee-producing areas, except China, Nepal and Papua New Guinea (CABI, 2016a). Although coffee is the favourite host plant of CBB, it can also develop on Fabaceae, Rubiaceae, Rosaceae, Malvaceae, Oleaceae and Vitaceae (Damon, 2000). The females begin to bore the coffee berries and the endosperm after about 60 and 120–150 days from flowering, respectively (Muñoz, 1989; Ruiz-Càrdenas and Baker, 2010). CBB gets inside the berry, digging a small visible hole at the apex (1 mm in diameter). The attack is visible also by the debris produced and accumulated outside the coffee berry by the pest. Dissecting the infested coffee berries, the beetle can be observed in the mesoderm between the two seeds or among the tunnels dug, depending on the maturity degree of the endosperm (Damon, 2000). The species follows the vegetative cycle of the plant and develops a number of yearly generations that vary from 2 in Ethiopia to 3–4 reported for Colombia and Tanzania (Baker, 1998;

Jaramillo *et al.*, 2009), probably due to the different environmental conditions (Damon, 2000). The insect spends almost the entire life inside the coffee berry, from eggs hatching up to the moment in which the fertilized female leaves the ripe berry to search for another one susceptible to be colonized.

Some cultural control techniques are suggested, especially to small producers, for their easy applicability and cheapness. All ripe coffee berries on trees and on the ground should be removed after the harvest and during interharvest period, because their persistence allows CBB to linger on crop and increase its infestation (Bustillo *et al.*, 1998; Jaramillo *et al.*, 2006). However, the method is expensive, laborious and can have a negative effect on parasitoids. In Kenya, Jaramillo *et al.* (2009) found that the density of the hymenopteran parasitoid *Prorops nasuta* Waterston on ground coffee berries was 90% higher than on berries from the trees and their destruction affected the population of the parasitoid. The same authors suggest gathering all coffee berries collected in a suitable container to allow the parasitoids to escape. Other cultural control methods include the cultivation of varieties resistant to the coffee leaf rust, so as not to damage *Beauveria bassiana* (Balsamo) Vuillemin when used to control CBB; to use large plant densities leaving two stems per tree, to favour the crop operations; to replace plants after the fifth harvesting year; to control the unwanted weeds (Benavides *et al.*, 2012).

For years, chemical control has been the exclusive control method used against coffee pests. Currently, insecticide treatments should be kept to a minimum, considering that: i) coffee is often grown in areas of exceptional ecological value; ii) they frequently fail to control CBB; iii) resistant strains may appear; iv) strong market restrictions are increasingly applied for insecticide residues on exporting coffee (Brun *et al.*, 1994; Botero and Baker, 2001; Benavides *et al.*, 2012). For crops in which flowering is sufficiently regular, it could be useful identifying the maximum bloom period and carry out the insecticidal treatment after about 8 weeks, when most females are outside in search of or beginning to penetrate coffee berries. However, the uneven maturity of the coffee berries and the fact that they remain hanging on the plants when not picked or on the ground when decayed, mean that it is not easy to identify this period (Baker, 1999a,b). Sampling the population and monitoring the flight activity of *H. hampei* are the first important steps of IPM. The so-called '30 tree-sampling procedure' appears to be the most efficient (Bustillo *et al.*, 1998; Aristizábal *et al.*, 2016). It consists in dividing the plantation in plots of about 5000 plants each, among which 30 plants will be chosen at random. On each plant, a central branch containing 30–100 developing coffee berries is selected, recording those with or without bores. Finally, on the total berries recorded, the percentage of attacked ones is calculated. In Colombia, the chemical intervention threshold is more than 2% of infested berries with 50% in which the females are still at the initial stage of penetration. In fact, any insecticide treatment would be ineffective if the insect has already reached the endosperm. The method makes it possible to localize the sites in the plantation with the highest concentration of *H. hampei*, in which control measures have to be started or intensified. An effective chemical treatment should be done within 5 days from sampling (Aristizábal *et al.*, 2012). Monitoring, through the use of traps, provides useful indications on seasonal flight activity of CBB. The most effective traps are coloured (red or white) and contain a mixture of ethanol/methanol (1:3 or 1:1 ratio) placed at 0.5–1.5 m high (Aristizábal *et al.*, 2015). The captures should be evaluated carefully, considering such factors as the weather conditions and the pest infestation level, that can alter the course and the number of catch (Borbón *et al.*, 2002; Uemura-Lima *et al.*, 2010).

Many parasites have been reported on *H. hampei* (Damon, 2000) and some of them have been used for biological control programmes. The hymenopteran parasitoids *Cephalonomia stephanoderis* Betrem, *Prorops nasuta* Waterston (Bethylidae), *Phymastichus coffea* LaSalle (Eulophidae) and

Heterospilus coffeicola Schmiedeknecht (Braconidae) are African species introduced in Central and South America for biological control programmes (Baker, 1998). Another species, *Chephalonomia hyalinipennis* Ashmead, is another parasitoid of CBB found in Mexico (Pérez-Lachaud, 1998). In Central and South America, several experimental field trials have been carried out on the three bethylid species, and the results agree regarding their limited impact on *H. hampei* population (Baker, 1998, 1999a,b; Damon, 2000). However, in African native areas, *C. stephanoderis* and *P. nasuta* are considered very effective in controlling the CBB populations (Damon, 2000; Jaramillo *et al.*, 2009). *P. coffea* appears to be the most promising natural enemy of the coffee berry borer (Jaramillo *et al.*, 2005). A ratio *P. coffea/H. hampei* of 1:5–1:10 is necessary for achieving more than 60% parasitism (Infante *et al.*, 2013). The species has been mass-reared and repeatedly released and has been established in North, Central and South America (Baker *et al.*, 2002). In Uganda, *H. coffeicola* is effective in CBB control, but the difficulties of its mass-rearing have hampered its use in extensive biological control programme so far (Jaramillo *et al.*, 2006; Kucel *et al.*, 2009). The studies carried out to date show that the parasitoids of *H. hampei* are effective only if integrated with other control measures and applied on large areas. Among the numerous entomopathogenic fungi infecting *H. hampei* (Vega *et al.*, 2009), *B. bassiana* is the most common and recommended in IPM programmes by some important coffee producers associations (Bustillo, 2002; Benavides *et al.*, 2012). However, its large-scale use by growers is limited mainly by its expensiveness, the need of high relative humidity and the occasional lack of efficacy (Jaramillo *et al.*, 2006). Research is making efforts to improve the efficacy of this fungus. As to this aspect, promising appear the results obtained, both in laboratory and in field trials, by mixing different strains of *B. bassiana* (Benavides *et al.*, 2012). Similarly, several entomopathogenic nematodes (e.g. *Steinernema feltiae* Filipjev and *Heterorabditis bacteriophora* Poinar) showed in laboratory trials a good activity against CBB, but their effectiveness in the field is to be further investigated (Molina and Lopez, 2002; Manton *et al.*, 2012).

A satisfactory control of *H. hampei* is still far to be achieved and further difficulties could take over. Jaramillo *et al.* (2009) stated that, as consequence of global warming, the pest could develop also in areas presently free due to the low temperatures prevalent (e.g. Ethiopia), or worsen its infestations. The future challenges of research on CBB control have several goals. Ceja-Navarro *et al.* (2015) discovered the microorganisms present in the gut of *H. hampei* and responsible for the degradation of caffeine, a substance otherwise toxic for the same insect. A method to inactivate these gut microbiotas may be a new option for pest control. A similar research regards the method to inactivate the proteobacterium *Wolbachia*, a microorganism involved in the typical unbalanced sex ratio in favour of female in *H. hampei* (Vega *et al.*, 2009). Further investigations are underway to explore the nature of kairomones involved in the coffee berry/CBB relationship, in order to improve the attractiveness of traps aimed at monitoring the insect populations (Ortiz *et al.*, 2004; Aristizábal *et al.*, 2015).

Other scolytid beetles can occasionally cause economic damage to coffee plantations, as the black twig borer, *Xylosandrus compactus* (Eichhoff), the brown twig borer, *X. morigerus* (Blandford), the granulate ambrosia beetle, *X. crassiusculus* (Motschulsky), and *X. discolour* (Blandford). They are cosmopolitan (except *X. discolour*, distributed only in Asia and Oceania) and polyphagous pests, known for the association with symbiotic ambrosia fungi and used as food for both adults and larval stages (Wood, 1982). *Xylosandrus* spp. attack shoots, twigs and small branches, on both suffering and healthy plants (Egonyu *et al.*, 2009). Damage is due to the digging activity of larvae and adults into the pith or wood to a depth of 1–3 cm and to the ambrosia fungi, which develop invading and occluding xylem vessels and impede vascular transport. Infested plants show wilting, necrosis, defoliation, branch dieback and a general debilitation

(Ngoan *et al.*, 1976; Greco and Wright, 2015). These beetles spend most of their life inside the internal tissues of coffee plants and this makes difficult their control with insecticides. Among the suggested practices, there is the use of a mixture of fungicide, against the ambrosia fungi, and systemic insecticides (1–2 treatments per month, depending on the level of infestation). The use of trapping lures helps to monitor the periods of flight. The most important cultural control practices are the cultivation of certified plants, the continuous visual inspections to detect eventually infested plants and the immediate destruction of twigs with ambrosia beetles. Moreover, it is fundamental to keep the plants in good vegetative state through proper pruning and fertilizing and to grow tolerant coffee varieties to *Xylosandrus* spp. (Burbano *et al.*, 2012; Greco and Wright, 2015; Kagezi *et al.*, 2015). As known, the control activity carried out by natural enemies is not effective (Greco and Wright, 2015).

Several coffee stem borers (Coleoptera Cerambycidae) can cause damage to coffee in various parts of the world (Vega *et al.*, 2007). Damage is caused by larvae that dig tunnels in the main stem and primary branch, causing withering and wilting symptoms to the attacked plants (Rhainds *et al.*, 2002). *Xylotrechus quadripes* (Chevrolat) is considered a serious pest in several Asian coffee-producing countries (Venkatesha and Divesh, 2012). In India, the most effective method is the stem hand-scrubbing, to smooth the surface so as to avoid egg-laying and removing any eggs and young larvae already present. Another method consists of covering the plants with a nylon net during the flight period of the cerambycid. However, the methods are labour-intensive and dangerous for plants, which may be damaged by scrubbing (Venkatesha and Divesh, 2012). Thapa and Lantinga (2016) found that a balanced fertilization, that keeps the coffee plants in a good vegetative vigour, is a good way to minimize damage by the coffee stem borers. No insecticide has proved completely effective on *X. quadripes* (Venkatesha and Divesh, 2012). Numerous hymenopteran parasitoids have been reported on the coffee stem borer, though they have low rates of parasitism (Venkatesha and Divesh, 2012). According to the literature, the most promising appears to be the braconid wasp *Allorhogas pallidiceps* (Perkins) (Prakasan, 1987) and the aulacid *Pristaulacus* sp. (Visitpanich, 1994) reported for India and Thailand, respectively.

15.3.3 Lepidoptera

The coffee leafminer, *Leucoptera coffeella* (Guérin-Méneville) (Lepidoptera, Lyonetiidae), is a tiny moth (2.5 mm long and a wingspan of 4–5 mm), with forewings white in colour and with a brown strip in the apical part. Although it is probably native of east Africa, currently it is widely distributed in all Central and South America (Green, 1984; Vega *et al.*, 2007), where the moth is considered a serious pest (Lomeli-Flores *et al.*, 2010). The larvae dig mines in the leaf mesophyll, causing necrotic lesions and reducing photosynthetic activity. The higher damage is recorded after long periods of dryness, in sparsely fertilized crops or those that have been debilitated by attack by other pests. Particularly during flowering, reduced photosynthetic activity can lead to important crop losses. Conversely, in rainy periods or in wetlands, the moth is less harmful, because water floods the mine and drowns the larvae (Vega *et al.*, 2007; Bustillo, 2008). *L. coffeella* has numerous natural enemies, which limit the moth populations below the damage threshold of 30% of attacked plants. The intensive use of insecticides has a severe impact on the numerous natural enemies of *L. coffeella* and can cause the emergence of resistant populations (Fragoso *et al.*, 2002; Lomeli-Flores *et al.*, 2009). The control of this leafminer can be achieved by keeping the coffee plants in good vegetative state through balanced fertilization, the adoption of appropriate cultivation techniques, the use of tolerant varieties and by minimizing the use of insecticides (Filho, 2006; Vega *et al.*, 2007). The sex pheromone traps contribute to monitor its flight activity and population density (Bacca *et al.*, 2012), whereas the use of pheromone for mating disruption needs further investigation (Ambrogi *et al.*, 2006).

15.3.4 Hemiptera

Several species of Heteroptera Pentatomidae are grouped as antestia bugs or variegated shield bugs and considered to be serious pests for African coffee crops (Cilas *et al.*, 2009; Mugo *et al.*, 2013). The most damaging species are *Antestiopsis intricata* (Ghesquiere & Carayon), *A. orbitalis* (Westwood) and *A. thumbergi* (Gmelin). Adults are bronze or brown in colour with orange patterns, and about 7 mm long. The nymphs look similar to the adults but lack wings. Damage is caused by feeding activity of both nymphs and adults of these species, by piercing on shoots, flower buds and especially coffee berries. On the latter, the early attacks induce abortion and premature loss; when pierced berries are fully grown, attacks produce the so-called 'potato taste', a defect that greatly lowers the quality of coffee (Le Pelley, 1968). Control is very difficult, because the damage threshold (1–3 bugs per tree) is very low. Regular pruning to favour the plant air circulation is recommended to expose the antestia bugs to abiotic and biotic natural enemies (Le Pelley, 1968). Insecticides (organophosphate or neem oil) should be applied only when the threshold of economic damage is reached. The use of *Beauveria bassiana* can represent a valid alternative in some areas (Nahajo and Beyisenge, 2012).

Among numerous scale insects that can infest coffee, *Planococcus citri* (Risso) (citrus mealybug), *P. kenyae* (Le Pelley) (coffee mealybug), *Ferrisia virgata* (Cockerell) (striped mealybug), *Coccus viridis* (Green) (green scale), *C. alpinus* De Lotto (soft green scale) and *Saissetia oleae* (Olivier) (black scale) are the species which are most frequently reported for their occasional damage to coffee plantations (Barrera, 2008; CABI, 2016b). Scale insects take advantage of the warm and wet environmental conditions in which coffee is cultivated. Damage is due to the large subtraction of sap and the production of sugary honeydew that falls on vegetation and on which sooty mould develops. On infested coffee plants, scale insects are visible on flower buds, apical shoots and inside clusters of berries. Control measures are based on regular pruning of trees to increase ventilation, enlarge plant spacing to prevent contact between canopies and the spread of scales among plant. The adoption of strategies to preserve natural enemies is strongly recommended. Scale insects are normally well controlled by numerous natural enemies (parasitoids and predators), and sudden infestations are sometimes a consequence of improper use of insecticides that suppress the populations of beneficial organisms, consequently advantaging those of pests. A monthly monitoring is necessary to detect the infested plants early. In Hawaii, it is advised that an infestation index be calculated for coffee crops of less than 2 ha, obtained by collecting 20 leaves from 10 plants randomly selected (Bittenbender and Smith, 2008). Upon reaching the damage threshold, it is suggested that mineral oil be added to an insecticide, and applied during the hours when the plants are not exposed to direct sunlight, not during flowering and, possibly, in the time of greatest presence of meander nymphs of first age (Barrera, 2008). Finally, it is important to control the ants that, attracted by the honeydew produced by scale insects, spread them over the crops.

15.4 Challenges of IPM on Tea, Cocoa and Coffee

Despite the fact that tea, cocoa and coffee are quite different crops, they share many problems and technical challenges when it comes to IPM. Coffee and cocoa are mainly produced by smallholder farmers in developing countries, usually with minimal education, low incomes, scarce access to information, technology and financial support. On the other hand, tea is rather diverse, with tea plantations undertaken mainly by tea estate owners, or realized in corporate or governmental plantations. However, the trend towards smallholders' cultivation is growing in the past few years also for this crop (FAO, 2015).

This gives an idea about the different realities existing for the three crops in terms of knowledge, economical resources, management and technology, though the difficulty to implement an IPM strategy often does not rely only on the education level or funding availability.

Farmers usually utilize chemical control because it is easier, less time-consuming and gives a visual response in the short term. At this stage, inaccurate dosage of the chemicals, poor crop coverage, wrong timing or calibration of the equipment are often the causes of failure (Asogwa and Dongo, 2009). This implies more sprays or overdosage. When switching to biological control approaches, which is one of the IPM pillars, there are several reasons why farmers do not tend to use biological control agents. The first one is that biocontrol is much slower to control pest populations if compared to pesticides, even if in the long term this strategy can pay. Another important

reason is that biological control is not appreciated by smallholder farmers due to the cost, limited conservation time and methods of distribution. The efficiency of commercial chains related to biological control agents is still rather poor in countries where the three crops are mainly cultivated. And again, the application of these agents requires relevant technical knowledge. The correct application timing (season, day temperature, etc.), while establishing when and how to integrate biocontrol with chemical sprays, are fundamental for a successful pest control.

In order to be successful, IPM approaches must be chosen to fit in the targeted cultural, social and agroecological setting (Moradi *et al.*, 2012). An interesting study carried out by Parsa *et al.* (2014) pointed out how one of the main obstacles for IPM adoption by farmers is the lack of training and support. If such an obstacle can be overcome, even if the farmers never fully adopt IPM strategies, they will gain enormous benefits in terms of scouting methodology, threshold of pests and chemical application guidelines. Moreover, farmers will be inspired to explore and utilize new alternative techniques.

References

Abate, T., van Huis, A. and Ampofo, J.K.O. (2000) Pest management strategies in traditional agriculture: an African perspective. *Annual Review of Entomology* 45, 631–659.

Adu-Acheampong, R., Sarfo, J.E., Appiah, E.F., Nkansah, A., Awudzi, G., Obeng, E., Tagbor, P. and Sem, R. (2015) Strategy for insect pest control in cocoa. *American Journal of Experimental Agriculture* 6, 416–423.

Ahmed, M., Saha, J.K. and Gazi, S.U. (1998) Economics of spraying endosulfan and synthetic pyrethroids against tea mosquito bug (*Helopeltis theivora* Waterh.). *Two and a Bud* 45, 11–14.

Ahmed, M., Paul, S.K. and Mamun, M.S.A. (2011) Field performance and economic analysis of some commonly used insecticides against tea mosquito bug, *Helopeltis theivora* W. *Bangladesh Journal of Agricultural Research* 36, 449–454.

Akpesse, A.A.M., Liabra, G.J., Boga, J.P., Coulibaly, T., Yapi, A. and Kouassi, K.P. (2014) Efficacy of the insecticide IMIDOR SL 200 (Neonicotinoid) against the mirids of cocoa (Theobroma cocoa variety Amelonado) in center Côte d'Ivoire. *Journal of Experimental Biology and Agricultural Sciences* 2 (6), 553–559.

Ambrogi, B.G., Lima, E.R. and Sousa-Souto, L. (2006) Efficacy of mating disruption for control of the coffee leaf miner *Leucoptera coffeella* (Guérin-Méneville) (Lepidoptera: Lyonetiidae). *BioAssay* 1, 8. Available at: https://www.bioassay.org.br/bioassay/article/download/43/76 (accessed 26 October 2017).

Amin, N., Salam, M. and Junaid, M. (2014) Isolation and identification of endophytic fungi from cocoa plant resistant VSD M. 05 and cocoa plant susceptible VSD M. 01 in South Sulawesi, Indonesia. *International Journal of Current Microbiology and Applied Science* 3, 459–467.

Aristizábal, L.F., Lara, O. and Arthurs, S.P. (2012) Implementing an integrated pest management for coffee berry borer in a specialty coffee plantation in Colombia. *Journal of Integrated Pest Management* 3. DOI:http://dx.doi.org/10.1603/IPM11006

Aristizábal, L.F., Jiménez, M., Bustillo, A.E., Trujillo, H.I. and Arthurs, S.P. (2015) Monitoring coffee berry borer, *Hypothenemus hampei* (Coleoptera: Curculionidae), populations with alcohol baited funnel traps in coffee farms in Colombia. *Florida Entomologist* 98, 381–383.

Aristizábal, L.F., Bustillo, A.E. and Arthurs, S.P. (2016) Integrated pest management of coffee berry borer: strategies from Latin America that could be useful for coffee farmers in Hawaii. *Insects* 7, 6. DOI:10.3390/insects7010006

Asogwa, E.U. and Dongo, L.N. (2009) Problems associated with pesticide usage and application in Nigerian cocoa production: a review. *African Journal of Agricultural Research* 4 (8), 675–683.

Awasom, J.A. (2011) Tea. *Journal of Agricultural & Food Information* 12, 12–22.

Azeez, O.M. (2015) Model on Integrated Pest Management (IPM) strategies for *Sahlbergella singularis* of cocoa in South West Nigeria: a review. *International Invention Journal of Agricultural and Soil Science* 3, 26–35.

Azhar, I. (1989) Towards the development of an integrated pest management system of *Helopeltis* in Malaysia. *Malaysian Agricultural Research and Development Institute Research Journal* 17 (1), 55–68.

Azhar, I. (1992) Role of black ants in cocoa pod borer natural control. *MAPPS Newsletter* 16 (4), 36–37.

Baah, F. and Garforth, C. (2008) Insights into cocoa farmers' attitudes in two districts of Ashanti, Ghana. *International Journal of Sustainable Development* 1, 8–14.

Babin, R., Ten Hoopen, G.M., Cilas, C., Enjalric, F., Gendre, P. and Lumaret, J.P. (2010) Impact of shade on the spatial distribution of *Sahlbergella singularis* in traditional cocoa agroforests. *Agricultural and Forest Entomology* 12, 69–79.

Bacca, T., Saraiva, R.M. and Lima, E.R. (2012) Captura de *Leucoptera coffeella* (Lepidoptera: Lyonetiidae) en trampas con feromona sexual y su intensidad de daño. *Revista Colombiana de Entomología* 38, 42–49.

Baker, P.S. (1984) Some aspects of the behaviour of the coffee berry borer in relation to its control in southern Mexico (Coleoptera, Scolytidae). *Folia Entomológica Mexicana (México)* 61, 9–24.

Baker, P.S. (1998) The biology, ecology and behaviour of the coffee berry borer and its parasitoids. *Second International Conference on Coffee Berry Borer*. Tapachula, Mexico.

Baker, P.S. (1999a) *The Coffee Berry Borer in Colombia; Final Report of the DFID-Cenicafé-CABI Bioscience IPM for Coffee Project (CNTR 93/1536A)*. Chinchinà (Colombia), DFID-CENICAFE.

Baker, P.S. (1999b) Colombian coffee IPM. *Biocontrol News and Information* 20, 72–73.

Baker, P.S., Jackson, J.A.F. and Murphy, S.T. (2002) *Natural Enemies, Natural Allies. Project Completion Report of the Integrated Management of Coffee Berry Borer Project.* CFC/ICO/02 (1998–2002). The Commodities Press, CABI Commodities, Egham, UK, and Cenicafé, Chinchinà, Colombia.

Bale, J.S., Van Lenteren, J.C. and Bigler, F. (2008) Biological control and sustainable food production. *Philosophical Transactions of the Royal Society of London B: Biological Sciences* 363, 761–776.

Banerjee, B. (1983) Arthropod accumulation on tea in young and old habitats. *Ecological Entomology* 8 (2), 117–123.

Barrera, J.F. (2008) Coffee pests and their management. In: Capinera, L. (ed.) *Encyclopedia of Entomology*. Springer Science & Business, Dordrecht, The Netherlands, 961–998.

Barthakur, B.K. (2011) Recent approach of Tocklay to plant protection in tea in North East India. *Science and Culture* 77, 381–384.

Bartlett, B.R. (1968) Outbreaks of two-spotted spider and cotton aphid follow pesticide treatment. I. Pest stimulation vs. natural enemy destruction as a cause of outbreaks. *Journal of Economic Entomology* 61, 297–303.

Benavides, P., Góngora, C. and Bustillo, A. (2012) IPM program to control coffee berry borer *Hypothenemus hampei*, with emphasis on highly pathogenic mixed strains of *Beauveria bassiana*, to overcome insecticide resistance in Colombia. In: Farzana Perveen (ed.) *Insecticides – Advances in Integrated Pest Management*. InTech, pp. 512–540.

Bisseleua, H.B.D., Yede and Vidal, S. (2011) Dispersion models and sampling of cacao mirid bug *Sahlbergella singularis* (Hemiptera: Miridae) on *Theobroma cacao* in Southern Cameroon. *Environmental Entomology* 40, 111–119.

Bisseleua, H.B.D., Fotio, D., Missoup, A.D. and Vidal, S. (2013) Shade tree diversity, cocoa pest damage, yield compensating inputs and farmers' net returns in West Africa. *PLOS ONE* 8 (3), e56115.

Bittenbender, H.C. and Smith, E.M. (2008) *Growing Coffee in Hawaii.* College of Tropical Agriculture and Human Resources, University of Hawaii at Manoa. Available at: www.ctahr.hawaii.edu/oc/freepubs/pdf/coffee08.pdf

Borbón, O., Mora, A.O., Oehlschlager, A.C. and González, L.M. (2002) Proyecto de Trampas, Atrayentes y Repelentes para el Control de la Broca del Fruto de Cafeto, *Hypothenemus hampei* (Ferrari) (Coleoptera: Scolytidae). Informe ICAFE, San José, Costa Rica.

Botero, J. and Baker, P.S. (2001) Coffee and biodiversity: a producer-country perspective. In: Baker, P.S. (ed.) *Coffee Futures: A Source Book of Some Critical Issues Confronting the Coffee Industry.* CAB International, Wallingford, UK, pp. 94–103.

Bowers, J.H., Bailey, B.A., Hebbar, P.K., Sanogo, S. and Lumsden, R.D. (2001) The impact of plant diseases on world chocolate production. *Plant Health Progress.* DOI:10.1094/PHP-2001-0709-01-RV

Brun, L.A., Marcillaud, C. and Gaudichon, V. (1994) Cross resistance between insecticides in coffee berry borer, *Hypothenemus hampei* (Coleoptera: Scolytidae) from New Caledonia. *Bulletin of Entomological Research* 84, 175–178.

Burbano, E.G., Wright, M.G., Gillette, N.G., Mori, S., Dudley, N., Jones, T. and Kaufmann, M. (2012) Efficacy of traps, lures, and repellents for *Xylosandrus compactus* (Coleoptera: Curculionidae) and other ambrosia beetles on *Coffea arabica* plantations and *Acacia koa* nurseries in Hawaii. *Environmental Entomology* 41, 133–140.

Bustillo, A.E.P (2002) *El manejo de cafetales y su relación con el control de la broca del café en Colombia.* Boletín Técnico No. 24. FNC – Cenicafé, Chinchiná, Colombia.

Bustillo, A.E.P. (2008) El minador de la hoja del cafeto, *Leucoptera coffeellum* (Lepidoptera: Lyonetiidae). In: Bustillo, A.E.P. (ed.) *Los insectos y su manejo en la caficoltura colombiana.* FNC – Cenicafé, Chinchiná, Colombia.

Bustillo, A.E.P., Cárdenas, R., Villalba, D., Benavides, P., Orozco, J. and Posada, F. (1998) *Manejo integrado de la broca del café,* Hypothenemus hampei *(Ferrari) en Colombia.* FNC – Cenicafé, Chinchiná, Colombia.

CABI (2016a) *Hypothenemus hampei.* In: *Invasive Species Compendium.* CAB International, Wallingford, UK. Available at: www.cabi.org/isc (accessed 5 April 2016).

CABI (2016b) *Planococcus kenyae.* In: *Invasive Species Compendium.* CAB International, Wallingford, UK. Available at: www.cabi.org/isc (accessed 6 April 2016).

Ceja-Navarro, J.A., Vega, F.E., Karaoz, U., Hao, Z., Jenkins, Z. *et al.* (2015) Gut microbiota mediate caffeine detoxification in the primary insect pest of coffee. *Nature Communications* 6, 7618. DOI:10.1038/ncomms8618

Chen, Z.. and Chen, X.F. (1989) An analysis on the world tea pest fauna. *Journal of Tea Science* 9, 13–22.

Chen, Z., Sun, J., Wu, G.Y.J., Jin, J.Z. and Zeng, M. (1997) Integrated management technology of citrus black spiny whitefly (*Aleurocanthus spiniferus* Quaintance). *Journal of Tea Science* 17, 15–20.

Chin, D. and Brown, H. (2008) Red-banded thrips on fruit trees. *Agnote.* Available at: www.nt.gov.au/dpifm/Content/File/p/Plant_Pest/719.pdf (accessed 7 April 2016).

Choudhary, T.C. (1999) Pesticide residues in tea. In: Jain, N.K. (ed.) *Global Advances in Tea Science.* Aravali Books, New Delhi, pp. 369–378.

Cilas, C., Bonyjou, B. and Decazy, B. (2009) Frequency distribution of *Antestiopsis orbitalis* Westwood (Hem., Pentatomidae) in coffee plantations in Burundi: implications for sampling techniques. *Journal of Applied Entomology* 112, 601–606.

Cocoasafe (2016) *Stem Borer.* Available at: www.cocoasafe.org/Resources/Datasheets/StemBorer.pdf (accessed 31 March 2016).

Collingwood, C.A. (1971) *Cocoa Capsids in West Africa: Report of International Capsid Research Team, 1965–1971.* Cocoa, Chocolate and Confectionery Alliance, London.

Cranham, J.E. (1961) The natural balance of pests and parasites on Ceylon tea, especially *Tortrix* and *Macrocentrus. Tea Quart* 32, 26–41.

Crowdy, S.H. (1947) Observation on the pathogenicity of *Calonectria rigidiuscula* (Berk. and Br.) Sacc. on *Theobroma cacao* L. *Annals of Applied Biology* 34, 45–59.

Curry, G.N., Koczberski, G., Lummani, J., Nailina, R., Peter, E., McNally, G. and Kuaimba, O. (2015) A bridge too far? The influence of socio-cultural values on the adaptation responses of smallholders to a devastating pest outbreak in cocoa. *Global Environmental Change* 35, 1–11.

Damon, A. (2000) A review of the biology and control of the coffee berry borer, *Hypothenemus hampei* (Coleoptera: Scolytidae). *Bulletin of Entomological Research* 90, 453–465.

Das, S.C. (1978) Cataloguing coccoids. *Two and a Bud* 25, 43–44.

Deka, M.K., Karan, S. and Hendique, R. (1998) Antifeedant and repellent effects of pongam (*Pongamia pinnata*) and wild sage (*Lantana camara*) on tea mosquito bug (*Helopeltis theivora*). *Indian Journal of Agricultural Science* 68, 274–276.

Denmark, H.A. and Wolfenbarger, D.O. (2010) Redbanded Thrips, *Selenothrips rubrocinctus* (Giard) (Insects: Thysanoptera: Thripidae). EENY-099 (IN256) (orig. pub. as DPI Entomology Circular No. 108). Florida Department of Agriculture and Consumer Services, Division of Plant Industry, Gainesville, Florida.

Dittrich, V., Ernst, G.H., Ruesch, O. and Uk, S. (1990) Resistance mechanisms in sweetpotato whitefly (Homoptera: Aleyrodidae) populations from Sudan, Turkey, Guatemala, and Nicaragua. *Journal of Economic Entomology* 83, 1665–1670.

Egonyu, J.P., Kucel, P., Kangire, A., Sewaya, F. and Nkugwa, C. (2009) Impact of the black twig borer on Robusta coffee in Mukono and Kayunga districts, central Uganda. *Journal of Animal and Plant Sciences* 3 (1), 163–169.

Elder, S.D., Lister, J. and Dauvergne, P. (2014) Big retail and sustainable coffee: a new development studies research agenda. *Progress in Development Studies* 14, 77–90.

Entwistle, P.F. (1972) *Pests of Cocoa*. Longman, Harlow, UK.

Falh, J.J., Queiroz-Voltan, R.B., Carelli, M.L.C., Schiavinato, M.A., Prado, A.K.S. and Souza, J.C. (2007) Alterations in leaf anatomy and physiology caused by the red mite (*Oligonychus ilicis*) in plants of Coffea arabica. *Brazilian Journal of Plant Physiology* 19, 61–68.

FAO (2014) *The State of Food and Agriculture: Innovation in Family Farming*. Available at: www.fao. org/3/a-i4040e.pdf (accessed 23 March 2016).

FAO (2015) *World Tea Production and Trade: Current and Future Development*. Available at: www. fao.org/3/a-i4480e.pdf (accessed 23 March 2016).

FAOSTAT (2015) *The Agricultural Production*. Available at: http://faostat3.fao.org/download/Q/*/E (accessed 23 March 2015).

FAOSTAT (2016) *Statistic Division*. Available at: http://faostat3.fao.org/download/Q/QC/E (accessed 23 March 2016).

FAS/USDA (2015) *Coffee: World Markets and Trade*. Available at: www.fas.usda.gov/data/coffee-world-markets-and-trade (accessed 15 March 2016).

Filho, O.G. (2006) Coffee leaf miner resistance. *Brazilian Journal of Plant Physiology* 18, 109–117.

Fragoso, D.B., Guedes, R.N.C., Picanto, M.C. and Zambolin, L. (2002) Insecticide use and organophosphate resistance in the coffee leaf miner *Leucoptera coffeella* (Lepidoptera: Lyonetiidae). *Bulletin Entomological Research* 92, 203–212.

Franco, R.A., Reis, P.R., Zacarias, M.S., Altae, B.F. and Barbosa, J.P.R.A.D. (2009) Influence of *Oligonychus ilicis* (McGregor, 1917) (Acari: Tetranychidae) infestation on the potential photosynthesis rate of coffee-plant leaves. *Arquivos do Instituto Biológico (São Paulo)* 76, 205–210.

Franco, R.A., Reis, P.R., Zacarias, M.S. and Oliveira, D.C. (2010) Influence of the webbing produced by *Oligonychus ilicis* (McGregor) (Acari: Tetranychidae) on associated predatory phytoseiid. *Neotropical Entomology* 39, 97–100.

Gathalkar, G.B. and Barsagade, D.D. (2016) Predation biology of weaver ant *Oecophylla smaragdina* (Hymenoptera: Formicidae) in the field of tasar sericulture. *Journal of Entomology and Zoology Studies* 4 (2), 7–10.

Ghorai, N. and Bera, S. (1999) Occurrence of an entomopathogenic fungus *Paecilomyces tenuipes* (Peck) Samson (Deuteromycetes: Moniliaceae) on the tea bunch caterpillar *Anadraca bipunctata* (Walker) (Lepidoptera: Bombycidae). *Science and Culture* 65, 323–324.

Ghorai, N., Kumar Raut, S. and Kumar Bhattachayya, A. (2010) Behavioural ecology of the tea pest, *Andraca bipunctata* (Lepidoptera: Bombycidae), in the Sub-Himalayan climate of Darjeeling (India). *Biological Letters* 47, 65–80.

Gillott, C. (2005) *Entomology*. Springer Science & Business Media, Dordrecht, The Netherlands.

Gogoi, I., Rahman, I. and Dolui, A.K. (2003) Antifeedant properties of some plant extracts against tea mosquito bug, *Helopeltis theivora* Waterhouse. *Journal of Entomological Research* 27, 321–324.

Gotsch, N. (1997) Cocoa crop protection: an expert forecast on future progress, research priorities and policy with help of the Delphi survey. *Crop Protection* 16, 227–233.

Greco, E.B. and Wright, M.G. (2015) Ecology, biology, and management of *Xylosandrus compactus* (Coleoptera: Curculionidae: Scolytinae) with emphasis on coffee in Hawaii. *Journal of Integrated Pest Management* 6 (1), 7. DOI:10.1093/jipm/pmv007

Green, S.D. (1984) A proposed origin of the coffee leaf-miner, *Leucoptera coffeella* (Guérin-Ménaville) (Lepidoptera: Lyonetiidae). *Bulletin of the Entomological Society of America* 30, 30–31.

Guastella, D., Lulah, H., Tajebe, L.S., Cavalieri, V., Evans, G.A. *et al.* (2015) Survey on whiteflies and their parasitoids in cassava mosaic pandemic areas of Tanzania using morphological and molecular techniques. *Pest Management Science* 71 (3), 383–394.

Gupta, M., Sharma, A. and Shanker, A. (2008) Dissipation of imidacloprid in orthodox tea and its transfer from made tea to infusion. *Food Chemistry* 106, 158–164.

Gurusubramanian, G., Rahman, A., Sarmah, M., Roy S. and Bora, S. (2008) Pesticide usage pattern in tea ecosystem, their retrospects and alternative measures. *Journal of Environmental Biology* 29, 813–826.

Hamasaki, R.T., Shimabuku, R. and Nakamoto, S.T. (2008) *Guide to insect and mite pests of tea (Camellia sinensis) in Hawaii.* CTAHR Publications. Available at: www.ctahr.hawaii.edu/freepubs (accessed 3 February 2016).

Hazarika, L.K., Puzari, K.C. and Wahab, S. (2001) Biological control of tea pests. In: Upadhyay, R.K., Mukerji, K.G. and Camola, B.P. (eds) *Biocontrol Potential and Its Exploitation in Sustainable Agriculture*, Vol. II. Kluwer Academic/Plenum Publishers, New York, pp. 159–180.

Hazarika, L.K., Bhuyan, M. and Hazarika, B.N. (2009) Insect pests of tea and their management. *Annual Review of Entomology* 54, 267–284.

Holman, J. (2009) *Host Plant Catalogue of Aphids: Palaearctic Region.* Springer, Berlin.

Hong, B.B. (1998) A list of viruses of tea pests discovered in tea garden in China. *Journal of Tea* 24, 82–84.

Infante, F., Castillo, A., Pérez, J. and Vega, F.E. (2013) Field-cage evaluation of the parasitoid *Phymastichus coffea* as a natural enemy of the coffee berry borer, *Hypothenemus hampei. Biological Control* 67, 446–450.

ICCO (2012) L'économie cacaoyère. *Le journal de L'Economie, Bulletin trimestriel de statistiques cacaoyères* XXXVIII (4).

ICCO (2013) *Quarterly Bulletin of Cocoa Statistics*, XXXIX (May) (4).

Jaggi, S., Sood, C., Kumar, V., Ravindranath, S.D. and Shanker, A. (2001) Leaching of pesticides in tea brew. *Journal of Agricultural and Food Chemistry* 44, 5479–5483.

James, D.G. and Price, T.S. (2002) Fecundity in two-spotted spider mite (Acari: Tetranychidae) is increased by direct and systemic exposure to imidacloprid. *Journal of Economic Entomology* 95, 729–732.

Jaramillo, J., Bustillo, A.E., Montoya, E.C. and Borgemeister, C. (2005) Biological control of the coffee berry borer *Hypothenemus hampei* (Coleoptera: Curculionidae) by *Phymastichus coffea* (Hymenoptera: Eulophidae) in Colombia. *Bulletin of Entomological Research* 95, 1–6.

Jaramillo, J., Borgemeister, C. and Baker, P. (2006) Coffee berry borer *Hypothenemus hampei* (Coleoptera: Curculionidae): searching for sustainable control strategies. *Bulletin of Entomological Research* 96, 223–233.

Jaramillo, J., Chabi-Olaye, A., Kamonjo, C., Jaramillo, A., Vega, F.E., Poehling, H.M. and Borgemeister, C. (2009) Thermal tolerance of the coffee berry borer *Hypothenemus hampei*: predictions of climate change on a tropical insect pest. *PLOS ONE* 4, e6487. DOI:10.1371/journal.pone.0006487

Jha, S., Bacon, C.M., Philpott, S.M., Méndez, V.E., Läderach, P. and Rice, R.A. (2014) Shade coffee: update on a disappearing refuge for biodiversity. *BioScience* 64, 416–428.

Kagezi, G., Kucel, B., Egonyu, J.P., Kyamanyna, S., Karungi, J.T. *et al.* (2015) A review of the status and progress in management research of the black coffee twig borer, *Xylosandrus compactus* (Heichhoff), in Uganda. In: *Proceedings of the 25th International Conference on Coffee Science*. ASIC, Paris, pp. 42–52.

Kawai, A. (1997) Prospects for integrated pest management in tea cultivation in Japan. *Japanese Agricultural Research Q* 31, 213–217.

Kucel, P., Kangire, A. and Egonyu J.P. (2009) Status and current research strategies for management of the coffee berry borer (*Hypothenemus hampei* Ferr) in Africa. *ICO Seminar on the Coffee Berry Borer*. Available at: www.ico.org/event_pdfs/cbb/presentations/Kangire%20NaCRRI.pdf (accessed 8 May 2016).

Lee, S.Y., Park, H., Boo, K.S., Park, K.T. and Cho, S. (2005) Molecular identification of *Adoxophyes honmai* (Yesuda) based on mitochondrial COI sequences. *Molecules and Cells* 19, 391–397.

Lehmann-Danzinger, H. (2000) Disease and pests of tea: overview and possibilities of integrated pest and disease management. *Journal of Agriculture in the Tropics and Subtropics* 101, 13–38.

Le Pelley, R.H. (1968) *Pests of Coffee*. Longmans, London.

Le Pelley, R.H. (1973) Coffee insects. *Annual Review of Entomology* 18, 121–142.

Liao, D.Q., Liao, G.W. and Qin, Y.J. (2007) Field efficacy of *Steinernema carpocapsae* (Nematoda: Steinernematidae) against the larvae of tea brown beetle, *Basilepta melanopus* Levevre (Coleoptera: Chrysomelidae). *Acta Phytophylactica Sinica* 34, 447–448.

Lomeli-Flores, R.J., Barrera, J.F. and Bernal, J.S. (2009) Impact of natural enemies on coffee leafminer *Leucoptera coffeella* (Lepidoptera: Lyonetiidae) population dynamics in Chiapas, Mexico. *Biological Control* 51, 51–60.

Lomeli-Flores, R.J., Barrera, J.F. and Bernal, J.S. (2010) Impacts of weather, shade cover and elevation on coffee leafminer *Leucoptera coffeella* (Lepidoptera: Lyonetiidae) population dynamics and natural enemies. *Crop Protection* 29, 1039–1048.

Mahob, R.J., Babin, R., ten Hoopen, G.M., Dibog, L., Hall, D.R. and Bilong Bilong, C.F. (2011) Field evaluation of synthetic sex pheromone traps for the cocoa mirid *Sahlbergella singularis* (Hemiptera: Miridae). *Pest Management Science* 67 (6), 672–676.

Mahob, R., Baleba, L., Yédé, D.L., Cilas, C., Bilong Bilong, C.F. and Babin, R. (2015) Spatial distribution of *Sahlbergella singularis* Hagl. (Hemiptera: Miridae) populations and their damage in unshaded young cacao-based agroforestry systems. *International Journal of Plants Animals and Environmental Science* 5 (2), 121–131.

Mamun, M.S.A. and Ahmed, M. (2011) Prospects of indigenous plant extracts in tea pest management. *Journal of Agricultural Research, Innovation and Technology* 1, 16–23.

Manton, J.L., Hollingsworth, R.G. and Cabos, R.Y.M. (2012) Potential of *Steinernema carpocapsae* (Rhabditida: Steinernematidae) against *Hypothenemus hampei* (Coleoptera: Curculionidae) in Hawaii. *Florida Entomologist* 95, 1194–1197.

Mao, H.X. and Kunimi, Y. (1991) Pupal mortality of the oriental tea tortrix, *Homona magnanima* Diakonoff (Lepidoptera: Tortricidae), caused by parasitoids and pathogens. *Japanese Journal of Applied Entomology and Zoology* 35, 241–245.

Mochizuki, F., Fukumoto, T., Noguchi, H., Sugie, H., Morimoto, T. and Ohtani, K. (2002) Resistance to a mating disruption composed of (Z)-11-tetradecenyl acetate in the smaller tea tortrix, *Adoxophyes honmai* (Yasuda) (Lepidoptera: Tortrididae). *Applied Entomology and Zoology* 37, 299–304.

Moguel, P. and Toledo, V.M. (1999) Review: biodiversity conservation in traditional coffee system of Mexico. *Conservation Biology* 13, 11–21.

Molina, J.P. and Lopez, J.C. (2002) Desplazamiento y parasitismo de entomonematodos hacia frutos infestados con la broca del café *Hypothenemus hampei* (Coleoptera: Scolytidae). *Revista Colombiana de Entomologia* 28, 145–151.

Moradi, P., Najafabadi, M.O. and Lashgarara, F. (2012) Identify the challenges in applying of integrated pest management (IPM) from farmers' perception. *Archives des Sciences* 65 (9), 300–305.

Mu, D., Cui, L., Ge, J., Wang, M.X., Liu, L.F. and Yu, X.P. (2012) Behavioural responses for evaluating the attractiveness of specific tea shoot volatiles to the tea green leafhopper, *Empoasca vitis. Insect Science* 19, 229–238.

Mugo, H.M., Kimemia, J.K. and Mwangi, J.M. (2013) Severity of antestia bugs, *Antestiopsis* spp. and other key insect pests under shaded coffee in Kenya. *International Journal of Science and Nature* 4, 324–327.

Mumford, J.D. (1986) Control of the cocoa pod borer (Acrocercops cramerella): a critical review. In: Rajaratnam, E. and Chew Poh Soon (eds) *Cocoa and Coconut: Progress and Outlook*. Incorporated Society of Planters, Kuala Lumpur.

Muñoz, R. (1989) Biological cycle and parthenogenetic reproduction of coffee berry borer *Hypothenemus hampei* Ferrari. *Turrialba* 39, 415–421.

Muraleedharan, N. (1992) Pest control in Asia. In: Wilson, K.C. and Clifford, M.N. (eds) *Tea: Cultivation to Consumption*. Chapman and Hall, London, pp. 375–411.

Muraleedharan, N. and Chen, Z.M. (1997) Pests and diseases of tea and their management. *Journal of Plantation Crops* 25, 15–43.

Muraleedharan, N., Sudarmani, D.N.P. and Selvasundaram, R. (2005) Bioecology and management of the red spider mite infesting tea in south India. In: *Proceedings of International Symposium on Innovation on Tea Science and Sustainable Development in Tea Industry*. China Tea Science Society, pp. 756–766.

Murphy, S.T. and Moore, D. (1990) Biological control of the coffee berry borer, *Hypothenemus hampei* (Ferrari) (Coleoptera, Scolytidae): previous programmes and possibilities for the future. *Biocontrol News and Information* 11 (2), 107–117.

Nadda, G., Eswara Reddy, S.G. and Shaker, A. (2013) Insect and mite pests of tea and their management. In: Ahuja, P.S., Gulati, A., Singh, R.D., Sud, R.K. and Boruah, R.C. (eds) *Science of Tea Technology*. Scientific Publishers, Rajasthan, India, pp. 317–333.

Nahajo, A. and Beyisenge, J. (2012) Biological control of coffee antestia bugs (*Antestiopsis lineaticolis*) by using *Beauveria bassiana*. *New York Science Journal* 5, 106–113.

Nakai, M. (2009) Biological control of Tortricidae in tea field in Japan using insect viruses and parasitoids. *Virologia Sinica* 24, 323–332.

Naranjo, S.E., Ellsworth, P.C., Frisvold, G.B. (2015) Economic value of biological control in integrated pest management of managed plant systems. *Annual Review of Entomology* 60, 621–645.

Narasimham, A.U. (1987) Scale insects and mealybugs on coffee, tea and cardamom and their natural enemies. *Journal Coffee Research* 17, 7–13.

Neilson, J. (2007) Global markets, farmers and the state: sustaining profits in the Indonesian cocoa sector. *Bulletin of Indonesian Economic Study* 43, 227–250.

Ngoan, N.D., Wilkinson, R.C., Short, D.E., Moses, C.S. and Mangold, J.R. (1976) Biology of an introduced ambrosia beetle, *Xylosandrus compactus*, in Florida. *Annals of the Entomological Society of America* 69, 872–876.

Ortiz, A., Ortiz, A., Vega, F.E. and Posada, F. (2004) Volatile composition of coffee berries at different stages of ripeness and their possible attraction to the coffee berry borer *Hypothenemus hampei* (Coleoptera: Curculionidae). *Journal of Agricultural and Food Chemistry* 52, 5914–5918.

Panda, H. (2011) *The Complete Book on Cultivation and Manufacture of Tea*. Asia Pacific Business Press, Delhi.

Panigrahi, A. (1997) Mycosis in the tea pest *Andraca bipunctata* Walker (Lepidoptera: Notondontidae). *Ecology, Environment and Conservation Paper* 3, 163–168.

Parsa, S., Morse, S., Bonifacio, A., Chancellor, T.C., Condori, B. *et al.* (2014) Obstacles to integrated pest management adoption in developing countries. *Proceedings of the National Academy of Sciences* 111 (10), 3889–3894.

Penman, D.R. and Chapman, R.B. (1988) Pesticide-induced mite outbreaks: pyrethroids and spider mites. *Experimental & Applied Acarology* 4, 265–276.

Pérez-Lachaud, G. (1998) A new bethylid attacking the coffee berry borer (Coleoptera:Scolytidae) in Chiapas (Mexico) and some notes on its biology. *Southwestern Entomologist* 23, 287–288.

Perfecto, I., Rice, R.A., Greenberg, R. and van der Voort, M.E. (1996) Shade coffee: a disappearing refuge for biodiversity. *BioScience* 46, 598–608.

Pohlan, H.A.J. and Janssens, M.J.J. (2012) Growth and production of coffee. In: *Soils, Plant Growth and Crop Production*, Vol. III, *Encyclopedia of Life Support Systems* (EOLSS). Developed under the auspices of UNESCO, EOLSS Publishers, Paris, France. Available at: www.eolss.net (accessed 5 May 2016).

Posada, F.J., Chaves, F.C., Gianfagna, T.J., Pava-Ripoll, M. and Hebbar, P. (2010) Establishment of the fungal entomopathogen *Beauveria bassiana* as an endophyte in cocoa pods (*Theobroma cacao* L.). *Revista UDCA Actualidad & Divulgación Científica* 13 (2), 71–78.

Prakasan, C.B. (1987) Biological control of coffee pests. *Journal of Coffee Research* 17 (1), 114–117.

Qin, D., Zhang, L., Xiao, Q., Dietrich, C. and Matsumura, M. (2015) Clarification of the identity of the tea green leafhopper based on morphological comparison between Chinese and Japanese specimens. *PLOS ONE* 10 (9), e0139202. DOI:10.1371/journal.pone.0139202

Rabindra, R.J. (2012) Sustainable pest management in tea: prospects and challenges. *Two and a Bud* 59, 1–10.

Rahman, A., Roy, S., Muraleedharan, N.N. and Phukan, A.K. (2014) Effect of potassium chloride and potassium sulphate on the efficacy of insecticides against infestation by *Helopeltis theivora* (Heteroptera: Miridae) in tea plantations. *International Journal of Tropical Insect Science* 34, 217–221.

Rattan, P.S. (1992) Pest and disease control in Africa. In: Wilson, K.C. and Clifford, M.N. (eds) *Tea: Cultivation to Consumption.* Chapman and Hall, London, pp. 331–352.

Rattan, P.S. (2005) Carpenter moth: its life history, damage, and control. *The Tea Research Foundation of Central Africa Newsletter* 145, 1–10.

Reis, P.R. and Chagas, S.J.R. (2001) Relacão entre a ataque do acaro/plano e da mancha/anular com indicadores da qualidade do café. *Ciência e Agrotecnologia* 25, 72–76.

Reis, P.R. and Teodoro, A.V. (2000) Efeito do oxicloreto de cobre sobre a reproducão do ácaro-vermelho do cafeeiro, *Oligonychus ilicis* (McGregor, 1917). *Ciência e Agrotecnologia* 24, 347–352.

Rhainds, M., Chin Chiew, L., Moli, Z. and Gries, G. (2002) Incidence, symptoms, and intensity of damage by three coffee stemborers (Coleoptera: Cerambycidae) in South Yunan, China. *Journal of Economic Entomology* 95, 106–112.

Rice, R.A. (1996) The coffee environment of Northern Latin America: tradition and change. In: Rice, R.A., Harris, A.M. and McLean, J. (eds) *Proceedings from the First Sustainable Coffee Congress. Smithsonian Migratory Bird Center.* Smithsonian Institution, Washington, DC, pp. 105–114.

Rice, R.A. and Greenberg, R. (2000) Cacao cultivation and the conservation of biological diversity. *AMBIO: A Journal of the Human Environment* 29 (3), 167–173.

Rosmana, A. (2010) Control of cocoa pod borer and Phytophthora pod rot using degradable plastic pod sleeves and a nematode, *Steinernema carpocapsae. Indonesian Journal of Agricultural Science* 11, 41–47.

Roy, S., Mukhopadhyay, A. and Gurusubramanian, G. (2009) Antifeedant and insecticidal activity of *Clorodendron infortunatum* Gaertn. (Verbenaceae) extract on tea mosquito bug, *Helopeltis theivora* Waterhouse (Heteroptera, Miridae). *Research on Crops* 10, 152–158.

Roy, S., Muraleedharan, N. and Mukhopadhyay, A. (2014) The red spider mite, *Oligonychus coffeae* (Acari: Tetranychidae): its status, biology, ecology and management in tea plantations. *Experimental and Applied Acarology* 63, 431–463.

Ruben, R. and Fort, R. (2012) The impact of fair trade certification for coffee farmers in Peru. *World Development* 40, 570–582.

Ruiz-Càrdenas, R. and Baker, P. (2010) Life table of *Hypothenemus hampei* (Ferrari) in relation to coffee berry phenology under Colombian field conditions. *Scientia Agricola* 67, 658–668.

Saha, D., Mukhopadhyay, A. and Bahadur M. (2012) Genetic diversity of *Empoasca flavescens* Fabricius (Homoptera: Cicadellidae), an emerging pest of tea from sub-Himalayan plantations of West Bengal, India. *Proceedings of the Zoological Society* 65, 126–131.

Saripah, B. (2014) Control of cocoa pod borer using insecticides and cocoa black ants. *Malaysian Cocoa Journal* 14–22.

Sarma, P.V. (1979) Possibilities of integrated control of major pests of tea in India. *PANS* 25, 237–245.

Sarmah, M. and Kumar Bhola, R. (2015) Bio-efficacy of *Acorus calamus* extracts against tea mosquito bug, *Helopeltis theivora* Waterhouse. *International Journal of Recent Scientific Research* 6, 7638–7641.

Shahabuddin, A.A. and Pasaru, F. (2013) Biological control of cocoa pod borer (*Conopomorpha Cramerella* Snell.) on cocoa plantation for maintaining cocoa production in Central Sulawesi, Indonesia. *Malaysian International Conference*, Malaysian Cocoa Board. Malaysian Plant Protection Society, Kuala Lumpur, pp. 62–68.

Sharma, D.C. and Kashyap, N.P. (2002) Impact of pesticidal spray on seasonal availability of natural predators and parasitoids in tea ecosystem. *Journal of Biological Control* 16, 31–35.

Shi, L.-Q., Zeng, Z.-H., Huang, H.-S., Zhou, Y.-M., Vasseur, L. and You, M.-S. (2015) Identification of *Empoasca onukii* (Hemiptera Cicadellidae) and monitoring of its populations in the tea plantations of South China. *Journal of Economic Entomology* 108, 1025–1033.

Sivapalan, P. (1999) Pest management in tea. In: Jain, N.K. (ed.) *Global Advances in Tea Science.* Aravali Books, New Delhi, pp. 625–646.

Sonwa, D.J., Coulibaly, O., Weise, S.F., Adesina, A.A. and Janssens, M.J. (2008) Management of cocoa: constraints during acquisition and application of pesticides in the humid forest zones of southern Cameroon. *Crop Protection* 27(8), 1159–1164.

Sosan, M.B. and Akingbohungbe, A.E. (2009) Occupational insecticide exposure and perception of safety measures among cacao farmers in southwestern Nigeria. *Archives of Environmental & Occupational Health* 64 (3), 185–193.

Srikumar, K.K. and Bhat, P.S. (2012) Field survey and comparative biology of tea mosquito bug (*Helopeltis* spp.) on cashew (*Anacardium occidentale* Linn.). *Journal of Cell and Animal Biology* 6 (14), 200–206.

Sudhakaran, R. and Muraleedharan, N. (2006) Biology of *Helopeltis theivora* (Hemiptera: Miridae) infesting tea. *Entomon* 31, 165–180.

Sudoi, V., Cheramgoi, E., Langat, J.K., Kamunya, S.M. and Wachira, F.N. (2011) Screening of Kenyan tea clones for susceptibility to mite attack at different ecological zones. *International Journal of Current Research* 3, 328–332.

Sundararaju, D. and Sundara Babu, P.C. (1999) Helopeltis spp. (Heteroptera: Miridae) and their management in plantation and horticultural crops of India. *Journal of Plant Crops* 27, 155–174.

Szczepaniec, A., Creary, S.F., Laskowsky, K.L., Nyrop, J.P. and Raupp, M.J. (2011) Neonicotinoid insecticide Imidacloprid causes outbreaks of spider mites on elm trees in urban landscapes. *PLOS ONE* 6, e-20018. DOI:10.1371/journal.pone.0020018

Takafuji, A., Ozawa, A. and Nemoto, H. (2000) Spider mites of Japan: their biology and control. *Experimental and Applied Acarology* 24, 319–335.

Takagi, K. (1978) Trap for monitoring adult parasites of tea. *Japanese Agricultural Research Quarterly* 12, 99–103.

Takahashi, R. and Todo, Y. (2014) The impact of a shade coffee certification program on forest conservation: a case study from a wild coffee forest in Ethiopia. *Journal of Environmental Management* 130, 48–54.

Takahashi, M., Nakai, M., Nakanishi, K., Sato, T., Hilton, S., Winstanley, D. and Kunini, Y. (2008) Genetic and biological comparisons of four nucleopolyedrovirus isolates that are infectious to *Adoxophyes honmai* (Lepidoptera: Tortricidae). *Biological Control* 46, 542–546.

Thapa, S. and Lantinga, E.A. (2016) Infestation by coffee white stem borer, *Xylotrechus quadripes*, in relation to soil and plant nutrient content and associated quality aspects. *Southwestern Entomology* 41, 331–335.

TRI (2003) *Live-wood termites of low grown tea and their management.* Advisory Circular no. 3, 1–5. Tea Research Institute of Sri Lanka, Talawakelle, Sri Lanka.

Tuelher, E.S., Venzon, M., Guedes, R.N.C. and Pallini, A. (2014) Toxicity of organic-coffee-approved products to the southern red mite *Oligonychus ilicis* and to its predator *Iphiseiodes zuluagai*. *Crop Protection* 55, 28–34.

Uemura-Lima, D.H., Ventura, M.U., Mikami, A.Y., da Silva, F.C. and Morales, L. (2010) Response of coffee berry borer, *Hypothenemus hampei* (Ferrari) (Coleoptera: Scolytidae), to vertical distributions of methanol: ethanol traps. *Neotropical Entomology* 39, 930–933.

Vacante, V. (2016) *The Handbook of Mites of Economic Plants: Identification, Bio-ecology and Control.* CAB International, Wallingford, UK.

van Mele, P., Cuc, N.T.T. and van Huis, A. (2001) Farmers' knowledge, perceptions and practices in mango pest management in the Mekong Delta, Vietnam. *International Journal of Pest Management* 47(1), 7–16.

Vega, F.E. (2004) Coffee berry borer *Hypothenemus hampei* (Ferrari) (Coleoptera: Scolytdae). In: Capinera, J.L. (ed.) *Encyclopedia of Entomology*, Vol. I. Kluwer Academic Publishers, Dordrecht, The Netherlands, pp. 575–576.

Vega, F.E., Posada, F. and Infante, F. (2007) Coffee insects: ecology and control. In: Pimentel, D. (ed.) *Encyclopedia of Pest Management*. Taylor & Francis, London. DOI:10.1081/E-EPM-120042132

Vega, F.E., Infante, F., Castillo, A. and Jaramillo, J. (2009) The coffee berry borer, *Hypothenemus hampei* (Ferrari) (Coleoptera: Curculionidae): a short review, with recent findings and future research directions. *Terrestrial Arthropod Reviews* 2, 129–147.

Venkatesha, M.G. and Divesh, A.S. (2012) The coffee white stemborer *Xylotrechus quadripes* (Coleoptera: Cerambycidae): bioecology, status and management. *International Journal of Tropical Insect Science* 32, 177–188.

Visitpanich, J. (1994) The parasitoid wasps of the coffee stem borer, *Xylotrechus quadripes* Chevrolat (Coleoptera, Cerambycidae) in Northern Thailand. *Japanese Journal of Entomology* 62, 597–606.

Walgama, R.S. and Pallemulla, R.M.T.D. (2005) The distribution of shot hole borer *Xyleborus fornicatus* Eichh. (Coleoptera: Scolytidae), across tea growing areas in Sri Lanka: a reassessment. *Sri Lanka Journal of Tea Science* 70, 105–120.

Wickremasinghe, R.L. and Thirugnanasuntheram, K. (1980) Biochemical approach to the control of *Xyleborus fornicatus* (Coleoptera: Scolytidae). *Plant Soil* 55, 9–15.

Wood, L. (1982) The bark and ambrosia beetles of North and Central America (Coleoptera: Scolytidae), a taxonomic monograph. *Great Basin Naturalist Memoirs* 6, 1–1356.

World Cocoa Foundation (2003) *Discovery Learning about Cocoa.* Available at: www.worldcocoafoundation.org/wp-content/uploads/files_mf/vos2003.pdf (accessed 15 April 2016).

Ye, G.-Y., Xiao, Q., Chen, M., Chen X.-X., Yuan, Z.-J., Stanley, D.W. and Hu, C. (2014) Tea: biological control of insect and mite pests in China. *Biological Control* 68, 73–91.

Zeiss, M.R. and den Braber, K. (2001) *Tea Integrated Pest Management Ecological Guide.* E-book, CIDSE, Vietnam.

Zeng, M.S., Liu, F.J., Wang, D.F. and Wu, G.Y. (2012) An experimental demonstration of comprehensive control of *Basilepta melanopus* Lefevre. *Fujian Journal of Agricultural Sciences* 8, 847–852.

16 Integrated Insect Pest Management in Tropical Forestry

Nitin Kulkarni*

Tropical Forest Research Institute, Jabalpur, India

16.1 Tropical Forests

The forest areas between the Tropics of Cancer (23.5°N latitude) and Capricorn (23.5°S latitude), occupying 10% of the total land mass of the world above sea level, is known as Tropical Forest. It is characterized by distinct seasonality, 12 h day-length, average temperature range being 20–25°C, multi-layered canopy for little light penetration, highly diverse flora and resulting rich fauna (Sambaraju *et al.*, 2016). Owing to the great diversity existing in the tropics, it is an enormous task to discuss forest insect pests and their management in the whole tropical forest, but efforts to cover some important available examples are made in this chapter. There is about 1700 million ha of geographical area of tropical forest (FAO, 1993), under the political barriers of more than 90 countries of North America, central America, South America, the Caribbean, central Africa, east Africa, west Africa and southeast Asia, including India, under the tropical region (Redublo, 2016). Africa, Morocco and Tunisia in the north and Lesotho and Swaziland in the south are the only nations that are not in the tropics. The Middle East has four: Yemen, which is entirely in the tropics, and parts of Saudi Arabia, Oman and the United Arab Emirates. All these have unique forest areas with diverse forest species.

16.2 Major Tree Species

The tropical forests have much greater species variety than any other forests in the world and thus are the biodiversity hotspots. A hectare of Malaysian rainforest may contain 180 kinds of trees, compared to only 10 in temperate forests. While some trees, like eucalypts, poplars, mahagonies, teak, *sal*, rosewoods and okoumes, provide valuable timber and pulpwood, others provide products like nuts, fruits, rubber, rattans and other non-timber forest produce (NTFP) (WWF, 2016). Major forest species growing in the tropics are listed in Table 16.1, along with a description of their insect pests and origin.

16.3 Overview of the Major Forest Insect Pests in Tropics

The entomofauna existing in the canopies of tropical forests of the world (Erwin, 1983) has been estimated to be as high as 30 million living species (Anon., 2016). However, this estimate includes beneficial and insect pest fauna. In any forest ecosystem, a population outbreak of insects and pathogens is a major biotic disturbance (FAO, 2007; Sambaraju *et al.*, 2016). This is apart from their essential role in forest ecosystem

---

* Corresponding author e-mail: kulkarnin@icfre.org; kulkarni_n27@hotmail.com

Table 16.1. An updated list of major insect pests damaging tropical trees. (Adapted from Wylie and Speight (2012).)

Host trees	Insect species	Common name	Insect group	Country	Mode of attack and impact	Reference
Acacia auriculiformis Benth.	*Hypothenemus dimorphus* Wood & Bright	Stem-boring scolytinae beetles	Coleoptera: Scolytidae	Malaysia	Seedling mortality	Nair, 2007
Acacia crassicarpa A. Cunn ex. Benth	*Platypus* spp.	Ambrosia beetle	Coleoptera: Platypodidae	Sabah	Stem boring up to 50 holes per tree with associated back stain	Thapa, 1991
Acacia cunninghamii Hook.	*Aesiotes notabilis* Pascoe	Pine bark weevil	Coleoptera: Curculionidae	Queensland, Australia	Attack tubed seedlings, tunnelling of main roots, chewing woody tissue/bark, later forming cocoons at or near ground level	Elliott *et al.*, 1998
Acacia mangium Willd.	*Xylosandrus crassiusculus* (Mochut'skii)	Stem-boring scolytinae beetles	Coleoptera: Scolytidae	Malaysia	Seedling mortality	Varma and Swaran, 2009
	Coptotermes curvignathus (Holmgren)	White ants	Isoptera: Rhinotermitidae	Indonesia	Causes death of seedlings	Nair and Sumardi, 2000
	Xytrocera festiva Pascoe	Stem borer	Coleoptera: Cerambycidae	Indonesia	Attacks small branches which often dry up and break	
	Xylosandrus sp. and *Xyleborus fornicates* Eichh.	Pinhole borer	Coleoptera: Scolytidae	Indonesia	Attacks small branches which often dry up and break	
	Helopeltis theivora Waterhouse and other *Helopeltis* spp.	Mosquito bug	Hemiptera: Miridae	Indonesia	Damages young seedlings	
	Pteroma plagiophleps Hamp.	Bagworms	Lepidoptera: Psychidae	Indonesia	Defoliates leaves	
	Valanga nigricornis Burm.	Grasshopper	Orthoptera: Acrididae	Thailand; Indonesia	Defoliation in nursery	
Acacia mearnsii De Wild.	Various species	White grubs	Coleoptera: Scarabaeidae	South Africa	2–30% mortality of seedlings due to root feeding in nursery	Govender, 2007

Acacia nilotica L.	*Microcerotermes minor* Holm.; *Odontotermes* spp. and *Microtermes obesi* Holm.	Termites	Isoptera: Termitidae	India	Root and stem damage in seedlings leading to death	Thakur, 1992
Acacia nilotica L. *Azadirachta indica* A. Juss.	*Holotrichia serrata* Fab.; *H. consanguinea* Blanch.	White grubs	Coleoptera: Scarabaeidae	India	Mortality in seedlings	Ali and Chaturvedi, 1996
Acacia senegal (L.) Willd. and *Acacia tortilis* (Forsk.) Hayne	*Myllocerus* spp.	Weevil	Coleoptera: Curculionidae	Thar Desert of India	Damage to nursery plants	Vir and Parihar, 1993
Ailanthus excelsa Roxb.	*Atteva fabriciella* Swed.	Webworm	Lepidoptera: Yponomeutidae	India	Larvae web around the tender leaves in seedlings and large trees and defoliate voraciously; also bores inside the terminal growing shoot tips	Joshi, 1992; Senthilkumar and Murugesan, 2015
Ailanthus triphysa (Dennst.) Alsten; *A. excelsa* Roxb.	*Eligma narcissus* (Cram.)	Noctuid caterpillar	Lepidoptera: Noctuidae	India	100% defoliation of nursery stock	Roonwal, 1990; Sivaramakrishnan and Remadevi, 1996
Albizia lebbek (L.) Benth.	Several unnamed	Pod/ seed borer	(Coleoptera: Curculionidae)	India	50–70% of seeds damaged in pods	Harsh and Joshi, 1993
Albizia lebbek (L.) Benth.; *Eucalyptus tereticornis* Sm.	*Agrotis ipsilon* Hüfnagel	Cutworm	Lepidoptera: Noctuidae	Brazil; China; India	10–30% shoot cutting incidence	Pang, 2003
Avicennia marina (Forssk.) Vierh.	*Nephopterix syntaractis* Turner	Mangrove caterpillar	Lepidoptera: Pyralidae	Hong Kong	Over 75% defoliation in the mangrove species	Anderson and Lee, 1995
Avicennia sp.	*Pteroma plagiophleps*	Bagworm	Lepidoptera: Psychidae	Indonesia	Foliage feeder on the mangrove seedlings	Nair and Sumardi, 2000
	Acanthopsyche sp.	Bagworm	Lepidoptera: Psychidae	Indonesia	Foliage feeder on the mangrove seedlings	
Bruguiera spp.	*Pteroma plagiophleps* Hamp.	Bagworm	Lepidoptera: Psychidae	Indonesia	Foliage feeder on the mangrove seedlings	Nair and Sumardi, 2000
	Acanthopsyche sp.	Bagworm	Lepidoptera: Psychidae	Indonesia	Foliage feeder on the mangrove seedlings	

continued

Table 16.1. *continued.*

Host trees	Insect species	Common name	Insect group	Country	Mode of attack and impact	Reference
Buchanania lanzan Spreng.	*Plocaderus obesus* Gahan	Stem borer	Coleoptera: Cerambycidae	India	40–50% damage to plant due to damaged internal tissues	Meshram, 2009
Casuarina spp.	*Microcerotermes minor* Holmgren; *Odontotermes* spp. and *Microtermes obesi* Holmgren	Termites	Isoptera: Termitidae	India	Root and stem damage in seedlings leading to death	Thakur, 1992
Cedrela odorata L. (Cigar box cedar)	*Anurogrillus* spp.	Grasshopper	Orthoptera: Gryllidae	Cuba	Seedling damage in nursery	Castilla *et al.*, 2002
	Hypsipyla grandida Zeller	Mahogany shoot borer	Lepidoptera: Pyralidae	Costa Rica	77% of trees attacked after 84 weeks	Newton *et al.*, 1998
Cedrus deodara (Roxb.) G. don.f.	*Scolytus major* Stebbing	The bark borer	Coleoptera: Scolytidae	Northwestern India	Serious pest of forests	Thapa and Singh, 1986
Cordia gerascanthus Linn. (Cordia wood)	*Anurogrillus* spp.	Grasshopper	Orthoptera: Gryllidae	Cuba	Seedling damage in nursery	Castilla *et al.*, 2002.
Dalbergia sissoo Roxb.	*Aristobia horridula* Hope	Longhorn beetle	Coleoptera: Cerambycidae	Nepal	Causes 43–97% mortality of young trees due to stem girdling/boring	Dhakal *et al.*, 2005
	Myllocerus spp.	Weevil	Coleoptera: Curculionidae	Bihar, India	Damage to nursery plants	Sah *et al.*, 2007
Dendrocalamus strictus (Roxb.) Nees., *D. giganteus* Munro, *Bambusa nutans* Wall ex. Munro, *B. vulgaris* Schrad. ex. Wendl. (bamboos)	*Pyrausta (Crypsiptya) coclesalis* Walk.	Bamboo leaf roller	Lepidoptera: Pyralidae	India	Larvae binds and rolls leaves and defoliates in nurseries and plantations	Joshi, 1992; Senthilkumar and Murugesan, 2015
	Alcidodes crassus Poscoe and *A. ludificator* Singh	Shoot tunneller	Coleoptera: Curculionidae	India	Adult feeds on twigs and leaves. Lays eggs in galleries in green shoot of seedlings in nurseries causing die-back	
	Holotrichia consanguinea Blanchard	White grubs	Coleoptera: Scarabaeidae	India	Grubs feed on roots and rootlets inside the soil and cause seedling death in forest nurseries	

	Species	Common name	Order: Family	Location	Damage	Reference
	Odontotermes microdentatus Roonwal and Sen-Sarma, *O. obesus* Rambur	Termites	Isoptera: Termitidae	India	These are mound building termite species, attacking roots of germinating seedlings, rhizomes below ground level and cause death in forest nurseries	
Eucalyptus spp.	*Strepsicrates* spp.	Leaf roller	Lepidoptera: Tortricidae	Ghana; Sabah	Causes 50% defoliation in nursery beds	Wagner *et al.*, 1991; Chey, 1996
	Helopeltis spp.	Tea mosquito bug	Hemiptera: Miridae	Indonesia	Bugs suck sap and damages plants in nursery	Nair, 2007
	Alcides sp.	Shoot borer	Coleoptera: Curculionidae	Indonesia	Bores the shoots sometimes killing the plants in nursery	
	Zeuzera coffeae Nietner	Red borer	Lepidoptera: Cossidae	Indonesia	Causes die-back of shoots in samplings	
	Coptotermis testaceus Linn.	Subterranean termite	Isoptera: Termitidae	Brazil	Feeds on roots in nurseries, killing 10–70% seedlings	Wilcken *et al.*, 2002
	Microtermes spp. *Odontotermes* spp.	Subterranean termites	Isoptera: Termitidae	Ethiopia; Africa; India	Up to 100% of transplants killed due to stem and root severance	Cowie *et al.*, 1989
	–	Scale insects and mealybugs	Homoptera: Coccidae	Ghana	They are serious pests on seedlings in forest nurseries	Wagner *et al.*, 2008
	Frankliniella occidentalis (Pergande)	Western flower thrips	Thysanoptera: Thripidae	Andhra Pradesh, India	Suck sap from tips and leaves, resulting in reduced leaf size, pale narrow laminae, stunting, distortion, wilting of foliage and dropping in nurseries and young plantations	Kulkarni, 2010a
Eucalyptus grandis W. Hill ex. Maiden	*Leptocybe invasa* Fisher & La Salle	Chalcid gall wasp	Hymenoptera: Eulophidae	Vietnam	Causes 65–92% incidence of shoot galls	Thu *et al.*, 2009

continued

Table 16.1. *continued.*

Host trees	Insect species	Common name	Insect group	Country	Mode of attack and impact	Reference
Eucalyptus tereticornis Sm.; *Dalbergia sissoo* Roxb.; *Syzigium cumini* (L.) Skeels.; *Peltophorum ferrugineum* Benth; *Santalus album* Q.F. Zheng	Species not known	Cockchafer swarms	Coleoptera: Melolonthinae	South India	Causes 70–80% defoliation by beetles	Thakur and Sivaramkrishanan, 1991
Eucalyptus urophylla S.T. Blake	*Cephisus siccifolius* Walker	Spittle bug	Hemiptera: Cicadellidae	Brazil	Feeds sap from leaves and shoots leading to shoot deformation in 99% seedlings	Ribeiro *et al.*, 2005
Eucalyptus spp., *Eucalyptus urophylla* S.T. Blake; *E. camaldulensis* Dehnh.	*Leptocybe invasa* Fisher & La Salle	Chalcid gall wasp	Hymenoptera: Eulophidae	Vietnam; Thailand; India; Argentina	Damage is more severe in younger seedlings	Mendel *et al.*, 2004; Dell *et al.*, 2008; Thu *et al.*, 2009; Botto *et al.*, 2010; Jhala *et al.*, 2009; Senthilkumar and Murugesan, 2015
Falcataria moluccana (Miq.) Barn. & Grim.	*Pteroma plagiophelps* Hampson	Bagworm	Lepidoptera: Psychidae	Southwest India	Death of 25% of trees and 17% severely damaged after 2.5 years repeat defoliation	Nair and Mathew, 1992
Gmelina arborea Roxb.	*Alcidodes ludificator* Marshal	Gamari weevil	Coleoptera: Curculionidae	Northeast, Central and South India	Damage to plants of nurseries	Joshi, 1992; Senthilkumar and Barthakur, 2009; Senthilkumar and Murugesan, 2015
	Eupterote geminate Wlk.	Gmelina defoliator	Lepidoptera: Eupteroptidae	Central and South India	Moves in gregarious forms and defoliates seedlings and trees	Joshi, 1992; Senthilkumar and Murugesan, 2015
	Prioptera punctipennis Vagener	–	Coleoptera: Chrysomelidae	South India	Larvae and adults defoliate leaves voraciously	Senthilkumar and Murugesan, 2015

Host plant	Insect pest	Common name	Order: Family	Country	Damage	Reference
	Tingis beesoni Drake	–	Homoptera: Tingidae	Central and South India	Aggregate in large numbers on stem and branches and suck sap from larger veins, making leaves spotted, discoloured and withered, causing shoot die-back	
	Xyleutes ceramica Walk.	Bee-hole borer	Lepidoptera: Cossidae	Sabah	7–12% damage due to stem boring	Gotoh *et al.*, 2003
	Zonocerus elegans Thunb.	Grasshopper	Orthoptera: Pyrgomorphidae	India	Feeding on tender leaves	
	Scale insects and mealybugs		Homoptera: Coccidae	Ghana	They are serious pests on forest nursery plants	Wagner *et al.*, 2008
	Xyleborus fornicates Eichh.	Stem-boring scolytinae beetles	Coleoptera: Scolytidae	India	Seedling mortality	Mathew, 2005
	Calopepla leayana Latr.	Gmelina leaf feeder	Coleoptera: Chrysomelidae	India	The larvae and adults devour the leaves completely, leaving only midribs	Senthilkumar and Murugesan, 2015
Hevea spp.	*Atta sexdens* (Latreille)	Leaf-cutting ant	Hymenoptera: Formicidae	Brazil	71.3% of seedlings were attacked	Calil and Soares, 1987
Hevea brasiliensis Müll.Arg.	*Anomala* spp.; *Leucopholis* spp.; *Holotrichia* spp.	White grubs	Coleoptera: Scarabaeidae	Bangladesh	Root feeder	Baksha ,1990
Khaya spp.	*Xylosandrus compactus* (Eichh.)	Bark beetle	Coleoptera: Scolytidae	India	60–70% bark boring in seedlings and young saplings	Meshram *et al.*, 1993b
Khaya anthotheca (Welw.) C.DC.	*Hypsipyla robusta* (Moore)	Mahogany shoot borer	Lepidoptera: Pyralidae	Ghana	Shoots of up to 88% of trees bored after 2 years	Ofori *et al.*, 2007
Khaya senegalensis (Desr.) A. Juss	Scale insects and mealybugs	Sap suckers	Homoptera: Coccidea	Ghana	They are serious pests on seedlings in forest nurseries	Wagner *et al.*, 2008
Leucaena leucocephala Lam.; *L. pallida* Britton & Rose	*Heteropsylla cubana* Crawford	Leucaena Psyllid	Hemiptera: Psyllidae	Australia	Sap sucker reduces production by 50–79%	Palmer *et al.*, 1989; Bray and Woodroffe, 1991; Mullen and Shelton, 2003

continued

Table 16.1. *continued.*

Host trees	Insect species	Common name	Insect group	Country	Mode of attack and impact	Reference
Milicia excelsa Welw.	*Phytolyma lata* Walker	Gall psyllid	Hemiptera: Homotomidae	Ghana	Over 40% mortality due to galls according to provenance	Ofori and Cobbinah, 2007
Moringa pterygosperma Lam.	*Noorda blitealis* Walker	Moringa leaf feeder	Lepidoptera: Pyralidae	India	100% mortality due to severe defoliation and tunnelling of shoots	Kulkarni *et al.*, 2003
Paraserianthes falcataria (L.) Nielsen	*Xystrocera festiva* Thompson; *X. globosa* Oliv.	Sengon borer (Albizia borer)	Coleoptera: Cerambycidae	Indonesia	Serious pest of trunks; *X. globosa* is a minor pest	Nair and Sumardi, 2000
	Pteroma plagiophleps Hampson	Small bagworm	Lepidoptera: Psychidae		Occasionally serious pest as leaf feeder and damage on bark surfaces	
	Eurema blanda Boisd.	Yellow butterfly caterpillar	Lepidoptera: Pieridae		Occasional pest as defoliator	
	Indarbela quadrinotata Walk.	Bark caterpillar	Lepidoptera: Indarbelidae		Damages bark	
	Xylosandrus morigerus Bland.	Scolytid beetle	Coleoptera: Scolytidae		Bores twigs	
Pinus caribaea Morelet	*Dendroctonus frontalis* Zimmerman	Southern pine beetles	Coleoptera: Scolytidae	Belize	Bores bark causing 90% tree mortality	Snyder *et al.*, 2007
P. caribaea Morelet; *P. cubensis* Grise; *P. maestrensis* Bisse	*Atta insularis* (Guérin-Méneville)		Hymenoptera: Formicidae	Cuba	Polyphagous pest of forest nurseries	Castilla *et al.*, 2002
P. elliotii var. *elliotii* Engelm.	*Heteronychus arator* Fab.	African black beetle	Coleoptera: Scarabaeidae	Australia	Seedling damage due to defoliation	Brown and Wylie, 1990
Pinus massoniana Lamb.	*Dendrolimus* spp.	Pine caterpillar	Lepidoptera: Lasiocampidae	South China	30–80% defoliation	Wang *et al.*, 2007
P. massoniana Lamb.	*Anomala cupripes* Hope.	Scarab beetle	Coleoptera: Scarabaeidae	Hong Kong	Defoliation due to feeding	Browne, 1968
Pinus merkusii Jungh. & de Vriese	*Dioryctria rubella* Hamp.	Tusampitch moth	Lepidoptera: Pyralidae	Indonesia	Causes shoot die-back and stem distortion	Nair and Sumardi, 2000
	Miliona basalis Walker	Pine looper	Lepidoptera: Geometridae			

	Nesodiprion biremes (Konow)	Pine sawfly	Hymenoptera: Diprionidae			
	Coptotermes sp.	Termites	Isoptera: Rhinotermitidae		Feeds on root collar	
	Several species of white grubs	White grubs	Coleoptera: Scarabaeidae		Feeds on roots of saplings	
Pinus oocarpa Schiede ex. Schltdl.	*Dioryctria* spp.	Pine shoot borer	Lepidoptera: Pyralidae	Indonesia	Bores shoot/ bark with 70% mortality of young trees	Intari and Ruswandy, 1986
Pinus radiata D.Don.; *P. taeda* L. and *Pinus* spp.	*Sirex noctilio* F.	Sirex wood wasp	Hymenoptera: Siricidae	Queensland	Toxic mucus and the fungus kills the host tree; major outbreaks of *Sirex* in SE South Australia-SW Victoria between 1987 and 1989, wasp killed over 5 million *Pinus radiata* trees	Ramsden, 2013
Pinus taeda L.	*Lygus lineolaris* (Pal. de Beau.)	Sap-sucking mirid and lygaeid bugs	Hemiptera	USA	50% damage to seedlings	Dixon and Fasulo, 2009
Pisonia grandis R.Br.	*Pulvinaria urbicola* Cock.	Softscale insect	Hemiptera: Coccidae	Coral Sea	16 ha trees destroyed due to sap sucking on leaves and stems	Greenslade, 2008
Populus spp.	*Eucosma glaciate* (Mey.)	Poplar shoot borer	Lepidoptera: Eucosmidae	Western Himalayas of India	Larvae tunnel the tender apical shoots of young poplars in seedlings, causing shoot death and forking	Sharma *et al.*, 2005
Populus spp.	*Condylorrhiza vestigialis* (Guenée)	Poplar caterpillar	Lepidoptera: Crambidae	South Brazil	50–100% defoliation on 2-year-old trees	Castro *et al.*, 2009
Populus spp.	*Chrysomela populi* (Linn.)		Coleoptera: Chrysomelidae	India; Pakistan	Eggs laid in clusters on lower leaf surfaces, larvae feed gregariously on seedlings	Khan and Ahmad, 1991; Sharma *et al.*, 2005
Populus spp.	*Nodostoma waterhousie* Jacoby		Coleoptera: Chrysomelidae	India; Pakistan	10–60% leaf feeding in seedlings	Singh and Singh, 1995
Pterocarpus marsupium (Roxb.) and *Albizzia odoratissima* (L.f.)	*Spanioneura* spp., *Arytaina* spp.		Hemiptera: Psyllidae	Kerala, India	Feeding causes pouch galls in nursery, results in leaf crinkling and severe seedling stunting in 40% of seedlings	Mathew, 2005

continued

Table 16.1. *continued.*

Host trees	Insect species	Common name	Insect group	Country	Mode of attack and impact	Reference
Rhizophora mangle L.	*Oiketicus kirbyi* Guild.	Bagworm	Lepidoptera: Psychidae	Ecuador	80% of foliage defoliated over 1200 ha of mangrove	Gara *et al.*, 1990
	Coccotrypes rhizophorae (Hop.)	Mangrove bark beetle	Coleoptera: Scolytidae	Panama	Bores bark of mangrove causing 72% tree mortality	Sousa *et al.*, 2003
Rhizophora stylosa Griff.	*Doratifera stenosa*	Nettle caterpillar	Lepidoptera: Limacodidae	Australia	Insect causes annual defoliation of 30–40% in mangrove hosts	Duke, 2002
Rhizophora spp.; *Bruguiera gymnorhiza* (L.) Lam.	*Chionaspis* sp.	Scale insect	Homoptera: Coccoidea	Indonesia	Sucks sap from mangrove species	Nair and Sumardi, 2000
Rhizophora spp.	*Aulacaspis marina* T. & W.	Scale insect	Homoptera: Coccoidea	Indonesia	Sucks sap from the mangrove seedlings	
	Zeuzera conferta Walker	Bee-hole borer	Lepidoptera: Cossidae	Indonesia	Bores stem of mangrove species	
	Xyleborus spp.	Scolytid borer	Coleoptera: Scolytidae	Indonesia	Bores twigs of mangrove species	
Samanea saman (Raintree)	*Anurogrillus* spp.	Grasshopper	Orthoptera: Gryllidae	Cuba	Seedling damage in nursery	Castilla *et al.*, 2002
Santalum album L.	*Zeuzera coffeae* Neitner	Santalum borer	Lepidoptera: Cossidae	India	Larvae tunnel the heartwood of trees and degrade values of timber	Senthilkumar and Murugesan, 2015
	Letana inflate	Grasshopper	Orthoptera: Tettigoniidae		Defoliating in seedlings in nursery; lay eggs in slits	Remadevi *et al.*, 2005
	Ceroplastes actinoformis Green and *Inglisia bivalvata* Green	Coccids insect	Homoptera: Coccidae		Causes mortality of sandalwood nurseries	
Shorea robusta Roth.	*Sitophilus rugicollis* (Casey)	Sal seed weevil	Coleoptera: Curculionidae	India	Infestation originates with seed fall on forest floor and continues in storage with potential of 100% seed damage; damaged seeds are completely hollow and lose economic value	Khatua, 1999; Choubey *et al.*, 2004, 2008, 2013a

	Hoplocerambyx spinicornis Newm.	Sal heartwood borer	Coleoptera: Cerambycidae	India	Heavy tunnelling in tree trunks; repeated epidemics every 18–20 years reported; kills millions of trees in every epidemic	Roonwal, 1978; Joshi *et al.*, 2006; Kulkarni *et al.*, 2007a, 2015
Shorea spp.; *Hopea odorata* Roxb.	*Nanophyes shoreae* Marshall; *Alcidodes dipterocarpi* Marshall	Seed weevil	Coleoptera: Curculionidae	Indonesia	Seeds are damaged by borer	Nair and Sumardi, 2000
Shorea javanica Koord. & Valeton.	*Mucanum* sp.	Plant bug	Hemiptera: Pentatomidae	Indonesia	Insects suck sap on seedlings in nurseries	
	Lawana candida Fab.	Cicada	Homoptera: Cicadidae	Indonesia	Occasional pest on older plants	
Shorea spp. and *Hopea* spp.	*Calliteara cerigoides* Walk.	Hairy caterpillar	Lepidoptera: Lymantriidae	Indonesia	Serious pest as defoliator	
Sonneratia apetala Buch.	*Zeuzera conferta* Walk.	Stem-boring moth	Lepidoptera: Cossidae	Bangladesh	Average 32% trees attacked in mixed and 51% in pure plantations	Wajihullah *et al.*, 1996
Swienenia macrophylla King; *Khaya senegalensis* Desr.	*Xyleborus compactus* Eichh.	Black twig borer	Coleoptera: Scolytidae	Sri Lanka	Attack resulted in complete failure of some nurseries	Mannakkara and Alawathugoda, 2005
Swietenia humilis Zuce.	*Hypsipyla grandella* Zeller	Mahogany shoot borer	Lepidoptera: Pyralidae	Honduras	Shoot borer with 44% incidence	Goulet *et al.*, 2005
Tabebuia angustata Britt.	*Anurogrillus* spp.	Grasshopper	Orthoptera: Gryllidae	Cuba	Damage to seedlings	Castilla *et al.*, 2002
Tectona grandis Linn. f.	*Schizonycha ruficollis* Fab.	White grub	Coleoptera: Scarabaeidae	India	While adults defoliate other species, grubs cause 14–42% nursery seedling destroyed due to root feeding	Kulkarni *et al.*, 2007b
	Hypsipyla robusta Moore	Teak shoot borer	Lepidoptera: Pyralidae	India	Caterpillars bore into the tips and shoots and cause apical side branching in young planted forests of several species of high-quality timber species of Meliaceae and Verbenaceae, including *Swietenia macrophylla, Toona ciliate, Cedrella* spp.	Joshi, 1992; Senthilkumar and Murugesan, 2015

continued

Table 16.1. *continued.*

Host trees	Insect species	Common name	Insect group	Country	Mode of attack and impact	Reference
	Myllocerus discolor (Bohe.)	Weevil	Coleoptera: Curculionidae	India	Grub bores into the green shoots making tunnels and hollows seedlings leading to mortality	Kulkarni *et al.*, 2007c
	–	Scale insects and mealybugs	(Homoptera: Coccidae)	Ghana	They are serious pests on forest nursery plants	Wagner *et al.*, 2008
	Apate monachus F. and *A. tenebrans* Hoff.	–	Coleoptera: Bostrychidae	Africa	Bores the stems	Wagner *et al.*, 2008
	Xyleutes ceramicus Walk.	Bee-hole borer	Lepidoptera: Cossidae	Sabah	Cause 10–65% stem boring	Gotoh *et al.*, 2003
	Holotrichia rustica Burm.	White grubs	Coleoptera: Scarabaeidae	India	While adults feed on other tree species, grubs cause 20% mortality in seedlings due to root feeding	Kulkarni *et al.*, 2009
	Hyblaea puera Cramer	Teak defoliator	Lepidoptera: Hyblaeidae	Southwest India; Indonesia	Defoliation causes 44% cumulative loss of volume increment in 5 years; 50% loss of height increment	Nair *et al.*, 1985; Joshi *et al.*, 2001; Nair, 2007
	Holotrichia serrata F.	White grubs	Coleoptera: Scarabaeidae	Sri Lanka	While adults feed on tree species, grubs cause 20% mortality in seedlings due to root feeding	Bandara, 1990
	Eutectona machaeralis Walker	Teak skeletonizer	Lepidoptera: Pyralidae	India; known as *Paliga demastisalis* in Indonesia	100% seedling infestation in nursery	Joshi *et al.*, 2001; Basalingappa and Ghandi, 1994
	Valanga nigricornis	Grasshopper	Orthoptera: Acrididae	Indonesia	Feeds on foliage	Nair and Sumardi, 2000
	Alcidodes ludificator Singh and *A. crassus* Poscoe	–	Coleoptera: Curculionidae	India	Small beetle causes die-back or death of seedlings in nurseries due to galleries made in the green shoots in nursery	Nair, 2001; Senthilkumar and Murugesan, 2015

Host	Insect pest	Common name	Order: Family	Distribution	Damage	Reference
	Neotermes tectonae Damm	Inger-inger	Isoptera: Kalotermitidae	Indonesia	Unique pest of teak in Indonesia damaging trunk/ stem	Nair and Sumardi, 2000
	Xyleutes ceramica Walker	Bee-hole borer	Lepidoptera: Cossidae		Bores trunks	
	Xyleborus destruens	Ambrosia beetle	Coleoptera: Scolytidae		Bores trunk/ stem	
	Zeuzera coffeae Neitner	Red borer	Lepidoptera: Cossidae	India; Indonesia	Bores saplings	Joshi, 1992; Nair and Sumardi, 2000; Senthilkumar and Murugesan, 2015
	Dihammus cervinus (Hope)	Stem borer	Coleoptera: Cerambycidae	India	Adults feed on bark of 2–8-year-old saplings and grubs cause formation of bulging canker all around the stem at or above ground level causing breakage in saplings	Joshi, 1992
	Dichocrocis punctiferalis (Guenee)	Seed borer	Lepidoptera: Pyralidae	India	Major pest of flowering shoots and young fruits of teak affecting seed production	Joshi, 1992; Senthilkumar and Murugesan, 2015
Tectona grandis L.f. and many other forest tree species	*Aleurodicus dispersus* Russell; *Acaudaleyrodes rachipora* Singh	Spiralling whitefly and Babul whitefly	Hemiptera: Aleyrodidae	India	Causes leaf chlorosis in seedlings and plantations, shedding and reduced growth rate, attracts mould development on leaves and affects photosynthesis	Sundararaj and Dubey, 2005
Toona ciliate M. Roem.	*Hypsipyla robusta* Moore	Shoot-boring moth	Lepidoptera: Pyralidae	Thailand	Shoots of 20–60% trees bored depending upon height class	Cunningham and Floyd, 2006
Triplochiton scleroxylon K. Schum	*Diclidophlebia* spp., *D. eastopi* Vondracek, *D. harrisoni* Osisanya	White aphids	Homoptera: Psyllidae	Nigeria; Ghana; Ivory Coast	Eggs deposited on principal leaf veins or edges of young leaves hatch in 8 days; sap sucking causes peripheral leaf curling, rolling and yellowing, causing die-back, stunted growth, copious branching	Wagner et al., 2008

processes, such as facilitating removal of weakened trees, decomposition of dead or decaying trees, nutrient cycling, biodiversity maintenance, and forest regeneration and succession (Ayres and Lombardero, 2000) during their non-epidemic or endemic period. Thus, while their presence is essential, their outbreaks result either in tree mortality or growth reduction. While aerial surveys, remote sensing and GIS technologies play a major role in collecting these insect infestation data in forests (Fierke et al., 2011), actual observations through ground-based monitoring plots is still considered more accurate and thus is still simultaneously practised in many parts of the world, particularly for site-specific chronological changes due to insect infestations at initial stages without crown death for closer monitoring.

Insect pests infesting forest trees at various stages, namely, inflorescence, fruits, seeds on standing trees and storage (Kulkarni and Joshi, 1998), seedlings in forest nurseries (Thakur, 2000; Kulkarni and Paunikar, 2009; Kulkarni et al., 2006), plantations and natural forests (Shukla and Joshi, 2001; Kulkarni, 2014) in India have earlier been compiled. Nair (2001) compiled information on insect pests infesting forestry species in Indonesia. With a global perspective, Wylie and Speight (2012) have recently compiled an excellent book describing numerous insect species. Discussing all these species is beyond the scope of this chapter. However, an updated list of major insect pests in the tropics on some economically important forestry species is presented in Table 16.1, which gives an overview of the diversity of forest insect pests, together with their countries of occurrence.

16.4 Examples of IPM of Forestry Insect Pests

The widespread concern about the negative impacts of pesticides in developing countries (Atreya et al., 2011), besides feasibility of application of conventional methods in forests (Kulkarni, 2014), development of ecofriendly and biorational methods, their logical integration, coupled with education and training has become more relevant today. However, only a few limited examples of IPM as a programme or the individual IPM components, are available in the world in forestry, as compared to the agricultural counterparts (Wylie and Speight, 2012), possibly due to various reasons like complexity in generating base-line field data, long gestation-period crops and resulting longer life cycles of the insect pests, and feasibility of application. Efforts are made here to briefly discuss a few examples detailed by Wylie and Speight (2012), with some updated recent information.

16.4.1 Masson pine caterpillar in pine plantations in Vietnam and China

Pine plantations in Vietnam, comprising *Pinus massoniana* Lamb., *P. merkusii* Jungh. & de Vriese, *P. kesiya* Royle ex Gordon and *P. caribaea* Morelet, face serious growth losses due to defoliation by the Masson pine caterpillar, *Dendrolimus punctatus* (Walker) (Lepidoptera: Lasiocampidae). The effect has been more pronounced in *P. massoniana*, a species indigenous to China (Huang et al., 2008). In the case of severe infestation, extending to thousands of hectares, young trees may even die. According to Billings (1991), trees growing in sites to which they are not best suited, were affected the most and thus demonstrates that such areas unsuitable to the species may be left only for native species. IPM of this insect pest in Vietnam centred on exchange of insects collected from infested pine with a kilogram of rice per day. The available information on the IPM of this insect pest is summarized and given below.

1. Monitoring and prediction through traps (Zhang et al., 2001), sex attractant pheromones (Zhang et al., 2003), use airborne video to characterize defoliation intensities (Chen et al., 1997), predict damage intensity using satellite data (Zhang et al., 2005), use economic threshold of 5 larvae per tree (Li et al., 2007).

2. Chemical control using aerial spraying of pyrethroid insecticides (Zhang *et al.*, 1998), reduce use of synthetic insecticides (Li *et al.*, 2007).

3. Silvicultural manipulations like enhancing plant diversity in pine stands (Peng *et al.*, 1996), using different species, i.e. *P. elliottii* Engelm. instead of *P. massoniana* (Liang *et al.*, 2008), mixing pines with cedar or broad-leaved tree species (Wang *et al.*, 2005).

4. Biological control through parasitoids by conservation of understorey vegetation to enhance populations of natural enemies (CABI, 2010), rear egg parasitoids (*Trichogramma dendrolimi* Matsumura) in laboratory and release in stands (Xu *et al.*, 2006).

5. Biological control through pathogens by applying fungi, such as *Metarhizium anisopliae* (Metchnikoff) Sorokin in warm and dry conditions (Song, 1997) or *Beauveria bassiana* (Bals.-Criv.) Vuill. in cooler and more humid conditions (Wang *et al.*, 2004), bacteria, such as *Bacillus thuringiensis* Berliner var. *kurstaki* (*Bt.*) (Chen *et al.*, 1997), or cytoplasmic polyhedrosis virus (CPV).

6. Use egg parasitoids to disseminate cytoplasmic polyhedrosis virus (CPV) (Sun and Peng, 2007).

7. Use transgenic *Pinus* species expressing *Bt.* CRY1Ac gene (Tang and Tian, 2003), mass trapping of adults with light traps (Zhao *et al.*, 1998).

8. Use pupae as food for ethnic minorities (He *et al.*, 1998).

16.4.2 Insect pests of eucalypt plantations in Brazil

Out of more than 500 insect pest species known to infest eucalyptus plantation in Brazil, most important among them are three Lepidoptera (Zanuncio *et al.*, 2003, 2006, 2009), two Coleoptera (Wilcken *et al.*, 2008; Flechtmann *et al.*, 2001), besides Isoptera, Hymenoptera and Hemiptera (Wylie and Speight, 2012).

As with all IPM systems, preventive measures should be used to combat insect pests in the eucalypt plantations. Aims were: (i) detection of primary pest outbreaks; (ii) outbreak evaluations; (iii) analysis of results; and (iv) definitions of control strategies. All this is possible only through periodical surveys, before outbreaks take a hold. Among Lepidopteran, *Thyrinteina arnobia* Stoll also had various potentially important parasitoids (Zanuncio *et al.*, 2009). However, efficacy of these biological control agents depends largely on climate and host-tree factors on lepidopteran population dynamics (Wylie and Speight, 2012).

Similarly, IPM was also employed for the management of psyllids *Ctenarytaina spatulata* Taylor and *Glycaspis brimblecambei* Moore in eucalypt plantation of Brazil (Santana and Burckhardt, 2007; Silva *et al.*, 2010). This included components like planting tolerant species, avoiding water stress condition by site selection, avoiding continuous eucalypt plantations which facilitate the spread of the pest, maintaining natural vegetation around plantations to promote biological control, monitoring pest densities with sticky traps, manual collections by beating foliage and branches, monitoring pest densities using egg counts related to subsequent numbers of damaging nymphs, encouraging native natural enemies such as spiders and ladybirds, establishing self-perpetuating populations of exotic parasitoids throughout eucalypt regions of Brazil and reducing reliance on insecticides (Wylie and Speight, 2012).

16.4.3 Insect pests of agroforestry on poplars in northwest India

A few important examples from subtropical regions have also been included in this chapter, since forestry in such areas has so many similarities and problems to that in the true tropics, like Punjab in northwest India.

Poplars, *Populus deltoides* W. Bartram ex Humphry Marshall, are planted on a large scale as part of agroforestry system in Uttar Pradesh, Himachal Pradesh, Punjab and Kashmir in India. This tree is attacked by the defoliators *Clostera fulgurita* Walker and *C. restitura* Walker (Lepidoptera: Notodontidae) coupled with other stem and shoot borers and root feeders in nursery and stands (Singh *et al.*, 2004; Sangha, 2011). Singh *et al.* (2004) has recommended various IPM measures for managing these pests, like: pruning and burning attacked branches, and storm damaged and old stumps, and removing alternative hosts like *Morus* spp. and *Prunus* spp. to manage longhorn stem borer, *Apriona cinerea* Chevrolat; pruning and burning affected branches with foliar spray of insecticide (dimethoate) in May/June and August/September against lepidopteran shoot borer, *Eucosma glaciata* (Meyrick); use of tolerant clone of *P. deltoides* and spray Clostera NPV; release of egg parasitoids (*Trichogramma* spp.); encourage predatory bugs in stands; spray neem oil after rains and monitor pest populations in alternative winter hosts like other poplars and *Salix* spp., etc. to manage lepidopteran defoliator *Clostera cupreata* Butler, *C. fulgurita*, *C. restitura* and *Phalantha phalantha* (Drury); encourage predatory ladybird beetles to manage aphids, *Eriosoma* spp.; manipulation in weeding time of the nursery beds; dress soil with slow release insecticidal formulations against white grubs, *Holotrichia* spp.; apply insecticides in water or dust to soil or base of tree to manage termite, *Odontotermes* spp. and *Coptotermes* spp. (Wylie and Speight, 2012).

16.4.4 *Phoracantha* species in the Americas, Africa and Europe

The eucalyptus cerambycids longhorn beetles, *Phoracantha semipunctata* (Fabr.) and *P. recurva* Newman, are widespread pests of eucalyptus in tropical, subtropical and even Mediterranean regions of the world (Paine *et al.*, 2011). IPM measures can be applied, like tree care and reduction of stressful conditions (including avoiding moisture deficits and unnecessary damage during pruning), avoidance of susceptible tree species, removal of sources of adult beetles by maintaining sanitation through

any means, uninfested cut logs to be split and debarked, biological control using egg and larval parasitoids; careful assessments of damage thresholds for the pest have also to be made (Paine *et al.*, 1995; Wylie and Speight, 2012).

16.4.5 Sal seed borer, *Sitophilus rugicollis* in India

Sal (*Shorea robusta* Gaertn. f.) is one of the most important natural forest tree species after teak, which is a plantation crop in India (Anon., 1972; Tewari, 1995). It occupies about 10% of the total forest cover of the country (ISFR, 2015). Sal fruits/seeds are eaten after roasting (only in times of food scarcity) (Anon., 1972). They are mainly valued for yielding fatty oil and oil cake which are used in various ways, i.e. cooking and lighting by tribes, soap making after blending with other soft oils, in chocolate and confectionery after refining (Anon., 1972). Its principal use is in soap making. Owing to the above economical importance, trade in sal seed was nationalized in the states of Madhya Pradesh, Chhattisgarh and Bihar as an economic activity, which generated large-scale employment in rural areas. Seeds are collected by hand picking during the seed setting months. The sal forest area is vast and the seed collection period is very short, i.e. four weeks. During rains, sal seeds require immediate storage in the areas of purchase (Tewari, 1995). Some studies have been reported regarding the oil content in seeds being affected by storage. Gautam *et al.* (1975) have also observed that available oil content (14%) in seeds does not diminish even if the sal fruits are stored at ambient room temperature for one year; instead, the fatty acid content of the extracted oil increases from 3 to 13.6%. As against the above fact, the extracted sal fat has shown general deterioration on storage, and hence in practice sal seeds are stored before extraction of oil instead of extracted oil (Sharma and Jain, 1981).

Among 16 insect pests of inflorescence/fruits/seeds on standing tree and 5 insect species feeding on seeds in storage, *Sitophilus* (*Calandra*) *rugicollis* Casey is the most important and a true storage pest (Khatua and Chakrabarti, 1990; Choubey, 2004). This insect has been reported to feed on the cotyledons and embryo of the sal seed. During early stages of the infestation, the seed looks sound and healthy from outside, but as larval feeding increases with time the whole inside of the seed is eaten away. Such seeds are filled with larval excreta and a powdery substance. At the time of emergence, one or more small circular holes on the surface of seeds are visible (Choubey *et al.*, 2013a).

Choubey *et al.* (2012) recommend direct fumigant effect of neem (*Azadirachta indica* A. Juss) product Multineem® and admixture of neem leaf powder at the ratios 10:100 and 5:100 proved best, with 43.76 and 31.66% reduction in emergence of adults against untreated seed kept as a control. Neem oil proved most effective as oviposition deterrent as evident reduction in these treatments over control, i.e. 69.44, followed by methanolic leaf extract of *Annona squamosa* Linn. with mean reduction over control being 63.88%. Multineem®, karanj oil and castor oil proved next with reduction of 53.32, 47.77 and 54.99%, respectively.

In case of severe incidence, seed treatment can be applied with deltamethrin, cypermethrin and malathion (Choubey *et al.*, 2013b), with persistent effect up to 5 weeks (Choubey, 2004). When methyl parathion 2G Dust was used, it proved significantly effective over control, when adults were released on freshly treated seeds. The highest treatment ratio, i.e. 1.5:100, continued to kill 100% of adults even after the seventh week. However mortality declined further, to only 24.13%, during the ninth week. Dichlorvos and Carbon tetrachloride caused respectively, 100.0% and 80.32% mortality of developmental stages inside the seeds when seed stock was fumigated with 0.5 ml/600 cm^3 dose. Chloroform required 2.0 ml/600 cm^3 for a similar effect and Carbon Disulphide and Naphthalene proved significantly less effective. This investigation was unique in the way that, unlike previous reports, it investigated penetration effect of treatments on different developing stages inside the seeds. Carbon tetrachloride killed eggs laid inside the seeds at and above 1.0 ml/600 cm^3, along with a direct effect on the adult in direct exposure to fumigation with persistent effect until the seventh week (Choubey *et al.*, 2013c).

16.4.6 White grubs in forest nurseries in India

White grubs of scarabaeid beetles (Coleoptera: Scarabaeidae) are the major reported pests on a variety of forest nursery stock (Mathur, 1960; Browne, 1968). In India, nearly 20 species belonging to 10 genera of white grub species are known as pests on agricultural crops (Musthak Ali, 2001); whereas 5 species of 2 genera, *Holotrichia* and *Schizonycha*, are pests on forestry crops in nurseries (Garg *et al.*, 2005; Kulkarni *et al.*, 2007b). The active species of major concern in teak (*Tectona grandis* Linn. f.) nurseries are *H. serrata* Fabr., *H. consanguinea* Blanchard and *H. reynaudi* Blanchard (syn. *H. insularis* Brenske) (Oka and Vaishampayan, 1979; Meshram *et al.*, 1993a; Thakur, 2000; Garg *et al.*, 2005). These species cause heavy losses in teak seedlings in India (Fig. 16.1).

Management efforts have mainly depended on chemical treatment of beds for controlling grub population in the soil. Only very recently Kulkarni *et al.* (2007b, 2009) have reported detailed bioecology of the three scarab species (*Holotrichia rustica* Burmeister, *H. mucida* Gyllenhall and *Schizonycha ruficollis* Fabr.) in a central Indian teak nursery with the aim to develop an Integrated Pest Management (IPM) model, including the prediction of the attack (Kulkarni, 2010b), for the management of scarab pests in teak nurseries. The IPM guideline is based on the following time-bound recommendations (Kulkarni, 2010b; Kulkarni *et al.*, 2016).

1. Deep plough of the nursery plot before bed preparation to expose and kill the insect stages inside the soil.

Fig. 16.1.   Damage caused by the Scarabaeid *Holotrichia* spp. to teak (*Tectona grandis*) seedlings: top left, teak nursery in central India with seedlings on raised soil beds to obtain root-shoots for plantations; top right, missing seedlings due to mortality caused by root damage by white grubs; bottom left, larvae; bottom right, adult. (Photos: N. Kulkarni)

2.   Monitoring and surveillance of the immature (grubs) and adult (chafer beetle) population by wilting in seed beds or checking the presence on identified host trees/ bushes/plants after the dusk or feeding symptoms.

3.   Monitoring of temperature and atmospheric RH for predicting emergence of beetles from previous year's soil beds (Kulkarni, 2010b).

4.   Record locality-specific field biology of the scarabaeid pest.

5.   Install standard light trap for trapping of adults.

6.   Identify host plants preferred by the scarabaeid pest growing naturally in and around the forest nursery area and retain them selectively as trap plants by removing them from the places where it is difficult to monitor.

7. Treatment of trap plants with 0.03% imidacloprid, which may be repeated after 15–20 days.

8. Just one week after the record of beetle emergence, treatment of nursery beds with phorate 10% G @ 200 g bed^{-1} to be repeated after a month with marketed product of *Beauveria bassiana* and/or *Metarhizium anisopliae.*

9. Introduction of native entomopathogenic nematodes (Kulkarni *et al.*, 2008).

10. Manual collection of white grubs from nursery beds by shallow digging around wilting seedlings and killing the grubs by dipping in kerosene.

11. To train nursery staff to preserve natural predatory elaterid grub, *Lenelater fuscipes* Fabr., which is usually inadvertently killed by the staff as a pest (Kulkarni *et al.*, 2013).

16.4.7 Management of *Sal* heartwood borer in India

Sal (*Shorea robusta*) heartwood borer (*Hoplocerambyx spinicornis* Newman) (Coleoptera: Cerambycidae) is one of the most dreaded and difficult-to-manage insect pests of the tropical region. During its annual life cycle, adults emerge just after the first heavy monsoon rain and lay eggs on the underside of the bark. After hatching, newly hatched grubs bore into the tree trunk of large sal trees in India. Many of them are thrown out due to the defensive resin secretion; but those that enter become established and feed on softwood and then heartwood, making large tunnels inside the tree trunk and emerging again as adults during the next rainy season (Roonwal, 1978), governed primarily by environmental factors (Kulkarni *et al.*, 2007a) (Fig. 16.2). The attacked tree with multiple tunnels in the heartwood succumbs to the injuries, gradually dies within a year and has value merely as fuel wood

(Joshi *et al.*, 2006). The insects have a history of reaching epidemic proportions every 15–20 years (Joshi *et al.*, 2006), killing millions of trees (Sambaraju *et al.*, 2016). The last epidemic was during 1996 to 2001 in Mandla Forest Division, in Madhya Pradesh State in central India, which resulted in removal of a million damaged/ attacked dead *sal* trees from the forest floor to maintain forest hygiene (Sambaraju *et al.*, 2016). Recently, the sal forests of Dindori Forest Division of Chhattisgarh State in central India have seen reoccurrence (Kulkarni *et al.*, 2015) in some new areas, not reported earlier in India. The beetle stage is the only free-moving stage of the insect for employing management options. The management options mainly concentrate on silvicultural options, baiting by natural tree logs through kairomonal attraction (Kulkarni *et al.*, 2004b) and identification, sensitization and conservation of natural enemies (Roychoudhury *et al.*, 2013) with other associated essential components.

1. Monitoring and surveillance of pest population by laying out permanent sample plots.

2. Periodical assessment of state of infestation on a scale/grade of 8 from dead (Cat. I) to healthy (Cat. VIII) (Roonwal, 1978).

3. Maintaining forest hygiene by removing dead and near to dead (Cat. I and II) and stumps (Cat. VI) from the site and logging to areas 4–5 km away from the sal forests to prevent emerging beetles from reverting to sal forests and laying eggs on healthy trees (Joshi *et al.*, 2006).

4. If transportation of logs is not possible due to inapproachability of the area or any other management issue, then fumigation of logs with fumigant before onset of monsoon rains to kill beetles inside their tunnels.

5. Baiting of beetles emerging from affected stands of other categories in the sal forests by standard trap-tree method. This is done by using wind-fallen, broken, unhealthy logs of 1 m and beating them to separate the bark. The process allows oozing out of the kairomonal compounds to attract beetles after dusk, which are caught

Fig. 16.2. Damage caused by the Cerambycid trunk borer, *Hoplocerambyx spinicornis*, known commonly as sal borer in *sal* (*Shorea robusta*) forest in central India: top left, view of a dead tree due to infestation by the borer; top right, view of wood dust around the tree, indicating heavy infestation, beetle and grub in inset; bottom left, inspection of the timber depot by the team where affected logs have been stacked; bottom right, section of a branch showing internal galleries by the beetle. (Photos: N. Kulkarni)

before sunrise (Joshi *et al.*, 2006; Sambaraju *et al.*, 2016).

6. The baiting of beetles is done by the community. The forest dwelling people are involved in laying out tree-traps at the rate of 1 trap/ha and paid for each beetle's head at the rate prescribed.

7. Training and demonstrations of these measures to the front-line are performed by forest staff in the affected areas.

8. Re-assessment of status of incidence during the next summer and in between to evaluate effect of the operations.

16.4.8 IPM of teak defoliator complex in India

Teak (*Tectona grandis*), the most valued timber tree species in India and abroad, is defoliated by successional or overlapping incidences of two oligophagous pests: the teak defoliator, *Hyblaea puera* (Cramer) (Lepidoptera: Hyblaeidae) (which consumes the whole leaf leaving only the midrib), and the teak skeletonizer, *Eutectona machaeralis* (Walker) (Lepidoptera: Pyralidae) (which consumes only mesophyl tissues, making the leaf look like a skeleton). However, the pests do not cause mortality in the nursery as in plantations and natural forests, but reportedly affect 65% incremental growth in plantation in a cumulative period of 5 years (Nair *et al.*, 1996), and 55% in forest nurseries in a year (Shukla and Joshi, 2001; Kulkarni *et al.*, 2004a). Management options ranging from identifying resistant teak clones (Roychoudhury, 2001), chemical applications, to the classical biological control have been experimented against these defoliators (Shukla and Joshi, 2001).

Sudheendrakumar *et al.* (1988) and Mohammad Ali *et al.* (1992) isolated and field-tested Nuclear Polyhedrosis Viruses (NPV) from the larvae of teak defoliator *H. puera* (HpNPV) at Nilambur (Kerala, India), with significant results within 3 days (Nair *et al.*, 1998). The freeze-dried wettable powder formulation with activated charcoal as UV protectant formulation not only provided more than 70% mortality in larvae, but protected trees also exhibited 41% more basal area increment (Nair *et al.*, 1996). This freeze-dried formulation of HpNPV is now a commercial product named Hybcheck®, with recommended field dose of 1×10^{10} POBs/ha to target third instar larvae. It may not be economical to target much older larvae, as it may incur a higher dose of the virus. This biopesticide is a new addition to the repertoire of measures undertaken in tropical forestry to reduce environmental hazards (Sajeev *et al.*, 2007; Sudheendra Kumar and Biji, 2009). However, the use of this product has been limited to Kerala State, possibly because of differences in environmental conditions in other parts of the country.

In a few earlier reports from south India, successful parasitization of *H. puera* eggs by *Trichogramma embryophagum* (Hartig) and *T. dendrolini* under laboratory conditions (Sudheendrakumar *et al.*, 1995) and *T. chilonis* Ishii under field conditions (Patil and Naik, 1997) was reported. In central India, successful classical biological control of teak defoliator complex began with collection, identification of native population of *Trichogramma raoi* Nagaraja (Yousuf *et al.*, 2004) and led to the development of TFRI-TRICHOCARD (Joshi *et al.*, 2007). Demonstration trials with TFRI-TRICHOCARD introduction of *T. raoi* alone in teak plantations or forests @ 1.25 lakh wasps per ha prevented 50% incidence of the defoliator complex. The following components can be listed out for management of teak defoliator complex.

1.   Planting of relatively resistant teak clones (Roychoudhury, 2001).
2.   Monitoring and surveillance of the defoliation incidence and intensity of attack.
3.   Introduction of *T. raoi* in batches from July onwards, so as to release a total of 1.25 lakh wasps per ha in plantation and natural forests.
4.   Spraying with *Bacillus thuringiensis* var. *karstaki* in Teak Seed Orchards (TSOs) or Clonal Seed Orchards (CSOs) and Seed Production Areas (SPAs) using a high-power sprayer to immediately check the infestation of inflorescence by the skeletonizer larvae (Shukla and Joshi, 2001).
5.   Post-introduction assessment of incidence.
6.   Supplementing the population in the subsequent years for better and prolonged effects.

16.4.9 Sirex woodwasp in Australia

Sirex wood wasp (*Sirex noctilio* Fabr.), detected first in 2009 in Queensland, is an introduced pest that has been affecting *Pinus radiata* D. Don, *P. taeda* L. and some other *Pinus* species grown in plantations in Australia since its accidental introduction into the mainland in 1961. The larval stages depend on the presence of the decay fungus *Amylostereum areolatum* (Chaillet ex Fr.)

Boidin, with spores of this fungus being deposited with phytotoxic mucus as the female lays her eggs. The host tree dies because of the combined effect of toxic mucus and the fungus. Major outbreaks of *Sirex* have occurred in 1987 killing over 5 million *Pinus radiata* trees with a royalty value of AUS$10–12 million (Ramsden, 2013). Its detection again in 2009 led to the establishment of the National Sirex Coordination Committee. In early 2010, 44 female *Sirex* were intercepted in static panel traps with an intercept peak in mid-March and the biocontrol measures widely used throughout the southern states were initiated. The parasitic nematode *Beddingia siricidicola* Bedding, was manually introduced into artificially stressed (poisoned) 'trap' trees. These nematodes infect late instar larvae, rendering female *Sirex* sterile, while it still lays eggs or rather sacks of nematodes. This was supplemented with a small number of parasitoid wasps, *Ibalia leucospoides* (Hochenwarth) and *Megarhyssa nortoni* (Cresson). Using this method, 72 million nematodes were introduced in 2012 into 76 trap tree plots. The infection rates of emerged adults between sample plots ranged from 48 to 100% (Ramsden, 2013).

References

Ali, M.S. and Chaturvedi, O.P. (1996) Major insect pests of forest trees in North Bihar. In: Nair, K.S.S., Sharma, J.K. and Varma, R.V. (eds) *Impact of Diseases and Insect Pests in Tropical Forests.* Proceedings of the IUFRO Symposium, 23–26 November 1993, Peechi, India, pp. 464–467.

Anderson, C. and Lee, S.Y. (1995) Defoliation of the mangrove *Avicennia marina* in Hong Kong – causes and consequences. *Biotropica* 27 (2), 218–226.

Anon. (1972) The wealth of India, raw materials (Vol. IX: Rh–So). Publication and Information Directorate, Council of Scientific and Industrial Research, New Delhi, pp. 313–321.

Anon. (2016) Encyclopedia Smithsonian: numbers of insects (species and individuals). Smithsonian buginfo, Smithsonian Institution Information Sheet no. 18. Available at: https://www.si.edu/spotlight/buginfo/bugnos (accessed 26 October 2017).

Atreya, K., Sitaula, B.K., Johnsen, F.H. and Bajracharya, R.M. (2011) Continuing issues in the limitations of pesticide use in developing countries. *Journal of Agricultural and Environmental Ethics* 24, 49–62.

Ayres, M.P. and Lombardero, M.J. (2000) Assessing the consequences of climate change for forest herbivore and pathogens. *The Science of the Total Environment* 262, 263–286.

Baksha, M.W. (1990) Some major forest insect pests of Bangladesh and their control. Bulletin No. 1, Forest Entomology Series. Forest Research Institute, Chittagong, Bangladesh.

Bandara, G.D. (1990) Chemical control of cockchafer grub (*Holotrichia serrata*) in teak nurseries. *Sri Lanka Forester* 19, 47–50.

Basalingappa, S. and Ghandi, M.R. (1994) Infestation of the seedlings of *Tectona grandis* by the lepidopteran larvae of *Hapalia machaeralis* (Pyralidae) and *Hyblaea puera* (Hyblaeidae). *Journal of Ecobiology* 6, 67–68.

Billings, R.F. (1991) The pine caterpillar *Dendrolimus punctatus* in Vietnam – recommendations for integrated pest-management. *Forest Ecology and Management* 39, 97–106.

Botto, E.N., Aquino, D., Lolacono, M. and Pathauer, P. (2010) Presence of *Leptocybe invasa* Fisher & La Salle, the 'eucalyptus gall wasp' in Argentina. *IPM Newsletter* No. 16. Institute of Microbiology and Agricultural Zoology.

Bray, R.A. and Woodroffe, T.D. (1991) Effect of the leucaena psyllid on yield of *Leucaena leucocephala* cv Cunningham in south-east Queensland. *Tropical Grasslands* 25, 356–357.

Brown, B.N. and Wylie, F.R. (1990) Diseases and pests of Australian forest nurseries: past and present. Paper presented at the first meeting of IUFRO Working Party S2.07-09 (Diseases and Insects in Forest Nurseries). Victoria, Canada, 22–30 August 1990.

Browne, F.G. (1968) *Pests and Diseases of Forest Plantation Trees: An Annotated List of the Principal Species Occurring in the British Commonwealth.* Clarendon Press, Oxford, UK.

CABI (2010) *Forestry Compendium.* CAB International, Wallingford, UK.

Calil, A.C.P. and Soares, W.O. (1987) Damage caused by leaf-cutter ants (*Atta sexdens*) in rubber (*Hevea* spp.) nurseries. *Boletim da Faculdade de Ciencias Agrarias do Para, Brasil* No. 16, 23–30.

Castilla, R.A.L., Casanova, A.D., Rivero, C.G., Escoto, H.C. and Issasi, N.T. (2002) *Forest Nursery Pest Management in Cuba*. National Proceedings RMRS-P-24: Forest and Conservation Nursery Associations 1999, 2000 and 2001. USDA Forest Service, September.

Castro, M.E.B., Ribeiro, Z.M.A., Santos, A.C.B., Souza, M.L., Machado, E.B., Sousa, N.J. and Moscardi, F. (2009) Identification of a new nucleopolyhedrovirus from naturally-infected *Condylorrhiza vestigialis* (Guenee) (Lepidoptera: Crambidae) larvae on poplar plantations in South Brazil. *Journal of Invertebrate Pathology* 102, 149–154.

Chen, S., Chen, E., Suo, Q., Duan, Z., Hu, C., Chen, S.W. *et al.* (1997) *Dendrolimus* virus resources and their application in Yunnan. *Chinese Journal of Biological Control* 13, 122–124.

Chey, V.K. (1996) *Forest Pest Insects in Sabah*. Sabah Forest Record No. 15, Sabah Forest Department, Sandakan, Borneo.

Choubey, V. (2004) Studies on some control methods on insect pests associated with seeds of sal (*Shorea robusta* Gaertn. F.) in Madhya Pradesh. PhD thesis, Rani Durgawati University, Jabalpur, India.

Choubey, V., Kulkarni, N. and Bhandari, R. (2004) Incidence of seed insect pests of sal (*Shorea robusta* Gaertn. F.). *Indian Journal of Tropical Biodiversity* 12, 48–52.

Choubey, V., Kulkarni, N. and Bhandari, R. (2008) Status of sal (*Shorea robusta*) seed insect pests in six sal dominating regions of central Indian state of Madhya Pradesh. *Indian Journal of Forestry* 31, 595–598.

Choubey, V., Bhandari, R. and Kulkarni, N. (2012) Effect of botanicals on the seed weevil, *Sitophilus rugicollis* Casey. *Journal of Entomological Research* 36 (3), 259–262.

Choubey, V., Bhandari, R. and Kulkarni, N. (2013a) Life history and morphology of seed weevil, *Sitophilus rugicollis* Casey (Coleoptera: Curculionidae), infesting sal seeds in Madhya Pradesh. *Journal of Entomological Research* 37 (3), 259–267.

Choubey, V., Bhandari, R. and Kulkarni, N. (2013b) Contact toxicity of some selected sysnthetic pyrethroids on the seed weevil, *Sitophilus rugicollis* Casey (Coleoptera: Curculionidae), and environmental implications for conservation of forests. Presented to VIIIth National Conference, Recent Advances in Biodivesity Conservation, Biotechnology and Environmental Management Research, 19–20 April 2013, Govt New Science College, Rewa, India.

Choubey, V., Bhandari, R., Kulkarni, N. and Mishra, V.K. (2013c) Effect of chemical fumigants on emergence of sal, *Shorea robusta*, seed weevil, *Sitophilus (Calandra) rugicollis* Casey. *National Journal of Life Science* 10 (2), 183–187.

Cowie, R.H. and Wood, T.G. (1989) Damage to crops, forestry and rangeland by fungus-growing termites (Termitidae: Macrotermitinae) in Ethiopia. *Sociobiology* 15, 139–153.

Cunningham, S.A. and Floyd, R.B. (2006) *Toona ciliata* that suffer frequent height-reducing herbivore damage by a shoot-boring moth (*Hypsipyla robusta*) are taller. *Forest Ecology and Management* 225, 400–403.

Dell, B., Hardy, G. and Burgess, T. (2008) Health and nutrition of plantation eucalypts in Asia. *South Forest* 70, 131–138.

Dhakal, L.P., Jha, P.K. and Kjaer, E.D. (2005) Mortality in *Dalbergia sissoo* (Roxb.) following heavy infection by *Aristobia horridula* (Hope) beetles: Will genetic variation in susceptibility play a role in combating declining health? *Forest Ecology and Management* 218, 270–276.

Dixon, W.N. and Fasulo, T.R. (2009) Tarnished plant bug, *Lygus lineolaris* (Palisot de Beauvois) (Insecta: Hemiptera: Miridae). EENY-245. Institute of Food and Agricultural Sciences Extension, University of Florida, Gainesville, Florida.

Duke, N.C. (2002) Sustained high levels of foliar herbivory of the mangrove *Rhizophora stylosa* by a moth larva *Doratifera stenosa* (Limacodidae) in north-eastern Australia. *Wetlands Ecology and Management* 10, 403–419.

Elliott, H.J., Ohmart, C.P. and Wylie, F.R. (1998) *Insect Pests of Australian Forests: Ecology and Management*. Inkata Press, Melbourne, Australia.

Erwin, T.L. (1983) Tropical forest canopies: the last biotic frontier. *Bulletin of the Entomological Society of America* 29, 14–19.

FAO (1993) *The Challenges of Sustainable Forest Management: What Future for the World's Forests?* Chapter 5: Forest Management Options, FAO Corporate Document Repository. Available at: www.fao.org/docrep/T0829E/T0829E04.htm (accessed 27 March 2016).

FAO (2007) *Forest Health & Biosecurity Working Papers: Overview of Forest Pests.* Working Paper FBS/32E. FAO, Rome.

Fierke, M., Nowak, D. and Hofstetter, R. (2011) Seeing the forest for the trees: forest health monitoring. In: Castello, J.D. and Teale, S.A. (eds) *Forest Health: An Integrated Perspective.* Cambridge University Press, Cambridge, UK, pp. 321–343.

Fletchmann, C.A.H., Ottati, A.L.T. and Berisford, C.W. (2001) Ambrosia and bark beetles (Scolytidae: Coleoptera) in pine and eucalypt stands in southern Brazil. *Forest Ecology and Management* 142, 183–191.

Gara, R.I., Sarango, A. and Cannon, P.G. (1990) Defoliation of an Ecuadorian mangrove forest by the bagworm, *Oiketicus kirbyi* Guilding (Lepidoptera: Psychidae). *Journal of Tropical Forest Science* 3, 181–186.

Garg, V.K., Kulkarni, N. and Meshram, P.B. (2005) Influence of weather parameters on the incidence of white grub, *Holotrichia serrata* Fabr., on teak seedlings in Chhindwara Forest Divisions. *Indian Journal of Tropical Biodiversity* 13, 48–51.

Gautam, A., Rao, S.D. and Thirumala (1975) Spoilage of sal fruit on storage. *Journal of Oil Technology Association of India* 7 (4), 111.

Gotoh, T., Kotulai, J.R. and Matsumoto, K. (2003) Stem borers of teak and yemane in Sabah, Malaysia, with analysis of attacks by the teak beehole borer (*Xyleutes senegal* Wlk.). *Jarq – Japan Agricultural Research Quarterly* 37, 253–261.

Goulet, E., Rueda, A. and Shelton, A. (2005) Management of the mahogany shoot borer, *Hypsipyla grandella* (Zeller) (Lepidoptera: Pyralidae), through weed management and insecticidal sprays in 1- and 2-year-old *Swietenia humilis* Zucc. plantations. *Crop Protection* 24, 821–828.

Govender, P. (2007) Status of seedling establishment pests of *Acacia mearnsii* De Wild. (Mimosaceae) in South Africa. *South African Journal of Science* 103, 141–148.

Greenslade, P. (2008) Climate variability, biological control and an insect pest outbreak on Australia's Coral Sea islets: lessons for invertebrate conservation. *Journal of Insect Conservation* 12, 333–342.

Harsh, N.S.K. and Joshi, K.C. (1993) Loss assessment of *Albizia lebbek* seeds due to insect and fungus damage. *Indian Forester* 119, 932–939.

He, J., Zhang, R., Tong, Q., Lu, N., Niu, J., He, J.Z. *et al.* (1998) Investigation on using pine caterpillars (*Dendrolimus* spp.) as food in Yunnan minor nationality areas. *Forest Research* 11, 396–401.

Huang, L.L., Wang, G.H., He, Z. and Ge, F. (2008) Effect of pine foliage damage on the incidence of larval diapauses in the pine caterpillar *Dendrolimus punctatus* (Lepidoptera: Lasiocampidae). *Insect Science* 15, 441–445.

Intari, S.E. and Ruswandy, H. (1986) Preferences of *Dioryctria* sp. to three species of pines. *Buletin Penelitian Hutan* 479, 44–48.

ISFR (2015) *India State of Forest Report.* Forest Survey of India, Ministry of Environment, Forests and Climate Change, Govt of India. Available at: http://fsi.nic.in/isfr-2015/isfr-2015-tree-cover.pdf (accessed 26 October 2017).

Jhala, R.C., Chauhan, N.R., Patel, M.G. and Bharpoda, T.M. (2009) Infestation of invasive gall inducer, *Leptocybe invasa* Fisher & La Salle (Hymenoptera: Eulophidae) in nurseries of *Eucalyptus* in Middle Gujarat, India. *Insect Environment* 14 (4), 191–192.

Joshi, K.C. (1992) *Handbook of Forest Zoology and Entomology.* Oriental Enterprise, Dehradun, India.

Joshi, K.C., Roychoudhury, N., Kulkarni, N. and Sambath, S. (2001) *Implementation Completion Report of World Bank Forestry Research, Education and Extension Project (FREEP), entitled Entomology (04).* Tropical Forest Research Institute, Jabalpur, India.

Joshi, K.C., Roychoudhury, N., Kulkarni, N. and Sambath, S. (2006) Sal heartwood borer in Madhya Pradesh. *Indian Forester* 132, 799–808.

Joshi, K.C., Sambath, S., Yousuf, M., Chander, S., Roychoudhury, N. and Kulkarni, N. (2007) Evaluation of *Trichogramma* spp. to minimize the attack of teak leaf skeletonizer. *Indian Forester* 133 (4), 527–533.

Khan, M.A. and Ahmad, D. (1991) Comparative toxicity of some chlorinated hydrocarbon insecticides to poplar leaf beetle, *Chrysomela populi* L. (Coleoptera: Chrysomelidae). *Indian Journal of Forestry* 14, 42–45.

Khatua, A.K. (1999) Sal seed infestation and adult emergence of *Sitophilus rugicollis* Casey (Coleoptera: Curculionidae) in Midnapur (West Bengal). *Indian Forester* 125, 1214–1223.

Khatua, A.K. and Chakrabarti, S. (1990) Life history and season activity of sal seed weevil, *Sitophilus* (Calandra) *rugicollis* Casey (Coleoptera: Curculionidae). *Indian Forester* 116 (1), 63–70.

Kulkarni, H.D. (2010a) Little leaf disease of eucalyptus by thrips, *Frankliniella occidentalis* (Pergande). *Karnataka Journal of Agriculture Sciences* 23 (1), 203–206.

Kulkarni, N. (2010b) Bioecology and management of white grub complex in teak forest nursery in India. In: Cram, M. (ed.) *Proceedings of the 7th Meeting of IUFRO Working Party 7.03.04: Diseases and Insects in Forest Nurseries.* Hilo, Hawaii, 13–17 July 2009. Available at: https://www.iufro. org/download/file/6253/4502/70304-hilo09_pdf/ (accessed 26 October 2017).

Kulkarni, N. (2014) Status of potential biocontrol components for integrated management of forest insect pests in India. In: Koul, O., Dhaliwal, G.S., Khokhar, S. and Singh, R. (eds) *Biopesticides in Sustainable Agriculture: Progress and Potential.* Science Publishers, New Delhi, pp. 389–419.

Kulkarni, N. and Joshi, K.C. (1998) Insect pests of forest tree seeds: their economic impact and control measures. *Journal of Tropical Forest Science* 10(4), 438–455.

Kulkarni, N. and Paunikar, S. (2009) Van ropvatika ani ropvananmadhye honare kitak ani roganche pradurbhav va tyavar vyavasthapan karyapustika [*Handbook on insect pests of forest nurseries and forest plantations and its management*]. *Technical Bulletin.*

Kulkarni, N., Meshram, P.B. and Joshi, K.C. (2003) Foliage feeder, *Noorda blitealis* (Walk.) on *Moringa pterygosperma* (Geart.) syn. *M. oleifera* (Lam.) and its control in nursery. *Journal of Tropical Forestry* 19(1–2), 79–82.

Kulkarni, N., Joshi, K.C. and Shukla, P.K. (2004a) Integrated Insect Pest Management of forest insect pests. In: Gujar, G.T. (ed.) *Contemporary Trends in Insect Science.* Campus Books, New Delhi, pp. 370–410.

Kulkarni, N., Tripathi, S. and Joshi, K.C. (2004b) Kairomonal activity of compounds isolated from bark of sal (*Shorea robusta* Gaert. f.) for attracting the sal heartwood borer, *Hoplocerambyx spinicornis* Newman (Coleoptera: Cerambycidae). *Indian Journal of Forestry* 27 (3), 321–325.

Kulkarni, N., Chandra, K., Singh, Ram Bhajan and Joshi, K.C. (2006) New species of white grubs, their menace on the seedlings of teak (Tectona grandis L.f.) and their management: an experimental study at Ramdongari Forest Nursery, Nagpur. In: Subramanian, K., Joshi, A.K., Mishra, A.K. and Kulkarni, M.M. (eds) *Proceedings of the Regional Workshop on Recent Advances in Teak Research and Management Practices in Central India,* Forest Develoopment Corporation of Maharashtra, Nagpur, India, pp. 184–189.

Kulkarni, N., Soni, K.K. and Joshi, K.C. (2007a) Biotic and abiotic factors responsible for outbreak of the sal heartwood borer, *Hoplocerambyx spinicornis* Newman, in Achanakmar-Amarkantak biosphere reserve. In: Joshi, K.C. and Mandal, A.K. (eds) *Research Needs for Achanakmar-Amarkantak Biosphere Reserve.* Tropical Forest Research Institute, Jabalpur, India, pp.115–126.

Kulkarni, N., Chandra, K., Wagh, P.N., Joshi, K.C. and Singh, R.B. (2007b) Incidence and management of white grub, *Schizonycha ruficollis* on seedlings of teak (*Tectona grandis* Linn.f.). *Insect Science* 14, 411–418.

Kulkarni, N., Chandra K., Singh, R.B. and Joshi, K.C. (2007c) Record of grey weevil, *Myllocerus discolor* (nr.) Boheman (Coleoptera : Curculionidae) grub infesting roots of teak (*Tectona grandis* L. f.) seedlings in Maharashtra. *Indian Journal of Forestry* 30 (3), 291–294.

Kulkarni, N., Paunikar, S., Hussaini, S.S. and Joshi, K.C. (2008) Entomopathogenic nematodes in insect pest management of forestry and plantations crops: an appraisal. *Indian Journal of Tropical Biodiversity* 16 (2), 155–166.

Kulkarni, N., Paunikar, S., Joshi, K.C. and Rogers, J. (2009) White grubs, *Holotrichia rustica* and *Holotrichia mucida* (Coleoptera: Scarabaeidae), as pests of teak (*Tectona grandis* L. f.) seedlings. *Insect Science* 16, 519–525.

Kulkarni, N., Paunikar, S. and Singh, R.B. (2013) Record of Elatirid, *Agrypnus fuscipes* (Fabricius) from Maharashtra and potentiality as predator on white grubs of *Holotrichia rustica. Indian Journal of Tropical Biodiversity* 21(1–2), 143–144.

Kulkarni, N., Roychoudhury, N., Chandra, S., Singh, R.S. and Das (2015) First report on incidence of sal borer (*Hoplocerambyx spinicornis*) in Korba, Kabirdham and Bhanupratappur Forest Division, Chhattisgarh. *Vaniki Sandesh*, 7 (3–4), 23–31.

Kulkarni, N., Ahmad, M., Paunikar, S. and Barve, S. (2016) Van ropaniyon me shwet illiyo ke ekeekrut nashikeet prabandhan hetu margdarshika [*Handbook for Integrated Pest Management of white grubs in forest nurseries*]. *Technical Bulletin* (in Hindi). Tropical Forest Research Institute.

Li, G., Zan, Q., Zhao, S., Xiao, Y., Wang, Y., Xu, H. *et al.* (2007) Biological characters of *Ptyomaxia* sp. and the toxicity and effectiveness of *Bacillus thuringiensis* against its larvae. *Chinese Journal of Applied and Environmental Biology* 13, 50–54.

Liang, Y.-Y., Li, H.-G., He, Z. and Ge, F. (2008) Effect of alternative hosts on the growth and development of masson pine caterpillar, *Dendrolimus punctatus*. *Chinese Bulletin of Entomology* 45(5), 739–743.

Mannakkara, A. and Alawathugoda, R.M.D. (2005) Black twig borer *Xylosandrus compactus* (Eichhoff) (Coleoptera: Scolytidae) damage for forest tree species. *Sri Lanka Forester* 28, 65–75.

Mathew, G. (2005) Nursery pest problems on some native tree species in Kerala province, India. *Working Papers of the Finnish Forest Research Institute* 11, 31–36.

Mathur, R.N. (1960) Pests of teak and their control. *Indian Forest Record* (N.S.) Entomology, 10, Manager of Publication, Government of India Press, New Delhi.

Mendel, Z., Protasov, A., Fisher, N. and La Salle, J. (2004) Taxonomy and biology of *Leptocybe invasa* gen. & sp. n. (Hymenoptera: Eulophidae), an invasive gall inducer on *Eucalyptus*. *Australian Journal of Entomology* 43, 51–63.

Meshram P.B. (2009) Stem borer *Plocaederus obesus* Gahn (Coleoptera: Cerambycidae) as a pest of *Buchanania lanzan* (Spreng). *World Journal of Zoology* 4(4), 305–307.

Meshram, P.B., Joshi, K.C. and Sarkar, A.K. (1993a) Efficacy of some insecticides against the white grubs, *Holotrichia insularis* Brenske, in teak nursery. *Annals of Forestry* 1, 196–198.

Meshram, P.B., Husen, N. and Joshi, K.C. (1993b) A new report of ambrosia beetle, *Xylosandrus compactus* Eichhoff (Coleoptera: Scolytidae), as a pest of African mahogany, *Khaya* sp.. *Indian Forester* 120, 58–61.

Mohammad Ali, M.I., Varma, R.V. and Sudheendrakumar, V.V. (1992) Microbial pathogens associated with forest insect pests in Kerala. In: Ananthakrishnan, T.N. (ed.) *Emerging Trends in Biological Control of Phytophagous Insects*. IBH Publishing Co. Pvt Ltd, Oxford and New Delhi, pp. 131–141.

Mullen, B.F. and Shelton, H.M. (2003) Psyllid resistance in Leucaena. Part 2: Quantification of production losses from psyllid damage. *Agroforestry Systems* 58, 163–171.

Musthak Ali, T.M. (2001) Biosystematics of phytophagous Scarabaeidae – an Indian overview. In: Sharma, G., Mathur Y.S. and Gupta, R.B.L. (eds) *Indian Phytophagous Scarabaeidae and Their Management*. Agrobios, India, pp. 5–47.

Nair, K.S.S. (2001) *Pest Outbreaks in Tropical Forest Plantation: Is there a Greater Risk for Exotic Tree Species?* Centre for International Forestry Research, Jakarta.

Nair, K.S.S. (2007) *Tropical Forest Insect Pests: Ecology, Impact and Management.* Cambridge University Press, Cambridge, UK.

Nair, K.S.S. and Mathew, G. (1992) Biology, infestation characteristics and impact of the bagworm *Pteromaplagiophleps* Hamps in forest plantations of *Paraserianthes falcataria*. *Entomon* 17(1–2), 179–180.

Nair, K.S.S. and Sumardi (2000) Insect pests and diseases of major plantation species. In: Nair, K.S.S. (ed.) *Insect Pests and Diseases of Indonesian Forests*. Centre for International Forestry Research (CIFOR), Bogor, Indonesia, pp. 21–44.

Nair, K.S.S., Sudheendrakumar, V.V., Varma, R.V. and Chacko, K.C. (1985) *Studies on the Seasonal Incidence of Defoliators and the Effect of Defoliation on Volume Increment of Teak*. Research Report, Kerala Forest Research Institute No. 30.

Nair, K.S.S., Sudheendrakumar, V.V., Varma, R.V., Chako, K.C. and Jayaraman, K. (1996) Effect of defoliation by *Hyblaea puera* and *Eutectona machaeralis* on volume increment of teak. In: Nair, K.S.S., Sharma, J.K. and Varma, R.V. (eds) *Proceedings of the IUFRO Symposium on Impact of Diseases and Insect Pests in Tropical Forests*. KFRI, Peechi & FAO, FORSPA, Bangkok, pp. 257–273.

Nair, K.S.S., Varma, R.V., Sudheendrakumar, V.V., Mohanadas, K. and Mohamed-Ali, M.I. (1998) Management of the teak defoliator (*Hyblaea puera*) using nuclear polyhedrosis virus (NPV). Kerala Forest Research Institute, Peechi, India.

Newton, A.C., Cornelius, J.P., Mesen, J.F., Corea, E.A. and Watt, A.D. (1998) Variation in attack by the mahogany shoot borer, *Hypsipyla grandella* (Lepidoptera: Pyralidae), in relation to host growth and phenology. *Bulletin of Entomological Research* 88, 319–326.

Ofori, D.A. and Cobbinah, J.R. (2007) Integrated approach for conservation and management of genetic resources of *Milicia* species in West Africa. *Forest Ecology and Management* 238, 1–6.

Ofori, D.A., Opuni-Frimpong, E. and Cobbinah, J.R. (2007) Provenance variation in *Khaya* species for growth and resistance to shoot borer *Hypsipyla robusta*. *Forest Ecology and Management* 242 (2–3), 438–443.

Oka, A.G. and Vaishanpayan, S.M. (1979) White grubs menace in teak nurseries in Maharashtra. *Second All India Symposium on Biology and Ecology*, April 1979, Bangalore, India.

Paine, T.D., Millar, J.G. and Hanks, L.M. (1995) Integrated program protects trees from eucalyptus longhorned borer. *California Agriculture* 49, 34–37.

Paine, T.D., Steinbauer, M.J. and Lawson, S.A. (2011) Native and exotic pests of eucalyptus: a worldwide perspective. *Annual Review of Entomology* 56, 181–201.

Palmer, B., Bray, R.A., Ibrahim, T.M. and Fulloon, M.G. (1989) The effect of the leucaena psyllid on the yield of *Leucaena leucocephala* cv. Cunningham at four sites in the tropics. *Tropical Grasslands* 23, 105–107.

Pang, Z. (2003) Insect pests of eucalypts in China. In: Turnbull, J.W. (ed.) *Eucalypts in Asia*. Proceedings of an international conference held in Zhanjiang, Guangdong, People's Republic of China, 7–11 April 2003. ACIAR Proceedings No. 111, pp. 183–184.

Patil, S.U. and Naik, M.I. (1997) Evaluation of *Trichogramma chilonis* Ishii – an egg parsitoid against teak defoliator, *Hyblaea puera* Cramer. *Indian Journal of Forestry* 20, 183–186.

Peng, J.W., Zhou, S.J., Jiang, Y. and Deng, X.H. (1996) Improve forest structure to heighten its resistance to Masson's pine caterpillar. In: Nair, K.S.S., Sharma, J.K. and Varma, R.V. (eds) *Impact of Diseases and Insect Pests in Tropical Forests*. Proceedings of the IUFRO Symposium, Peechi, India, 23–26 November 1993, pp. 373–378.

Ramsden, M. (2013) *Sirex* in Queensland. *News Bulletin, Entomological Society of Queensland* 41(2), 24–25.

Redublo, M.M. (2016) *List of Tropical Countries*. Available at: www.scribd.com/doc/76928345/List-of-Tropical-Countries#download (accessed 25 March 2016).

Remadevi, O.K., Nagaveni, H.C. and Muthukrishnan, R. (2005) Pests and diseases of sandalwood plants in nurseries and their management. *Working Papers of the Finnish Forest Research Institute* 11, 69–75.

Ribeiro, G.T., Mendonca, M.D., de Mesquita, J.B., Zanuncio, J.C. and Carvalho, G.S. (2005) Spittlebug *Cephisus siccifolius* damaging eucalypt plants in the state of Bahia, Brazil. *Pesquisa Agropecuaria Brasileira* 40, 723–726.

Roonwal, M.L. (1978) The biology, ecology and control of the sal heartwood borer, *Hoplocerambyx spinicornis*: a review of recent work. *Indian Journal of Forestry* 1 (1), 21–34.

Roonwal, M.L. (1990) Field-ecological observations on the *Ailanthus excelsa* defoliator, *Eligma narcissus indica* (Lepidoptera: Noctuidae), in Peninsular India. *Indian Journal of Forestry* 13, 81–84.

Roychoudhury, N. (2001) Tree resistance to insects: a novel approach of forest insect management. In Shukla, P.K. and Joshi, K.C. (eds) *Recent Trends in Insect Pest Control to Enhance Forest Productivity*. Tropical Forest Research Institute, Jabalpur, Madhya Pradesh, India, pp. 28–60.

Roychoudhury, N., Kulkarni, N. and Das, A.K. (2013) Sal heartwood borer and its natural enemies. *Vaniki Sandesh* 4 (3), 1–7.

Sah, S.B., Ali, M.S. and Mandal, S.K. (2007) Efficacy of some conventional insecticides against *Myllocerus curvicornis* on shisham in nursery. *Asian Journal of Horticulture* 2(1), 90–91.

Sajeev, T.V., Sudheendrakumar, V.V., Mahiba Helen, S., Meera, C.S., Bindu, T.N. and Bindu, K.J. (2007) Hybcheck – the biopesticide for managing teak defoliator (*Hyblaea puera* Cramer): an announcement. In: Bhat, K.M., Balasundaran, K.V., Muralidharan, E.M. and Thulasidhas, P.K. (eds) *Processing and Marketing of Teak Wood Products of Planted Forests*. Kerala Forest Research Institute, Peechi, India.

Sambaraju, K., DesRochers, P., Rioux, D., Boulanger, Y., Kulkarni, N. *et al.* (2016) Forest ecosystem health and biotic disturbances: perspectives on indicators and management approaches. In: Larocque, G.R. (ed.) *Ecological Forest Management Handbook*. CRC Press, Boca Raton, Florida, pp. 460–502.

Sangha, K.S. (2011) Evaluation of management tools for the control of poplar leaf defoliators (Lepidoptera: Notodontidae) in northwestern India. *Journal of Forestry Research* 22, 77. DOI:10.1007/s11676-011-0129-0

Santana, D.L.W. and Burckhardt, D. (2007) Introduced *Eucalyptus* psyllids in Brazil. *Journal of Forest Research* 12, 337–344.

Senthilkumar, N. and Barthakur, N.D. (2009) *Alcidodes ludificator* Faust: a serious insect pest of nursery and young plantations of *Gmelina arborea* (Roxb.) in northeastern India. *Current Biotica* 2 (4), 493–494.

Senthilkumar, N. and Murugesan, S. (2015) *Insect Pests of Important Tree Species in South India and Their Management Information.* An e-book published by Institute of Forest Genetics and Tree Breeding (IFGTB), Coimbatore (TN), India.

Sharma, A., Verma, T.D. and Sood, A. (2005) Some important insect pests of poplars in the western Himalayas and their management. *Indian Forester* 131(4), 553–562.

Sharma, B.K. and Jain, P.P. (1981) Studies on the effect of period of collection and storage of Sal (Shorea robusta Gaertn. f.) seed kernel and its oil. *Indian Forester* 107(5), 316.

Shukla, P.K. and Joshi, K.C. (eds) (2001) *Recent Trends in Insect Pest Control to Enhance Forest Productivity.* Tropical Forest Research Institute, Jabalpur (MP), India.

Silva, J.O., Oliveira, K.N., Santos, K.J., Espirito-Santo, M.M., Neves, F.S. and Faria, M.L. (2010) Effects of landscape structure and eucalyptus genotype on the abundance and biological control of *Glycaspis brimblecombei* Moore (Hemiptera : Psyllidae). *Neotropical Entomology* 39 (1), 91–96.

Singh, A.P. and Singh, K.P. (1995) Damage evaluation of *Nodostoma waterhousie* Jacoby (Coleoptera: Chrysomelidae), on different clones, provenances and species of poplar in Himachal Pradesh. *Indian Journal of Forestry* 18, 242–244.

Singh, K., Rana, B.S. and Singh, R.P. (2004) Biomass and productivity of an age series of three cottonwood clones (*Populus deltoids*) in central Himalayan tarai region, India. *Journal of Tropical Forest Science* 16, 384–395.

Sivaramakrishnan, V.R. and Remadevi, O.K. (1996) Insect pests in forest nurseries in Karnataka, India, with notes on insecticidal control of psyllid on *Albizia lebbeck.* In: Nair, K.S.S., Sharma, J.K. and Varma, R.V. (eds) *Impact of Diseases and Insect Pests in Tropical Forests.* Proceedings of the IUFRO Symposium, 23–26 November 1993, Peechi, India, pp. 460–463.

Snyder, A.I., Bower, N.W. and Snyder, M.A. (2007) Susceptibility of coastal populations of Caribbean pine to attack by southern pine beetle in Belize. *Ecological Society of America Annual Meeting Abstracts.*

Song, Z. (1997) Preliminary study on using *Metarhizium anisopliae* to control *Dendrolimus punctatus. Journal of Fujian College of Forestry* 17(2), 107–109.

Sousa, W.P., Quek, S.P. and Mitchell, B.J. (2003) Regeneration of *Rhizophora mangle* in a Caribbean mangrove forest: interacting effects of canopy disturbance and a stem-boring beetle. *Oecologia* 137, 436–445.

Sudheendrakumar, V.V. and Biji, C.P. (2009) Biocontrol of the teak defoliator, *Hyblaea puera*: prospects and constraints. *Envis Forestry Bulletin,* 9(1), 8–12.

Sudheendrakumar, V.V., Mohamed Ali, M.I. and Verma, R.V. (1988) Nuclear polyhedrosis virus of the teak defoliator *Hyblaea puera. Journal of Invertebrate Pathology* 51, 307–308.

Sudheendrakumar, V.V., Jalali, S.K. and Singh, S.P. (1995) Acceptance of the teak defoliator Cramer (Lepidoptera, Hyblaeidae) by two exotic species of *Trichogramma* (Hymenoptera: Trichogrammatidae). *Journal of Biological Control* 9(1), 43–44.

Sun, X.-L and Peng, H.-Y. (2007) Recent advances in biological control of pest insects by using viruses in China. *Virologica Sinica* 22(2), 158–162.

Sundararaj, R. and Dubey, A.K. (2005) Potential of plant products for the management of whiteflies in nurseries. *Working Papers of the Finnish Research Institute* 11, 65–68.

Tang, W. and Tian, Y.C. (2003) Transgenic loblolly pine (*Pinus taeda* L.) plants expressing a modified delta-endotoxin gene of *Bacillus thuringiensis* with enhanced resistance to *Dendrolimus punctatus* Walker and *Crypyothelea formosicola* Staud. *Journal of Experimental Botany* 54, 835–844.

Tewari, D.N. (1995) *A Monograph of Sal (Shorea robusta Gaertn. f.).* International Book Distributors, Dehradun, India.

Thakur, M.L. (2000) *Forest Entomology.* Sai Publishers, Dehradun, India.

Thakur, R.K. (1992) Termites as pests of fodder and fuelwood plantations in South India. *Journal of Tropical Forestry* 8, 96–98.

Thakur, R.K. and Sivaramakrishnan, V.R. (1991) A note on the insect pest problems in Kundapur and Mangalore Forests, Karnataka (South India). *Myforest* 27, 187–190.

Thapa, R.S. (1991) Studies on widespread attack of ambrosia beetle, *Platypus* sp., in provenance trial plots of *Acacia crassicarpa* in Sipitang District of Sabah, Malaysia. Breeding Technologies for

Tropical Acacias. *Proceedings of an International Workshop*, Tawau, Sabah, Malaysia, 1–4 July 1991. ACIAR Proceedings No. 37, pp. 31–34.

Thapa, R.S. and Singh, P. (1986) Large-scale mortality of deodar trees by the bark borer, *Scolytus major* Stebbing (Scolytidae, Coleoptera), in Kulu Forest Division, Himachal Pradesh. *Indian Forester* 112, 392–398.

Thu, P.Q., Dell, B. and Burgess, T.I. (2009) Susceptibility of 18 eucalypt species to the gall wasp *Leptocybe invasa* in the nursery and young plantations in Vietnam. *Science Asia* 35, 113–117.

Varma, R.V. and Swaran, P.R. (2009) Pest problems in root-trainer raised forest nurseries of selected species in Kerala. *Indian Journal of Forestry* 32(3), 375–377.

Vir, S. and Parihar, D.R. (1993) Insect pests of *Acacia senegal* (L.) Willd. and *Acacia tortilis* (Forsk) Hyne in the Thar Desert of India. *Myforest* 29, 105–111.

Wagner, M.R., Atuahene, S.K.N. and Cobbinah, J.R. (1991) *Forest Entomology in West Tropical Africa: Forest Insects of Ghana*. Kluwer Academic Publishers, Dordrecht, The Netherlands.

Wagner, M.R., Cobbinah, J.R. and Bosu, P.P. (2008) *Forest Entomology in West Tropical Africa: Forest Insects of Ghana*, 2nd edn. Springer, Dordrecht, The Netherlands.

Wajihullah, A.K.M., Islam, S.S., Rahman, F. and Das, S. (1996) Reduced attack of keora (*Sonneratia apetala*) by stem borer in mixed-species plantation in coastal Bangladesh. *Journal of Tropical Forest Science* 8, 476–480.

Wang, C., Fan, M., Li, Z. and Butt, T.M. (2004) Molecular monitoring and evaluation of the application of the insect-pathogenic fungus *Beauveria bassiana* in southeast China. *Journal of Applied Microbiology* 96 (4), 861–870.

Wang, G., Liu, X., Ge, F. and Li, Z. (2005) Population dynamics of *Dendrolimus punctatus* in different pine forest. *Shengtaizue Zazhi* 24(4), 355–359.

Wang, Y., Solberg, S., Yu, P., Myking, T., Vogt, R.D. and Du, S. (2007) Assessments of tree crown condition of two Masson pine forests in the acid rain region in south China. *Forest Ecology and Management* 242, 530–540.

Wilcken, C.F., Raetano, C.G. and Forti, L.C. (2002) Termite pests in *Eucalyptus* forests of Brazil. *Sociobiology* 40, 179–190.

Wilcken, C.F., de Oliveira, N.C., Sartorio, R.C., Loureiro, E.B., Bezerra Jr, N. and Rosado Neto, G.H. (2008) *Gonipterus scutellatus* Gyllenhal (Coleoptera: Curculionidae) occurrence in *Eucalyptus* plantations in Espirito Santo State, Brazil. *Arqivos do Instituto Biologico Sao Paulo* 75, 113–115.

WWF (2016) *Building a Future in which People Live in Harmony with Nature*. World Wide Fund for Nature (World Wildlife Fund). Available at: http://awsassets.wwf.org.za/downloads/wwf_annual_review_2013_web.pdf (accessed 26 October 2017).

Wylie, F.R. and Speight, M.R. (2012) *Insect Pests in Tropical Forestry*, 2nd edn. CAB International, Wallingford, UK.

Xu, Y.-X., Sun, X.-G., Han, R.-D. and He, Z. (2006) Parasitoids of *Dendrolimus punctatus* in China. *Chinese Bulletin of Entomology* 43 (6), 767–773.

Yousuf, M., Joshi, K.C. and Kulkarni, N. (2004) *Trichogramma*: the noble egg parasitoids and their potential in biological control of insect pests. In: Gujar, G.T. (ed.) *Contemporary Trends in Insect Science*. Campus Books, New Delhi, pp. 307–369.

Zanuncio, J.C., Zanuncio, T.V., de Freitas, F.A. and Pratissoli, D. (2003) Population density of Lepidoptera in a plantation of *Eucalyptus urophylla* in the State of Minas Gerais, Brazil. *Animal Biology* 53(1), 17–26.

Zanuncio, T.V., Zanuncio, J.C., de Freitas, F.A., Pratissoli, D., Sediyama, C.A.Z. and Maffia, V.P. (2006) Main lepidopteran pest species from an eucalyptus plantation in Minas Gerais, Brazil. *Revista de Biologia Tropical* 54(2), 553–560.

Zanuncio, J.C., Pires, E.M., Almado, R.D., Zanetti, R., Monne, M.A., Pereira, J.M.M. and Serrao, J.E. (2009) Damage assessment of host plant records of *Oxymerus basalis* (Dalman, 1823) (Crambycidae: Cerambycinae: Trachyderini) in Brazil. *Coleopterists Bulletin* 63, 179–181.

Zhang, A.B., Tan, S.J., Gao, W., Tu, J.B., Wang, R., Hao, Q., Cheng, L.S. and Chen, L.M. (2001) Primary studies on monitoring *Dendrolimus punctatus* with sex pheromone in Qianshan county, China. *Entomological Knowledge* 38, 223–226.

Zhang, A.B., Wang, Z.J., Tan, S.J. and Li, D.M. (2003) Monitoring the masson pine moth, *Dendrolimus punctatus* (Walker) (Lepidoptera: Lasiocampidae) with synthetic pheromone-baited traps in Qianshan County, China. *Applied Entomology and Zoology* 38 (2), 177–186.

Zhang, Y., Ban, X., Chen, P., Feng, R., Ji, R. and Xiao, Y. (2005) *Dendrolimus* spp. damage monitoring by using NOAA/AVHRR data. *Yingyong Shengtai Xuebao* 16(5), 870–874.
Zhang, Z.F., Huang, B.R., Lin, Q.Y., Cao, X.J. and Cao, G.M. (1998) A study on use of the Petrel 650B type light aircraft to control insect pests. *Journal of Nanjing Forestry University* 22, 56–60.
Zhao, M.F., Yu, F.Q. and Li, G.C. (1998) Preliminary report on control of *Dendrolimus punctatus* Nun by trap lamp. *Journal of Zhejiang Forestry Science and Technology* 18, 49–51.

Index